Ring Transformations
of Heterocycles

ORGANIC CHEMISTRY

A SERIES OF MONOGRAPHS

EDITORS
ALFRED T. BLOMQUIST
Department of Chemistry, Cornell University, Ithaca, New York

HARRY WASSERMAN
Department of Chemistry, Yale University New Haven, Connecticut

Ring Transformations of Heterocycles

of Heterocycles

H. C. VAN DER PLAS
*Landbouwhogeschool, Wageningen,
The Netherlands*

Volume 2

1973
ACADEMIC PRESS
London and New York
A Subsidiary of Harcourt Brace Jovanovich, Publishers

ACADEMIC PRESS INC. (LONDON) LTD.
24–28 Oval Road
London NW1

US edition published by
Academic Press Inc.
111 Fifth Avenue
New York, New York 10003

Library of Congress Catalog Card Number: 71-170748
ISBN: 0-12-711702-4

Printed in Great Britain by
William Clowes & Sons Limited
London, Colchester and Beccles

Preface

Although some excellent books have been published in the last few years covering the general field of heterocyclic chemistry and a number of papers have appeared on specific aspects of the subject, the object of this work is to provide a guide to perhaps one of the most fascinating properties of heterocyclic systems, that is the ease with which they can be converted into other heterocycles, in many cases by quite simple procedures.

It is the author's belief that the synthesis of heterocycles by ring interconversion is becoming of increasing importance as an alternative to the classical method using either a precursor containing the required ring system or an alicyclic intermediate as starting substance.

In order to give a reasonably comprehensive treatment of the diverse ring interconversions which occur in three-, four-, five-, six- and seven-membered heterocycles containing oxygen, sulphur and nitrogen, it has been decided to publish this work in two volumes. In Volume 1, three-, four- and five-membered heterocycles are discussed, and Volume 2 is concerned with six- and seven-membered heterocycles. The transformations in each ring system are presented systematically within the framework of ring contraction, ring retention and ring expansion. Where feasible the scope of application of the various interconversions and the possibility of synthesis has been considered; also the mechanisms of the proposed reactions have been extensively discussed.

The conversion of heterocycles into homocycles has not been considered as this would require separate treatment outside the scope of the present work.

An attempt has been made to cover the literature available on the subject up to the end of 1970; however, the task has been a difficult one in that ring transformations tend to be inadequately indexed in journals of abstracts and therefore it has, unfortunately, been necessary to rely rather heavily on secondary references.

I am greatly indebted to Professor Dr H. J. den Hertog who initially inspired my interest in heterocyclic chemistry. The preparation of this work would not have been possible without the encouragement that he gave to me during the progress of the research and the many invaluable discussions we had on numerous topics of mutual research interest.

I am grateful to Mr B. H. J. van Amersfoort and also the typing department of our Agricultural University for preparing the manuscript. Mr H. Jongejan

v

dedicatedly assisted me in drawing the structure diagrams, checking the references, reading the manuscript and compiling the subject indexes.

I hope that the work discussed will prove a useful contribution to this new but rapidly developing aspect of heterocyclic chemistry and that it will encourage chemists to study the fundamental aspects of the ring interconversions described.

Wageningen   H. C. VAN DER PLAS
March 1973

Contents

4
Ring Transformations of Six-Membered Heterocycles

I. RING TRANSFORMATIONS OF SIX-MEMBERED HETEROCYCLES CONTAINING ONE HETEROATOM

A. Ring transformations of six-membered heterocycles containing one oxygen atom

1. RING CONTRACTIONS INTO FIVE-MEMBERED HETEROCYCLES

a. *Pyrylium salts into dihydrofurans, pyrazolines, pyrazoles and isoxazolines; dihydropyrans into tetrahydrofurans; pyrones into pyrazoles; flavylium salts into arylisoxazolines*

Substituted pyrylium salts have a strong tendency to react with nucleophilic reagents and with appropriate nucleophiles ring transformations can occur. Since there are a number of methods[1-4a] by which substituted pyrylium salts can readily be prepared (these include the reaction of pyrones with nucleophilic reagents, dehydration of 1,5-diketones, the treatment of α-β unsaturated methylketones with carboxylic acid derivatives[3,4] and the more recent method of diacylation of olefins[4]) pyrylium salts are excellent starting substances for conducting ring transformations.

In these ring transformation reactions the attack of the nucleophile takes place mainly in position 2; nmr data as well as molecular orbital calculations on related thiopyrylium salts have firmly indicated that position 2 (6) is indeed the most electron-deficient position[7].

Most of these ring transformations are used for the conversion of pyrylium salts into other six-membered heterocycles (see sections I.A.2.a and I.A.2.b); however, a few ring contraction reactions with pyrylium salts have been reported.

Mild treatment of the pyrylium salts (4.1) with hydrogen peroxide in an acidic medium yields the 3,5-dialkylfurylketones (4.5)[8,9,16]. This method has been elegantly used for the preparation of 3,5-di(methyl-*d3*)-2-acetylfuran (4.6)[10]. It can be assumed that the hydrogen peroxide initially attacks the α-position of the pyrylium salt with formation of the intermediary 2*H*-pyran hydroperoxide (4.2). This peroxide is decomposed further by the acid into the cation (4.3) which is in equilibrium with the open-chain compound (4.4). Because this step is irreversible it is promoted by energy gained through the formation of the aromatic five-membered ring[7].

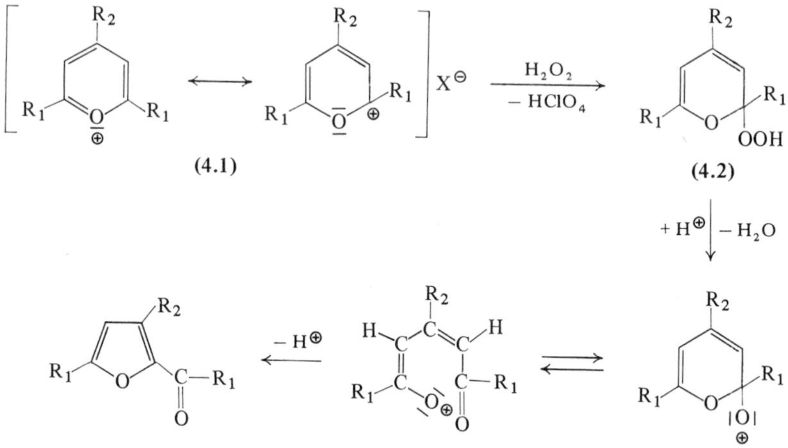

(4.1)   (4.2)

(4.5)   (4.4)   (4.3)

R_1	CH_3	CH_3	C_2H_5	CH_3	$CH(CH_3)_2$	$C(CH_3)_3$
R_2	CH_3	C_2H_5	CH_3	$CH(CH_3)_2$	CH_3	CH_3

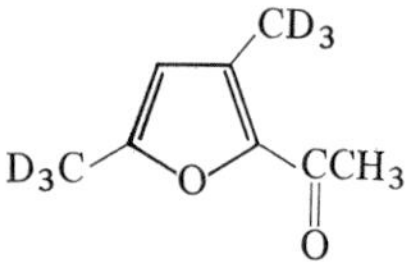

(4.6)

It has recently been found that oxidation in air of 2,4,6-triphenylpyrylium oxide (4.6a) in darkness[11a] gives ring contraction into the unsaturated γ-lactone (4.6c) (36 per cent). This compound is also formed during a benzophenone sensitized irradiation at 366 nm. It has been proposed that the dibenzoyl compound (4.6b) is intermediate in this conversion and that it gives (4.6c) after the ring closure and rearrangement as indicated [11b].

Oxidative ring contraction has also been reported[11c] to take place when 2-ethoxy-3,4-dihydro-2*H*-pyrans are treated with 1 equivalent of *m*-chloroperbenzoic acid in chloroform or methylene chloride, yielding 2-ethoxy-2-alkanoyltetrahydrofurans. Similarly, hydrolysis of 2,4,5,6-tetraphenylpyrylium perbromide gives ring contraction into 2-benzoyl-3,4,5-triphenylfuran[11].

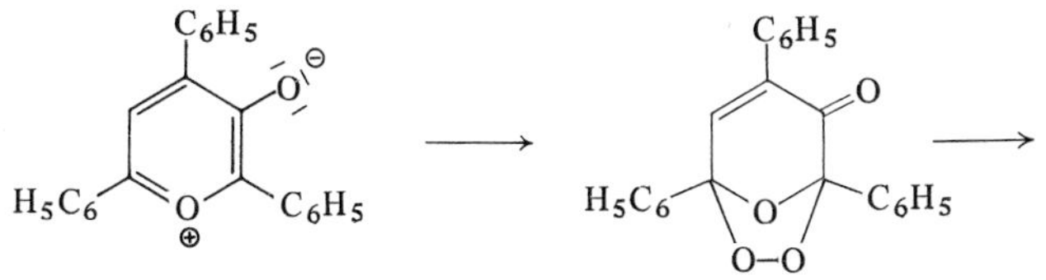

(4.6a)

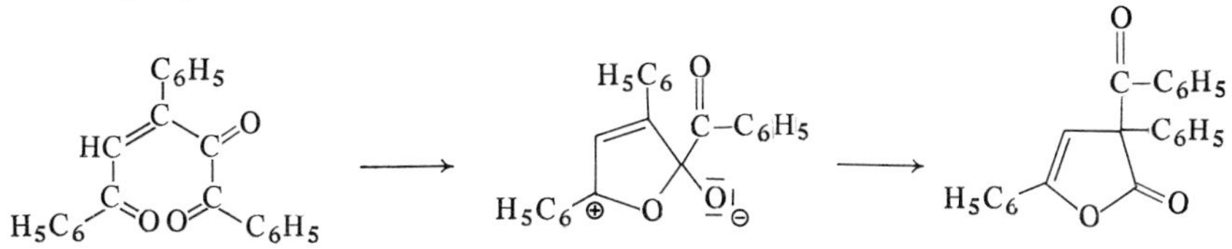

(4.6b)   (4.6c)

Hydroxylamine has also been found to be capable of bringing about ring contraction; thus, 2,6-bis(aryl 1-substituted)-4-(aryl 2-substituted) pyrylium fluoroborates (4.7) can be converted into 3-(aryl 1-substituted)-5-(aryl 2-substituted)-5-(aryl 1-COCH$_2$ substituted)Δ^2-isoxazolines (4.9)[12,13]. The intermediate is the monooxime (4.8) of the pseudobase and this isomerizes easily into (4.9). Isoxazoles (4.10) are obtained by the loss of substituted acetophenones from (4.9) in the presence of acid. It is of interest that 2,4,6-tri-t-butylpyrylium fluoroborate is not converted by hydroxylamine in ethanol into an isoxazoline derivative but into 3,5-di-t-butyl-2-furylketone (4.5) (R$_1$ = R$_2$ = t-C$_4$H$_9$)[14].

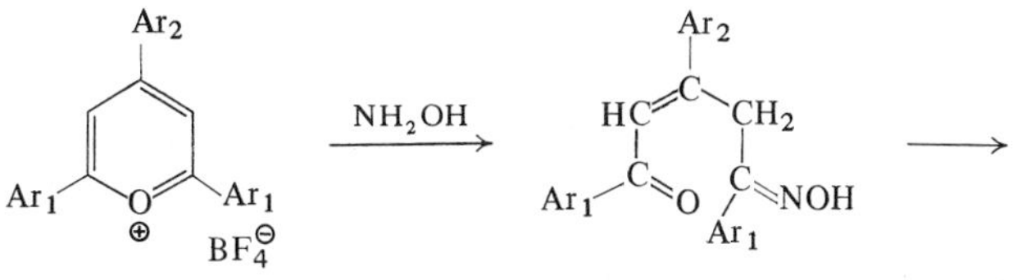

(4.7)   (4.8)

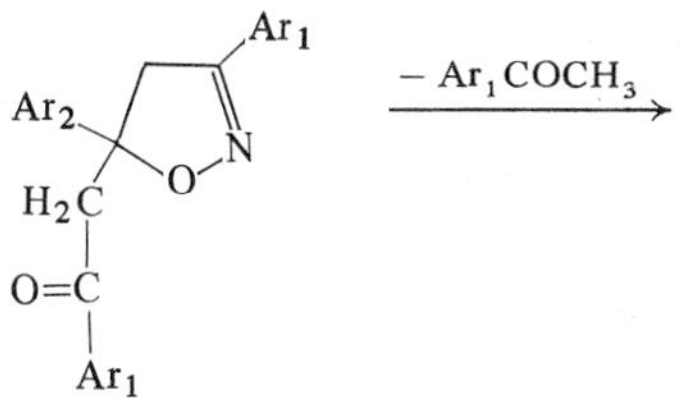

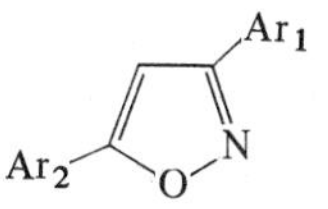

(4.9)   (4.10)

Ar$_1$	C$_6$H$_5$	p-BrC$_6$H$_4$	C$_6$H$_5$	C$_6$H$_5$	p-CH$_3$C$_6$H$_4$
Ar$_2$	C$_6$H$_5$	C$_6$H$_5$	p-ClC$_6$H$_4$	m-NO$_2$C$_6$H$_4$	p-OCH$_3$C$_6$H$_4$

Very recently, ring contraction has been reported[14a] when the flavylium salt (4.10a) is oximinated, yielding the Δ^3-isoxazoline (4.10d). In accordance with the mechanism proposed above[13], it has been advanced that the flavylium salt may undergo a direct nucleophilic attack at C-2, giving the Δ^3-isoxazoline via the intermediates (4.10b) and (4.10c).

(4.10a)   (4.10b)

(4.10c)   (4.10d)

Ar = p-OCH$_3$C$_6$H$_4$

It has been reported that the pyrylium salts (4.1) (R = CH$_3$) are converted into N-aminopyridinium salts by reaction with equimolar amounts of hydrazine (see section I.A.2.a), but that in the presence of an excess of hydrazine, ring contraction into pyrazolines or pyrazoles occurs[15]. Ring contractions into the Δ^2-pyrazolines (4.10f) have also been observed when 2,4,6-trisubstituted pyrylium salts (4.10e) react with phenylhydrazine[12,13] or benzenesulphono-hydrazide[15a]. A mono-hydrazone is probably intermediate in these ring contractions, since it has been found that similar intermediates can be isolated in certain conditions (see section I.A.3.a). In fact, this reaction of pyrylium salts with hydrazines is rather complicated as appears from a recent report on the reaction of (4.10e) with methylhydrazine[15b] according to which, besides ring contraction into Δ^2-pyrazolines and the corresponding pyrazoles, ring expansion into 1,2-diazepines and formation of N-aminopyridinium salts occurs. Similar reactions have been reported[15c] with phenylhydrazine. Moreover, in the reaction of (4.7) with phenylhydrazine, the interesting formation of pyrazolo[2,3-a]quinolines (4.10g) was observed. It is worthy of mention that the ring contraction of pyrylium salts into Δ^2-pyrazolines has its counterpart in the ring expansion of (4.10f) (R = SO$_2$C$_6$H$_5$, R$_1$ = R$_2$ = R$_3$ = C$_6$H$_5$) with perchloric acid into the pyrylium salt (4.10e) (R$_1$ = R$_2$ = R$_3$ = C$_6$H$_5$).

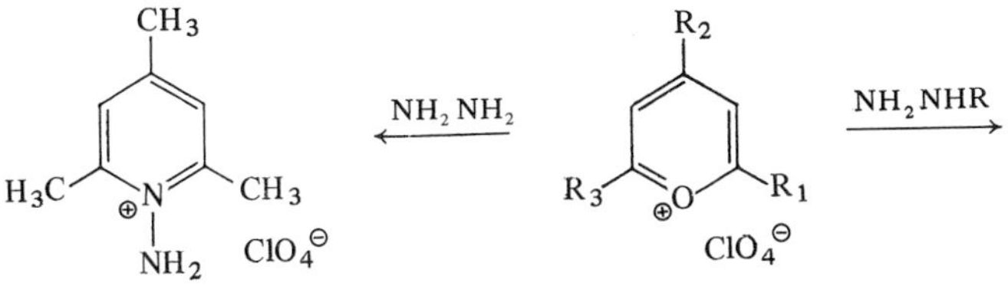

(4.10e)

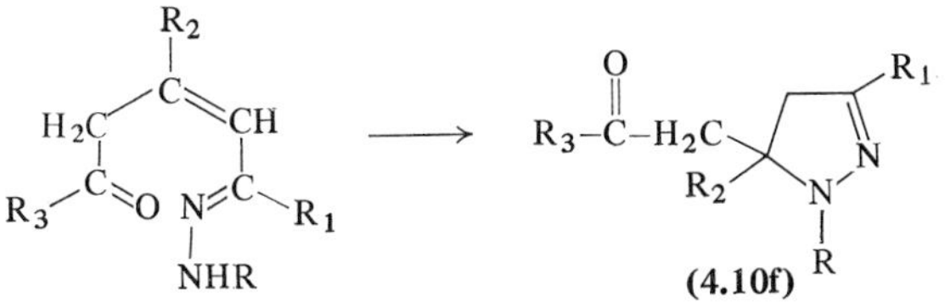

(4.10f)

$R = C_6H_5, SO_2C_6H_5$
$R_1 = CH_3, C_6H_5$
$R_2 = H, C_6H_5$
$R_3 = CH_3, C_6H_5$

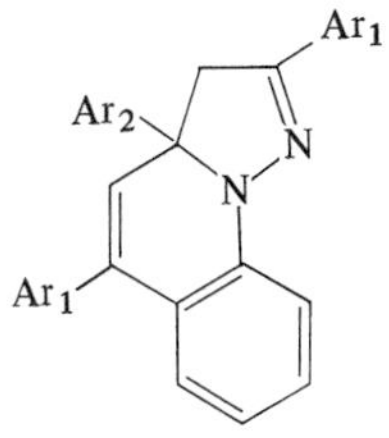

(4.10g)

Ar_1	Ar_2
C_6H_5	C_6H_5
$p\text{-}BrC_6H_4$	C_6H_5
C_6H_5	$p\text{-}CH_3C_6H_4$

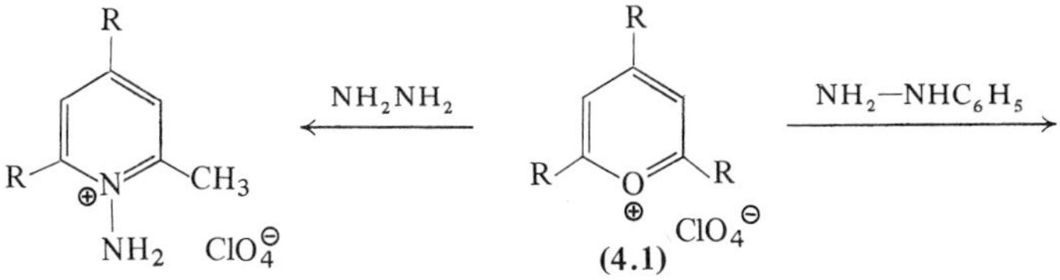

(4.1)

$R = CH_3$

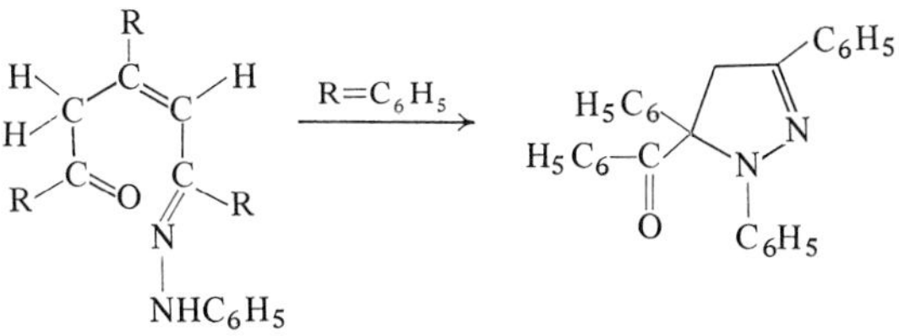

b. *Pyrans into furans; coumarins into benzofurans and benzisoxazoles; iso-*
coumarins into phthalides; chromones into phenylpyrazoles and phenylis-
oxazoles

In correlation with the reported conversion of pyrylium salts into furans (see
section I.A.1.a), appropriately substituted pyrones and their benzo derivatives
can be converted into furans and benzofurans respectively, under hydrolytic
conditions. Base treatment of bromodehydroacetic acid (4.11) thus gives the
furan-3-carboxylate (4.12)[17], 3-bromocoumalic acid (4.13) gives the furan-3,5-
dicarboxylate (4.14)[18], and several 4,5,6-triaryl-3-bromo-2-pyrones (4.15) yield
the furan-2-carboxylates (4.16)[19].

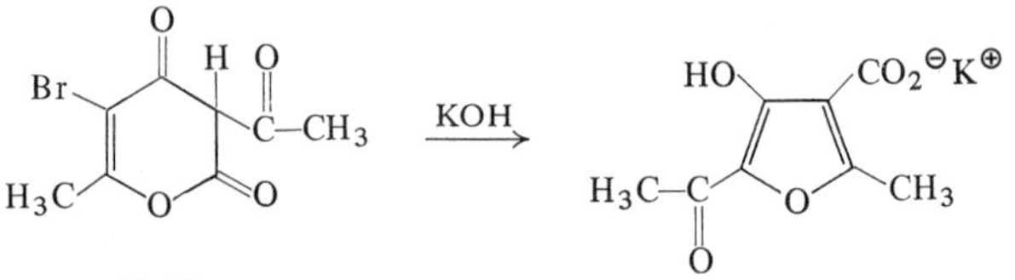

(4.11)   (4.12)

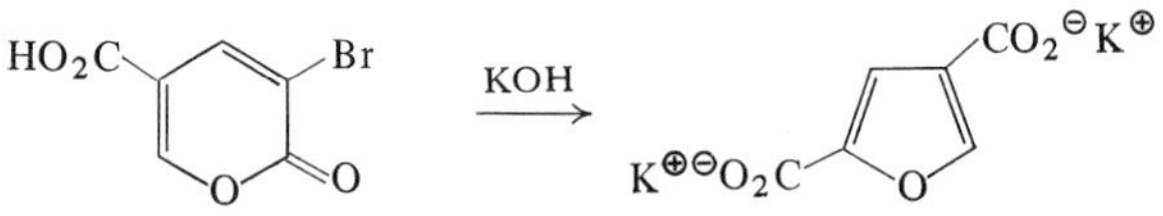

(4.13)   (4.14)

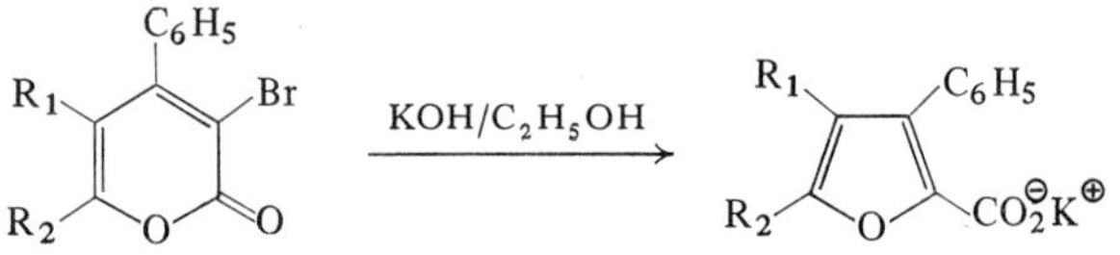

(4.15)   (4.16)

$R_1 = C_6H_5; p\text{-}OCH_3C_6H_4$
$R_2 = C_6H_5; p\text{-}BrC_6H_4; p\text{-}CH_3C_6H_4; p\text{-}OCH_3C_6H_4$

From the 4,6-dimethyl-3-bromocoumalic acid[20] (not from the ester)[21] with
evolution of carbon dioxide the 2,4-dimethylfuran-3-carboxylate is formed, and
from 3-bromocoumarin (4.17) the benzofuran-2-carboxylate (4.19)[22] is
obtained. It is clear that cleavage of the lactone bond is the primary step in these
reactions; this bond fission is followed by ring closure via an internal attack of
the nucleophilic moiety with bromide elimination (4.17) → (4.18) → (4.19).
Interestingly, benzofuran-2-carboxylic acid (4.19) has also been reported to be
formed in high yield when 4-chlorocoumarin (4.20) is treated with a hot
solution of sodium hydroxide in aqueous dioxane[23]. This ring transformation is
supposed to be brought about via a base-catalysed β-elimination of halogen
halide in the open-chain product and cyclization of the o-hydroxyphenyl-
propiolic acid formed.

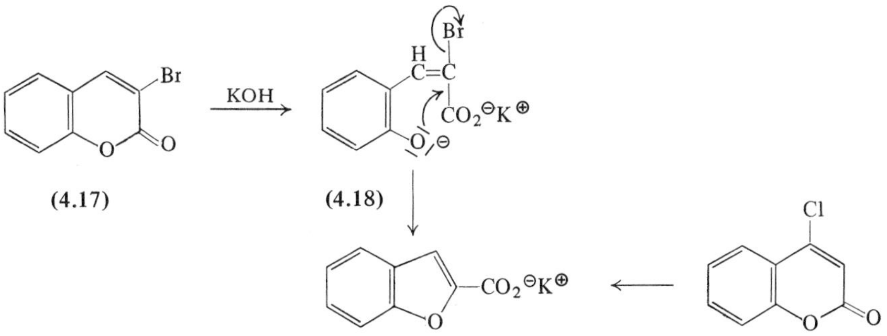

(4.17) $\xrightarrow{\text{KOH}}$ (4.18) $\downarrow$

(4.19) $-CO_2^{\ominus}K^{\oplus}$ $\leftarrow$ (4.20)

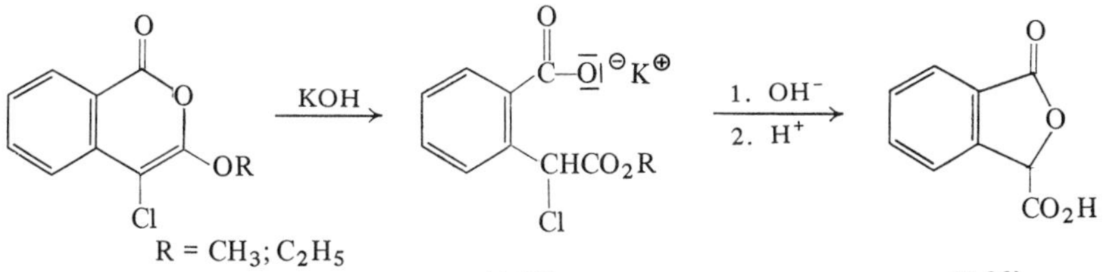

(4.21) $R = CH_3; C_2H_5$ $\xrightarrow{\text{KOH}}$ (4.22) $\xrightarrow[\text{2. H}^+]{\text{1. OH}^-}$ (4.23) CO_2H

In the isocoumarin series, it has been found that in an alkaline medium the 3-alkoxy-4-chloroisocoumarins (4.21) give ring contraction into the phthalide carboxylic acid (4.23) via the open-chain intermediate (4.22)[24]. Recently it has been shown that with aqueous sodium hydroxide 7-methylpyrano [4,3-b] pyran-2,5-dione (4.24) (R = H) is converted into the benzene dicarboxylic acid (4.26), while the same reaction with the corresponding 3-bromo derivative (4.24) (R = Br) gives ring contraction into the furan-2,4-dicarboxylic acid (4.28)[25]. In this reaction the intermediate is the diendiol (4.25) or the diketone; thus the mechanism proceeds with ring closure into (4.27) by elimination of hydrogen bromide. Acid hydrolysis leads to loss of the acetyl group.

Pyrolysis of 2-pyrone at 900-1000°C in a stream of nitrogen leads to ring contraction into furan (besides the formation of propyne and allene). The yield of furan is, however, low (15 per cent)[26]. On the other hand, pyrolysis of coumarin (4.29) under the same conditions gives, surprisingly, a very high yield (85 per cent) of benzofuran (4.30). A similar reaction has been reported for xanthone (4.31); at 860°C a 10 per cent yield of dibenzofuran (4.32) is obtained[27].

It has been reported[28] that during irradiation of 2,6-dimethyl-4-pyrone (4.33) ring contraction into 4,5-dimethyl-2-furaldehyde (4.35) takes place as well as formation of the photodimer (4.34)[29]. The juxtaposition of both methyl groups might suggest a photochemical route via compound (4.36).

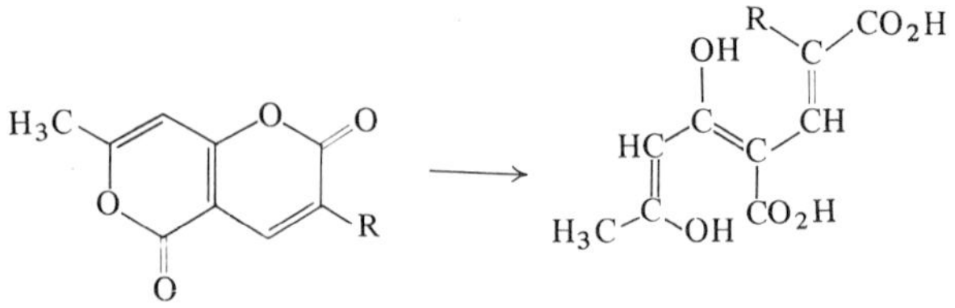

(4.24)

(4.25)

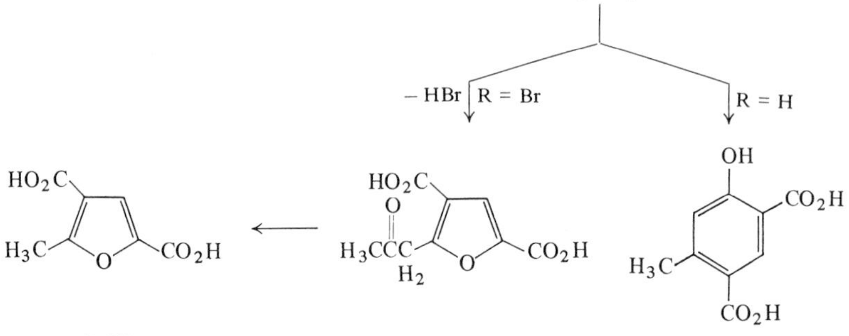

$- HBr$ $R = Br$

$R = H$

(4.28)

(4.27)

(4.26)

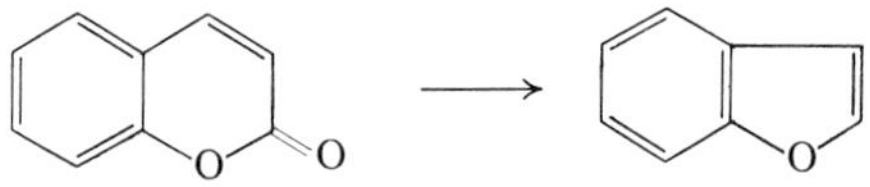

(4.29)

(4.30)

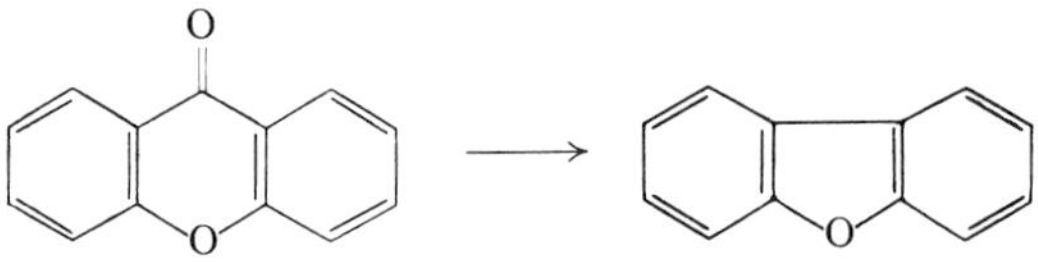

(4.31)

(4.32)

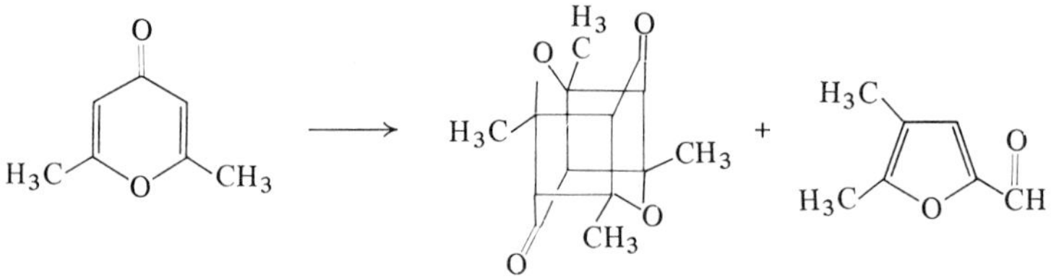

(4.33)

(4.34)

(4.35)

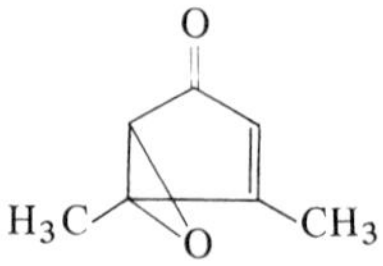

(4.36)

The acid-catalysed acyl lactone rearrangement[30], a useful general method for converting substituted α-acyl-γ- or -δ-lactones into tetrahydrofuran or tetrahydropyran derivatives, has been found not to take place with the β-acyllactone, 4-acetyl-3,4-dihydrocoumarin (4.37). Treatment with $3N$ hydrochloric acid leads to ring contraction into (2-methylbenzofuryl) acetic acid (4.38)[31].

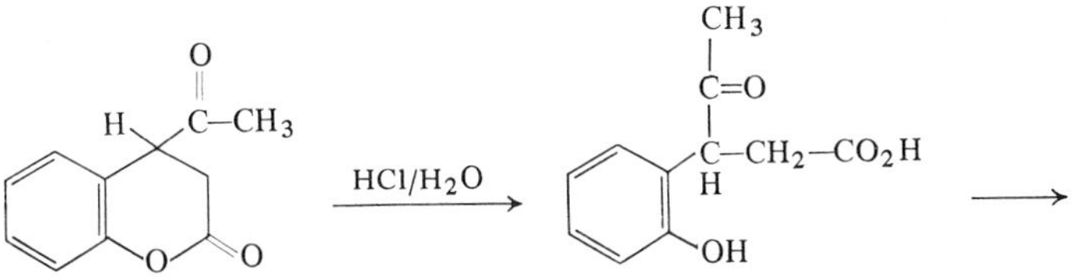

(4.37)

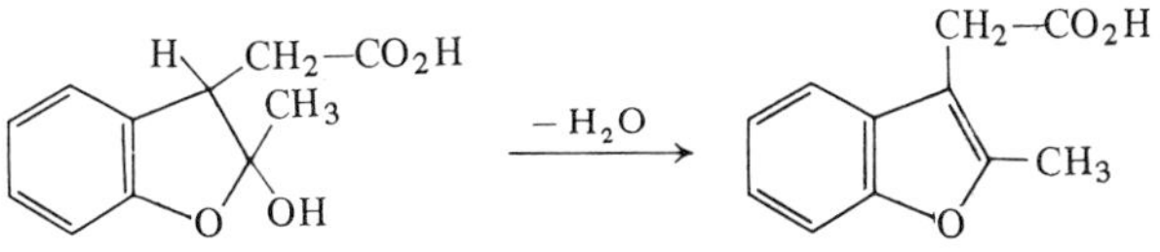

(4.38)

Ring transformations have also been reported in reactions of chromone derivatives with hydrazine and reactions of chromones, coumarins and flavone with hydroxylamine. If 2-methylchromone (4.39) is treated with hydrazine at 150°C or boiling acetic acid, 3-methyl-5-(o-hydroxyphenyl) pyrazole (4.41) is obtained[32,33]. Since the 4-carbonyl group in chromone is generally inert towards the usual ketonic reagents, it seems reasonable to assume that in this ring transformation reaction the first step is the attack of the phenylhydrazine at position 2, yielding the open-chain intermediate (4.40) which recyclizes into

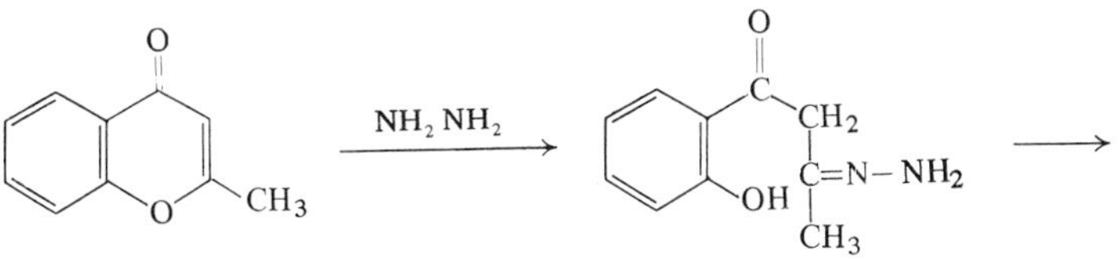

(4.39)

(4.40)

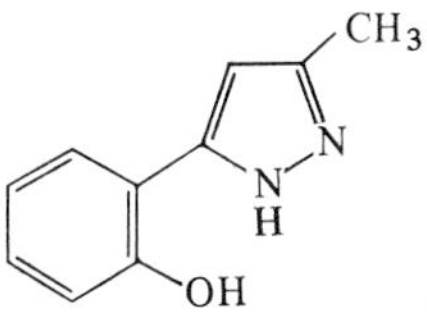

(4.41)

(4.41). Similarly, 4-methoxy-2-methyl-4-thionchromone[34] and 2,3-dimethyl-4-thionchromone[35] when treated with phenylhydrazine and alkali give 3-methyl-5-(2-hydroxy-4-methoxyphenyl)-1-phenylpyrazole and 5-(o-hydroxyphenyl)-3,4-dimethyl-1-phenylpyrazole respectively (and not as has been previously suggested[36] a chromone phenylhydrazone).

During hydrazinolysis of virnagin, khellol and schellin[37] and 3-acyl-chromones[34,38], the same type of ring transformation was found to take place.

When hydroxylamine was used instead of hydrazine, analogous reactions were reported. Thus, flavone (4.42) (R = C_6H_5) with hydroxylamine in pyridine gives 3-(o-hydroxyphenyl)-5-phenylisoxazole[35] (4.43) (R = C_6H_5) and 2-methyl-chromone (4.42) (R = CH_3) gives the corresponding isoxazole (4.43) (R = CH_3)[33]. A similar ring contraction was also reported with 2,6-dimethyl-3-acetyl chromone[33a]. From the benzopyranyl[3,2-c]pyrandione (4.44) the isoxazole (4.45) [or its isomer (4.46)] is obtained[39]. Reaction of 4-hydroxycoumarin (4.47) with an aqueous solution of hydroxylamine also gives rearrangement into an isoxazole ring, 1,2-benzisoxazole-3-acetic acid (4.48) being obtained[40], and not, as previously claimed[39], 4-hydroxylaminocoumarin (4.49).

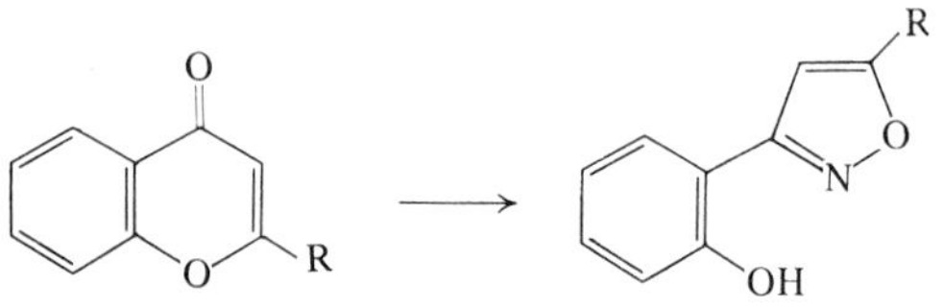

(4.42)   (4.43)

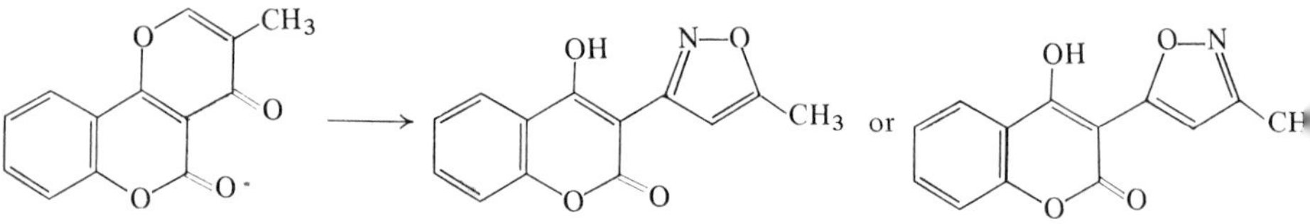

(4.44)   (4.45)   (4.46)

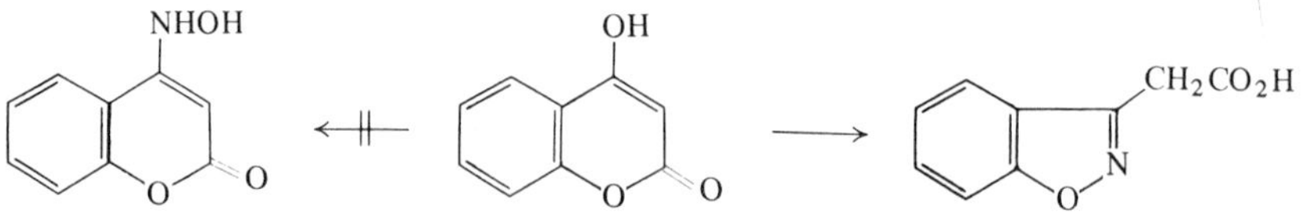

(4.49)   (4.47)   (4.48)

c. *Dihydroflavonols, isoflavons and flavanones into coumaranones; benzodi-
hydropyrans into benzodihydrofurans*

The action of alkali on dihydroflavonols can lead to ring contraction into either
3-benzylidenecoumaran-2-ones or 2-benzylidenecoumaran-3-ones. Which of
these coumaranones is formed appears dependent on the reaction conditions
(alkali concentration, temperature, reaction time). Under mild conditions
usually the coumaran-3-ones are obtained, while under more strenuous conditions the formation of coumaran-2-ones is favoured. Since these ring
contractions have not always been recognized immediately, for a long time there
has been considerable confusion in the literature about the structure of the
products obtained when dihydroflavonols are treated with alkali. If dihydro-
quercetin (4.50) is methylated by adding portions of dimethyl sulphate and
aqueous potassium hydroxide[41] alternately, no penta- and tetramethylethers,
which have the dihydroflavonol structure[42], were obtained, but compounds
were yielded which were later recognized to be the 2-benzylidenecoumaran-3-
one derivatives (4.51) and (4.52)[41]. The tetramethylcoumaranone (4.51) could

(4.50)   (4.51)

+

(4.52)

also be obtained by treating the corresponding tetramethylether (4.53) with
alkali. The formation of both coumaran-3-one derivatives is explained by an
initial formation of the anion (4.54) of the tetramethylether (4.53) which
subsequently yields the open-chain diketone (4.55). Ring closure into (4.51)
occurs by an internal ketalization-type reaction. This conversion has a reversible
character, since in further experiments it was shown that (4.51), when boiled
with an aqueous solution of methanolic potassium hydroxide, is converted into
3-benzylidenecoumaran-2-one (4.57)[43] and not into a coumaran-3-one derivative as suggested before; this result can reasonably be explained by reformation

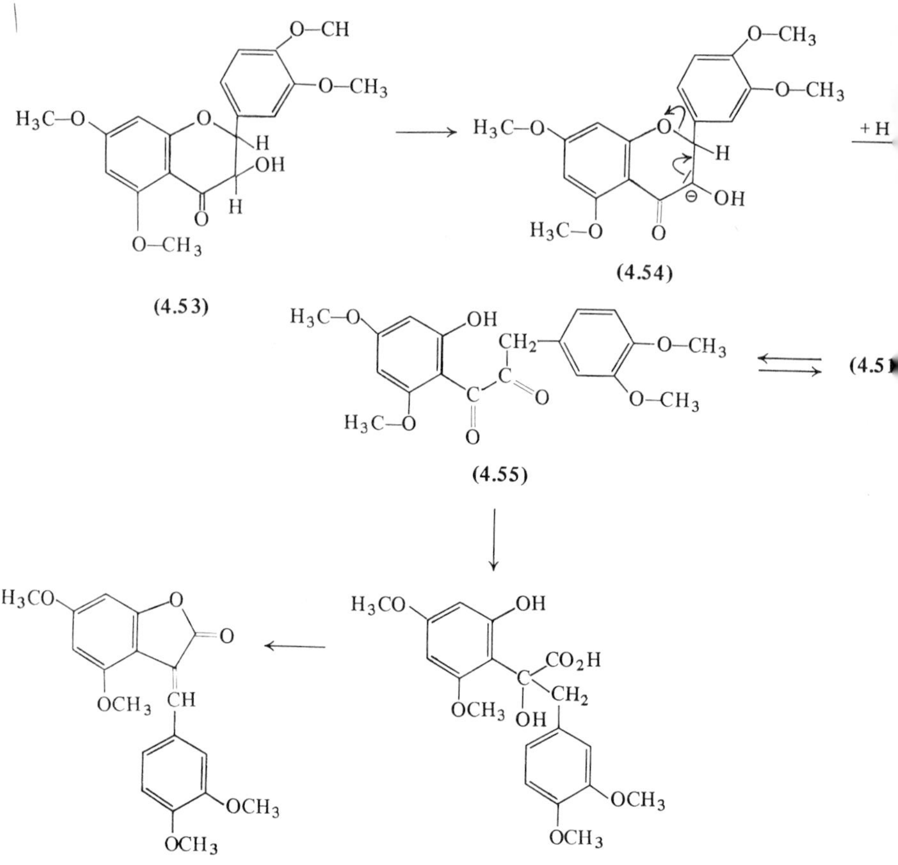

(4.53)

(4.54)

(4.55)

(4.57)

(4.56)

of the 1,2-diketone (4.55), a benzilic acid rearrangement into the α-hydroxy-carboxylic acid (4.56), and dehydrative cyclization into (4.57). Other examples of this ring contraction are the conversion of the trimethyl ether of dihydroflavonol, that is fustin (4.58) (R_1 = R_2 = H), into the 3-benzylidene-coumaran-2-one (4.59) (R_1 = R_2 = H)[44,45], the conversion of the tetramethyl ether of dihydroflavonol, i.e. (4.58) (R_1 = H, R_2 = OCH$_3$) and the pentamethyl ether (4.58) (R_1 = R_2 = OCH$_3$) respectively into the corresponding coumaran-2-ones (4.59)[46,47]. In all these ring contractions a benzilic acid rearrangement is apparently involved.

From the results discussed above it is clear that, if during a reaction in a basic medium, intermediates with a dihydroflavonol structure are formed, ring contraction into coumaranones is one of the side reactions (or possibly main reaction) which can occur. So it has appeared that the reduction of

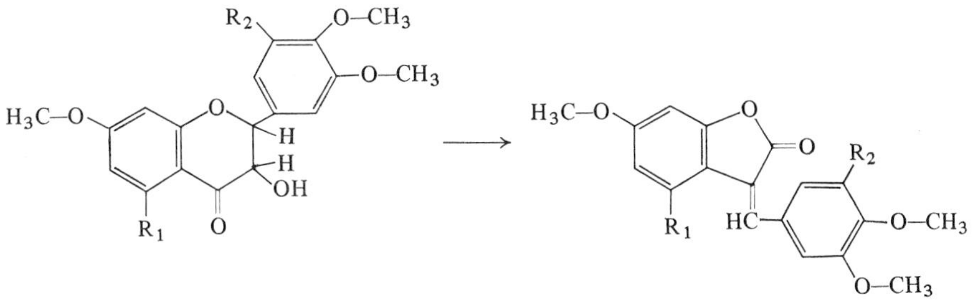

(4.58)

(4.59)

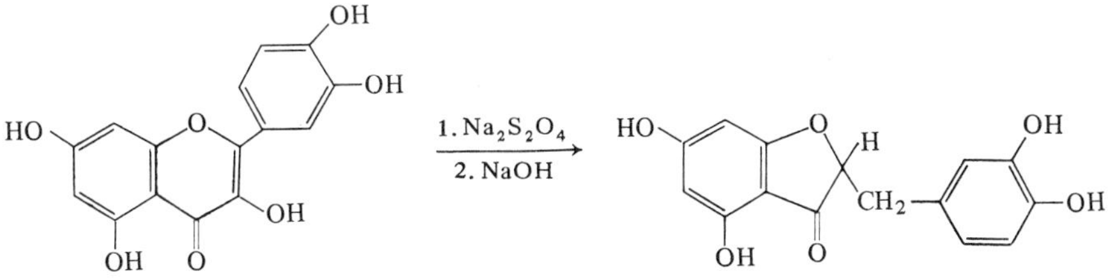

1. $Na_2S_2O_4$
2. NaOH

(4.60)

(4.61)

quercetine (4.60) by an alkaline solution of sodium hydrosulphite not only yields the normal dihydro compound but also the 2-benzylcoumaran-3-one (4.61)[48,49].

Ring contraction was also reported when malacacidine tetramethyl ether (4.62) was treated with periodic acid and the normal work-up procedure was applied. Besides 3-hydroxy-3,4-dimethoxybenzaldehyde[50,51], a product was formed which was proved to be the coumarone (4.65)[50]. Probably via the dialdehyde (4.63) and an internal aldol condensation (4.64) is formed; this is deformylated and dehydrated into (4.65) by successive alkaline and acid treatment during the extraction. Treatment of a compound similar to (4.62), that is the 3',4',7 trimethoxy-3-flavanol (4.66), with concentrated acid also results in a ring contraction yielding the coumarone derivative (4.67)[52]. Ring contraction of the isoflavone derivatives (4.68) has recently been reported in reactions with dimethylsulphoxonium methylides[53,53a]. It leads to the formation of the 2-vinylcoumaran-3-one derivative (4.72) with, as main products, the cyclopropane derivatives (4.70). 2,6-Dimethylisoflavone only gives the corresponding cyclopropane derivative. It has been suggested that the reaction takes place in two stages. In the first a nucleophilic attack by the ylide takes place yielding the resonance-stabilized enolate anion (4.69a) ↔ (4.69b) which either gives (4.70) or undergoes β elimination into the phenolate anion (4.71). The coumaran-3-one (4.72) is obtained by an allylic displacement reaction. These results indicate that cyclopropane formation probably does not occur by a

concerted process. It has also been found that *cis*-3-bromoflavanone (4.73) undergoes ring contraction when treated with dimethylsulphoxonium methylide[54]. In
this reaction 2,3-methanoflavanone (4.75) (in a 32 per cent yield) and 2-
phenylcyclopropyl-1-spiro-2-coumaranone (4.76) were formed and not the
expected cyclopropylspiro-3-flavanone (4.74). The formation of a coumaranone

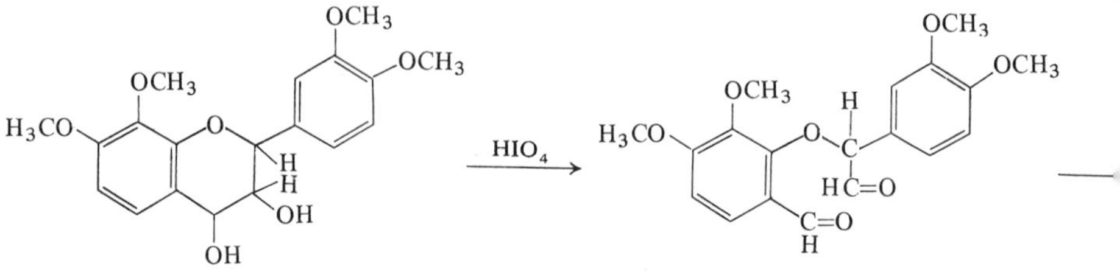

(4.62)   (4.63)

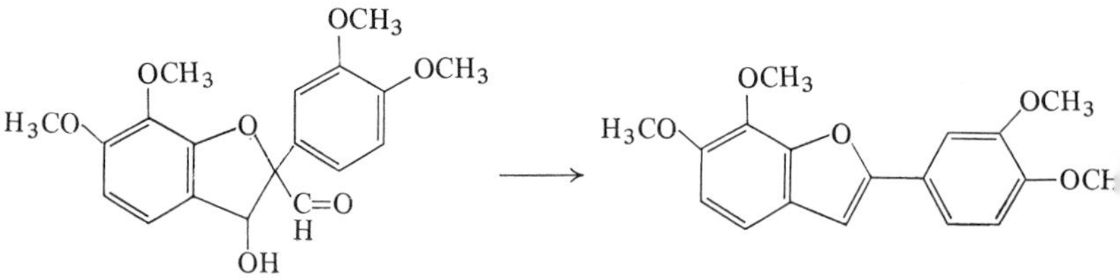

(4.64)   (4.65)

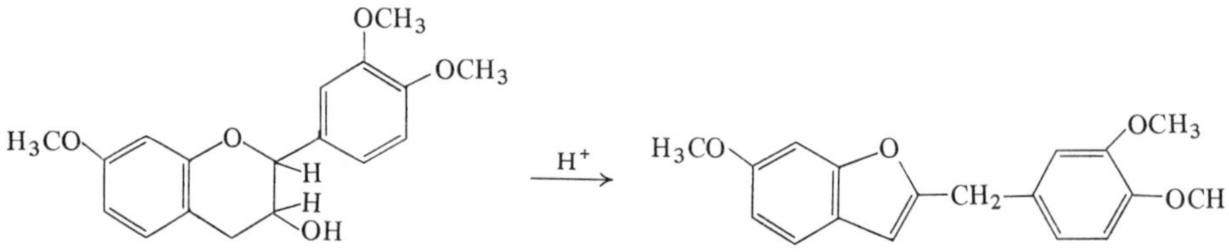

(4.66)   (4.67)

derivative has also been reported when 3-chloroflavone (4.77) is refluxed with a
1 per cent alcoholic potassium hydroxide solution. After acidification of the
reaction mixture 2-benzoylcoumaranone (4.79) is obtained[55]. The ring
contraction occurs via the intermediate (4.78).

A very unusual ring contraction has been found to occur during reaction of
the benzopyranobenzofuran derivatives (4.80) with diethylaniline containing
p-toluene sulphonic acid[56]. Two different isomeric coumaranocoumarans (4.81)
and (4.82) respectively, are obtained; the formation of these compounds is
dependent on the presence of the substituents R in the benzene ring. This
reaction is unique in that respect that the ring contraction does not involve any
loss or displacement of the existing functional groups.

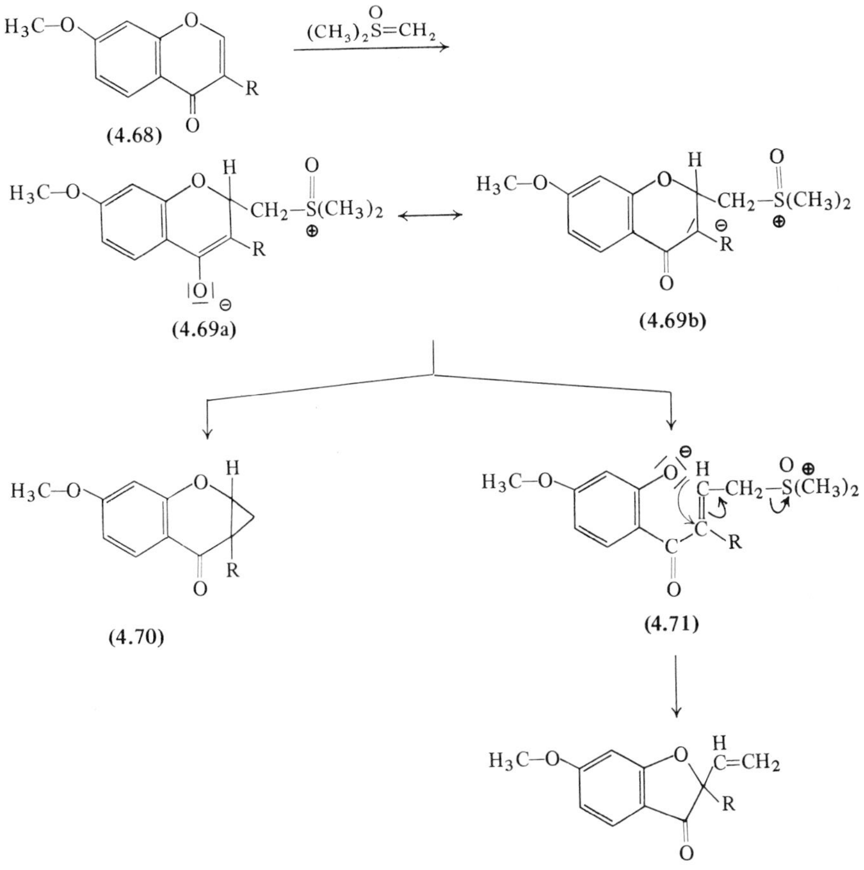

(4.68)

(4.69a)   (4.69b)

(4.70)   (4.71)

(4.72)

$R = C_6H_5;\ 2'\text{-OHC}_6H_4;\ 2'\text{-OCH}_3C_6H_4;\ 2'\text{-OH-4}'\text{-OCH}_3C_6H_3$

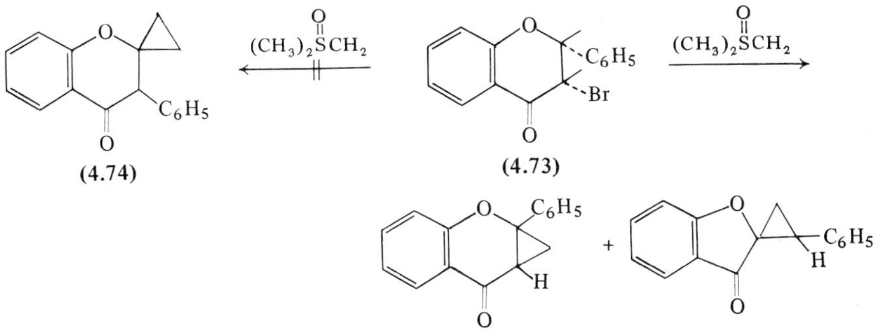

(4.74)   (4.73)

(4.75)   (4.76)

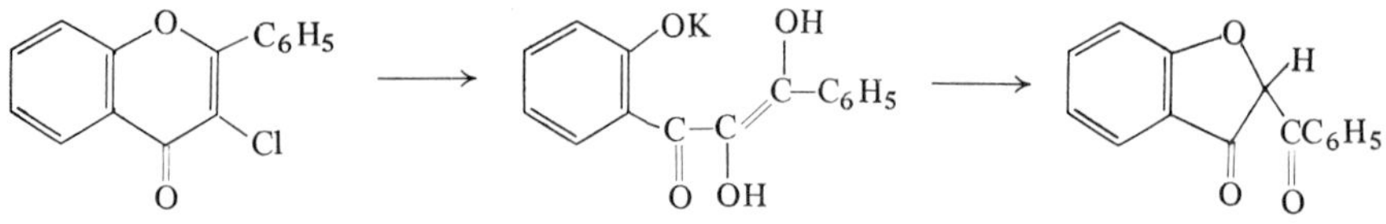

(4.77)   (4.78)   (4.79)

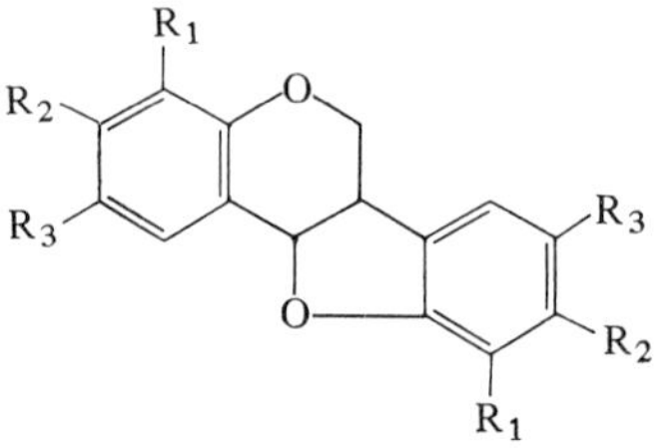

(4.80)

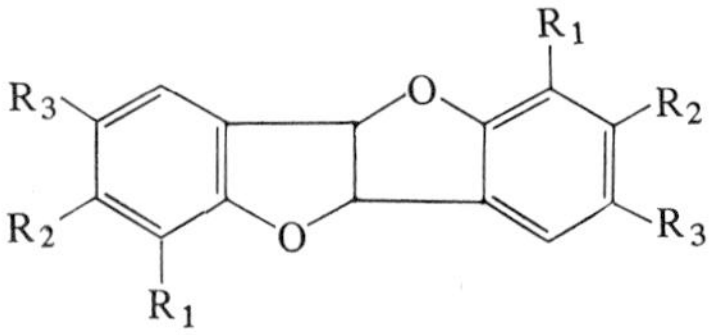

(4.81) 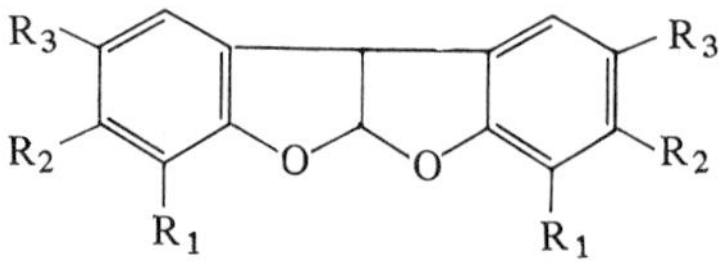   (4.82)

R_1	R_2	R_3	(4.81)	(4.82)
H	H	H	–	+
Cl	H	H	+	–
O–CH$_3$	H	H	+	–
H	H	Cl	+	–
H	H	CH$_3$	–	+
H	H	Br	+	–
H		phenylene	+	–

d. *Miscellaneous examples of ring contraction*

2,2,6,6-Tetramethyl-4-hydroxy-2,3-dihydro-[6*H*]-3-pyrone (4.83) has been reported to be converted into the tetrahydrofuran-3-carboxylate (4.84) when treated with alkali[57]. Although it has not been proved, it seems highly probable that this ring transformation occurs via an internal benzilic acid rearrangement.

Oxidation of the perhydro *cis*-3-hydroxycoumarin (4.85) with alkali dichromate leads to ring contraction into 2-oxo-7-oxa-bicyclo[0,3,4]nonane (4.86)[58].

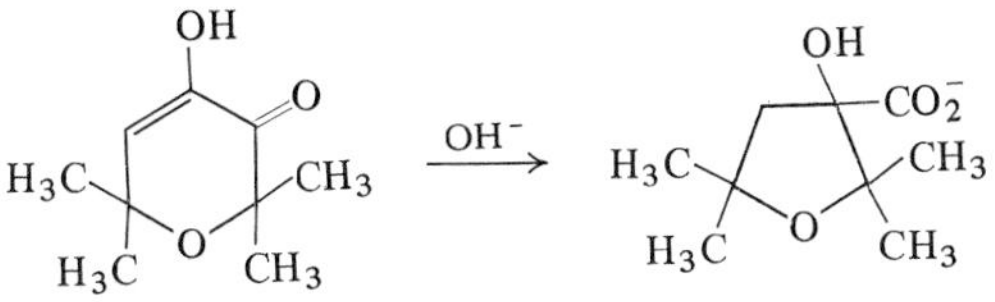

(4.83)   (4.84)

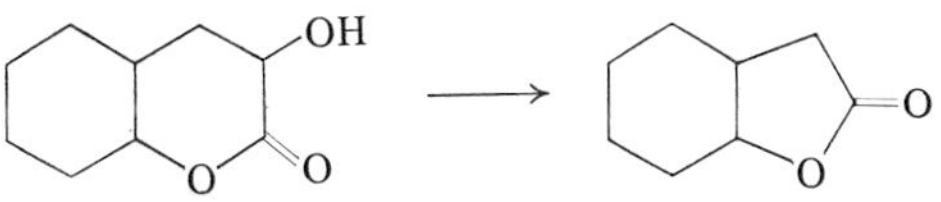

(4.85)   (4.86)

The ozonide (4.88) obtained from the indenone (4.87) by treatment with ozone decomposes at low temperature in about two days into a product which has been proved[59] to be the phthalide (4.89).

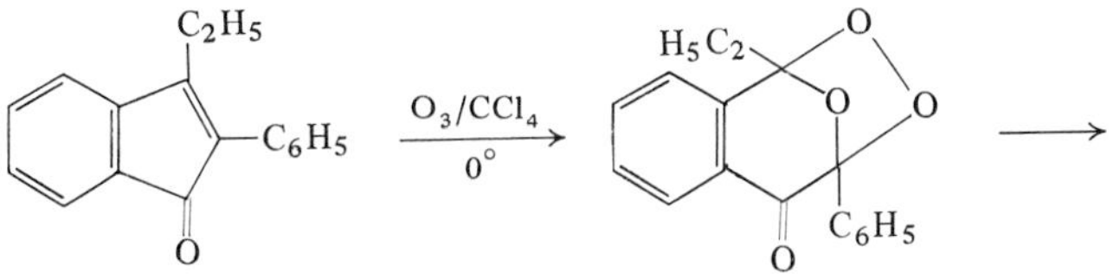

(4.87)   (4.88)

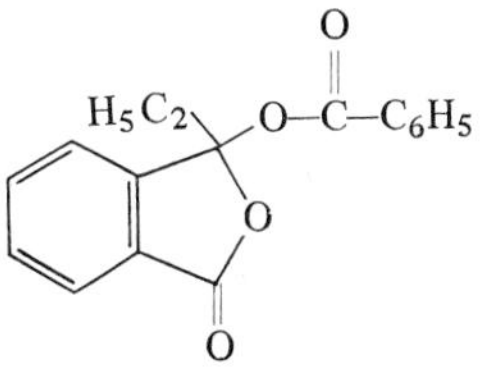

(4.89)

2. RING TRANSFORMATIONS INTO OTHER SIX-MEMBERED HETERO-CYCLES

a. *Pyrylium salts into pyridines and pyridinium salts*

As already pointed out in section I.A.1.a, pyrylium salts have a strong tendency to react with nucleophiles. The carbonium-oxonium resonance stabilization in the 6π-electron system of the pyrylium salts might lead one to predict that the nucleophilic attack could take place either on position 2 or on the vinylogous position 4; however, an attack on position 2 is more favoured[5,6]. The route by which nucleophilic reagents ($R\text{-}AH_2$) react with the pyrylium salts is shown in the following scheme:

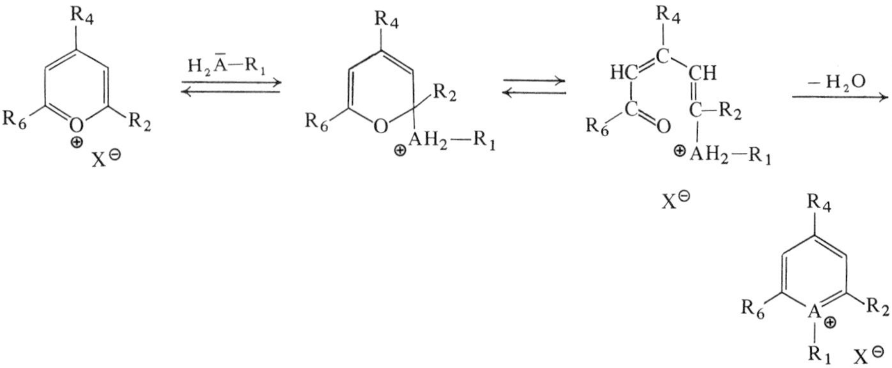

When using ammonia or primary amines as nucleophiles (R_1-AH_2 = NH_3 or R_1NH_2) the ring transformation of pyrylium salts into pyridines or pyridinium salts can be achieved. This conversion is usually carried out by mild heating with an aqueous solution of ammonia, ammonium carbonate, or ammonium acetate with an alcoholic ammonia solution, or by leading dry ammonia into a warm suspension of the pyrylium salts in t-butanol. This ring transformation is of great interest since the reaction has a very broad scope and the required pyrylium salts can be readily prepared (see section I.A.1.a). That this method has a general application is amply illustrated by the numerous differently substituted pyridines which can be obtained by it. The pyridines, which are prepared from the corresponding pyrylium salts, are listed in Tables 4.1 to 4.5.

Table 4.1 *Mono- and disubstituted pyridines*

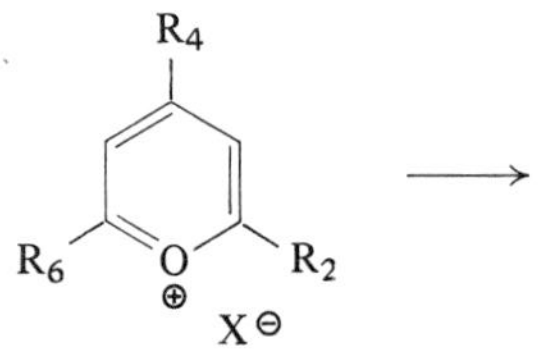

R_2	R_4	R_6	Reference
C_6H_5	H	C_6H_5	60-65
4-$OCH_3C_6H_4$	H	4-$OCH_3C_6H_4$	61
4-$OC_2H_5C_6H_4$	H	4-$OC_2H_5C_6H_4$	61
3,4-$(OCH_3)_2C_6H_3$	H	3,4-$(OCH_3)_2C_6H_3$	61
CH_3	H	CH_3	62
C_2H_5	H	C_2H_5	62
CH_3	CH_3	H	65
H	C_6H_5	C_6H_5	67
H	CH_3	H	64
H	OCH_3	H	66

Table 4.2 *2,4,6-Trialkyl substituted pyridines*

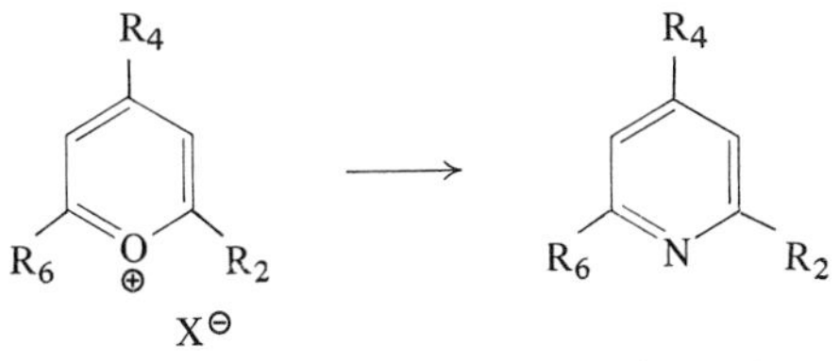

$R_2 = R_4 = R_6 = $ alkyl

R_2	R_4	R_6	Reference
CH_3	CH_3	CH_3	65, 68-71
CH_3	CH_3	C_2H_5	69, 70
CH_3	C_2H_5	CH_3	69, 70, 72, 73
C_2H_5	CH_3	C_2H_5	69
CH_3	$i\text{-}C_3H_7$	CH_3	69
C_2H_5	C_2H_5	C_2H_5	69
$n\text{-}C_3H_7$	CH_3	$n\text{-}C_3H_7$	69
$i\text{-}C_3H_7$	CH_3	$i\text{-}C_3H_7$	69
$n\text{-}C_4H_9$	CH_3	$n\text{-}C_4H_9$	69
$i\text{-}C_4H_9$	CH_3	$i\text{-}C_4H_9$	69
$t\text{-}C_4H_9$	CH_3	$t\text{-}C_4H_9$	69
$R_2 = R_6 = (CH_2)_{10}$	CH_3	$R_2 = R_6 = (CH_2)_{10}$	74
$t\text{-}C_4H_9$	$t\text{-}C_4H_9$	$t\text{-}C_4H_9$	14

Table 4.3 *2,4,6-Triaryl substituted pyridines*

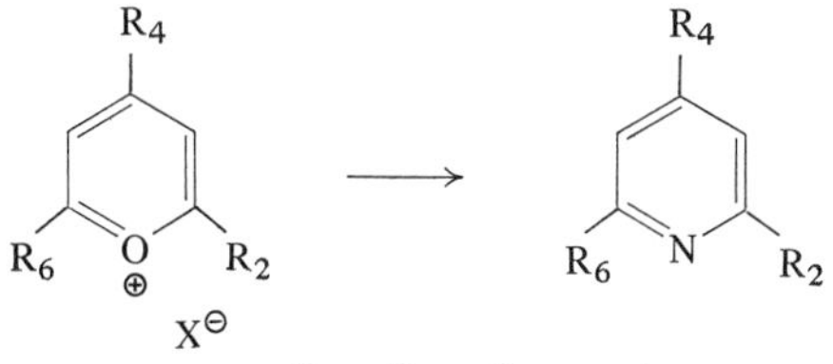

$R_2 = R_4 = R_6 = $ aryl

R_2	R_4	R_6	Reference
$4\text{-}BrC_6H_4$	$4\text{-}BrC_6H_4$	$4\text{-}BrC_6H_4$	75
$4\text{-}OCH_3\text{-}3\text{-}CH_3C_6H_3$	C_6H_5	C_6H_5	75
$4\text{-}OH\text{-}3\text{-}CH_3C_6H_3$	C_6H_5	C_6H_5	75
$2\text{-}OH\text{-}4\text{-}CH_3C_6H_3$	C_6H_5	$4\text{-}OCH_3C_6H_4$	76
$2\text{-}OH\text{-}4\text{-}CH_3C_6H_3$	C_6H_5	C_6H_5	76
$2\text{-}OH\text{-}4\text{-}CH_3C_6H_3$	C_6H_5	$2\text{-}OHC_6H_4$	76
$2\text{-}OH\text{-}4\text{-}OCH_3C_6H_3$	$4\text{-}OCH_3C_6H_4$	C_6H_5	76
$2\text{-}OH\text{-}4\text{-}OCH_3C_6H_3$	C_6H_5	$4\text{-}OCH_3C_6H_4$	76
$2\text{-}OH\text{-}4\text{-}OCH_3C_6H_3$	$4\text{-}OCH_3C_6H_4$	$4\text{-}OCH_3C_6H_4$	76

Table 4.3—*continued*

R_2	R_4	R_6	Reference
$3\text{-}NO_2C_6H_4$	C_6H_5	C_6H_5	77
C_6H_5	$3\text{-}NO_2C_6H_4$	C_6H_5	77, 78
C_6H_5	3-indolyl	3-indolyl	79
C_6H_5	$1\text{-}CH_3\text{-}2\text{-pyrrolyl}$	$1\text{-}CH_3\text{-}2\text{-pyrrolyl}$	79
C_6H_5	5-triphenylmethyl-2-thienyl	5-triphenylmethyl-2-thienyl	79
2-thienyl	C_6H_5	C_6H_5	79
2-furyl	C_6H_5	C_6H_5	79
C_6H_5	$4\text{-}FC_6H_4$	C_6H_5	80
C_6H_5	$4\text{-}ClC_6H_4$	C_6H_5	80
C_6H_5	$4\text{-}BrC_6H_4$	C_6H_5	80
$4\text{-}NO_2C_6H_4$	$4\text{-}NO_2C_6H_4$	$4\text{-}NO_2C_6H_4$	80
C_6H_5	C_6H_5	C_6H_5	78, 81-86, 89, 90
C_6H_5	$4\text{-}NO_2C_6H_4$	C_6H_5	78, 82
$4\text{-}OCH_3C_6H_4$	$4\text{-}NO_2C_6H_4$	$4\text{-}OCH_3C_6H_4$	82
C_6H_5	$4\text{-}N(CH_3)_2C_6H_4$	C_6H_5	78, 84, 87, 88
$4\text{-}OCH_3C_6H_4$	$3\text{-}NO_2C_6H_4$	$4\text{-}OCH_3C_6H_4$	78
$4\text{-}OCH_3C_6H_4$	$4\text{-}N(CH_3)_2C_6H_4$	$4\text{-}OCH_3C_6H_4$	78
$\alpha\text{-}C_{10}H_7$	$4\text{-}NO_2C_6H_4$	$\alpha\text{-}C_{10}H_7$	78
C_6H_5	$4\text{-}CH_3C_6H_4$	C_6H_5	83, 84, 86
C_6H_5	$4\text{-}OCH_3C_6H_4$	C_6H_5	84, 89
C_6H_5	$4\text{-}ClC_6H_4$	C_6H_5	84
C_6H_5	$4\text{-}N(C_2H_5)_2C_6H_4$	C_6H_5	84
C_6H_5	$4\text{-}CH_3C_6H_4$	$4\text{-}CH_3C_6H_4$	84
$4\text{-}CH_3C_6H_4$	$4\text{-}CH_3C_6H_4$	$4\text{-}CH_3C_6H_4$	84
$4\text{-}CH_3C_6H_4$	$4\text{-}OCH_3C_6H_4$	$4\text{-}CH_3C_6H_4$	84
$4\text{-}ClC_6H_4$	C_6H_5	$p\text{-}ClC_6H_4$	84, 89
$4\text{-}ClC_6H_4$	$2\text{-}ClC_6H_4$	$4\text{-}ClC_6H_4$	84
$4\text{-}BrC_6H_4$	C_6H_5	$4\text{-}BrC_6H_4$	84
$\alpha\text{-}C_{10}H_7$	C_6H_5	C_6H_5	86
$\alpha\text{-}C_{10}H_7$	C_6H_5	$\alpha\text{-}C_{10}H_7$	86
$\beta\text{-}C_{10}H_7$	C_6H_5	C_6H_5	86, 89
$\beta\text{-}C_{10}H_7$	C_6H_5	$\beta\text{-}C_{10}H_7$	86
$4\text{-}C_6H_5\text{-}C_6H_4$	C_6H_5	C_6H_5	86, 89
$4\text{-}C_6H_5\text{-}C_6H_4$	C_6H_5	$\beta\text{-}C_{10}H_7$	86
$4\text{-}BrC_6H_4$	C_6H_5	C_6H_5	86
$C_6H_5CH = CH$	C_6H_5	C_6H_5	86
C_6H_5	$C_6H_5CH = CH$	C_6H_5	86
$4\text{-}ClC_6H_4$	C_6H_5	$4\text{-}ClC_6H_4$	89
$4\text{-}CH_3C_6H_4$	C_6H_5	$4\text{-}CH_3C_6H_4$	89
$4\text{-}OCH_3C_6H_4$	C_6H_5	C_6H_5	89
$4\text{-}OCH_3C_6H_4$	$4\text{-}OCH_3C_6H_4$	C_6H_5	89
$4\text{-}OCH_3C_6H_4$	C_6H_5	$4\text{-}OCH_3C_6H_4$	89
$4\text{-}OCH_3C_6H_4$	$4\text{-}OCH_3C_6H_4$	$4\text{-}OCH_3C_6H_4$	89

Table 4.4 *2,4,6-Trialkyl(aryl) substituted pyridines*

R = alkyl, aryl

R_2	R_4	R_6	Reference
CH_3	C_6H_5	CH_3	68-70, 97
CH_3	$3,4-(OCH_3)_2C_6H_3$	$3,4-(OCH_3)_2C_6H_3$	91
CH_3	$2,5-(OCH_3)_2C_6H_3$	$2,5-(OCH_3)_2C_6H_3$	91
CH_3	$3,4-(OCH_3)_2C_6H_3$	$3,4-(OC_2H_5)_2C_6H_3$	91
CH_3	$4-BrC_6H_4$	$4-BrC_6H_4$	91
CH_3	$4-NO_2C_6H_4$	$4-NO_2C_6H_4$	91
CH_3	$\alpha-C_{10}H_7$	$\alpha-C_{10}H_7$	91
C_2H_5	C_6H_5	C_6H_5	91, 92
C_3H_7	C_6H_5	C_6H_5	91
C_3H_7	$4-OCH_3C_6H_4$	$4-OCH_3C_6H_4$	91
CH_3	$4-CH_3C_6H_4$	$4-CH_3C_6H_4$	93
CH_3	C_6H_5	C_6H_5	85, 92, 94
CH_3	CH_3	C_6H_5	69
C_2H_5	C_6H_5	C_2H_5	69
C_6H_5	CH_3	C_6H_5	69, 81
C_6H_5	C_2H_5	C_6H_5	69

A more detailed study of these conversions has revealed that in the reaction of 2,4,6-triphenylpyrylium perchlorate with ammonia, besides 2,4,6-triphenyl-pyridine, a product is formed, to which, based on the results of ultraviolet and infrared spectroscopic examinations the imino enol structure (4.90) was assigned[102]. A derivative of this imino enol with benzodioxoborole, i.e. (4.91) has been prepared. Since the compound (4.90) can be hydrated to give triphenylpyridine it provides evidence for the presence of open-chain inter-mediates during the ring transformation of pyrylium salts into pyridines[102]. Further evidence for the occurrence of an open-chain intermediate is provided by the finding that when an 4-acetylalkylpyrylium salt (4.92) is aminated the compound (4.95) is obtained and not the expected isomeric pyridine (4.94)[103]. In the open-chain intermediate (4.93) the carbonyl group attached to the aliphatic group apparently participates in the ring closure and not the carbonyl group attached to the aryl group.

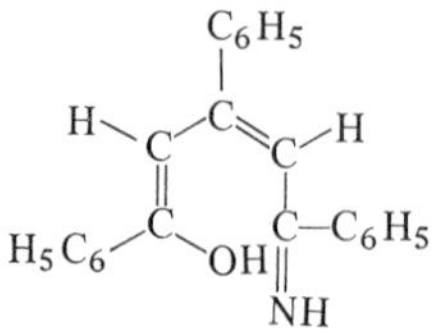

(4.90)

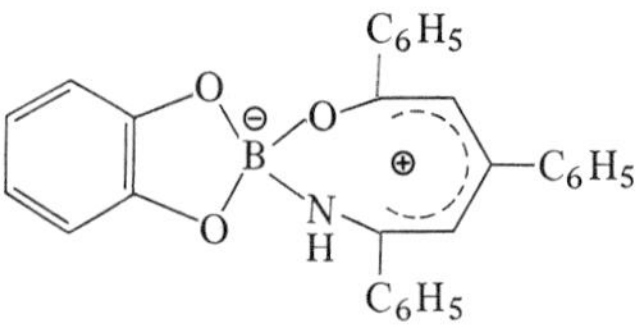

(4.91)

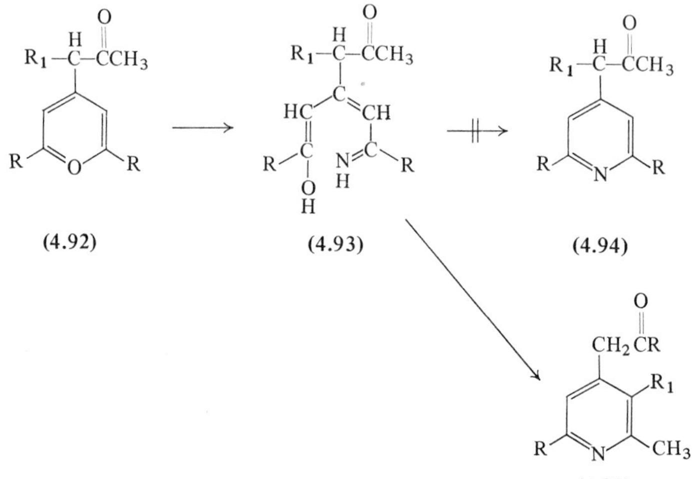

(4.92)   →   (4.93)   ─‖→   (4.94)

(4.95)

R	C_6H_5	p-BrC$_6$H$_4$
R_1	CH_3	$CH_2C_6H_5$

By this method it is possible to obtain with comparative ease certain pyridine derivatives which are difficult to prepare by other procedures; as examples we can cite the preparation of the pyridine derivative (4.96)[14], the polyalkene pyridines (4.97) and (4.98)[95], and the preparation of 2,4,6-tri(methyl-$d3$)pyridine (4.99) from the corresponding 2,4,6-tri(methyl-$d3$)pyrylium salt[10,104].

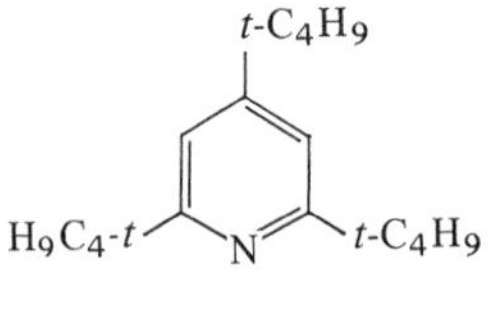

(4.96)

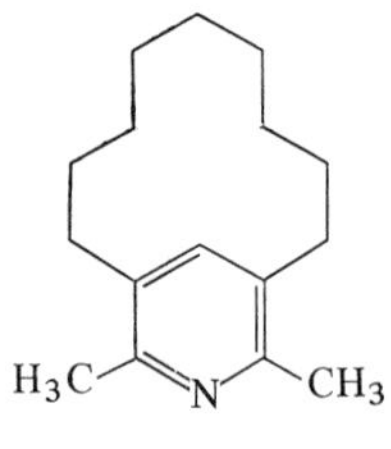

(4.97)

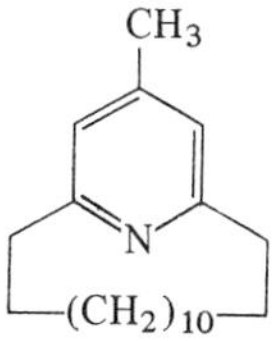

CH$_3$

—(CH$_2$)$_{10}$—

(4.98)

CD$_3$

D$_3$C   N   CD$_3$

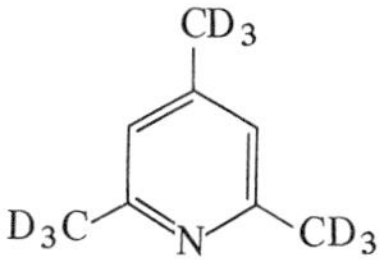

(4.99)

Its general applicability can further be demonstrated by the fact that these ring transformations are not confined only to monocyclic pyrylium salts. Bicyclic, tricyclic and polycyclic pyrylium salts also can be converted into the corresponding pyridines by this method; thus for example, the bicyclic 2,4-diphenyltetrahydrobenzopyrylium salt (4.100)[105,106], the substituted isobenzopyrylium salt (4.101)[67,107,108] and its tetrahydro derivative[99], the octahydroxanthylium salts (4.102)[109] and its dibenzo derivative (4.103)[110], the furopyrylium salt (4.104)[111] and the indolopyrylium salts (4.105)[112] can be converted into their corresponding pyridines. Recently m- and p-phenylene bispyrylium salts (4.106) have been prepared and then converted into the corresponding bispyridines[113-115].

The conversion of pyrylium salts into pyridines is not only restricted to a reaction with ammonia. The ring contraction can also be performed with primary aliphatic amines, aromatic amines or amino-acids, N-substituted pyridinium compounds being obtained.

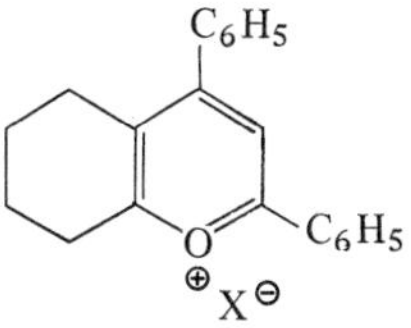

C$_6$H$_5$

O   C$_6$H$_5$
⊕ X⊖

(4.100)

R$_1$ —→   R$_3$

O⊕   X⊖
R$_2$

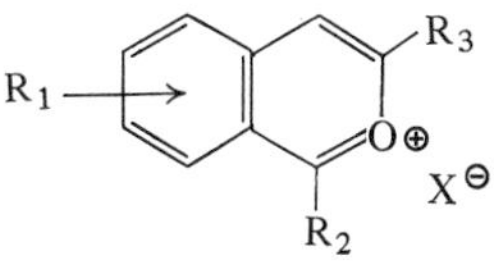

(4.101)

R$_1$ = OCH$_3$
R$_2$ = CH$_3$; C$_2$H$_5$; X–C$_6$H$_4$
R$_3$ = CH$_3$

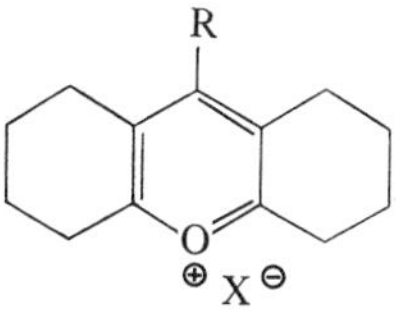

R

O
⊕ X⊖

(4.102)

R = CH$_3$; H

O
⊕ X⊖

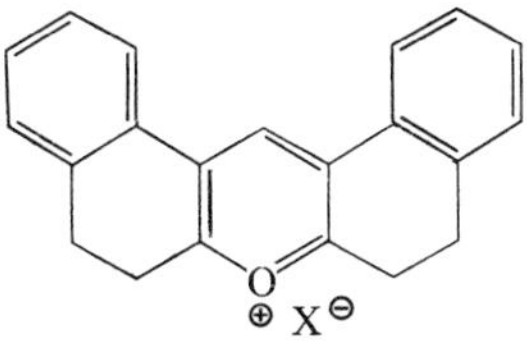

(4.103)

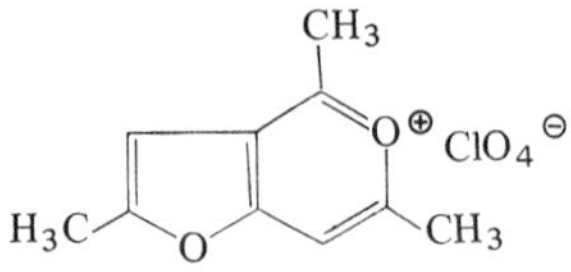
CH3
O⊕ ClO4⊖
H3C
O
CH3

(4.104)

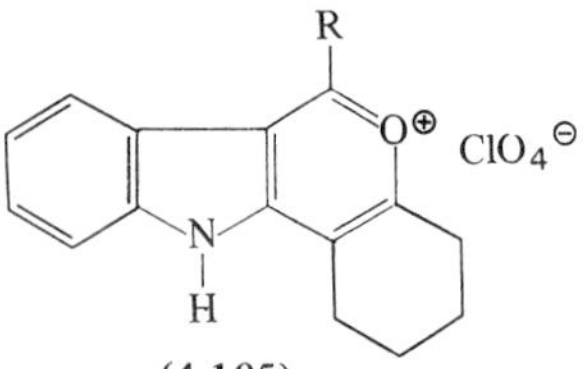
R
O⊕ ClO4⊖
N
H

(4.105)

R = CH_3; C_3H_7

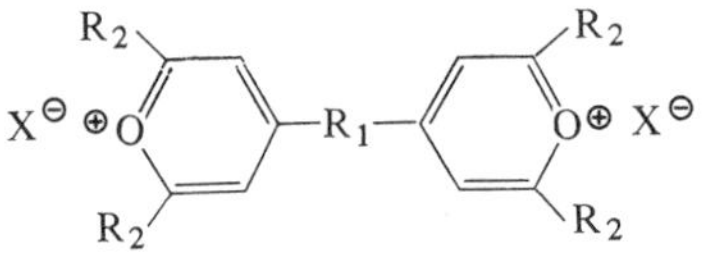
R2
R2
X⊖ ⊕O
R1
O⊕ X⊖
R2
R2

(4.106)

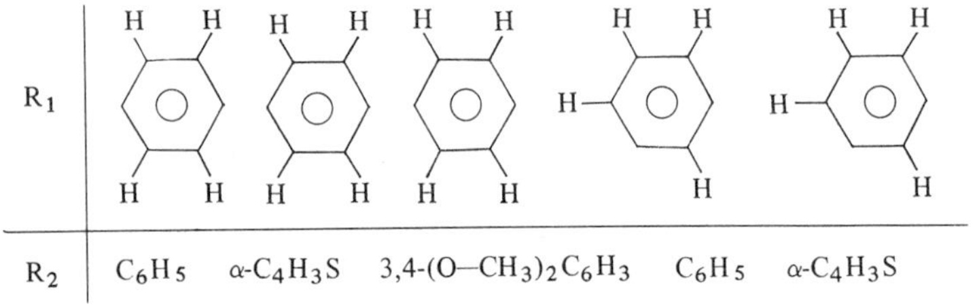
R1
H
R2
C6H5
α-C4H3S
3,4-(O—CH3)2C6H3
C6H5
α-C4H3S

It has been found that this method can successfully be used for the preparation of those quaternary pyridinium salts which are somewhat difficult to obtain by other methods. Also with semicarbazide, alkyl- and aryl-substituted hydrazines, or with hydroxylamine, the corresponding pyridinium compounds have been formed. A survey of the pyridinium salts obtained by this method is given in Table 4.6.

The behaviour of 2,4,6-triarylpyrylium salts with an excess of *t*-amines has been investigated and it has established that the pseudo bases (4.107) are formed. With an excess of secondary amines the red *N,N*-dialkylketodienamines (4.108) are obtained[127].

As a useful extension of the general applicability of this reaction, it has been found that the 4-alkoxy and 4-thiomethyl group present in pyrylium salts can be replaced by nucleophilic groups such as alkylamino or dialkylamino groups[130,131]. This means that in the reaction of (4.109) with an excess of primary amines replacement of the 4-alkoxy group as well as the oxygen of the pyrylium salt takes place, yielding the pyridinium salt (4.110)[130]. As it is not possible to bring about ring transformation with secondary amines (piperidine,

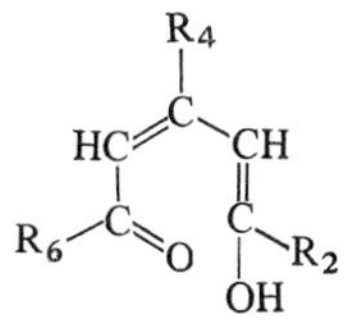

(4.107) 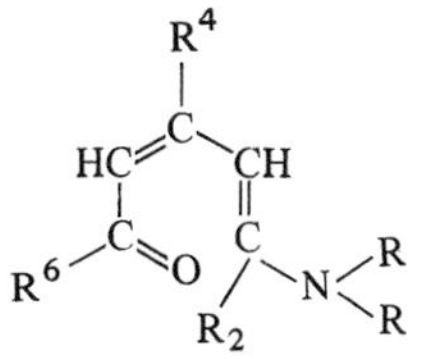   (4.108)

morpholine, or piperazine), only replacement of the 4-methoxygroup occurs yielding (4.111); a subsequent treatment of (4.111) with ammonia gives the pyridine derivative (4.112).

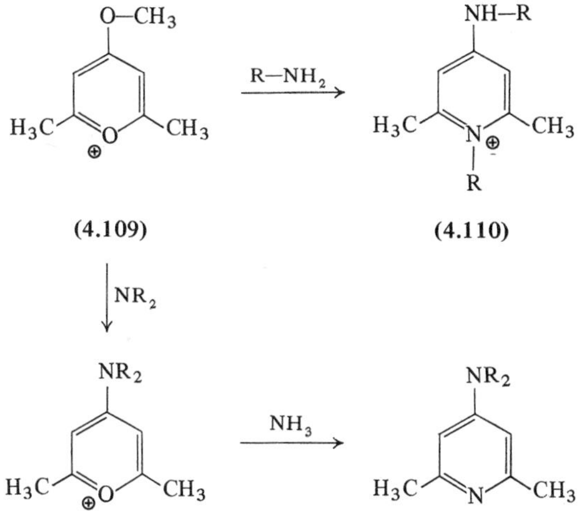

(4.109)   (4.110)

$R = CH_3; C_6H_5$

(4.111)   (4.112)

b. *Pyrylium salts into thiapyrylium salts, phosphorines and phosphapyrans*
In close correlation with the conversions described in the previous section, 2,4,6-triaryl- or tristyrylpyrylium salts (4.113) are found to be readily converted into the thiopyrylium salts (4.115) by treatment with disodium sulphide in aqueous acetone[132,133]. During the course of the reaction the mixture takes on a blue-red colour, and this is attributed to the presence of the intermediary ketothioenolate (4.114). This method can be used for the conversion of a large number of pyrylium salts into the corresponding thiapyrylium salts. A summary of these conversions is given in Table 4.7.

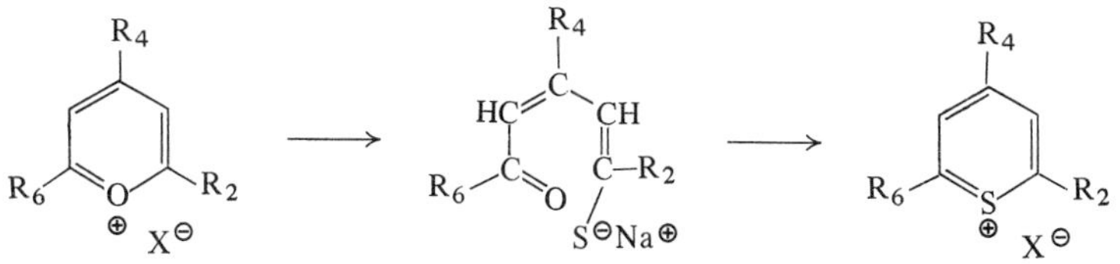

(4.113)   (4.114)   (4.115)

Table 4.5 *Tetra- and pentaalkyl(aryl)pyridines*

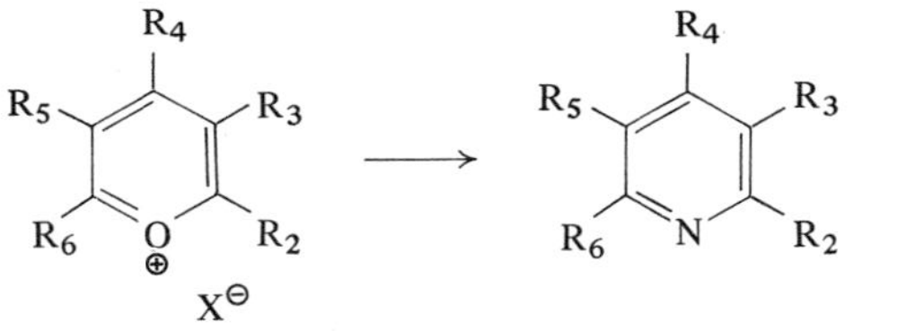

R = alkyl, aryl

R₂	R₃	R₄	R₅	R₆	Reference
CH$_3$	R$_3$ = R$_5$ = (CH$_2$)$_9$	H	R$_5$ = R$_3$ = (CH$_2$)$_9$	CH$_3$	95
C$_6$H$_5$	C$_6$H$_5$	C$_6$H$_5$	C$_6$H$_5$	C$_6$H$_5$	75
C$_6$H$_5$	H	C$_6$H$_5$	C$_6$H$_5$	C$_6$H$_5$	75, 86
p-BrC$_6$H$_4$	H	C$_6$H$_5$	C$_6$H$_5$	C$_6$H$_5$	75
CH$_3$	t-C$_4$H$_9$	CH$_3$	H	CH$_3$	65, 72, 98
CH$_3$	CH$_3$	CH$_3$	H	CH$_3$	70, 72, 73
CH$_3$	H	(CH$_2$)$_4$	·	CH$_3$	96, 99
C$_2$H$_5$	H	(CH$_2$)$_4$		C$_2$H$_5$	96, 99
CH$_3$	H	(CH$_2$)$_3$		CH$_3$	96, 99
C$_2$H$_5$	H	(CH$_2$)$_3$		C$_2$H$_5$	96, 99
C$_6$H$_5$	CH$_3$	C$_6$H$_5$	H	C$_6$H$_5$	86
C$_6$H$_5$	COC$_6$H$_4$	C$_6$H$_5$	H	C$_6$H$_5$	86

C_6H_5	H	C_6H_5				86
α-$C_{10}H_7$	H	C_6H_5				86
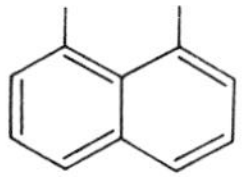		C_6H_5				86
CH_3	H	$[6,7(OCH_3)_2(CH)_4]$			CH_3	100
CH_3	H	$[6,7(OCH_3)_2(CH)_4]$			C_2H_5	100
CH_3	H	$[6,7(OCH_3)_2(CH)_4]$			p-$OCH_3C_6H_4$	100
CH_3	H	$(CH)_4$			CH_3	101
C_6H_5	H	$(CH)_4$			C_6H_5	101

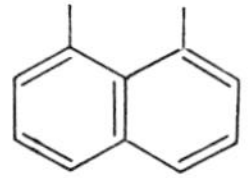

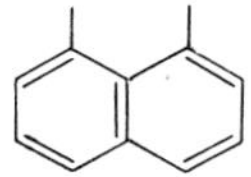

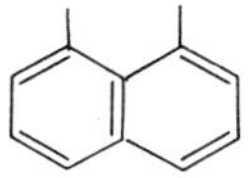

Table 4.6 *N-aryl (alkyl) pyridinium compounds*

$$R = alkyl, aryl$$

R_1	R_2	R_4	R_6	Reference
4-CH$_3$C$_6$H$_4$	CH$_3$	CH$_3$	CH$_3$	68, 1
4-CH$_3$C$_6$H$_4$	CH$_3$	C$_6$H$_5$	CH$_3$	68
C$_6$H$_5$	C$_6$H$_5$	C$_6$H$_5$	C$_6$H$_5$	117
4-CH$_3$C$_6$H$_4$	C$_6$H$_5$	C$_6$H$_5$	C$_6$H$_5$	117
2-CH$_3$C$_6$H$_4$	C$_6$H$_5$	C$_6$H$_5$	C$_6$H$_5$	117
4-OC$_2$H$_5$C$_6$H$_4$	C$_6$H$_5$	C$_6$H$_5$	C$_6$H$_5$	117
2-OCH$_3$C$_6$H$_4$	C$_6$H$_5$	C$_6$H$_5$	C$_6$H$_5$	117
4-OCH$_3$C$_6$H$_4$	C$_6$H$_5$	C$_6$H$_5$	C$_6$H$_5$	117
4-OHC$_6$H$_4$	C$_6$H$_5$	C$_6$H$_5$	C$_6$H$_5$	117
C$_6$H$_5$	CH$_3$	4-CH$_3$C$_6$H$_4$	4-CH$_3$C$_6$H$_4$	93
CH$_3$NC$_6$H$_5$	CH$_3$	4-CH$_3$C$_6$H$_4$	4-CH$_3$C$_6$H$_4$	93
C$_6$H$_5$	CH$_3$	C$_6$H$_5$	C$_6$H$_5$	94
CH$_2$C$_6$H$_5$	CH$_3$	C$_6$H$_5$	C$_6$H$_5$	94
CH$_3$NC$_6$H$_5$	CH$_3$	C$_6$H$_5$	C$_6$H$_5$	94
CH$_3$NH	CH$_3$	C$_6$H$_5$	C$_6$H$_5$	94, 1
C$_6$H$_5$NC$_6$H$_5$	CH$_3$	C$_6$H$_5$	C$_6$H$_5$	94
HNC$_6$H$_5$	C$_6$H$_5$	CH$_3$	C$_6$H$_5$	94, 1
HNC$_6$H$_5$	C$_6$H$_5$	C$_6$H$_5$	C$_6$H$_5$	94
HNC$_6$H$_5$	CH$_3$	C$_6$H$_5$	C$_6$H$_5$	119
HNC$_6$H$_5$	CH$_3$	4-OCH$_3$C$_6$H$_4$	4-OCH$_3$C$_6$H$_4$	120
HNC$_6$H$_4$-4CH$_3$	CH$_3$	4-OCH$_3$C$_6$H$_4$	4-OCH$_3$C$_6$H$_4$	120
HNC$_6$H$_4$-4Br	CH$_3$	4-OCH$_3$C$_6$H$_4$	4-OCH$_3$C$_6$H$_4$	120
HNC$_6$H$_5$[a]	CH$_3$	4-CH$_3$C$_6$H$_4$	4-CH$_3$C$_6$H$_4$	120
HNC$_6$H$_5$[b]	CH$_3$	4-BrC$_6$H$_4$	4-BrC$_6$H$_4$	120
HNC$_6$H$_5$[c]	CH$_3$	3-CH$_3$C$_6$H$_4$	3-CH$_3$C$_6$H$_4$	120
HNC$_6$H$_5$	CH$_3$	3-BrC$_6$H$_4$	3-BrC$_6$H$_4$	120
HNC$_6$H$_5$[d]	CH$_3$	4-ClC$_6$H$_4$	4-ClC$_6$H$_4$	120
HNC$_6$H$_5$[c]	CH$_3$	3-ClC$_6$H$_4$	3-ClC$_6$H$_4$	120
HNC$_6$H$_5$	C$_2$H$_5$	C$_6$H$_5$	C$_6$H$_5$	120
HNC$_6$H$_5$	C$_2$H$_5$	4-CH$_3$C$_6$H$_4$	4-CH$_3$C$_6$H$_4$	120
C$_6$H$_5$	C$_6$H$_5$	4-OHC$_6$H$_4$	C$_6$H$_5$	121
OH	CH$_3$	CH$_3$	CH$_3$	122,
OH	C$_2$H$_5$	CH$_3$	C$_2$H$_5$	122,
OH	HC(CH$_3$)$_2$	CH$_3$	CH(CH$_3$)$_2$	122
OH	C$_6$H$_5$	CH$_3$	C$_6$H$_5$	122
NH-CO(NH$_2$)	CH$_3$	CH$_3$	CH$_3$	122,
C$_6$H$_5$CH$_2$CHCO$_2$H	CH$_3$	CH$_3$	CH$_3$	124

Table 4.6—*continued*

R_1	R_2	R_4	R_6	Reference
$C_2H_5CHCO_2H$	CH_3	CH_3	CH_3	124
CH_3CHCO_2H	CH_3	CH_3	CH_3	124
$(CH_3)_2CHCHCO_2H$	CH_3	CH_3	CH_3	124
CH_3	CH_3	C_6H_5	C_6H_5	125
$CH_2COOC_2H_5$	CH_3	CH_3	CH_3	123
CH_2COOH	CH_3	CH_3	CH_3	126, 122
$CH_2COOC_2H_5$	C_6H_5	C_6H_5	C_6H_5	123
$CH_2(CH_2)_3CHNH_2COOH$	CH_3	CH_3	CH_3	123
$CH_2(CH_2)_3CHNH_2COOH$	CH_3	CH_3	CH_3	123
OH	C_2H_5	C_2H_5	C_2H_5	69
OH	CH_3	C_2H_5	CH_3	69
CH_2COOH	CH_3	CH_3	CH_3	69
CH_2COOH	CH_3	C_2H_5	CH_3	69
$NHCONH_2$	CH_3	C_2H_5	CH_3	69
CH_3	C_6H_5	C_6H_5	C_6H_5	127, 128
C_2H_5	C_6H_5	C_6H_5	C_6H_5	127
CH_2CH_2OH	C_6H_5	C_6H_5	C_6H_5	127
CH_3	$4\text{-}CH_3C_6H_4$	$4\text{-}CH_3C_6H_4$	$4\text{-}CH_3C_6H_4$	127
C_2H_5	$4\text{-}CH_3C_6H_4$	$4\text{-}CH_3C_6H_4$	$4\text{-}CH_3C_6H_4$	127
CH_3	$4\text{-}ClC_6H_4$	C_6H_5	$4\text{-}ClC_6H_4$	127
C_2H_5	$4\text{-}ClC_6H_4$	C_6H_5	$4\text{-}ClC_6H_4$	127
CH_3	C_6H_5	$4\text{-}N(CH_3)_2C_6H_4$	C_6H_5	88
CH_3	$4\text{-}N(CH_3)_2C_6H_4$	$4\text{-}N(CH_3)_2C_6H_4$	C_6H_5	88
CH_3	$4\text{-}N(CH_3)_2C_6H_4$	C_6H_5	$4\text{-}N(CH_3)_2C_6H_4$	88
CH_3	$4\text{-}N(CH_3)_2C_6H_4$	$4\text{-}N(CH_3)_2C_6H_4$	$4\text{-}N(CH_3)_2C_6H_4$	88
CH_3	$4\text{-}N(CH_3)_2C_6H_4$	$4\text{-}N(CH_3)_2C_6H_4$	C_6H_5	88
C_6H_5	CH_3	CH_3	CH_3	116
$4\text{-}ClC_6H_4$	CH_3	CH_3	CH_3	116
$4\text{-}BrC_6H_4$	CH_3	CH_3	CH_3	116
$4\text{-}JC_6H_4$	CH_3	CH_3	CH_3	116
$2\text{-}OHC_6H_4$	CH_3	CH_3	CH_3	116
$4\text{-}OHC_6H_4$	CH_3	CH_3	CH_3	116
$4\text{-}NH_2C_6H_4$	CH_3	CH_3	CH_3	116
$4\text{-}OHC_6H_4$	C_6H_5	C_6H_5	C_6H_5	129
$4\text{-}OHC_6H_4$	$4\text{-}CH_3C_6H_4$	C_6H_5	C_6H_5	129
$4\text{-}OHC_6H_4$	C_6H_5	$4\text{-}CH_3C_6H_4$	C_6H_5	129
$4\text{-}OHC_6H_4$	$4\text{-}CH_3C_6H_4$	C_6H_5	$4\text{-}CH_3C_6H_4$	129
$4\text{-}OHC_6H_4$	$4\text{-}C_2H_5C_6H_4$	$4\text{-}CH_3C_6H_4$	$4\text{-}C_2H_5C_6H_4$	129
$4\text{-}OHC_6H_4$	$4\text{-}i\text{-}C_5H_{11}C_6H_4$	C_6H_5	$4\text{-}i\text{-}C_5H_{11}C_6H_4$	129
$4\text{-}OHC_6H_4$	$4\text{-}ClC_6H_4$	C_6H_5	C_6H_5	129
$4\text{-}OHC_6H_4$	$4\text{-}ClC_6H_4$	C_6H_5	$4\text{-}ClC_6H_4$	129
$4\text{-}OHC_6H_4$	$4\text{-}OCH_3C_6H_4$	C_6H_5	C_6H_5	129
$4\text{-}OHC_6H_4$	C_6H_5	$4\text{-}OCH_3C_6H_4$	C_6H_5	129
$4\text{-}OHC_6H_4$	$4\text{-}OCH_3C_6H_4$	C_6H_5	$4\text{-}OCH_3C_6H_4$	129
$4\text{-}OHC_6H_4$	$4\text{-}OCH_3C_6H_4$	$4\text{-}OCH_3C_6H_4$	$4\text{-}OCH_3C_6H_4$	129
$4\text{-}OHC_6H_4$	$4\text{-}NO_2C_6H_4$	C_6H_5	C_6H_5	129
$4\text{-}OHC_6H_4$	C_6H_5	$4\text{-}NO_2C_6H_4$	C_6H_5	129
$4\text{-}OHC_6H_4$	$4\text{-}NO_2C_6H_4$	C_6H_5	$4\text{-}NO_2C_6H_4$	129

Table 4.6—*continued*

R_1	R_2	R_4	R_6	Reference
$4\text{-}OHC_6H_4$	CH_3	C_6H_5	C_6H_5	129
$4\text{-}OHC_6H_4$	C_6H_5	CH_3	C_6H_5	129
$4\text{-}OHC_6H_4$	$C(CH_3)_3$	C_6H_5	C_6H_5	129
$4[4\text{-}OHC_6H_4]C_6H_4$	C_6H_5	C_6H_5	C_6H_5	129
$2,6\text{-}(CH_3)_2\text{-}4\text{-}OHC_6H_2$	C_6H_5	C_6H_5	C_6H_5	129
$3,5\text{-}(CH_3)_2\text{-}4\text{-}OHC_6H_2$	C_6H_5	C_6H_5	C_6H_5	129
$3,5\text{-}(C_2H_5)_2\text{-}4\text{-}OHC_6H_2$	C_6H_5	C_6H_5	C_6H_5	129
$3,5\text{-}CH(CH_3)_2\text{-}4\text{-}OHC_6H_2$	C_6H_5	C_6H_5	C_6H_5	129
$3,5\text{-}C(CH_3)_3\text{-}4\text{-}OHC_6H_2$	C_6H_5	C_6H_5	C_6H_5	129
$\alpha\text{-}[5\text{-}OH]\text{-}C_{10}H_6$	C_6H_5	C_6H_5	C_6H_5	129
$\alpha\text{-}[4\text{-}OH]\text{-}C_{10}H_6$	C_6H_5	C_6H_5	C_6H_5	129
$3,5\text{-}(CH_2)_7\text{-}4\text{-}OHC_6H_2$	C_6H_5	C_6H_5	C_6H_5	129
$3,5\text{-}(CH_2)_9\text{-}4\text{-}OHC_6H_2$	C_6H_5	C_6H_5	C_6H_5	129
$3,5\text{-}(CH_2)_{12}\text{-}4\text{-}OHC_6H_2$	C_6H_5	C_6H_5	C_6H_5	129
$3,5\text{-}(C_6H_5)_2\text{-}4\text{-}OHC_6H_2$	C_6H_5	C_6H_5	C_6H_5	129
$CH_2C_6H_5$	CH_3	CH_3	CH_3	116
CH_3	CH_3	CH_3	CH_3	116
C_2H_5	CH_3	CH_3	CH_3	116
$n\text{-}C_3H_7$	CH_3	CH_3	CH_3	116
$n\text{-}C_4H_9$	CH_3	CH_3	CH_3	116
$i\text{-}C_5H_{11}$	CH_3	CH_3	CH_3	116
$cyclo\text{-}C_6H_{11}$	CH_3	CH_3	CH_3	116
$n\text{-}C_7H_5$	CH_3	CH_3	CH_3	116
$n\text{-}C_{18}H_{37}$	CH_3	CH_3	CH_3	116
	CH_3	CH_3	CH_3	129a
	CH_3	CH_3	CH_3	129a
	CH_3	CH_3	CH_3	129a
	CH_3	CH_3	CH_3	129a

[a] The N-anilino derivatives (p-Br, p-CH$_3$ and m-CH$_3$) with the same groups R_2, R_4 and R_6 are also prepared[120].

[b] The N-anilino derivatives (p-CH$_3$ and p-Br) with the same groups R_2, R_4 and R_6 are also prepared[120].

[c] The N-anilino derivatives (p-CH$_3$) with the same groups R_2, R_4 and R_6 are also prepared[120].

[d] The N-anilino derivatives (p-CH$_3$ and m-CH$_3$) are also prepared[120].

Table 4.7 *2,4,6-Trisubstituted thiapyrylium salts*

R_2	R_4	R_6
$_5H_5$	C_6H_5	C_6H_5
$_5H_5$	$4\text{-}OCH_3C_6H_4$	C_6H_5
$OCH_3C_6H_4$	C_6H_5	C_6H_5
$OCH_3C_6H_4$	C_6H_5	$4\text{-}OCH_3C_6H_4$
$OCH_3C_6H_4$	$4\text{-}OCH_3C_6H_4$	$4\text{-}OCH_3C_6H_4$
$OCH_3C_6H_4$	$4\text{-}OCH_3C_6H_4$	C_6H_5
$_,H_5$	$4\text{-}N(CH_3)_2C_6H_4$	C_6H_5
$N(CH_3)_2C_6H_4$	C_6H_5	C_6H_5
$N(CH_3)_2C_6H_4$	C_6H_5	$4\text{-}N(CH_3)_2C_6H_4$
$N(CH_3)_2C_6H_4$	$4\text{-}N(CH_3)_2C_6H_4$	$4\text{-}N(CH_3)_2C_6H_4$
$N(CH_3)_2C_6H_4$	$4\text{-}N(CH_3)_2C_6H_4$	C_6H_5
$_,H_5$	$CH_2{=}CHC_6H_4\text{-}4\text{-}N(CH_3)_2$	C_6H_5
$I_2{=}CH\text{-}C_6H_4\text{-}4\text{-}OCH_3$	C_6H_5	C_6H_5
$I_2{=}CH\text{-}C_6H_4\text{-}4\text{-}N(CH_3)_2$	C_6H_5	C_6H_5
$I_2{=}CH\text{-}C_6H_4\text{-}4\text{-}N(CH_?)_2$	$4\text{-}OCH_3C_6H_4$	$4\text{-}OCH_3C_6H_4$
$I_2{=}CH\text{-}C_6H_4\text{-}4\text{-}N(CH_3)_2$	C_6H_5	$CH_2{=}CH\text{-}C_6H_4\text{-}4\text{-}N(CH_3)_2$

Treatment of 4-methoxy-2,6-dimethylpyrylium perchlorate with potassium hydrosulphide in boiling methanol leads to replacement of the ring oxygen as well as the 4-methoxy group by sulphur[134].

The conversion of pyrylium salts (4.116) into phosphorus heterocycles (4.119) has recently been reported; phosphine[135], tris-hydroxymethyl-phosphine[14,136,137], tris-trimethylsilylphosphine[138] or phenylphosphine[139] were used as reagents. Although phosphine is only effective if catalysed by mineral acids, it is found to be the most versatile reagent; tris-hydroxymethyl-phosphine (PX_3) ($X = CH_2OH$) and tris-trimethylsilylphosphine (PX_3) ($X = Si(CH_3)_3$) are only capable of converting pyrylium salts containing bulky groups on positions 2, 4 and 6. Phosphine (PX_3) ($X = H$), however, can also react with pyrylium salts containing a methyl group in the 2- or 4-position. The various phosphorines obtained by these methods are listed in Table 4.8.

In this ring transformation the initial attack of the phosphorus-containing reagent has to take place at that position of the pyrylium ring which is most vulnerable to nucleophilic attack, that is position 2 (see section I.A.1.a). The

Table 4.8 *Polysubstituted phosphorines*

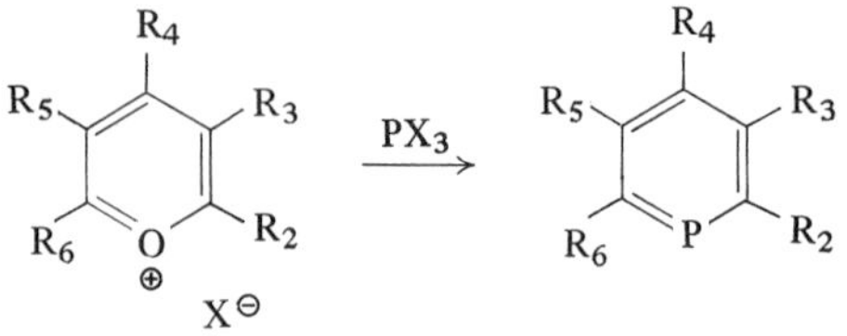

(4.116)   (4.119)

R_2	R_3	R_4	R_5	R_6	reagent PX_3	Refere
C_6H_5	H	C_6H_5	H	C_6H_5	X = Si$(CH_3)_3$ or H	135, 1
C_6H_5	H	C_6H_5	C_6H_5	C_6H_5	X = Si$(CH_3)_3$ or H	135, 1
C_6H_5	C_6H_5	C_6H_5	C_6H_5	C_6H_5	X = Si$(CH_3)_3$	138
4-$CH_3C_6H_4$	H	C_6H_5	H	4-$CH_3C_6H_4$	X = Si$(CH_3)_3$ or H	135, 1
C_6H_5	H	4-$OCH_3C_6H_4$	H	C_6H_5	X = Si$(CH_3)_3$ or H	135, 1
4-$OCH_3C_6H_4$	H	C_6H_5	H	4-$OCH_3C_6H_4$	X = Si$(CH_3)_3$	138
4-$OCH_3C_6H_4$	H	4-$OCH_3C_6H_4$	H	4-$OCH_3C_6H_4$	X = Si$(CH_3)_3$	138
t-C_4H_9	H	t-C_4H_9	H	t-C_4H_9	X = CH_2OH	14, 13
CH_3	H	C_6H_5	H	C_6H_5	X = H	138
C_6H_5	H	CH_3	H	C_6H_5	X = H	138

formation of the intermediate (4.117) becomes an irreversible step when (PX_3) (X = CH_2OH) is used as reagent, since in this reaction CH_2O is eliminated; on the other hand it is reversible with (PX_3) (X = H). Although (PX_3) (X = H) is a less voluminous reagent than, for example (PX_3) (X = CH_2OH or Si$(CH_3)_3$), its nucleophilicity is, however, low, making ring closure of the open-chain primary phosphine very unlikely. Thus it is for that purpose necessary to add acid to promote ring closure in this reaction.

When phenylphosphine reacts with 2,4,6-triphenylpyrylium salts in pyridine the hydrate of the 1-hydroxy-1,2,4,6-tetraphenyl derivative (4.121) and the

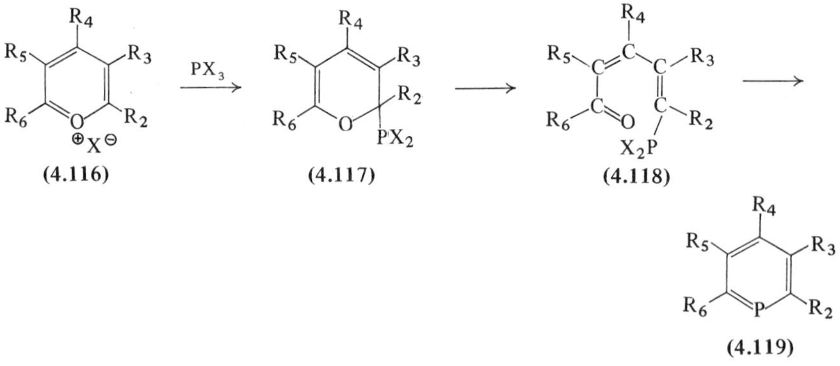

(4.116)   (4.117)   (4.118)

(4.119)

isomeric 2,3-dihydrophosphapyran oxide (4.122) are formed as main products[139]; the expected tetraphenylphosphorinium derivative (4.120) is not obtained. From a consideration of the mechanism it has been suggested that

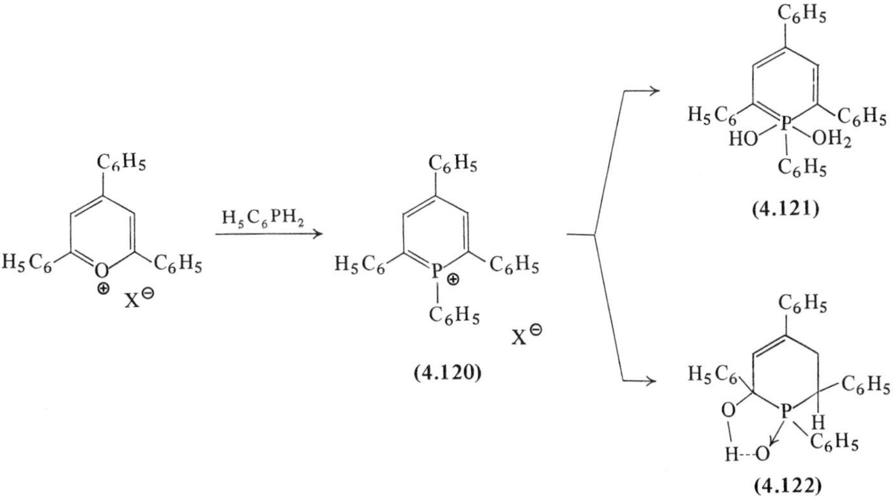

(4.121)

(4.120)

(4.122)

(4.120) is formed first and that this reacts further with a hydroxyl ion to form the phosphorabenzene (4.122a). Prototropic rearrangement of (4.122a) into (4.122b), followed by 1,4-addition of water, would give (4.122). However, the evidence provided by existing data cannot definitively exclude the possibility of an 1,2-addition of water yielding an isomer of (4.122)[139].

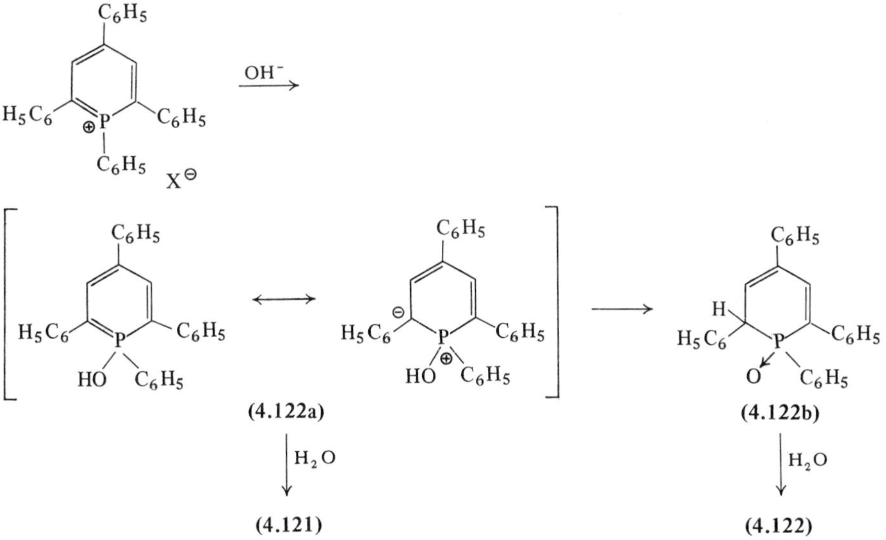

(4.122a)

(4.122b)

(4.121)

(4.122)

c. *Pyrans into pyridines, pyridazines and thiapyrans; tetra- and hexahydro-*
   coumarins into tetra- and hexahydroquinolines; isocoumarins into iso-
   carbostyrils

The aminolysis of derivatives of α-pyrone and γ-pyrone is found to be an important and most useful method for the preparation of 2-pyridones and 4-pyridones respectively. The reaction, which proceeds without difficulty is normally carried out with ammonia in aqueous or alcoholic solution. The reaction course essentially involves addition of ammonia or primary amines to the 2-position which in the case of the α-pyrone, leads to the intermediate (4.123); after ring opening into the aldehydoglutaconamide and subsequent dehydration, (4.124) (R = H) is yielded. Many examples of this ring transformation with substituted α- and γ-pyrones[140,141,141a] and with the corresponding pyranthiones[130,142-144] have been described. There are a few

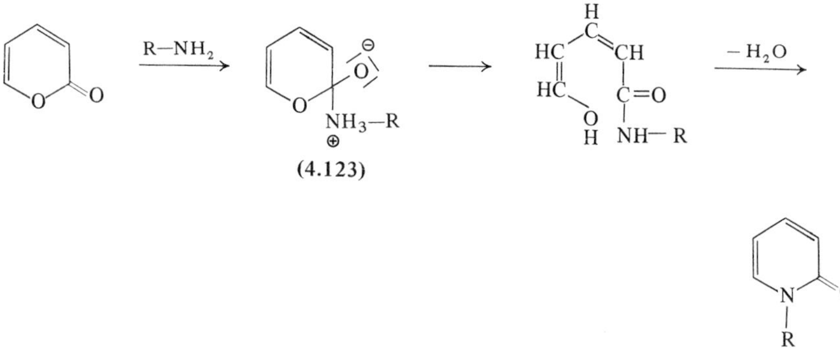

(4.123)

(4.124)

reviews dealing with this subject[2,145]. We mention the conversion of methyl-coumalate (4.125) (or acid) into 5-carboxypyridone-2 (4.126) (R = H) by treatment with ammonia[146-148], the corresponding reaction of (4.125) with primary amines to form (4.126) (R = $C_6H_5CH_2$; p-$OCH_3C_6H_4CH_2$; $C_6H_5CH(CH_3)$ and 4-$OCH_3C_6H_4CH_2CH_2$)[149], and the formation of 6-carboxy-2-pyridones from 2-pyrone-6-carboxylic acid[149]. In the reaction mixture obtained when methylcoumalate is treated with benzylamine, a crystalline product is found to be deposited, to which the structure (4.127) has been

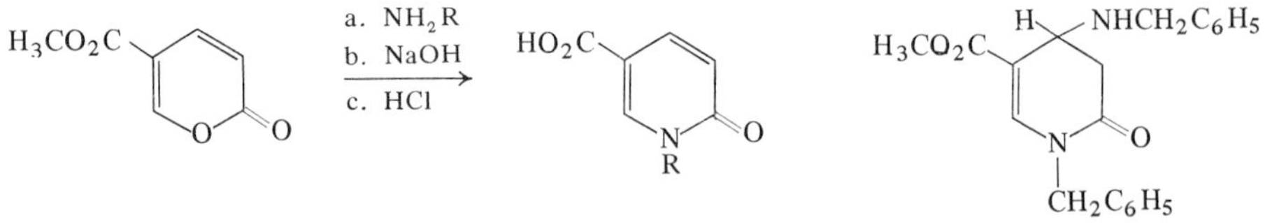

(4.125)   (4.126)   (4.127)

assigned. The formation of this adduct might give an indication of the course of the reaction. It has been suggested[145] that in this ring transformation first a 1:1 adduct is formed and subsequently a second mol of amine is added to the carbonyl group. Ring opening, dehydrative ring closure and finally aromatization by elimination of one mole of amine yields the pyridone.

Recently it has been reported that the tetrahydropyran-2-one (4.128), when heated with an acid at 140°C, decarboxylates into the N-phenyl-1,4,5,6-tetra-hydropyridine derivative (4.129)[150], and further that simple ethers derived from 2-hydroxy-3,4-dihydropyran, by treatment with either hydroxylamine in the presence of acid or with NH_3 in the vapour phase in the presence of a noble-metal catalyst[151], give pyridine[152,153].

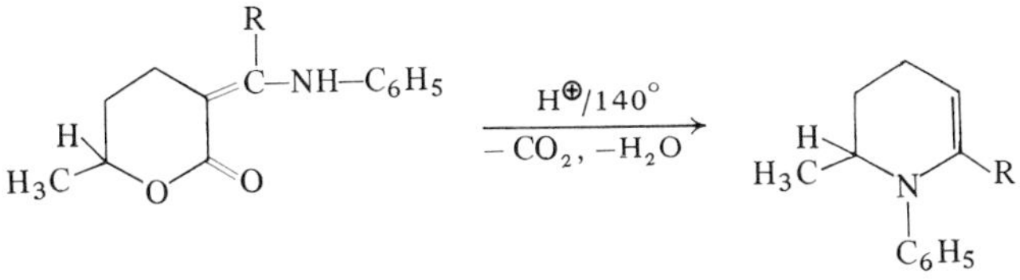

(4.128)   (4.129)

R = H; CH_3

For the reaction of γ-pyrone and its important derivatives, chelidonic acid (4.130), meconic acid (4.131), dehydroacetic acid (4.132), pyromeconic acid (4.133) and kojic acid (4.134) with ammonia or primary amines, we refer to the reviews cited above[2,145].

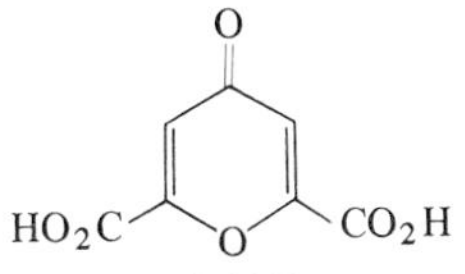

(4.130) 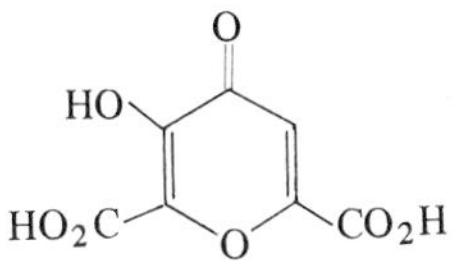   (4.131)

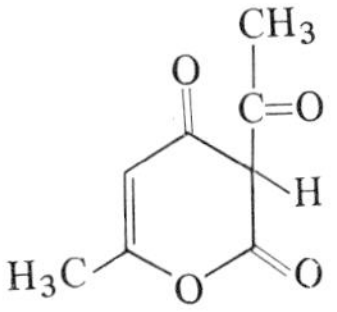

(4.132) 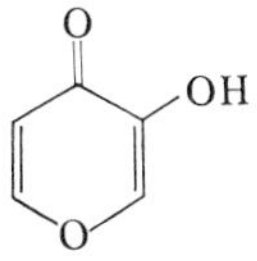   (4.133) 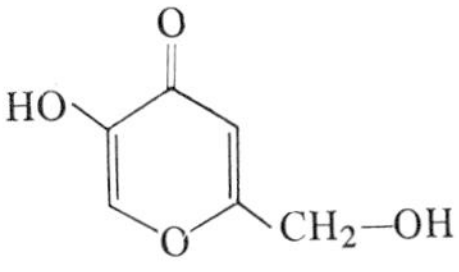   (4.134)

It is worth noting here that hydrazinolysis of some of these compounds were found to give a mixture of derivatives of pyridazine and pyrazoles and not as might have been expected N-aminopyridines. Thus, kojic acid (4.134) reacts with anhydrous hydrazine to give a mixture of 65 per cent 3,6-dihydroxy-methyl-4-oxo-1,4-dihydropyridazine (4.138)[154] and 21 per cent of the hydrazone of substituted hydroxymethylpyrazole (4.137)[154]; hence the pyrazole derivative (4.139) previously suggested is not obtained[155]. In both reactions there is intermediary formation of the open-chain compound (4.135) through an initial nucleophilic attack of the hydrazine group on position 2; this compound either cyclizes into the pyrazole acetaldehyde (4.136), which is subsequently converted into (4.137), or gives, via (4.135), ring closure into the pyridazine derivative (4.138). Similar mixtures of pyridazine and pyrazole derivatives were obtained from γ-pyrone[154,156], and from pyromeconic acid[154] and its 6-methyl derivative (allo maltol)[154].

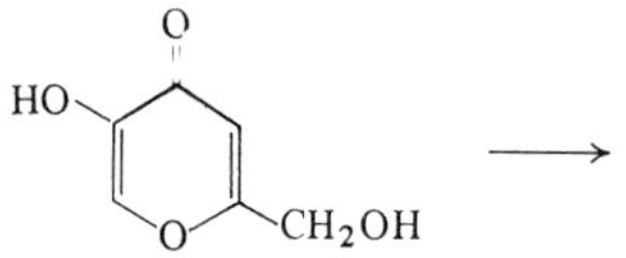

(4.134)

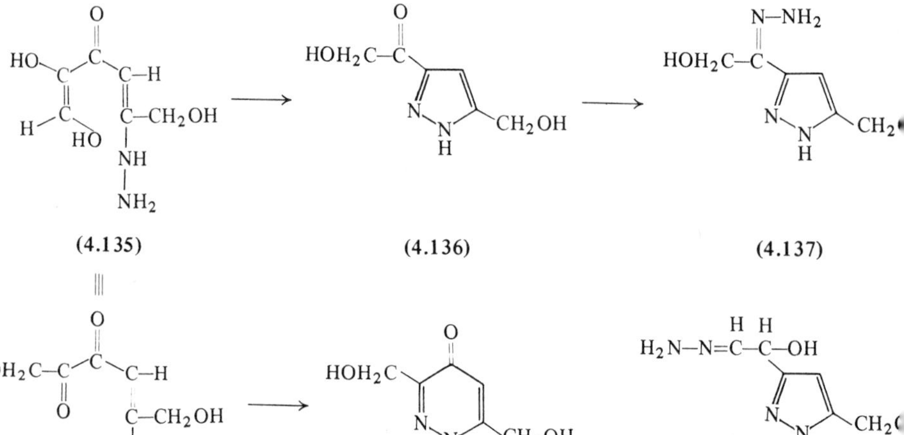

(4.135)   (4.136)   (4.137)

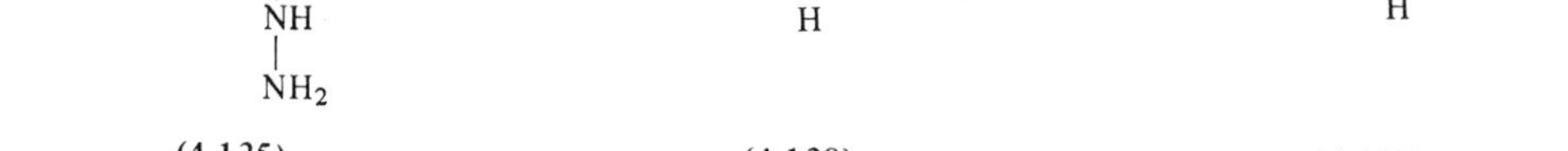

(4.135)   (4.138)   (4.139)

It is of interest that when methoxykojic acid (4.140) is treated with hydrazine, a mixture of the pyrazole (4.142) and the N-aminopyridine (4.141)[154,157] is yielded. As reported above, isomerization into a diketo compound is a necessary step in ring closure to a pyridazine derivative; that this isomerization cannot occur if a methoxy group is present in (4.143), which is presumed to be the intermediate in this reaction, accounts for the fact that a N-aminopyridine was obtained rather than a pyridazinone.

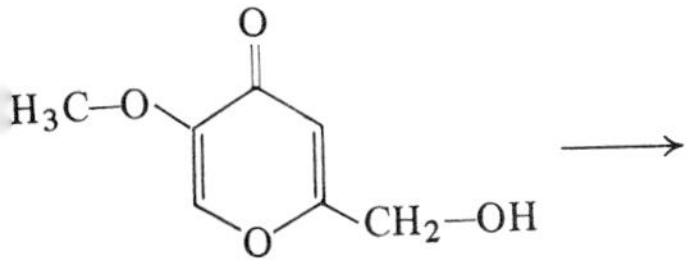

(4.140)

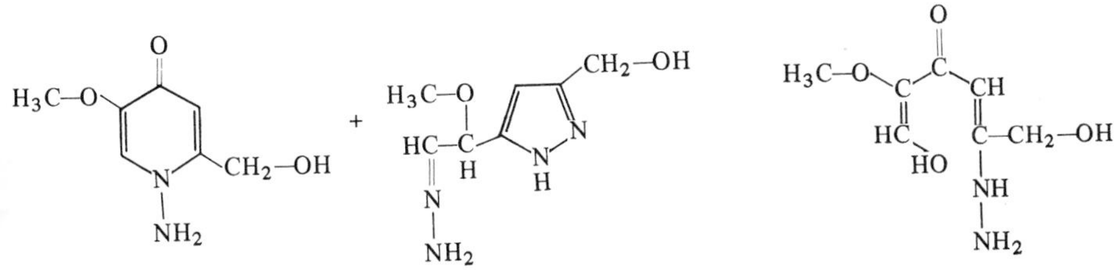

(4.141)   (4.142)   (4.143)

Replacement of oxygen by nitrogen does not always require the presence of ammonia or primary amines as a nitrogen source. Benzonitrile has very recently been used[158a] as a nitrogen source in the ring conversion of the pyrone (4.145a) into the pyridine-3-carboxylate (4.145b). Very drastic conditions (250 h!, 215°C) are necessary for the occurrence of this ring transformation. Although

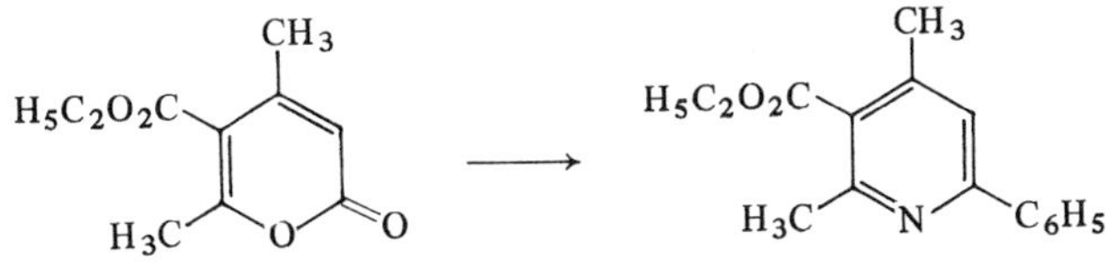

(4.145a)   (4.145b)

the mechanism is not clear, it would appear that intermediates formed by a homolytic bond breaking are involved. It has been reported that nitroso-pyromeconic acid (4.144) rearranges into the trihydroxypyridone-4 (4.145) during reduction with sulphur[158].

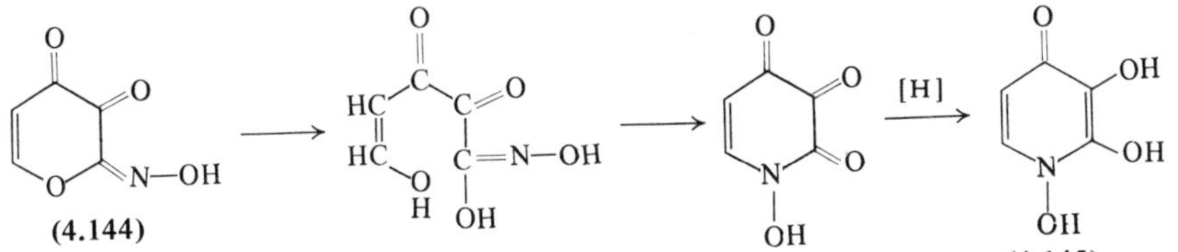

(4.144)   (4.145)

4-Methylene-4H-pyrans (4.146) have also been found to be converted with ammonia or with ammonium acetate in acetic acid into the corresponding pyridine derivatives[159-162]; with primary amines the same ring transformation has been observed. This method offers a valuable route for the synthesis of many otherwise not so easily obtainable pyridine derivatives. A very recent example is the conversion of the 4-dicyanomethylene-2-phenyl-4H-1-benzopyran (4.146a) into the benzopyrano[3,4-c]pyridine derivative (4.146b) by reaction with ammonia[162a]. With primary amines the same conversion occurs, but more vigorous reaction conditions are necessary. Then Dimroth rearrangements take place, leading to compounds with the structure (4.146c). The reaction course can be described as follows:

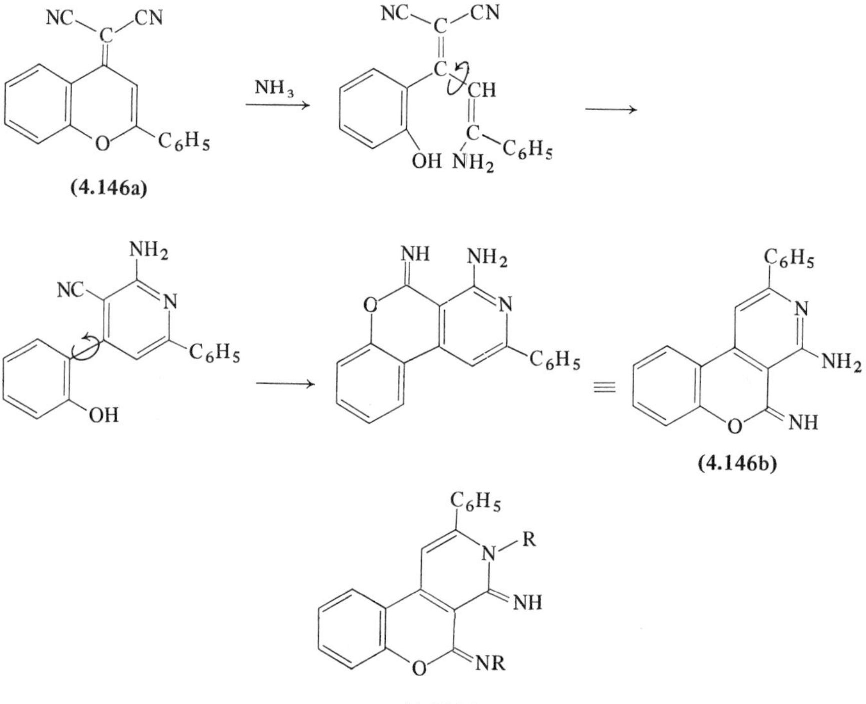

(4.146a)

(4.146b)

(4.146c)

$$R = CH_3; C_4H_9; C_6H_5CH_2$$

The 'oxygen-nitrogen exchange' also takes place when mono-, di- and tri-alkyltetrahydropyran-4-ones (4.147) are treated with an aqueous solution of ammonia or primary amines at about 80-90°C for a prolonged period[163-165]; the corresponding N-substituted 4-piperidones (4.149) are obtained. Thiapyran-4-ones (4.148) were formed when hydrogen sulphide[164,165] was used instead

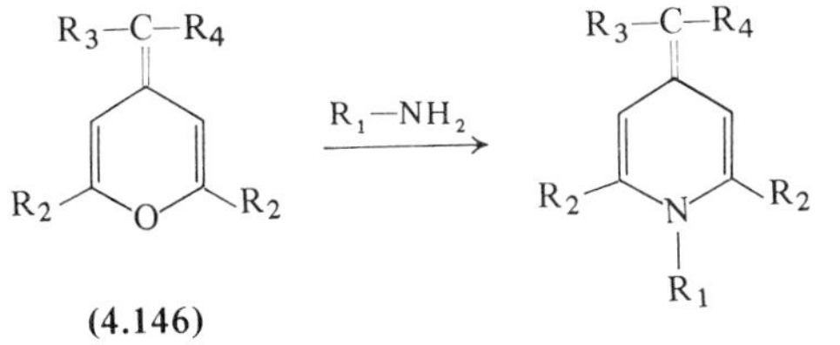

(4.146)

R_1	C_6H_5	$CH_2C_6H_5$	NH_2	$CH_2{-}C_6H_5$	NH_2
R_2	CH_3	CH_3	CH_3	CH_3	CH_3
R_3	CN	CN	CN	CN	CN
R_4	CN	CN	CN	$CO_2C_2H_5$	$CO_2C_2H_5$

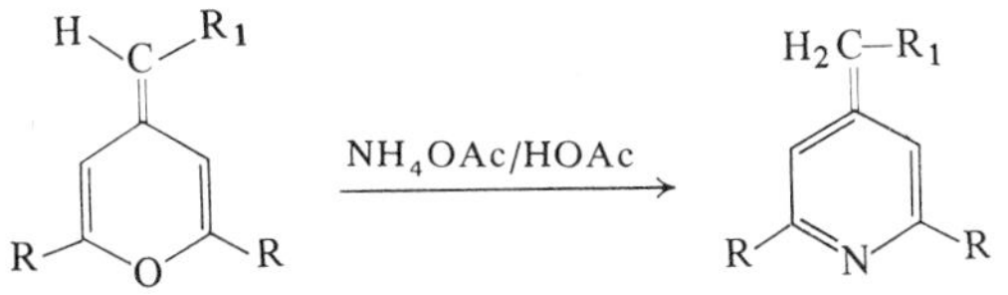

R	C_6H_5	C_6H_5	$4\text{-}BrC_6H_4$	$4\text{-}BrC_6H_4$	$4\text{-}NO_2C_6H_4$
R_1	C_6H_5	$4\text{-}BrC_6H_4$	C_6H_5	$4\text{-}BrC_6H_4$	C_6H_5

of primary amines. It has been found that all these ring systems are mutually interconvertible under the conditions described[164].

In the bicyclic series we mention the conversion of tetrahydrocoumarins[166] and hexahydrocoumarins[167] with ammonia and primary amines into the corresponding tetrahydro- and hexahydroquinoline derivatives, and the ring transformation of isocoumarins (4.150) into isocarbostyrils (4.151)[168-173]. All

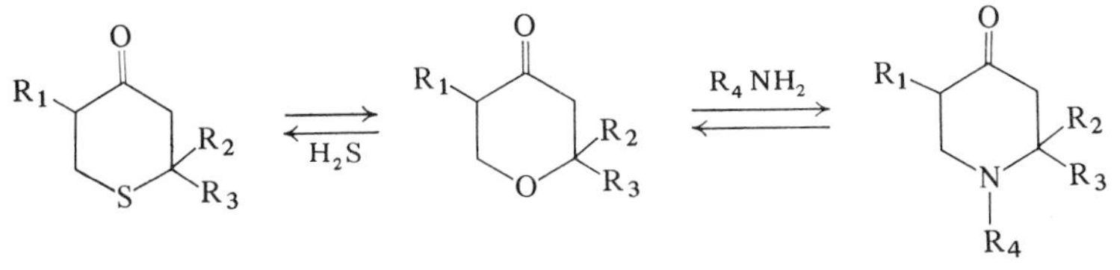

(4.148)   (4.147)   (4.149)

$$R_1 = H; CH_3. \; R_2 = H; CH_3; C_2H_5$$
$$R_3 = CH_3; C_2H_5; C_3H_7; i\text{-}C_4H_9$$
$$R_4 = H; CH_3; C_2H_5; C_4H_9$$

these reactions have been carried out using the normal procedures. With hydrazine 3,4-diphenylisocoumarin is converted into the corresponding N-aminoisocarbostyril[173a]. This reaction is in marked contrast to that observed when 3-phenylisocoumarin reacts with hydrazine giving ring expansion into the seven-membered 4-phenyl-2,5-dihydro-2,3-benzodiazepin-1-one[202] (see section I.A.3.b). This compound, however, can readily be converted into the corresponding N-aminoisocarbostyril by treatment with an acid[173a]. Isocarbostyrils are also formed from 1-thioisocoumarin and primary aromatic amines in the presence of a strong acid[174]. However, it appears that this reaction has little

(4.150)   (4.151)

value as a method of preparation since a rather complex reaction mixture is obtained, the composition of which is strongly dependent on the reaction conditions used.

The replacement of the ring oxygen in 4-pyrones and in 4-thiopyrones by sulphur can be accomplished by boiling these compounds with a solution of potassium hydrogen sulphide in water or in acetone[175-177]. With 4-thiopyrones this replacement occurs more easily[176,177]. The analogous conversion of the 2-n-propyltetrahydropyran-4-ones (4.152) into the corresponding tetrahydro-thiopyran-4-ones (4.153) can also be carried out by refluxing with a methanolic solution of hydrogen sulphide in the presence of the acid catalyst sulphuric acid[164,165].

(4.152)   (4.153)

R_1	H	CH_3
R_2	H	H

In the xanthene series, it has been possible in several cases to bring about the replacement of oxygen by both nitrogen and sulphur. Heating of fluorescein (4.155) with ammonia in water at 180°C gives the acridine (4.156)[178,179]. When (4.155) is treated with a basic solution of disodium sulphide, the thiofluorescein (4.154) is obtained. An important specific characteristic of fluorescein is exhibited in this reaction[179].

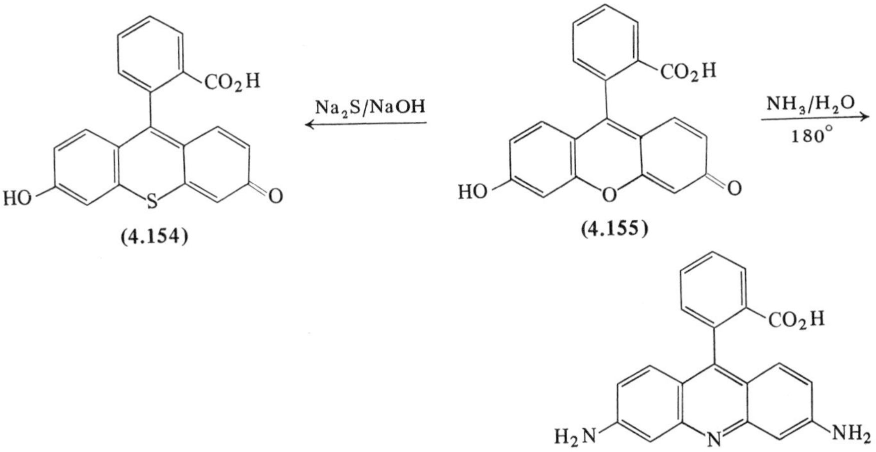

(4.154)　　　　**(4.155)**

(4.156)

3. RING EXPANSIONS OF SIX-MEMBERED HETEROCYCLES CONTAINING ONE OXYGEN ATOM INTO SEVEN-MEMBERED HETEROCYCLES

a. *Pyrans and pyrylium salts into oxepines; pyrylium salts into 1,2-diazepines*
The methods which are known to cause ring expansion in the homocyclic series include carbene addition, Beckmann rearrangement, Schmidt reaction and nitrosative deamination. Generally, they are equally successful for performing ring expansions in the heterocyclic field.

There are several papers dealing with ring expansion of pyrans achieved by using chlorocarbene or dichlorocarbene. Thus, when 2,3-dihydro-4H-pyran (4.157) is treated with dichlorocarbene (generated by the treatment of sodium trichloroacetate with alkali) and the reaction mixture is subsequently pyrolized, 6-chloro-2,3-dihydrooxepine (4.160) is formed[180]. Primarily the carbene is added to its acceptor giving the 2-oxa-7,7-dichloronorcarane (4.158) which, via the resonance-stabilized cyclic allylic carbonium ion (4.159), is converted into (4.160). Dichlorocarbene, which is generated from chloroform and potassium-t-butoxide, reacts with the 6-ethoxy derivative of (4.157), that is (4.161), to give 7-ethoxy-2-t-butoxy-2,5,6,7-tetrahydrooxepine (4.163)[181]. Apparently, in the decomposition of 2-oxanorcarane (4.162) attack of the t-butylate ion is more favoured than loss of hydrogen chloride. The corresponding conversion of (4.158) into 2-t-butoxy-3-chloro-2,5,6,7-tetrahydrooxepine has recently been reported[131a].

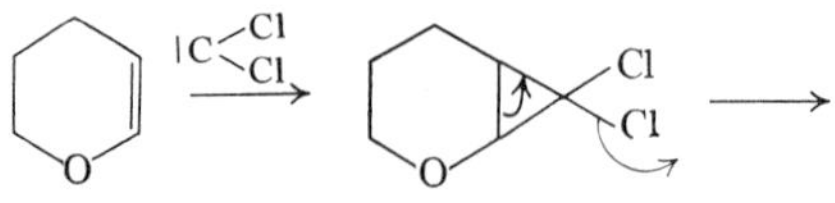

(4.157)   (4.158)

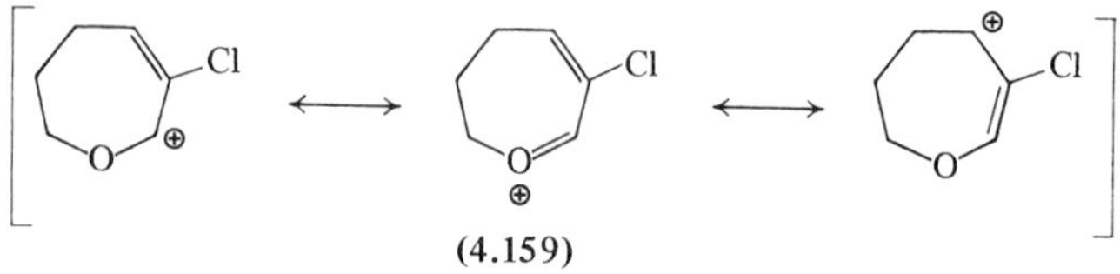

(4.159)

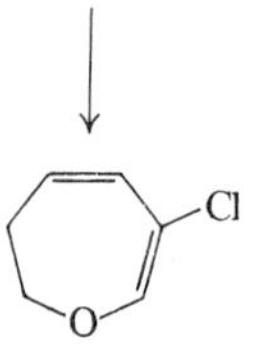

(4.160)

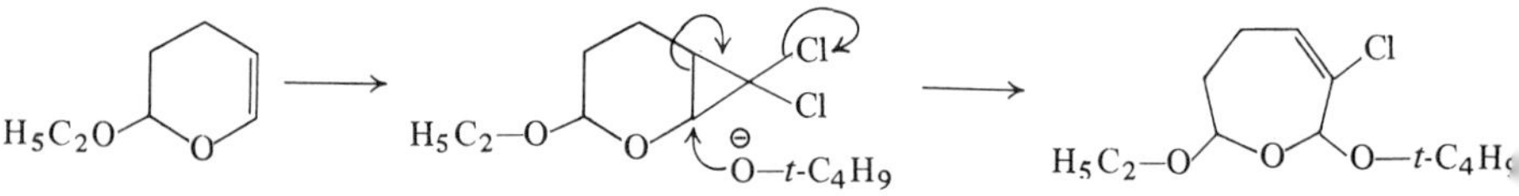

(4.161)   (4.162)   (4.163)

When (4.157) is treated with chlorocarbene instead of dichlorocarbene a racemic mixture of the *endo* and *exo* isomer (4.164) and (4.165) is formed[180]. It has been found that the *exo* isomer (4.165) is more easily pyrolized into 2,3-dihydrooxepine (4.166) in the presence of quinoline than the *endo* isomer (4.164). This is explained by the fact that in (4.165) neighbouring group

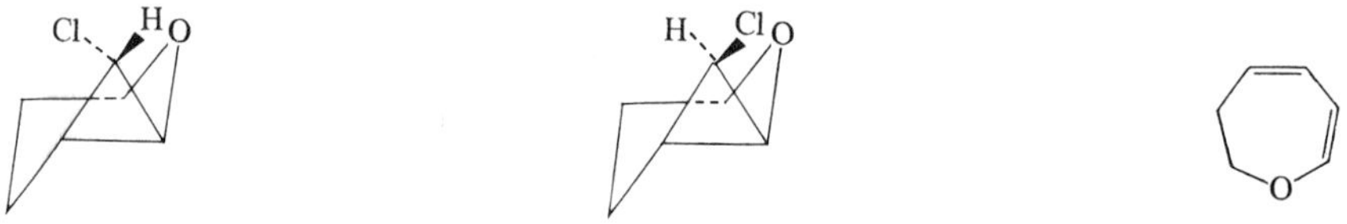

(4.164)   (4.165)   (4.166)

assistance of the *trans* oxygen atom favours the loss of the chlorine ion. From the point of view of synthesis a useful extension of this reaction is the conversion of (4.157) into 3-chloro-(2-chloroethoxy)tetrahydrooxepine (4.167) by dichlorocarbene and oxirane in the presence of tetraethylammonium salts[182]. It is now possible to convert this chloroethoxy compound (4.167) in a few steps into 2,3,4,5-tetrahydrooxepine (4.168) which can serve as a starting substance for a new ring expansion reaction: the preparation of eight-membered oxocine derivatives by the series of reactions described above. This tetrahydro-oxepine (4.168) has recently been prepared by gas-phase dehydration of 2-hydroxymethyltetrahydropyran (4.169) over a copper-chromium oxide catalyst; cyclopentane carboxaldehyde is a minor reaction product[183]. Close

(4.157) →

(4.158)   (4.167)   (4.168)

correlation with the mechanism reported for the conversion of 2-hydroxy-methyltetrahydrofuran into 5,6-dihydro-4H-pyran (see vol. 1. chapter 3, section I.A.4.a) would suggest the seven-membered cyclic carbonium ion (4.170), produced by the Wagner-Meerwein rearrangement, as reaction intermediate.

(4.169)   (4.170)   (4.168)

In the bicyclic series the addition of dibromocarbene to benzo[2H] pyran, yielding the adduct (gem-dibromocyclopropa[c] benzopyran), has been reported. Under the influence of methanol and silver ions this compound is converted into 3-bromo-2,3,4,5-tetrahydrobenzooxepin-4-one[184].

Where an attempt was made to introduce a bromine atom at the benzylic position of the benzofurobenzopyran (4.170a) by treatment with benzoyl peroxide and N-bromosuccinimide in tetrachloromethane, ring expansion leading to a benzoxepine derivative (4.170d) was observed[184a]. The reaction pathway has been described as a series involving the ready entry of bromine on the benzylic carbon: an intramolecular displacement of the bromine atom by the benzylic bond, yielding the stable spirocarbenium ion (4.170b), followed by proton loss and allylic migration, leading to ring expansion into (4.170c). In this compound the benzylic-allylic methylene group is highly susceptible to oxidation and is thus readily converted into the oxo group.

Diazomethane may also react with heterocyclic compounds giving ring expansion: tetrahydropyrone (4.171) can be converted into a mixture of the hexahydrooxepin-4-one (4.173) and the spiro epoxy compound (4.172)[185]; the xanthylium salt (4.174) is converted into dibenzo[b,f] oxepine (4.177)[186]. This last compound is also reported to be formed in the acid-catalysed rearrangement of the carbinol (4.175) with phosphorus pentoxide[187,188]. In both reactions the exocyclic carbonium ion (4.176) would appear to be the intermediate. The

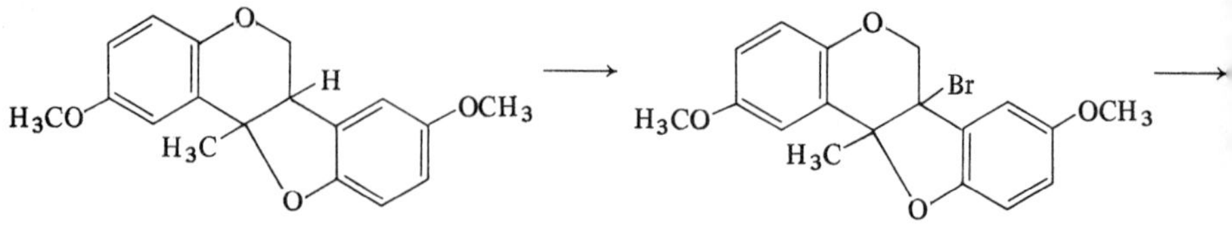

(4.170a)

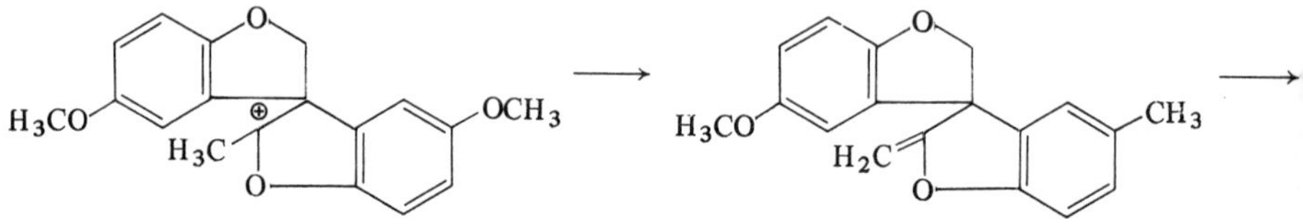

(4.170b)

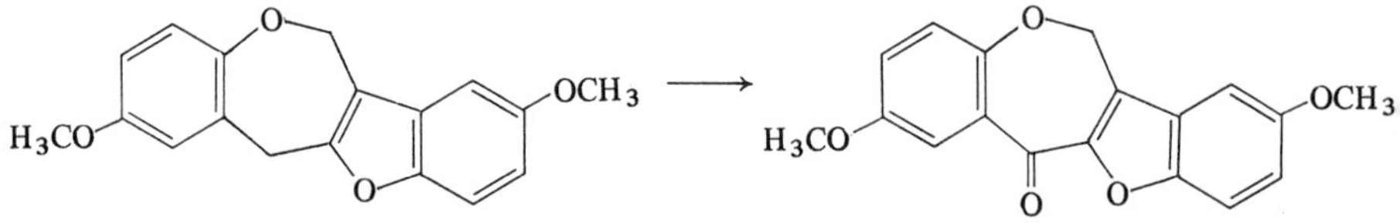

(4.170c)   (4.170d)

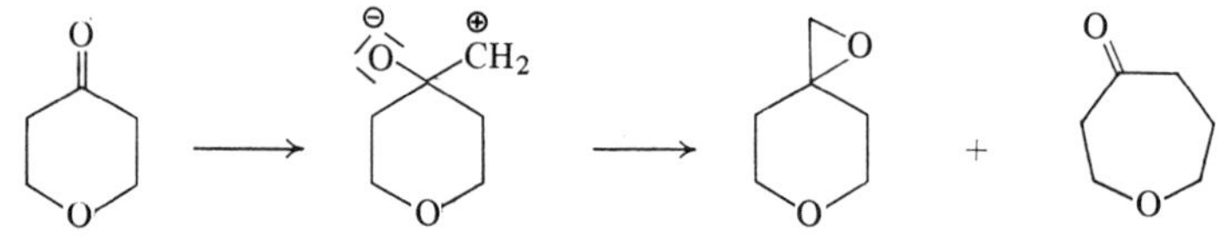

(4.171)   (4.172)   (4.173)

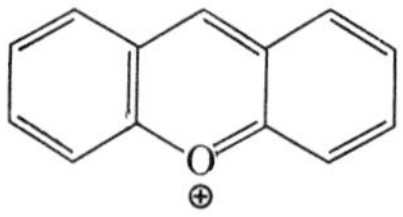

(4.174)

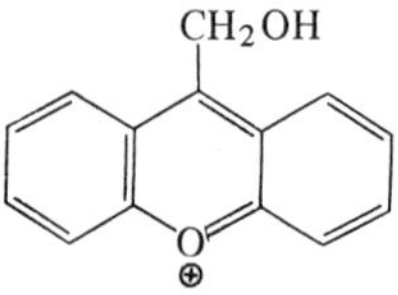

(4.175)

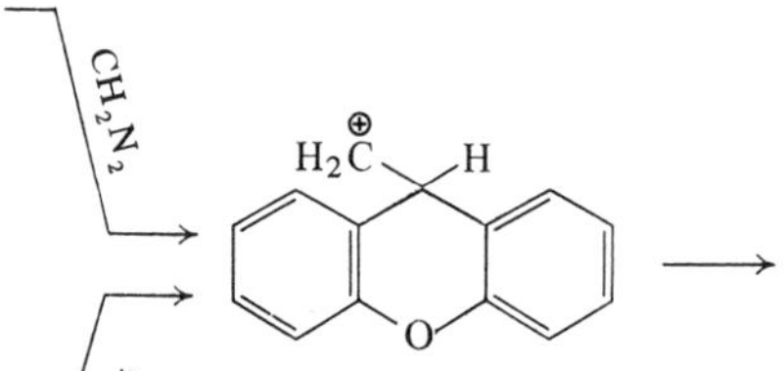

(4.176)

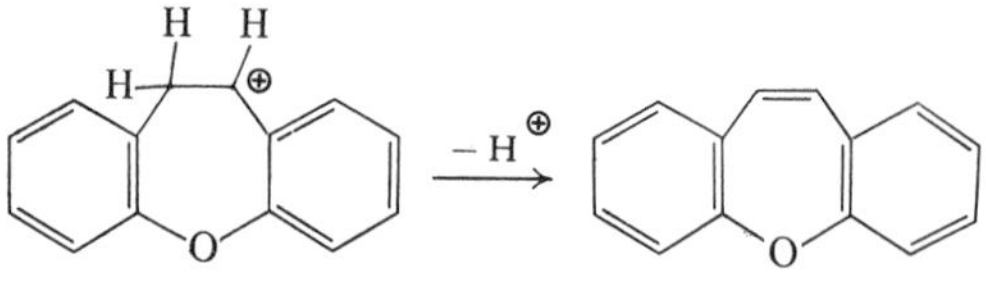

(4.177)

possibility cannot be excluded that the spiro compound (4.172) is the precursor in the formation of the hexahydrooxepin-4-one (4.173), as epoxychromanone-4 and epoxy-1-thiochromanone-4 are known to be convertible into 4-aryl-2,3,4,5-tetrahydro-1-benzoxepine-3,5-dione and into its corresponding 1-benzothiepine derivative respectively[189,190] by the action of concentrated sulphuric acid or with borontrifluoride-etherate in benzene or liquid sulphur dioxide. Very recently it has been reported[188a] that treatment of 3-acetyl-5,7-dimethyl-coumarin with diazomethane gives a pyrazoline derivative, which, under the influence of methanol, rearranges into a benzoxepine derivative.

Although nitrous acid deaminative ring transformations are frequently encountered with homocyclics, only a few reports are available on carbon skeletal rearrangements during deamination of heterocyclic compounds. The 2-amino-methyltetrahydropyran (4.178) gives, with nitrous acid, 2-hydroxyhexahydro-oxepine (4.179), together with the unrearranged 2-hydroxymethyltetrahydro-pyrans (4.180)[191]. Following the results obtained in the homocyclic series[192], the carbonium ions (4.181), (4.182) or perhaps the non-classical ones like (4.183) and (4.184), can be proposed as intermediates in this ring enlargement.

When a suspension of 2,4,6-triarylpyrylium salts (4.7) in ethanol reacts with hydrazine hydrate, 3,5,7-triaryl-4H-1,2-diazepines (4.186) are formed in nearly

(4.178)   (4.179)   (4.180)

(4.181)   (4.182)   (4.183)   (4.184)

(4.7)   (4.185)   (4.186)

$Ar_1 = C_6H_5; p\text{-}CH_3\text{-}C_6H_4; p\text{-}Br\text{-}C_6H_4$
$Ar_2 = C_6H_5; p\text{-}CH_3\text{-}C_6H_4; p\text{-}OCH_3\text{-}C_6H_4; p\text{-}Cl\text{-}C_6H_4; p\text{-}Br\text{-}C_6H_4;$
$\quad\quad p\text{-}NO_2\text{-}C_6H_4; m\text{-}NO_2\text{-}C_6H_4$

quantitative yield[8,193]. If the reaction is carried out in a two-phase system (ether–water) a colourless compound is found in the ethereal layer and this is believed to be the pseudo-base, the monohydrazone (4.185). Since (4.185) can be readily dehydrated into (4.186)[194], it is probable that it is the intermediary compound in this ring expansion reaction.

b. *Tetrahydropyrones into 1,4-oxazepines; chromanones and isoflavanones into 1,5-benzoxazepines; flavanones into 1,4-benzoxazepines; isocoumarins into benzodiazepines*

The ketoxime of (4.187) is converted via a Beckmann rearrangement into hexahydro-1,4-oxazepin-5-one (4.188) by the action of polyphosphoric acid[195]. The reported occurrence of this ring expansion reaction in thiapyrones will be discussed in section I.B.3.a.

(4.187) (4.188)

(4.189) (4.190) (4.191)

1,5-Benzoxazepines are found to be formed when the ketoximes of substituted isoflavanones (4.190) are reduced with lithium aluminium hydride in ether. 3-Aryl-2,3,4,5-tetrahydro-1,5-benzoxazepine derivatives (4.191) are obtained[196] and not the expected 4-aminoisoflavans (4.189). Similar rearrangements have also been reported[196a] to occur during LiAlH₄ reduction of 7-methoxychromanone oxime and 7-methoxy-3-methylchromanone oxime, yielding 2,3,4,5-tetrahydro-1,5-benzoxazepine derivatives. A more detailed investigation has shown that chromanone oximes *without* a C_2-substituent give only the rearrangement product, but that *with* a C_2-substituent a mixture of the normal reduction product (primary amine) and the rearrangement product is obtained; furthermore, it has been revealed that with a bulky substituent at C-2 the primary amine formation is increased at the cost of the benzoxazepine formation.

When flavanone (4.192) is treated with sodium azide in acetic acid or sulphuric acid[197], the 1,4-benzoxazepin-5[4*H*]-one (4.193) is formed. Thus, this Schmidt reaction does not lead to 2,3-dihydro-2-phenyl-1,5-benzoxazepin-4[5H]-one as previously reported[198].

Examples of ring expansions of six-membered oxygen heterocycles into diazepines include the conversion of isocoumarins (4.194) by hydrazine and phenylhydrazine into 1,2-dihydro-5H-2,3-benzodiazepines (4.195)[199-202], and

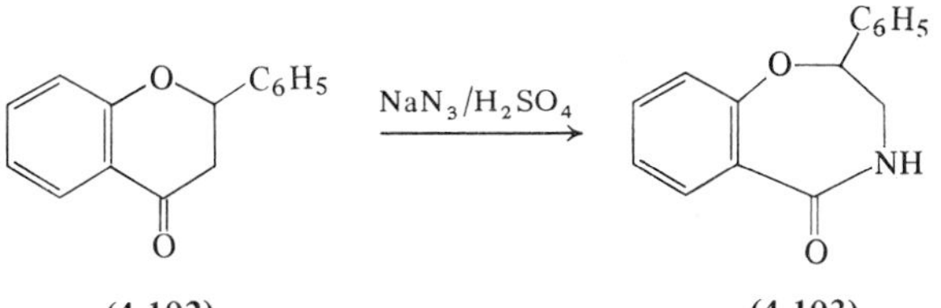

(4.192)   (4.193)

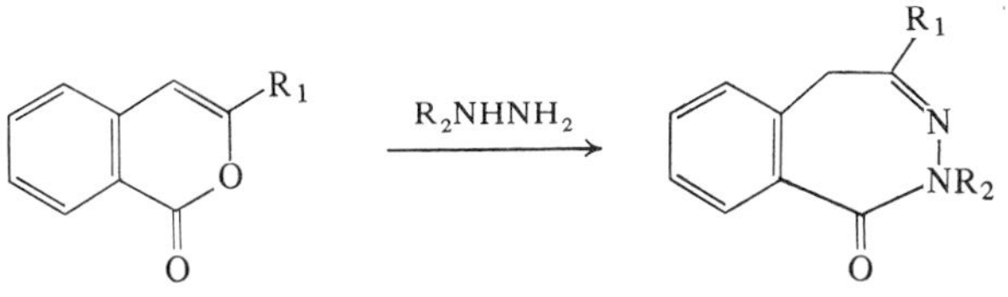

(4.194)   (4.195)

R_1	C_6H_5	m-CH$_3$C$_6$H$_4$	CH$_3$
R_2	H	H	C_6H_5

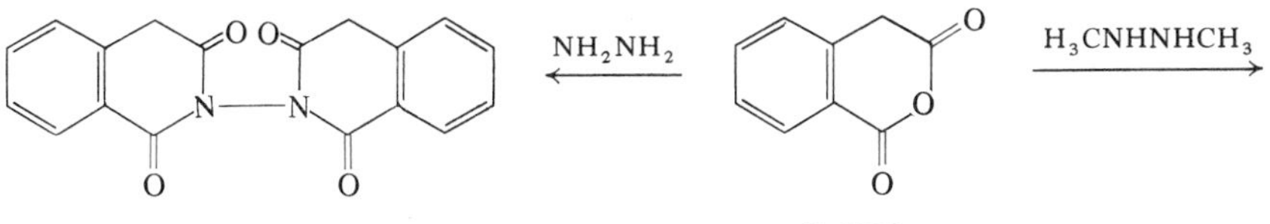

(4.196)   (4.197)

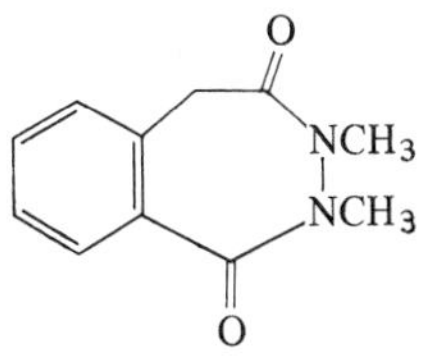

(4.198)

the conversion of homophthalic anhydride (4.197) into the 2,3-dimethylbenzo-diazepi,-1,4-dione (4.198)[203]. This ring expansion reaction has also been reported to occur with hydrazine[204], but it seems that this cannot be so because the product obtained (4.196) has been proved to be formed from a reaction between two moles of homophthalic anhydride and one mole of hydrazine[203].

It is interesting to note that it has been found that when 2,6-dimethyl-4-thiopyrone reacts with disodium sulphide in water or in aqueous acetone, and the reaction mixture is then acidified and aerated, 3,7-dimethyl-5H-1,2-dithiepin (4.199)[205,206] is formed; when 2,6-dimethyl-4-selenopyrone and disodium selenide are used the corresponding 1,2-diselena derivative (4.200) is obtained[207].

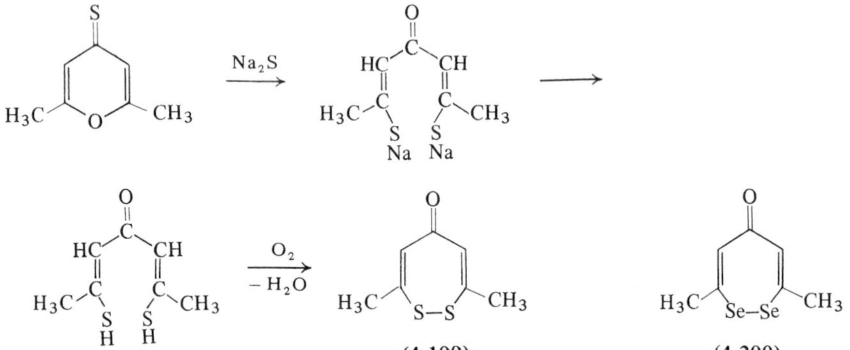

(4.199)   (4.200)

B. Ring transformations of six-membered heterocycles with one ring sulphur atom

1. RING CONTRACTIONS INTO FOUR- AND FIVE-MEMBERED HETEROCYCLES

a. *Tetrahydrothiopyrons into thietanes and thiolanes; thiapyrylium salts into thiophenes; thiochromans and isothiochromans into benzothiophenes*

Ring contractions of sulphur-containing six-membered heterocycles have been reported to be achieved by oxidation or reduction reactions and, more recently, by photolytic methods.

An interesting ring contraction occurs during photolysis of 5-methyl-2,3-dihydro(2H,6H)thiopyran-3-one (4.201) when the 2-isopropenyl-1-thietan-3-one (4.202) is yielded[208,209]. A similar compound, the thietanone-3 (4.204), has been postulated as intermediate during the light-induced isomerization of isothiachromanone-4 (4.203) into thiachromanone-3 (4.205)[209,210] (for further details see section I.B.2.a). Photolysis of tetrahydrothiapyrone-4 (4.206) in *t*-butyl alcohol gives, together with the formation of the open-chain compound *t*-butyl-4-thiahexanoate ring contraction into the thiolactone (4.209). It has been reasonably argued[210] that (4.209) is formed by an intramolecular electron transfer from sulphur to the excited carbonyl group

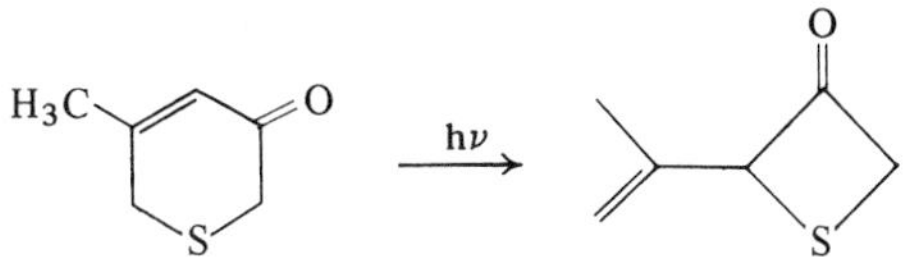

(4.201)   (4.202)

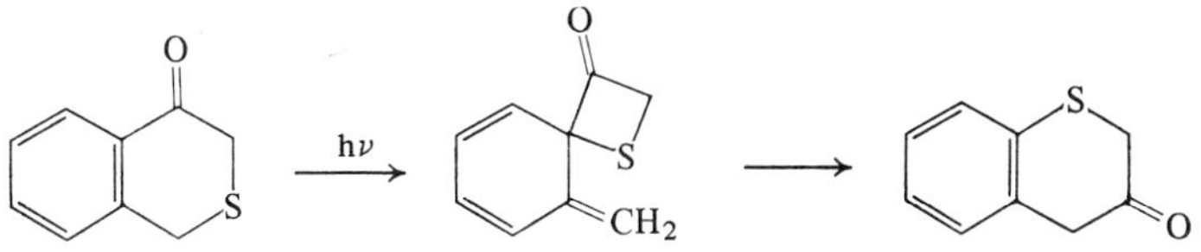

(4.203)   (4.204)   (4.205)

giving (4.207) and fragmentation of the dipolar sulphonium species (4.208) with formation of (4.209) and ethylene. The results so far obtained do not exclude the occurrence of the diradical (4.210) as intermediate. Both ring contraction and ring expansion have been observed during irradiation of tetrahydro-thiopyran-3-one (4.211) when 1,1,2-trichloro-1,2,2-trifluoroethane is used as solvent. Together with thiolan-2-one (4.212), thiepan-4-one (4.213) was formed,

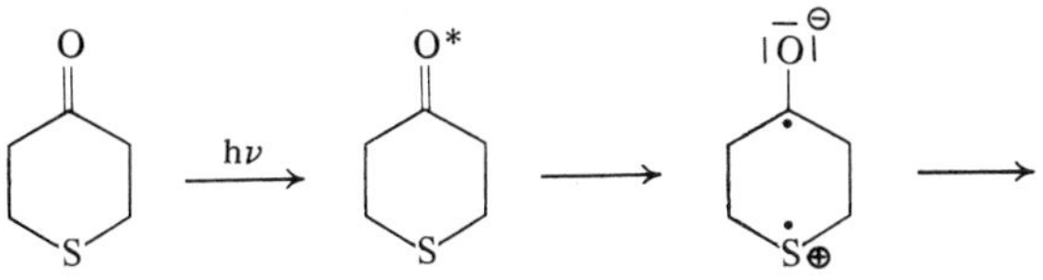

(4.206)   (4.207)

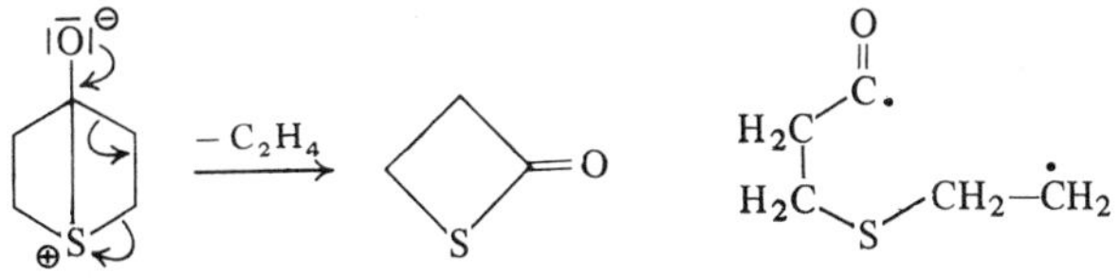

(4.208)   (4.209)   (4.210)

both products being obtained in yields of 6-8 per cent[211]. This ring contraction was not found to occur to any extent with the 2-methyl- and 6-methylderivative of (4.211). Irradiation in *t*-butyl alcohol did not give ring transformation but reduction to the corresponding tetrahydrothiopyranol occurred. Although it has been suggested that the reaction occurs via the ylid (4.214) in the presence of cyclohexene, no norcaran is yielded.

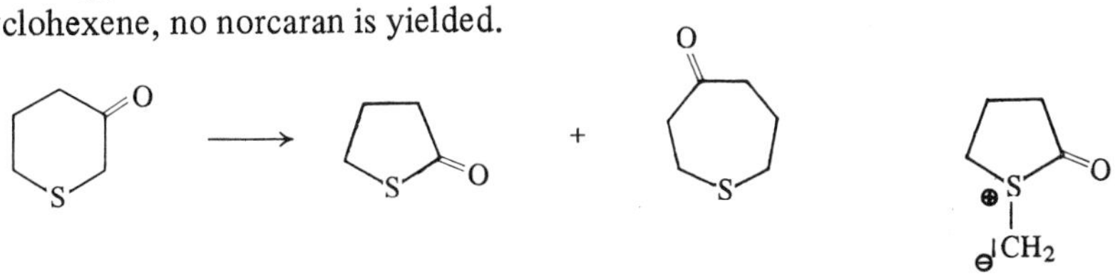

(4.211)   (4.212)   (4.213)   (4.214)

The Clemmensen reduction with zinc amalgam and hydrochloric acid[212,213] has been shown to be a very effective method for bringing about ring contraction of cyclic α-thiaketones. This excellent method is frequently used for performing ring contractions of cyclic α-aminoketones (see section I.C.1.a); when used for tetrahydrothiapyran-3-one (4.211) it gives formation of 2-methylthiolane (4.215)[212]. In the bicyclic series, isothiachromanone-4 (4.216) is converted by this method into 1-methyl-1,3-dihydroisothianaphthene (4.217)[213]. A correlation with the mechanism for the conversion of cyclic α-aminoketones has led to the suggestion that in the primary step hydrogenolysis of the Cα-S bond occurs.

Oxidation of thiapyrylium iodide by treatment with a suspension of MnO_2 in chloroform gave 2-thiophene aldehyde[214]. A remarkable ring contraction has been reported to occur when the *trans*-3-bromo-1-thio-4-chromanols (4.218) are heated in dioxane. Formation of the benzo[*b*]thiophene (4.219) with dehydration has been established[215]. In aqueous dioxane, the same ring contraction has

(4.211)   (4.215) 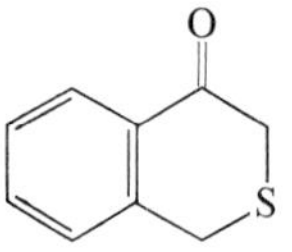   (4.216) 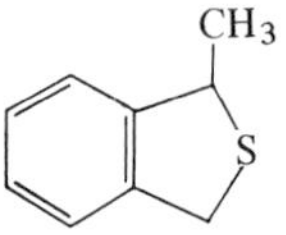   (4.217)

been found to occur, except that the 2-benzo[*b*]thienylmethanol (4.220) is formed from (4.218) (R = H) and the 2-vinylbenzothiophene (4.221) from (4.218) (R = CH_3). It is assumed that primarily the carbonium ion (4.222) is formed and that this is stabilized by neighbouring group participation of the thioether group (4.223).

When (4.223a) (R_1 = R_3 = H; R_2 = H; R_2 = OCH_3) was treated with dimethylsulphoxonium methylide at 50°C, a 1:1 mixture of the corresponding

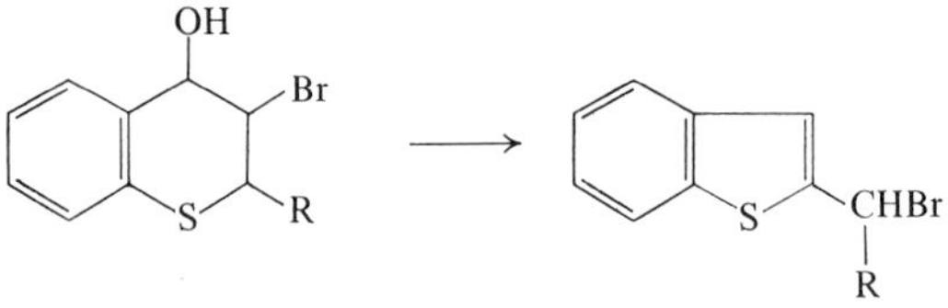

(4.218)   (4.219)

R = H; CH_3

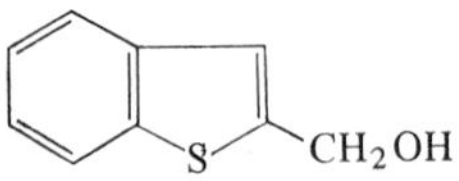

(4.220) 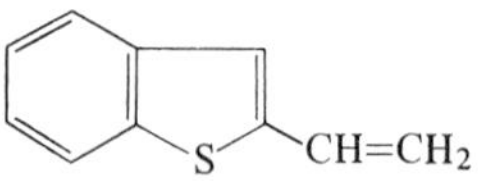   (4.221)

(4.218)

(4.222) (4.223) (4.219)

thianaphthene (4.223b) and the cyclopropylphenylketone (4.223c) ($R_1 = R_3 =$ H; $R_2 = OCH_3$) was obtained[215a]; at low temperature, or in the presence of an excess of ylide, the compound (4.223c) ($R_1 = R_3 = H$; $R_2 = OCH_3$) was the only product. The composition of the reaction mixture appears to be strongly dependent upon the nature of the substituents R_1, R_2 and R_3 and the applied temperature. The observed products obtained from (4.223a) ($R_1 = R_3 = H$, $R_2 = OCH_3$) were thought to be formed via a ring-opened zwitterion (4.223d), which, in the absence of a steric or kinetic factor, may undergo an internal cyclization to the thianaphthene precursor (4.223e); attack of the sulphide anion leads to ring closure and ring opening, yielding (4.223f).

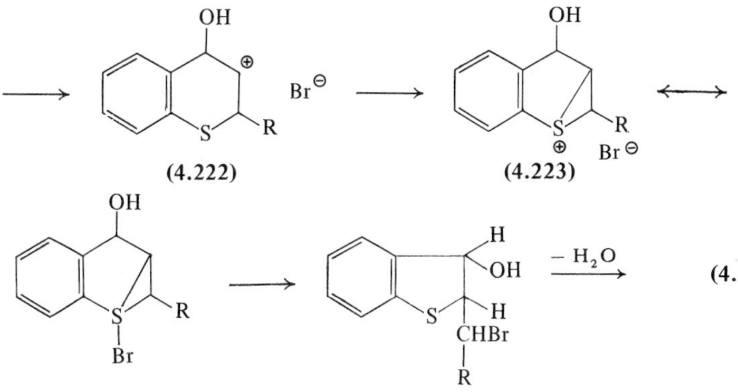

A ring contraction of a thiochroman-4-one system into a dihydro thianaphthene has been observed during the irradiation of the sulphoxides (4.223g) ($R_1 = C_6H_5$, $R_2 = H$, $R_3 = CH_3$) and (4.223g) ($R_1 = R_2 = CH_3$, $R_3 = H$)[215b]. It was found that the 3-position must remain unsubstituted since the photoreaction with these compounds only leads to the formation of substituted benzene

(4.223a) (4.223b) (4.223c)

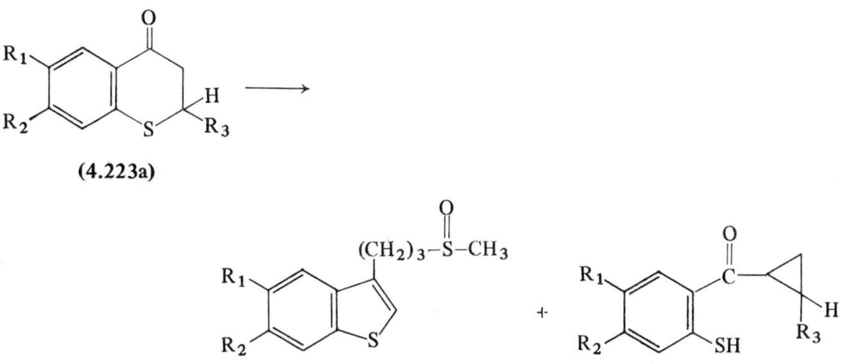

(4.223a)
$R_1 = R_3 = H;$
$R_2 = OCH_3$

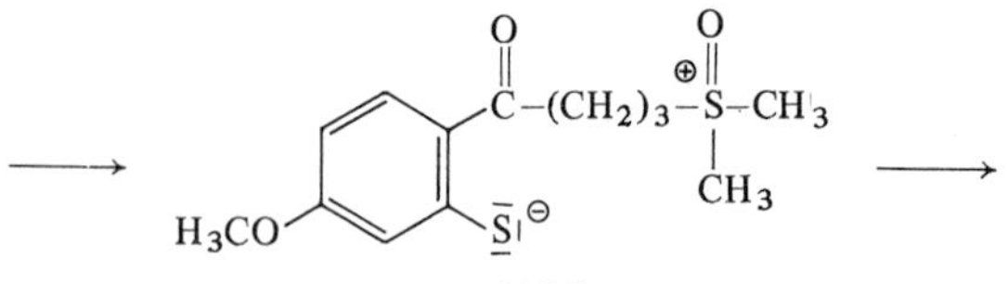

(4.223d)

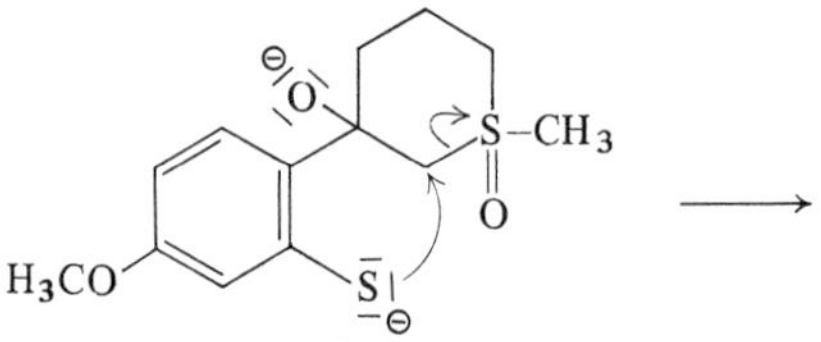

(4.223e)

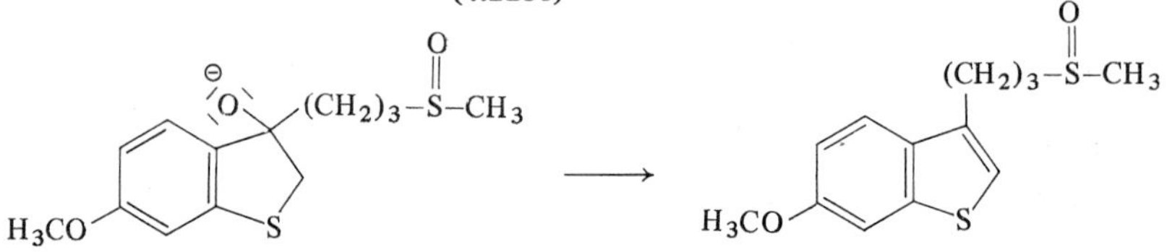

(4.233f)

derivatives. The mechanistic pathway tentatively suggested for this photo-induced ring contraction is a α-cleavage of the sulphoxide, formation of an intermediate sulphenic acid followed by ring closure as indicated.

An interesting ring contraction recently observed[215c] is the pyrolytic conversion of the enamine (4.223h) (160°C/0.00002 mm) into the isothianaphthene (4.223j). Although the exact mechanism is not certain, a thiiranium ion

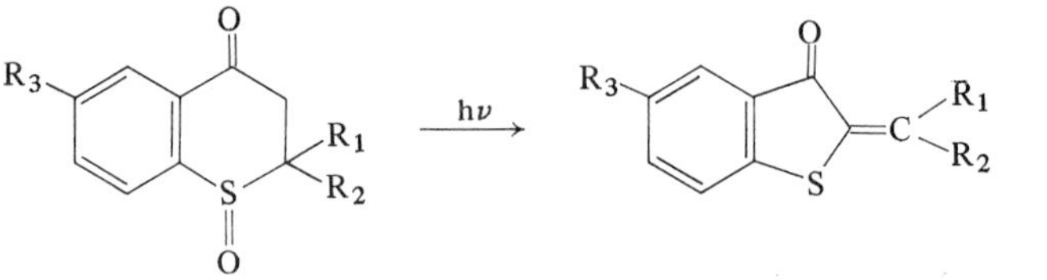

(4.223g)

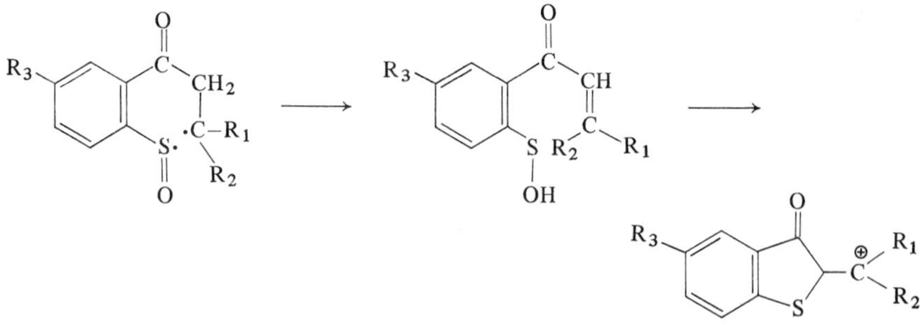

can be suggested as intermediate; it is thought to arise via an initial protonation of the enamine and participation of sulphur in stabilization of the resulting imminium form. Hydride transfer from a second molecule leads to (4.223i), which, on loss of pyrrolidine, yields (4.223j).

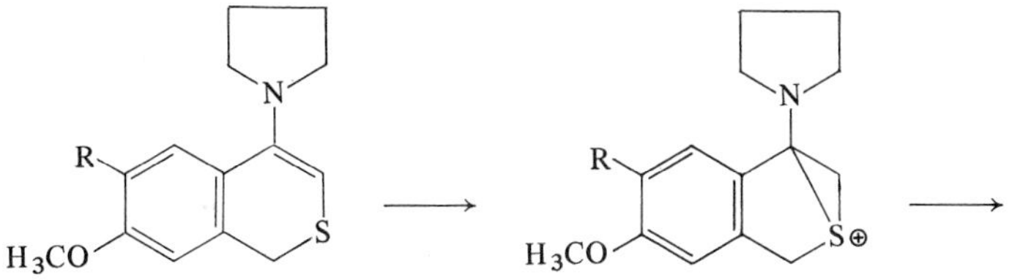

(4.223h)

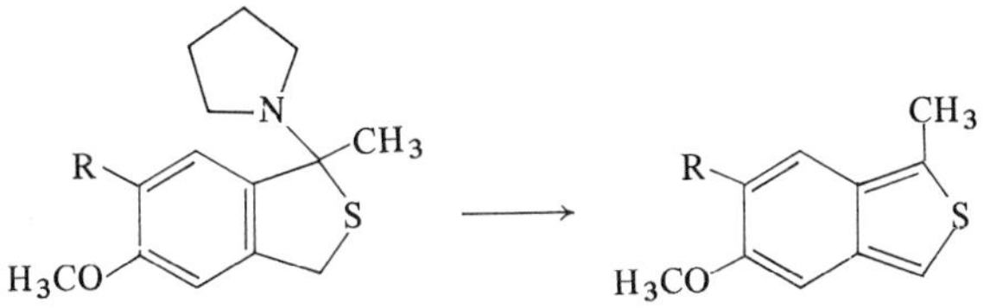

(4.223i)   (4.223j)

2. RING TRANSFORMATIONS INTO SIX-MEMBERED AND SEVEN-MEMBERED HETEROCYCLES

a. *Thiabenzenes into pyrylium salts; tetrahydropyrones into thiepanes; thiapyrylium salts into diazepines; isothiocoumarins into dihydroquinolines; isothiochromans into thiochromans*

When oxygen is bubbled through a freshly prepared violet-coloured solution of 1,2,4,6-tetraphenylthiabenzene (4.224) in ether, and then HCl gas is introduced, with simultaneous formation of phenylmercaptane a precipitate of 2,4,6-triphenylpyrylium salt (4.226) is yielded[216]. This ring transformation probably occurs via the peroxide (4.225), which is converted in the way indicated. This peroxide has not, however, been isolated.

In contrast to the extensively studied conversion of pyrones and pyranthiones into pyridones and thiopyridones respectively (see section I.A.2.c) the replacement of the ring sulphur in $2H$-thiopyran-2-ones by nitrogen through the action of ammonia on primary amines has scarcely attracted attention. Recently a few examples of these ring transformations have been reported.[217,218]. In the bicyclic series, the conversion of the hydrazone of 2-thio-3-arylisothiocoumarins (Ar = C_6H_5, p-$CH_3OC_6H_4$) into an N-aminoisoquinoline derivative by heating with an excess of N_2H_4 has been reported[218a].

An interesting ring interconversion was found during irradiation of the isothiochroman-4-ones (4.227) in cyclohexane which yielded the thiochroman-3-

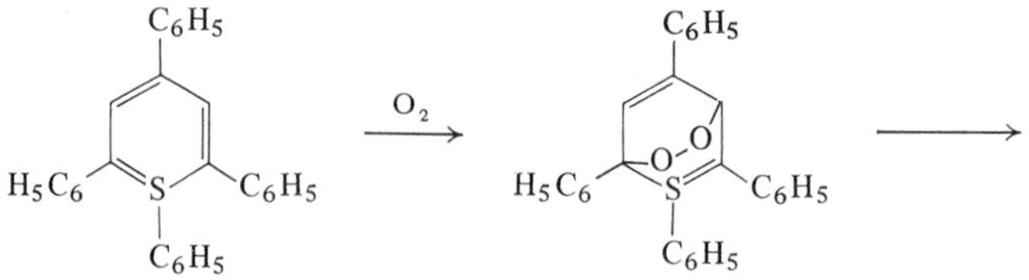

(4.224)   (4.225)

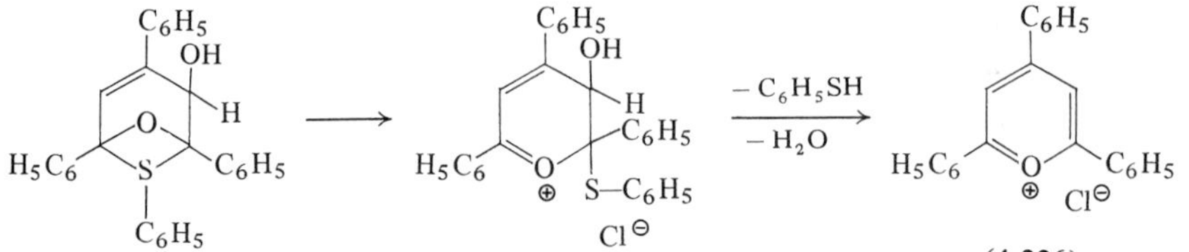

(4.226)

ones (4.228)[208]. From 3,3-dimethylisothiochroman-4-one (4.229), the corresponding compound (4.230) was formed. The ring transformation with the 8-methyl derivative (4.227) (R = CH_3, R_1 = H) clearly indicates that the sulphur atom in (4.228) is held to the same aryl carbon atom which is held to the carbonyl group in the starting substance (4.227). Moreover, it is clear that C_3 in the starting substance becomes C_2 in the photoproduct. It is suggested that the initial photoproduct is (4.231); this further rearranges into (4.228). Evidence for the existence of an intermediate with a similar structure, (4.202), has already been presented in section I.B.1.a.

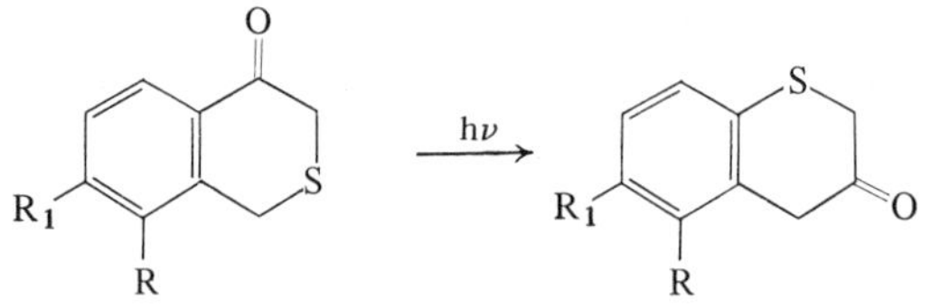

(4.227)   (4.228)

R	H	H	CH_3
R_1	H	OCH_3	H

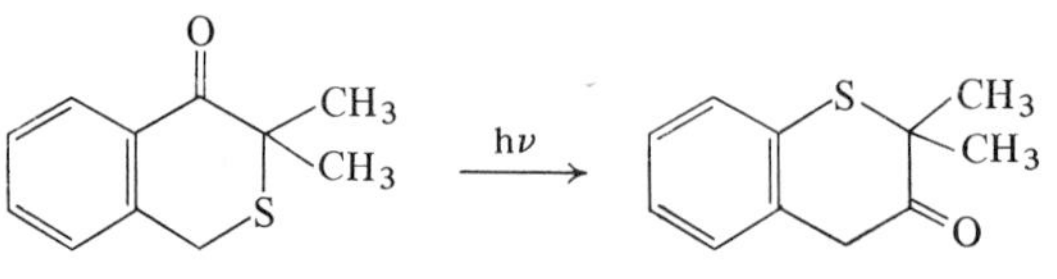

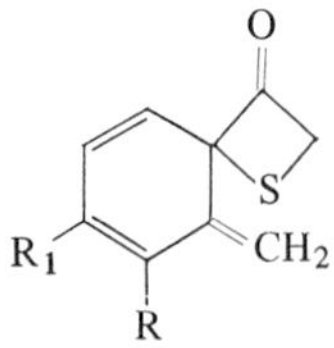

(4.229)   (4.230)   (4.231)

3. RING EXPANSIONS INTO SEVEN-MEMBERED HETEROCYCLES

a. *Tetrahydrothiapyrones into thiepanes and thiazepines; thiapyrylium salts into thiepines and 1,2-diazepines; thiochromans into benzothiazepines; isothiocoumarins into 2,3-benzodiazepines*

In correlation with the corresponding reaction of tetrahydropyrones (see section I.A.3.a), tetrahydrothiapyrone (4.232) can be converted with diazomethane into the seven-membered thiepan-4-one (4.233), together with the spirooxiran (4.234)[219]. The failure has been reported of attempts to achieve ring expansion with the sulphur heterocycles 2,3- and 3,4-thiochromenes using dichlorocarbene—a reagent found to be effective with the corresponding oxygen analogues (see section I.A.3.a)[220]. Only dichloromethyl norcarane derivatives are formed. The corresponding S-dioxide of (4.233), that is (4.236), is formed by rearrangement of the acetate salt of (4.235) which takes place by the action of nitrous acid at low temperature[219].

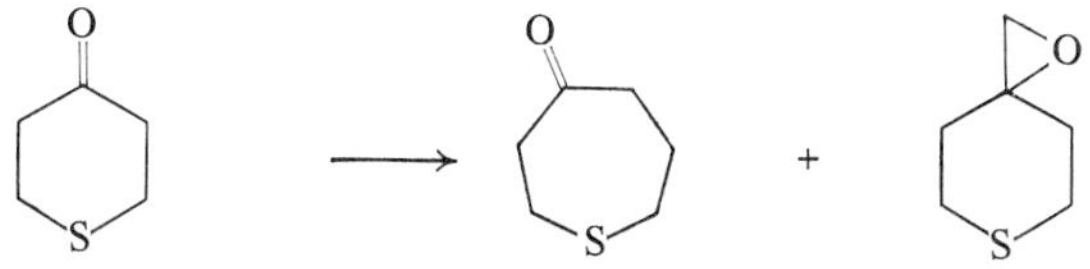

(4.232)   (4.233)   (4.234)

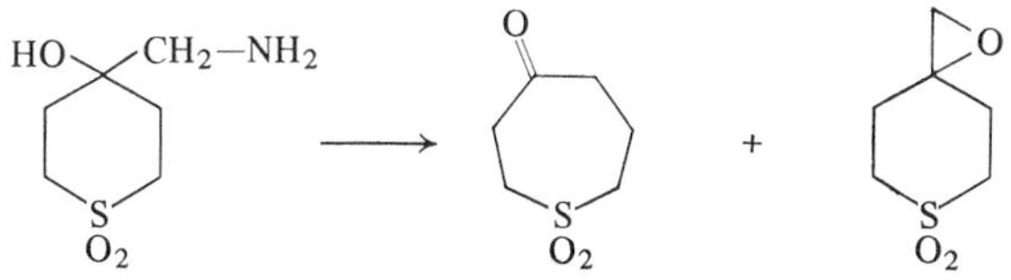

(4.235)   (4.236)

Closely analogous to the reported ring expansion of the dibenzopyrylium salts with diazomethane (see section I.A.3.a) is the conversion of thioxanthylium perchlorate (4.237) into dibenzo[*b,f*]thiepine (4.239)[186]. A more detailed investigation[221] of this homologenization reaction has revealed that, besides (4.239), many by-products are present and that all of them are derivatives of 10,11-dihydrodibenzo[*b,f*]thiepine. Solvolysis of the tosylate of 9-hydroxymethylthioxanthene (4.238) with 95 per cent formic acid also gives (4.239)[222]. In both reactions the carbonium ion (4.240) is the probable intermediate.

Beckmann rearrangement of the thioxanthone oxime (4.241) under the influence of polyphosphoric acid at 160°C gave the dibenzothiazepine

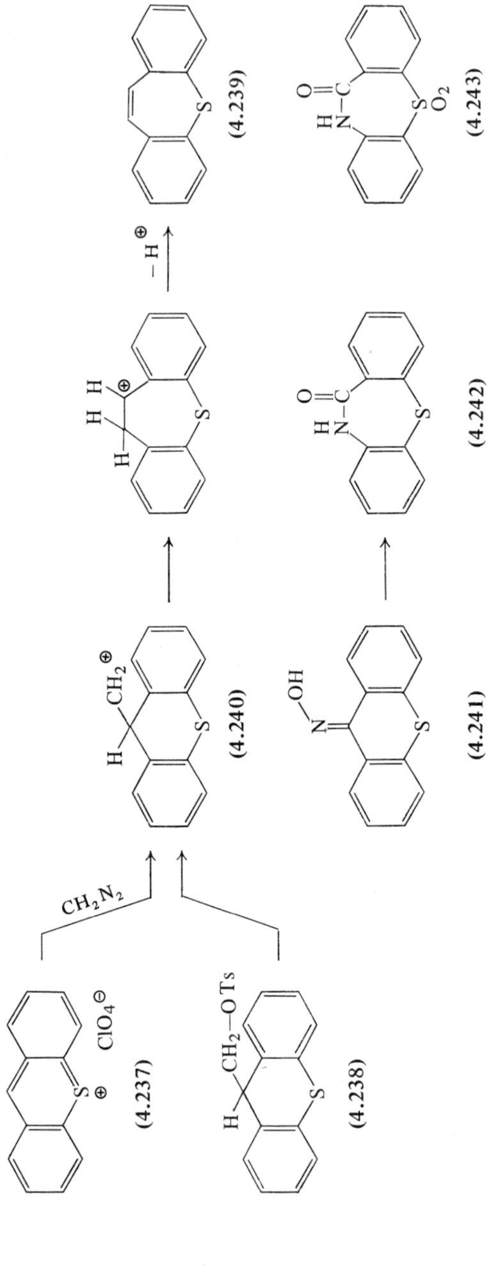

(4.237) (4.238) (4.240) (4.239) (4.241) (4.242) (4.243)

derivative $(4.242)^{223}$. Similarly, the corresponding S-dioxide of (4.241) gave (4.243).

The Schmidt reaction of 6,8-disubstituted 1-thiochroman-4-ones (4.243a) with HN_3 has been reported[223a] to yield 3-29 per cent 7,9-disubstituted-5-oxo-2,3,4,5-tetrahydro-1,4-benzothiazepines (4.243b) (R_1 = H, CH_3; R_2 = H, CH_3) and 18-84 per cent 7,9-disubstituted 4-oxo-2,3,4,5-tetrahydro-1,5-benzo-thiazepines (4.243c) (R_1 = H, CH_3, Cl or Br; R_2 = H, CH_3). Schmidt reaction of the S-dioxide of (4.243a) with HN_3 was found to give only the corresponding S-dioxides of (4.243c).

An interesting ring transformation involving a ring expansion with two carbon atoms has been reported[223b] to occur with isothiochroman-1-one (4.243d) in a reaction with vinyllithium; the 2H-benzo[e] thiocin-4-one (4.243e) was thereby obtained. For the mechanism of this reaction we refer to vol. 1, chapter 3, section I.B.2.b, in which an analogous ring expansion with thiophthalide is discussed.

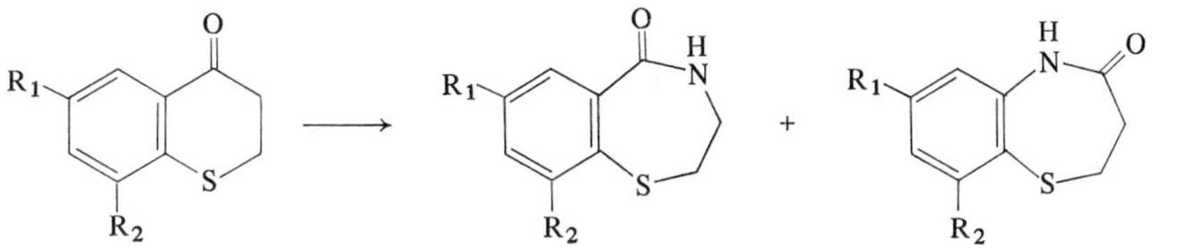

(4.243a)   (4.243b)   (4.243c)

Ring expansion of 2,4,6-triphenylthiapyrylium salt with hydrazine and methylhydrazine gives formation of 3,5,7-triphenyl-1,2-diazepines[8,193,224].

(4.243d)   (4.243e)

This conversion is similar to that reported for 2,4,6-triarylpyrylium salts (4.7) (see section I.A.3.a for the mechanism of the reaction). In analogy to the conversion of isocoumarins into 2,3-benzodiazepines (see section I.A.2.c), it has been found that isothiocoumarin gives an identical ring expansion[173a] when treated with hydrazine.

C. Ring transformations of six-membered heterocycles containing one nitrogen atom

1. RING CONTRACTIONS INTO FIVE-MEMBERED HETEROCYCLES

a. *Piperidines into pyrrolidines and pyrrolines; tetrahydropyridines into pyrazoles, pyrrolines, pyrazolines and isoxazolines*

The occurrence of a group of very important ring contractions has been reported with 3-chloro- and the 3-oxo-derivates of piperidine. When N-alkyl-3-chloro-piperidines (4.244) are treated with nucleophiles, ring contraction to 2-substituted pyrrolidines has been reported. With primary amines[225,226] the substituted 2-aminomethylpyrrolidines (4.245) are obtained, and with sodium alkoxides the 2-alkoxymethylpyrrolidines (4.246) are formed. With potassium cyanide[227] the corresponding 2-cyanomethyl compound (4.248) and with alkali[228] the 2-hydroxymethyl pyrrolidine (4.247) are yielded; with sodium diphenylacetonitrile[229] (4.249) is obtained. It has been proposed that in all these reactions neighboring group participation by nitrogen easily leads to an aziridinium intermediate (4.251)[225,228]; such an aziridinium-type compound is

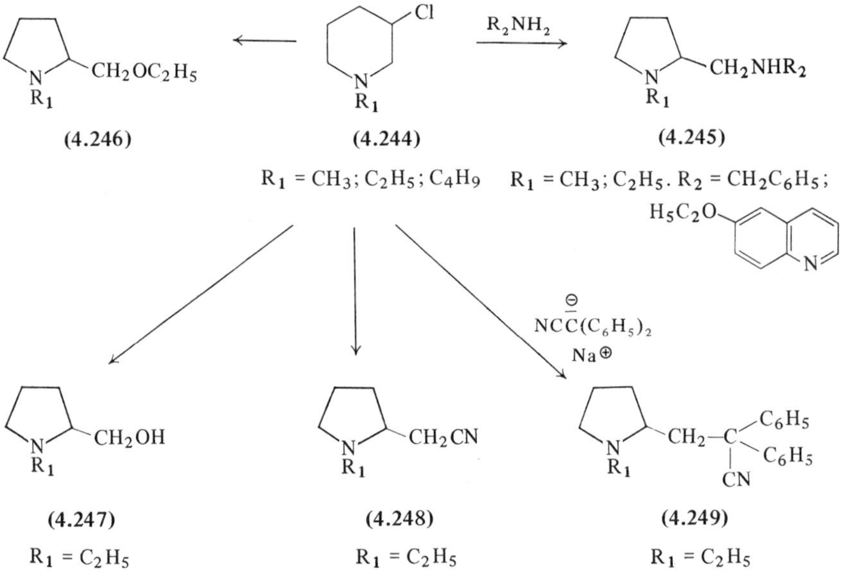

(4.246)   (4.244)   (4.245)

$R_1 = CH_3; C_2H_5; C_4H_9$ $R_1 = CH_3; C_2H_5. R_2 = CH_2C_6H_5;$

(4.247)   (4.248)   (4.249)

$R_1 = C_2H_5$ $R_1 = C_2H_5$ $R_1 = C_2H_5$

suggested as the probable intermediate in reactions of aliphatic β-haloethyl-amines with nucleophilic reagents[230,231]. The proposal has also been made that a similar intermediate is formed during the ring expansion of N-alkyl-2-chloromethylpyrrolidines into N-alkyl-3-chloropiperidines (for a more detailed discussion see vol. 1, chapter 3, section I.C.4.c). A further argument for the occurrence of an aziridinium salt as intermediate is provided by the fact that in the alkaline hydrolysis of optically active N-ethyl-3-chloropiperidine (4.250) a 95 ± 5 per cent retention of configuration was observed in the products formed, thereby ruling out a mechanism involving a dissociation into a carbonium ion intermediate.

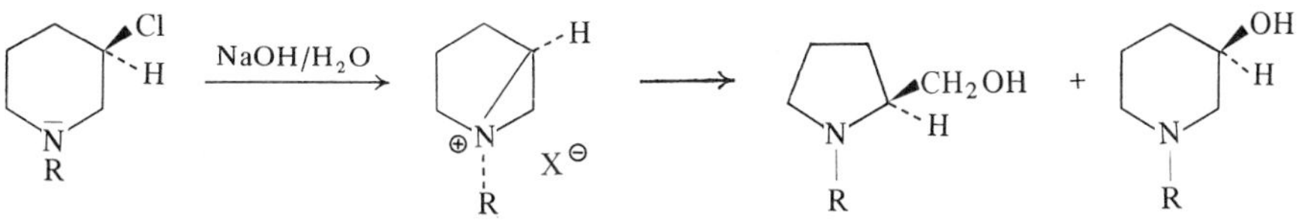

(4.250)   (4.251)

Similar ring contractions have been reported in reactions of N-alkyl-3-chloro-piperidines with the free acid 4-hydroxydiphenylacetic acid[232] and recently, during thermal decarboxylation of 3-piperidylphenothiazine-10-carboxylate (4.252), the pyrrolidinylmethylphenothiazine (4.253) was obtained[233]. It has been stated that in this reaction the product structures have been uncritically assigned.

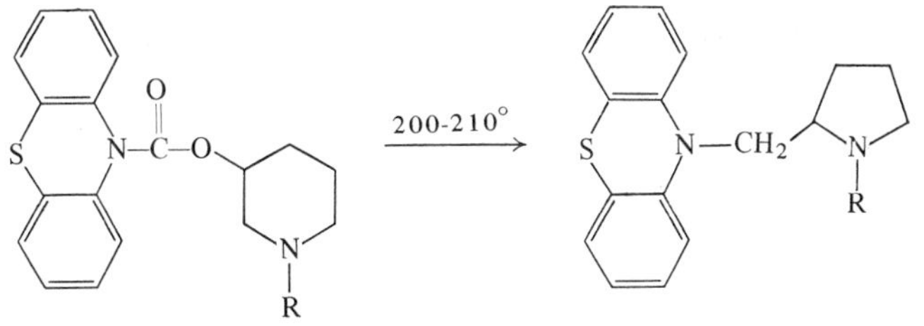

(4.252)   (4.253)

R = CH$_3$; C$_2$H$_5$

Reduction of cyclic α-aminoketones under Clemmensen conditions (and not by the Wolff–Kishner method) has been found a very useful method which, as well as leading to reduction, causes rearrangement into products containing a contracted heterocyclic ring. Treatment of a number of N-alkyl-3-piperidones (4.254) with zinc amalgam and hydrochloric acid, whether or not the 2-position is unsubstituted or carries one or more alkyl (aryl) groups, leads to the formation of 2-substituted pyrrolidines (4.255)[234-236]. By a process similar to that described above, the tetrahydro-3,3-dimethyl-5-oxopyridine-1-oxide (4.256) yields the Δ^1-pyrroline (4.257)[237]. An important key step for elucidating the mechanism of this reaction is the finding that Clemmensen reduction of

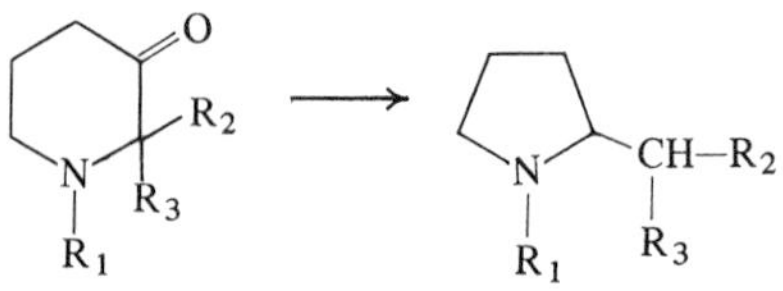

(4.254)   (4.255)

R_1	CH_3	CH_3	CH_3	CH_3	C_2H_5	C_6H_5
R_2	H	H	CH_3	CH_3	H	H
R_3	C_2H_5	H	CH_3	C_2H_5	C_6H_5	H

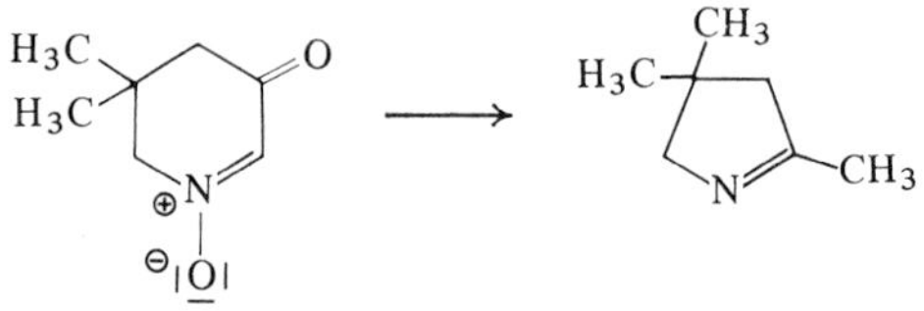

(4.256)   (4.257)

N-methyl-2-acetylpiperidine   (4.258)   gives N-methyl-7-aminoheptan-2-one (4.259), together with N-methyl-1-heptylamine (4.260)[238]. This result indicates that the vital step in this sequence of reactions is the hydrogenolysis of the N–C bond yielding (4.259) which is normally reduced to (4.260). Because of these results the mechanism of the transformation of a piperidine ring into the pyrrolidine ring can be assumed to take place as shown below.

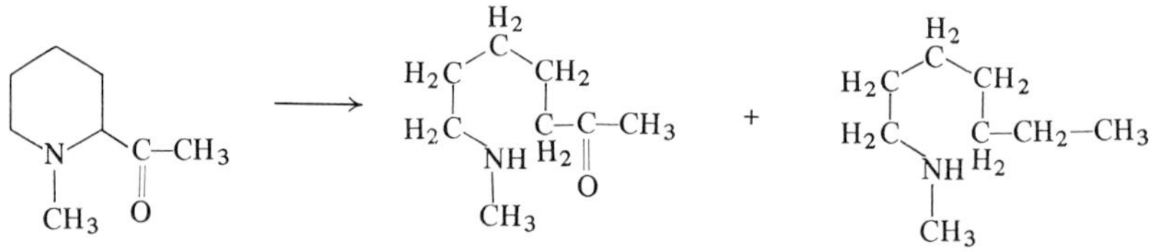

(4.258)   (4.259)   (4.260)

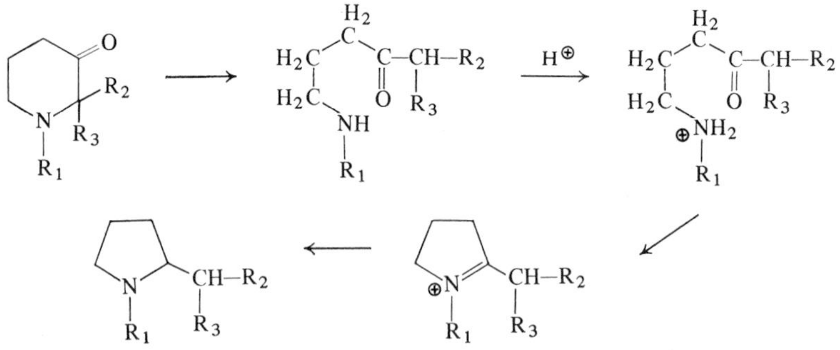

Further convincing evidence for the occurrence of an intermediate which is formed by this initial hydrogenolysis of the N–C bond was provided by the results obtained when the behaviour of the optically active ketone (4.261) towards Clemmensen reducing agents was investigated[239]. The product, N-methyl-2-s-butylpyrrolidine (4.263), was devoid of optical activity, indicating that in the intermediate (4.262) a rapid racemization takes place during refluxing with hydrochloric acid leading to loss of assymetry at the α-carbon atom.

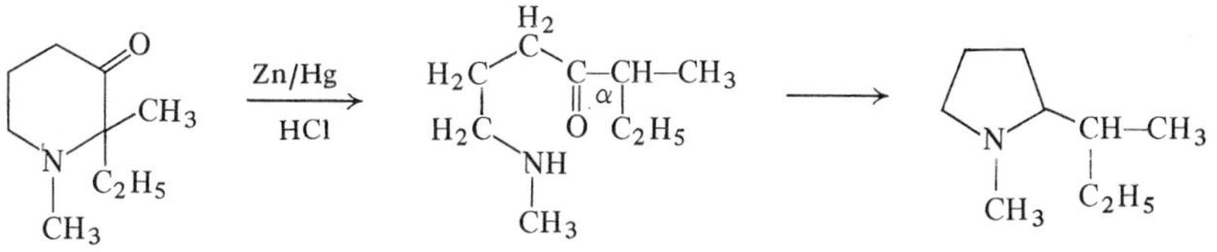

(4.261)   (4.262)   (4.263)

This reductive ring contraction has also been reported in the bicyclic series. Thus 1-ketoquinolizidine (4.264) rearranges under Clemmensen conditions into 1-aza-bicyclo[5.3.0] decane (4.265)[240,241]. In the bicyclic series it is important to know whether the ring containing the keto groups or the other six-membered ring in (4.264) contracts. The technique of introducing a "label" into each ring

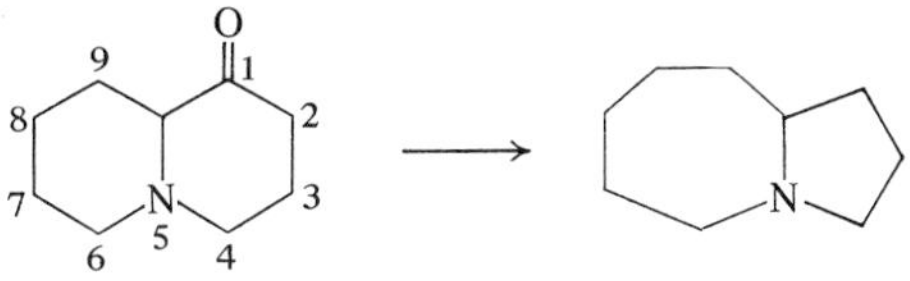

(4.264)   (4.265)

was used. On reduction of the 2-ketoquinolizidine (4.268) the compound (4.269) was obtained; (4.266) was formed on reduction of the 8-methyl derivative (4.267), clearly indicating that contraction occurred in the ring

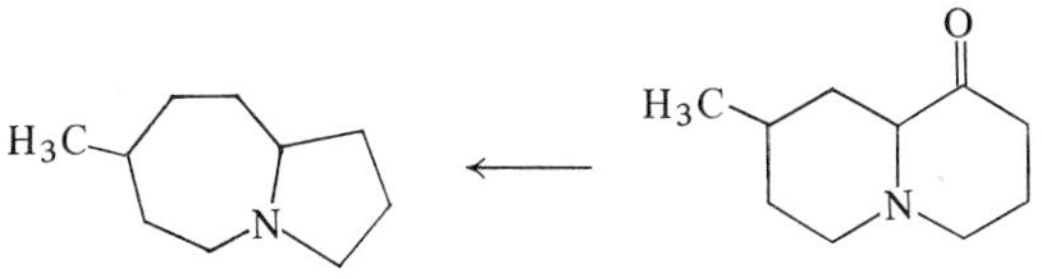

(4.266)   (4.267)

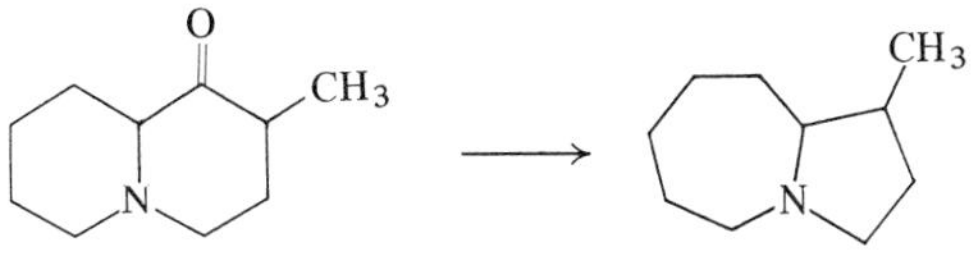

(4.268)   (4.269)

containing the keto group while there was enlargement of the other six-membered ring[242].

The following conversions take place by the mechanisms discussed above: benzoquinolizidine (4.270) into (4.271)[242]; the 1-ketoquinolizidine (4.272) into (4.273)[242]; 3-ketoquinolizidine (4.274) into a racemate of the stereo-isomers of (4.275)[243]; the 1-aroyl-1,2,3,4-tetrahydroisoquinolines (4.276) into the 2,3,4,5-tetrahydro-1H-3-benzazepines (4.277)[244]. Reduction of 1-veratroyl-2-methyl-octahydroisoquinoline with zinc and acetic acid also gives ring expansion into a benzazepine derivative[245]; the tetrahydro-6,13-diazanaph-thacene (4.279) yields (4.280)[246]. This last reaction is of considerable interest since instead of a C–C bond fission, a N–N bond fission is involved.

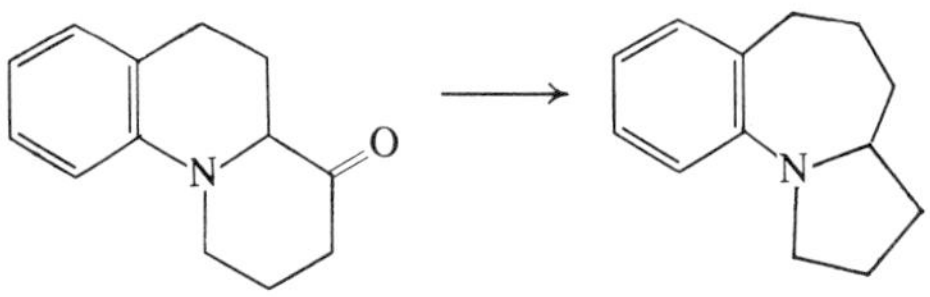

    (4.270)   **(4.271)** 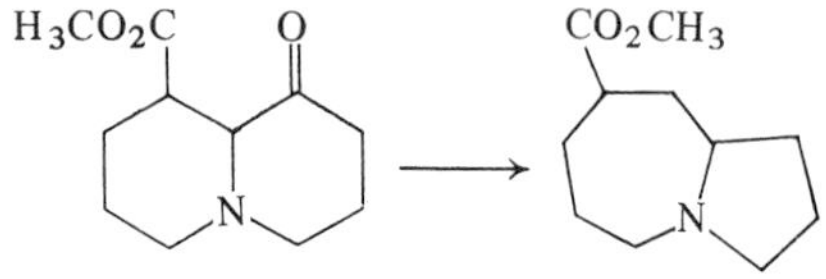   **(4.272)**   **(4.273)**

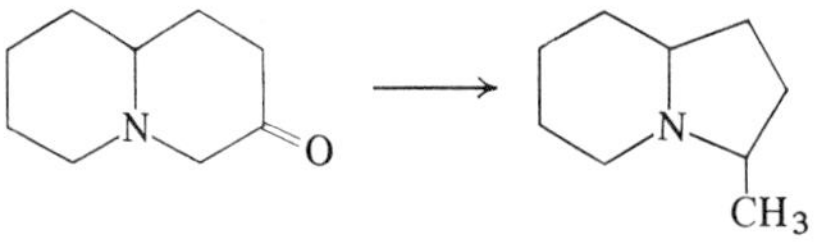

    (4.274)   **(4.275)** 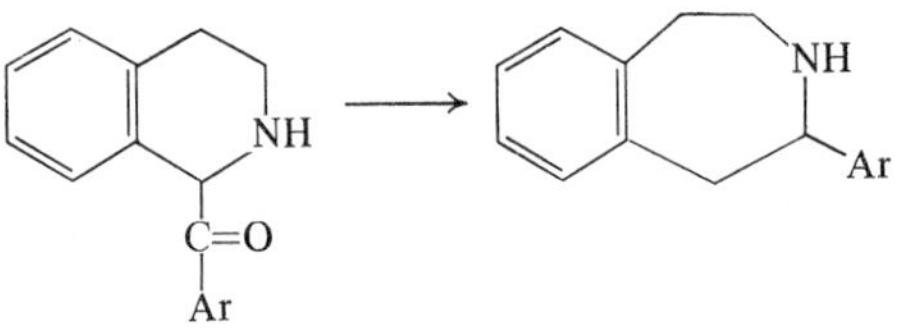   **(4.276)**   **(4.277)**

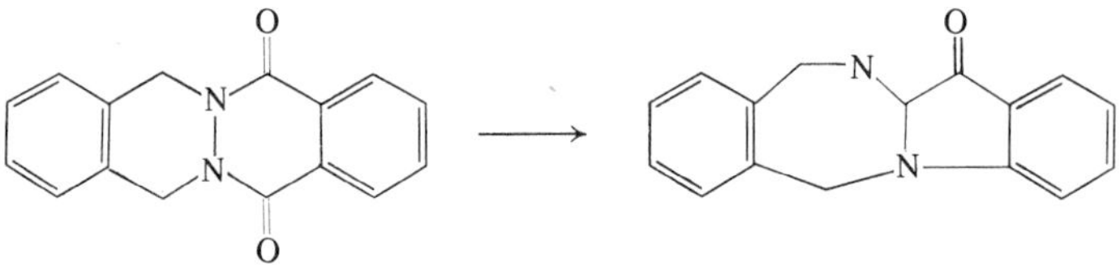

    (4.279)   **(4.280)**

When N-hydroxypiperidone-2 is treated with polyphosphoric acid at about 180-190°C, ring contraction into a five-membered ring is found to occur. This thermal ring contraction can take place either under decarboxylation or decarbonylation, depending on the stereochemical requirements of the groups or atoms which are eliminated. Whereas the N-hydroxylactam (4.281) loses CO and H_2O and yields the Δ^1-pyrroline (4.282), the *trans*-octahydroisocarbostyril (4.283) loses only CO_2 and yields the octahydroindole (4.284)[247]. These results suggest that elimination of CO may occur in a synchronous process requiring coplanarity of the bonds α and β and the C=O bond, while CO_2 is evolved in the case of compounds in which it is not possible for coplanar bonds to exist, for example, (4.283).

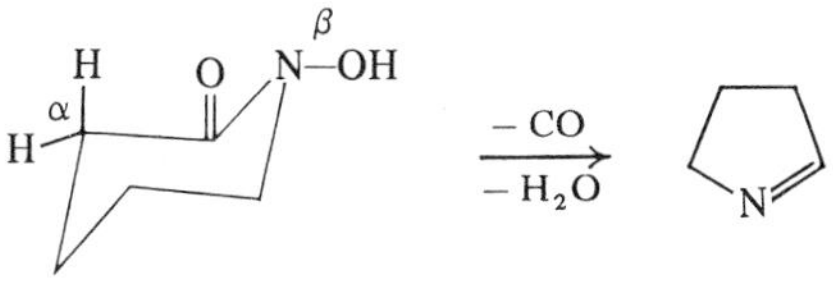

(4.281)   (4.282)

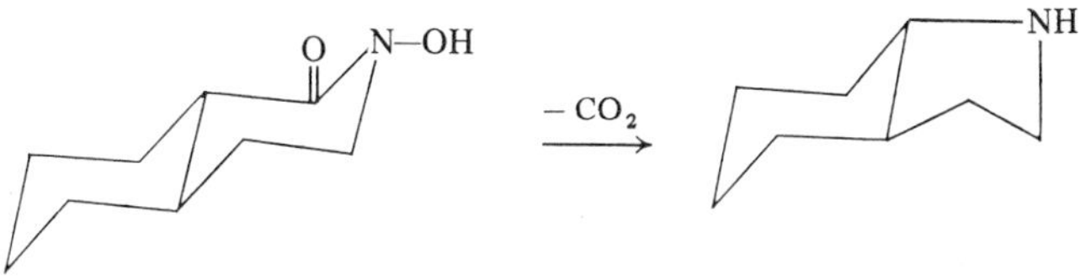

(4.283)   (4.284)

It has been established that tetrahydronicotinamide (4.285) reacts with compounds containing two nucleophilic centra to form ring contracted products. With hydrazine the pyridine ring is opened, yielding the Δ^3-pyrazolin-5-one (4.286); with hydroxylamine the Δ^2-isoxazolin-5-one (4.287) is obtained[248]. The formation can be depicted as follows:

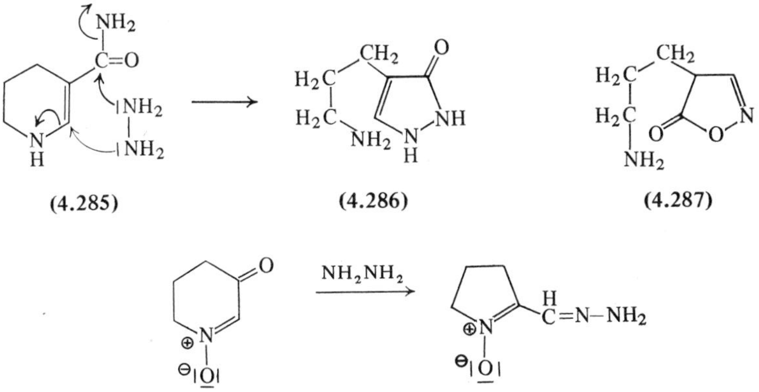

(4.285)   (4.286)   (4.287)

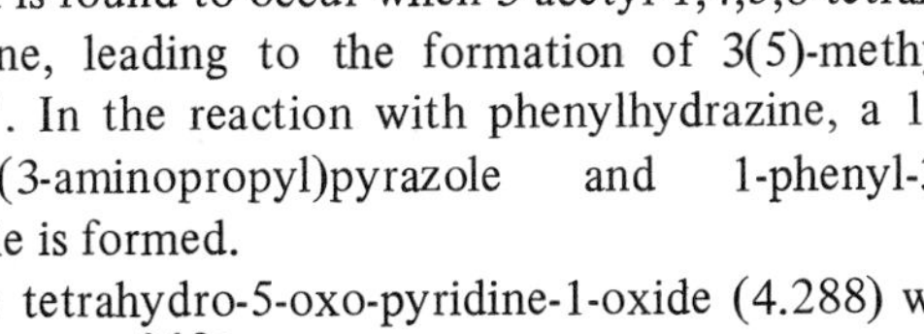

(4.288)   (4.289)

A similar reaction is found to occur when 3-acetyl-1,4,5,6-tetrahydropyridine reacts with hydrazine, leading to the formation of 3(5)-methyl-4-(3-amino-propyl)pyrazole[248b]. In the reaction with phenylhydrazine, a 1:1 mixture of 1-phenyl-5-methyl-4-(3-aminopropyl)pyrazole and 1-phenyl-3-methyl-4-(3-aminopropyl)pyrazole is formed.

Treatment of the tetrahydro-5-oxo-pyridine-1-oxide (4.288) with hydrazine gives the hydrazone (4.289)[248a].

b. *1,4-Dihydropyridines into pyrroles*

It has recently been reported that 3,5-bis(ethoxycarbonyl)-1,4-dihydro-2,6-dimethylpyridine-4-carboxylic acid (4.290) (R = $CO_2C_2H_5$) gives, on heating for 5 min at 240°C, a mixture of 3,5-bis(ethoxycarbonyl)-2,6-dimethylpyridine (4.291) (R = $CO_2C_2H_5$) and two ring-contracted products, 3-ethoxycarbonyl-4-(ethoxycarbonyl)1,4-dihydro-2,6-dimethylpyridine does not contain a carboxy 3-ethoxycarbonyl-2,5-dimethylpyrrole (4.292) (R = $CO_2C_2H_5$)[249]. 3,5-Bis-(ethoxycarbonyl)1,4-dihydro-2,6-di-methylpyridine does not contain a carboxy group on position 4 and does not therefore yield any of these products on heating; consequently it cannot be considered as an intermediate in this reaction. On the other hand the *N*-methyl derivative of (4.290) (R = $CO_2C_2H_5$) produced the corresponding *N*-methyl derivatives of (4.292) (R = $CO_2C_2H_5$) and (4.293) (R = $CO_2C_2H_5$). The suggestion has been made that the bicyclic intermediate (4.294) plays a central role in these ring contraction reactions. Heating the 3,5-dicyano compound (4.290) (R = CN) did not give products with a ring-contracted ring[249]; only products formed by an intramolecular reaction of the cyano group and carboxylic acid group and which thus still contained the

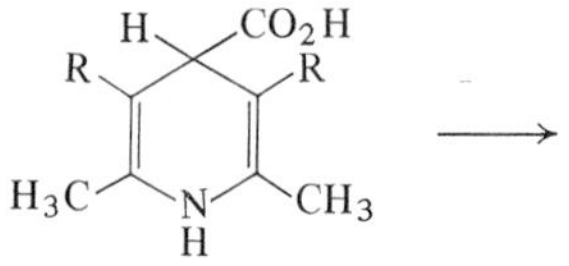

(4.290)

R = $CO_2C_2H_5$

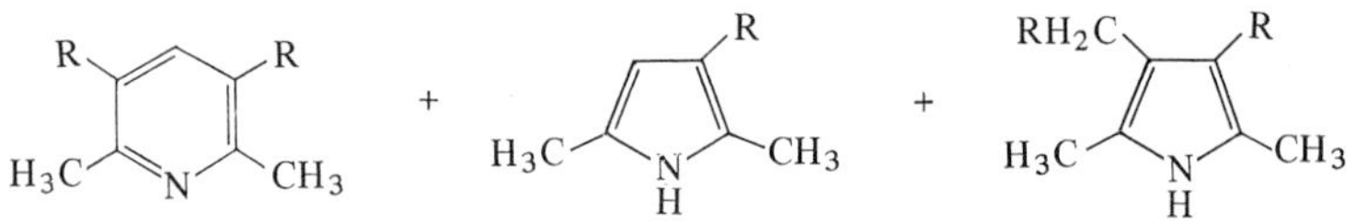

(4.291)   +   (4.292)   +   (4.293)

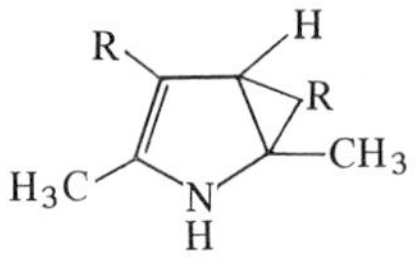

(4.294)

pyridine ring, were yielded. Treatment of the 4-chloromethyl-1,4-dihydropyridine derivatives (4.295) (R = H) with liquid ammonia[250], or of (4.295) (R = CH_3) with potassium *t*-butoxide[251], leads to ring contraction into the corresponding pyrroles (4.296). Since both ring contraction reactions very probably occur through the intermediary occurrence of seven-membered azepine derivatives, we refer to chapter 5, section I.B.1, in which these ring enlargements with ring contractions of 1,4-dihydropyridines are discussed in greater detail.

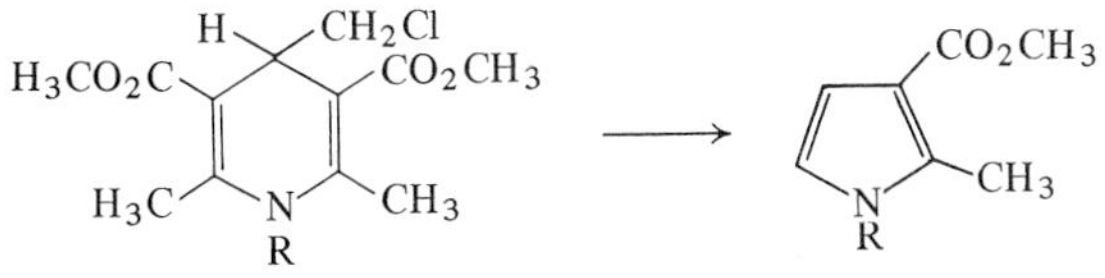

(4.295)   (4.296)

R = H; CH$_3$

c. *Pyridines into pyrroles and pyrrolidines; quinolines into indoles; phenyliso-quinolines into phthalimidines*

An important method for converting pyridines into pyrroles is the low-temperature ultraviolet irradiation of 3,4-chinon-3-diazides derived from pyridine[252,253], quinoline[253-255], 1,5- and 1,7-naphthyridine[253,256], and 6,7-benzoquinoline[254]. In all these reactions the carboxy derivatives derived from pyrrole, indole or azaindole respectively are obtained.

In the pyridine series, the 2,3-chinon-3-diazid [= 3-diazo-2(3*H*)pyridone (4.297)] as well as the 3,4-chinon-3-diazides [= 3-diazo-4(3*H*)pyridones (4.299)] are found to be photolysed into pyrrole-2-carboxylic acid (4.298) and the pyrrole-3-carboxylic acid (4.300) respectively[253]. This result is of considerable interest since the two examples illustrate that in these photoreactions carbon-carbon as well as carbon-nitrogen bond breaking processes can occur.

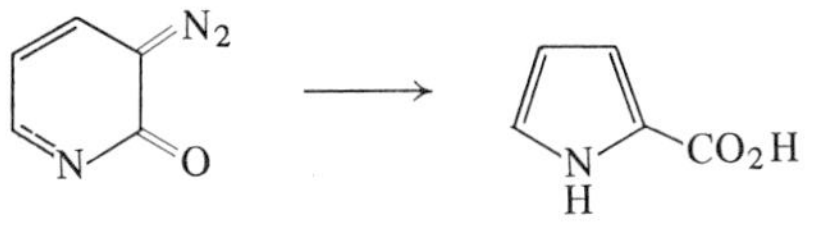

(4.297)   (4.298)

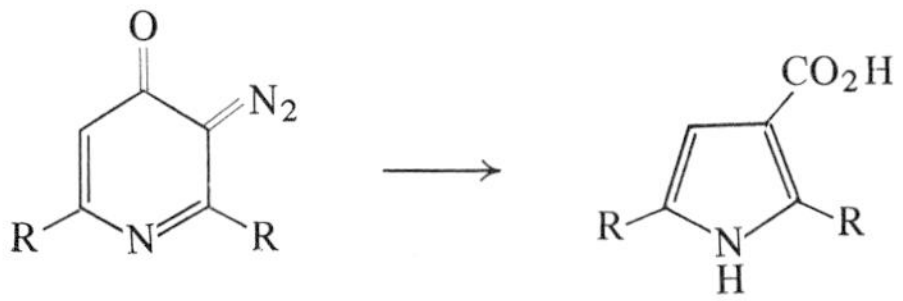

(4.299)   (4.300)

R = H; CH$_3$

Quite analogously, a number of 3-diazo-4(3*H*)quinolones (4.301) are converted into the indole-3-carboxylic acids (4.303)[253,254]. The carbon-to-carbon re-arrangement observed with the 3,4-chinon-3-diazides is closely related to the Wolff rearrangement of α-diazoketones into ketenes, which has been demon-strated by isotopic labelling[257] to be a carbon-skeleton rearrangement and not

(4.301)

(4.302)   (4.303)

R_1	H	H	H	H	H	$O-C_6H_5$	$SO_2-O-C_6H_5$	$-(CH)_4-$	Cl	CH_3
R_2	H	Cl	C_6H_5	CH_3	CO_2CH_3	H	H		H	H
R_3	H	H	H	H	H	H	H		H	$O-CH_3$

oxygen wandering[258]. It seems justified that in all these photo-induced ring contractions with o-chinon-diazides, a ketoketene, (4.302), is involved and this is further hydrolysed into an acid.

A very similar ring contraction has been reported[258a] during vapour-phase pyrolysis of the $1H$-benzotriazoles, $1H$-naphthotriazoles and $1H$-pyridotriazoles. It was found that on pyrolysis of $1H$-pyridotriazoles at $500°C$ 2-cyanopyrrole was obtained from (4.303a) and 3-cyanopyrrole was obtained from (4.303b); both ring contractions produced good yields. Both cyanopyrroles show a sigmatropic rearrangement when heated at $800°C$; they are thus interconvertible. It is of interest that an attempt to achieve the corresponding reaction of $1H$-triazolopyrimidine into cyanoimidazole failed. It is most probable that this contraction involves a Wolff rearrangement of the iminocarbene (4.303d) formed from the 1,3-biradical (4.303c), which is obtained initially. It has been assumed that ring contraction of the iminocarbene does not take place as a discrete stage in the reaction via the azirine (4.303e), although it carries the available orbitals of both carbon and nitrogen atoms, since this intermediate would certainly lead to a mixture of 1- and 2-cyanopyrrole and these have not been detected.

It has been reported[258b] that a recent attempt to prepare a hydroperoxide adduct of 1-methyl (or 1-benzyl)-3-carbamoylpyridiniumchloride (4.303f) by treatment with alkaline hydrogen peroxide, resulted in the formation of the hydroperoxide (4.303i), the end product of an extensive rearrangement. Although the yields are low (7-8 per cent), products of high purity are obtained.

(4.303a)   (4.303b)

(4.303c)   (4.303d)

(4.303e)

Treatment of 1-benzyl-3-acetylpyridinium chloride or the corresponding 3-bromo compound with alkaline hydrogen peroxide resulted in the formation of substituted benzenes. A possible mechanism in the formation of (4.303i) would involve the oxide (4.303g), formed from the hydroperoxide precursor, followed by solvolysis of the oxirane ring into the diol and subsequent ring opening, yielding the dialdehyde (4.303h). Ring closure and subsequent oxidation gives (4.303i). A similar oxidative ring contraction of dihydro-pyrazines into Δ^2-imidazolines is described in section II.D.1.b.

Ultraviolet-light induced ring contractions of mesoionic pyridinium compounds are being considered in an increasing number of publications. The photolysis of the pyridine-N-oxides (4.304)[259-262a,b] carried out in an inert solvent at 15°C with light that induces only n — π* transitions (>300 mµ) leads to the formation of pyrroles carrying a carbonyl group in position 2, i.e. (4.305); this reaction seems to be common in the monocyclic series[260]. It has been found that in the case of methylpyridine-N-oxides, besides ring contraction and deoxygenation, the formation of a pyridine, containing a hydroxy group in position 3, occurs[261-262a]. Thus 2-methylpyridine N-oxide (4.304) ($R_2 = R_1 =$ H, R = CH_3) gives, on irradiation, the pyrrole-2-carboxaldehyde (4.305) ($R_1 =$ $R_2 =$ H; R = CH_3), together with the corresponding 3-hydroxy- and 5-hydroxy-2-methyl-pyridine (4.306). From 2,6-lutidine-N-oxide, 2-acetyl-5-methylpyrrole (4.305) (R = $R_2 =$ CH_3; $R_1 =$ H) is obtained together with 3-hydroxy-2,6-lutidine (4.306) (R = $R_2 =$ CH_3; $R_1 =$ H)[262]. 2,3,6-Collidine gives similar products, but pyridine- and 4-methylpyridine-N-oxide do not give 3-hydroxypyridines[262a].

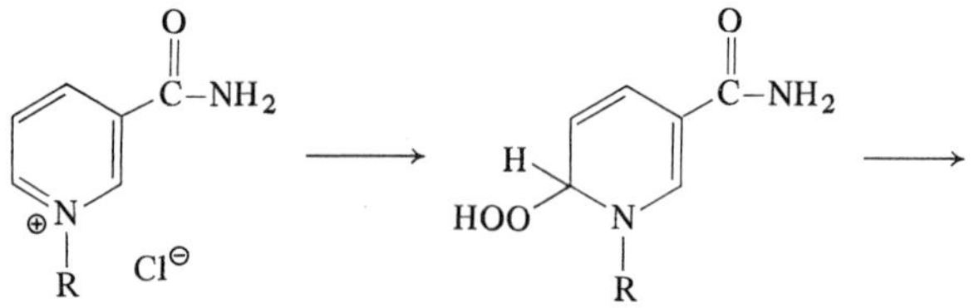

(4.303f)

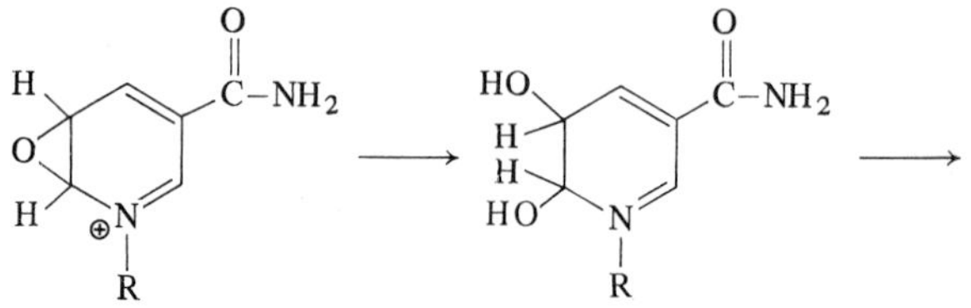

(4.303g)

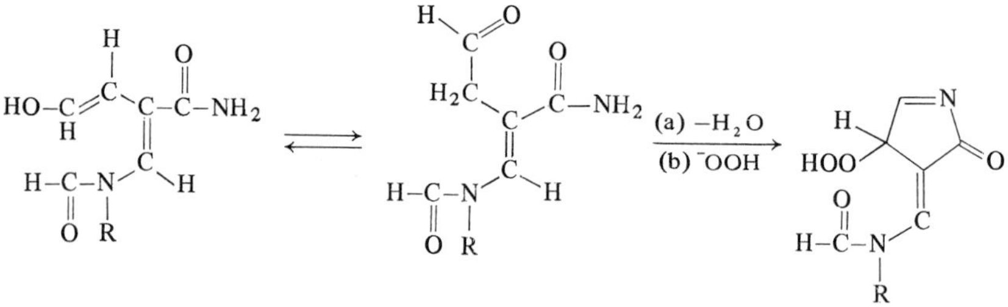

(4.303h)   (4.303i)

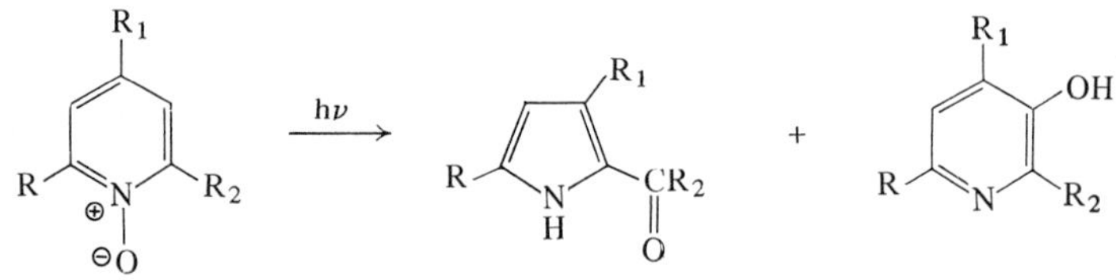

(4.304)   (4.305)   (4.306)

More insight into the course of these ring transformations was obtained when a solution of 2,6-dicyanopyridine-N-oxide (4.307) in dichloromethane was irradiated with light of >290 mμ. Besides deoxygenation into 2,6-dicyanopyridine and ring contraction into 5-cyano-2-pyrrolecarbonylcyanide (4.311), ring expansion into a seven-membered compound was observed; this compound was tentatively assigned the structure 2,4-dicyano-1,3-oxazepine (4.309)[263]. From the mechanism of these reactions, it can be postulated that in all these ring transformations the first step is the formation of a non-isolable, unstable oxaziridine (4.308), a reaction known to be characteristic for nitrones[264].

Direct evidence for the occurrence of this intermediate is the fact that when the irradiation of pyridine *N*-oxide is carried out in ethanol, besides ring contracted products, acetaldehyde is formed[265]. This indicates that during the irradiation a powerful oxidant[266], the oxaziridine, is formed which is able to oxidize the solvent ethanol into ethanal. Valence-tautomerization of the oxaziridine (4.308) yields partly the 1,3-oxazepine (4.309), partly the 1,2-oxazepine (4.310), which, however, is unstable and on further irradiation gives the pyrrole (4.311) via nitrogen–oxygen cleavage and recombination.

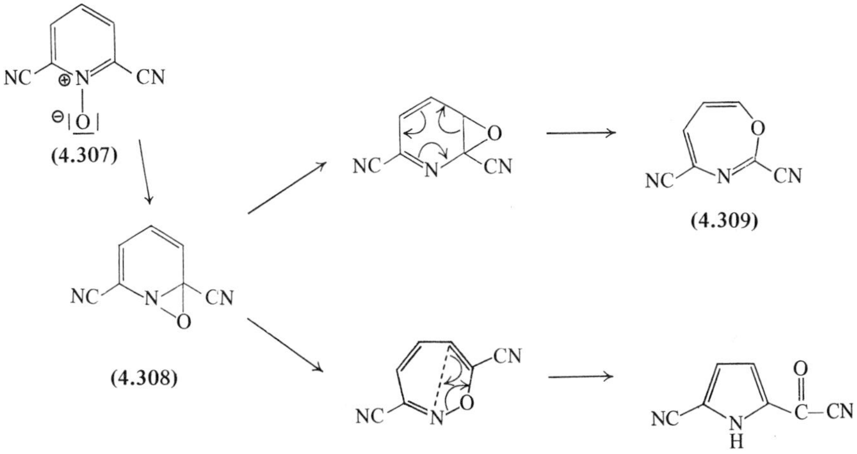

It is of interest that pyridine-*N*-oxide, when irradiated in the gas phase at 3261 Å (n-π* transition) and 2537 Å (π-π* transition), only undergoes deoxygenation[267].

A solution of the mesoionic pyridinium dicyanomethylide (4.312) (formally the analogue of pyridine-1-oxides)[259] in benzene is found to be isomerized exclusively into the dicyanovinylpyrrole derivative (4.313) when light of ~400 mμ is used[268]. With light of higher energy, besides (4.313), 7,7-dicyanonorcaradiene (4.314), formed by addition of dicyanocarbene to the solvent benzene, is obtained. The energy of the photons of the light of 400 mμ is apparently

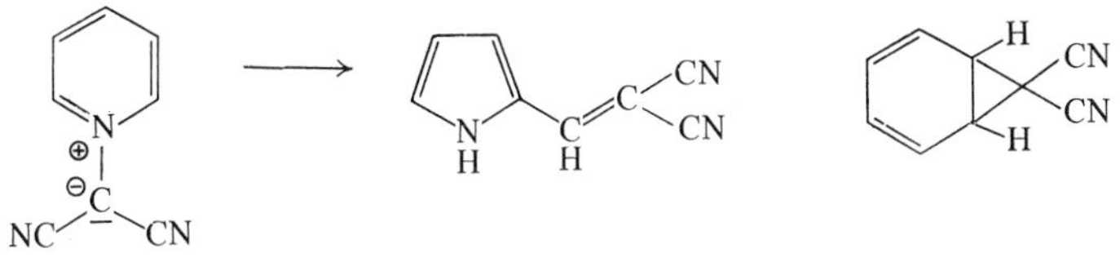

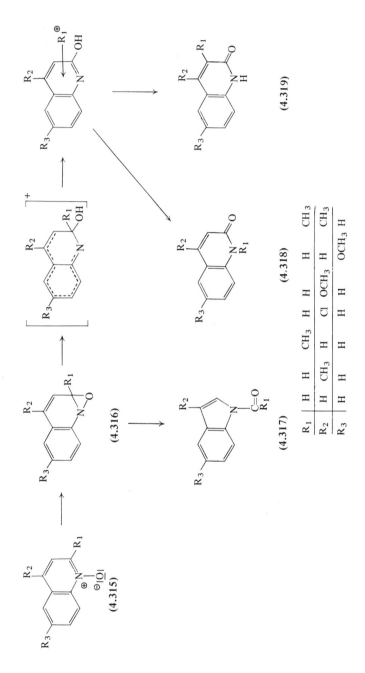

R_1	H	H	CH$_3$	H	H	H	CH$_3$
R_2	H	CH$_3$	H	Cl	OCH$_3$	H	CH$_3$
R_3	H	H	H	H	H	OCH$_3$	H

sufficiently low to allow only the weakest electronic transition and to prevent fission of the $C(CN)_2$ group from the pyridine ring.

Some quite fascinating rearrangements were observed during the irradiation of quinoline-N-oxides. Thus when an ethanolic solution of 2-methylquinoline-N-oxide (4.315) ($R_1 = CH_3$, $R_2 = R_3 = H$) is exposed to ultraviolet light there is formation of 1-methylcarbostyril (4.318) (16 per cent), 3-methylcarbostyril (4.319) (22 per cent) and the ring-contracted product N-acetylindole (4.317) (10 per cent)[269]. In order to investigate the scope of these rearrangements in more detail and to clarify further the mechanism of these ultraviolet-light induced ring transformations, the reaction has been extended to a variety of substituted quinoline-N-oxides[270]. In correlation with the monocyclic N-oxides, the oxaziridine species (4.316) is advanced as the primarily formed intermediate in all these conversions; this contention has recently been supported by a trapping experiment with amines[271]. When R_2 is a hydrogen atom or methyl group (both are characterized by a low affinity for electrons) in a protic solvent, a carbonium ion rearrangement of the substituents is suggested, as indicated in the reaction scheme[272].

It has been reported that photolysis of a solution of the 2-methylquinoline-N-oxides (4.320) in benzene or ether, which contains water, gives, besides N-acetylindoles, as photoproducts the N-acetyl-2-hydroxy-2,3-dihydroindoles (4.325) or the tautomeric oxo compound (4.324)[260,273,274]. For quinoline-N-oxide hydrate the same type of ring contraction has been reported[275]. The formation of (4.325) can be explained by heterocyclic cleavage of the oxaziridine species—possibly photochemically induced—yielding a resonance-stabilized dipolar ion (4.321), which, after a subsequent attack of water and carbon-carbon bond fission, gives the open-chain intermediate (4.323), which is in equilibrium with (4.324) (and/or its rotamer). Reinvestigation of the uv-induced rearrangements of quinoline-N-oxides without a substituent on position 2 (3-, 4-, 5-, 6-, 7- and 8-methylquinoline-N-oxides) in protic solvents at very low concentration (10^{-3} M) has indicated that under these conditions the compounds are mainly converted into the corresponding carbostyrils[276,277]. The great effect that the solvent and hydration has on the course of the reaction presents an interesting problem in the photochemistry of quinoline-N-oxides.

Very recently it was found that the 8-substituted quinolines (4.326) are converted into the 7-substituted oxindoles (4.328) when treated with hydrogen peroxide in acetic acid at 70-80°C for eight hours[278]. It has been proved that the reaction proceeds via the 3-hydroxy compound (4.327) and not via the corresponding N-oxide or by a previous ring contraction which would lead to an indole derivative which would subsequently be oxidized. As this ring contraction does not take place with the unsubstituted compound (4.326) (R = H) or the 8-methyl derivative (4.326) (R = CH_3), a structural requirement for the

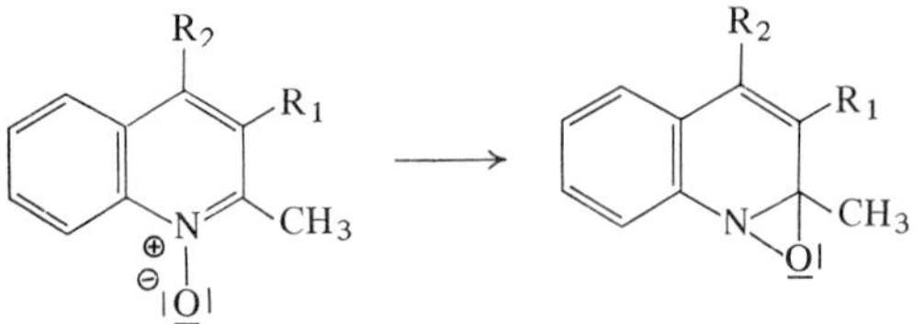

(4.320)

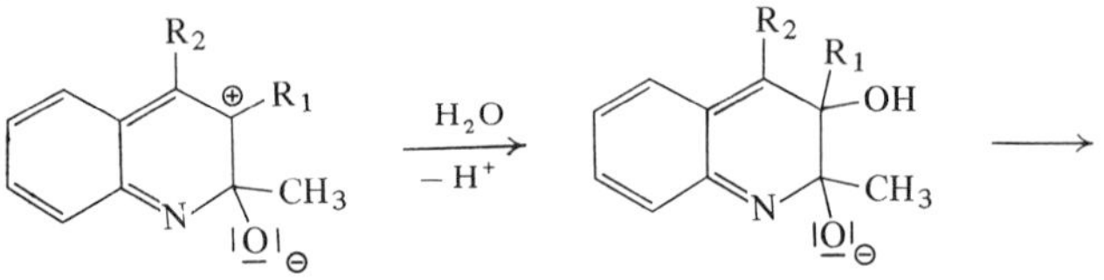

(4.321)   (4.322)

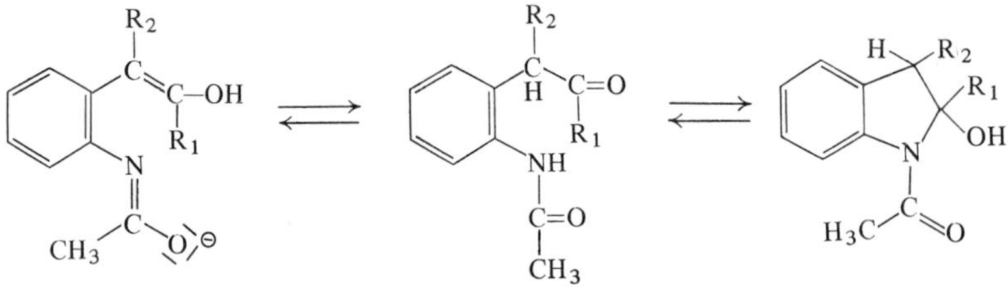

(4.323)   (4.324)   (4.325)

R$_1$	H	CH$_3$	H	CH$_3$
R$_2$	H	H	CH$_3$	CH$_3$

achievement of this reaction is apparently the presence of bulky and electron-withdrawing substituents.

Extension of these uv-induced rearrangements to the photochemistry of isoquinoline-N-oxide[270] and phenanthridine-N-oxides[279] has shown that neither of these compounds yield ring contracted products containing the indole or carbazole ring respectively; only the corresponding isocarbostryrils or phenanthridones were formed as main reaction products.

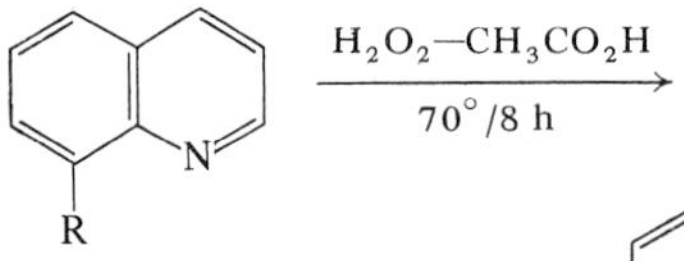

(4.326)

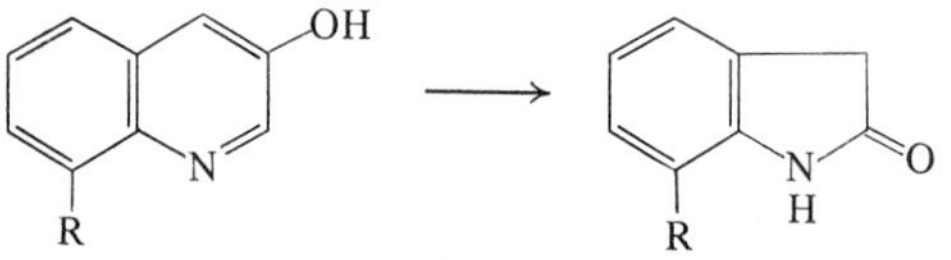

$$\xrightarrow[70°/8\ h]{H_2O_2-CH_3CO_2H}$$

(4.327)   (4.328)

R = NO$_2$; CO$_2$H; CO$_2$C$_2$H$_5$

An uv-induced ring contraction has also been reported to occur when an ethanolic solution of 3,5-bis(ethoxycarbonyl)-2,4,6-trimethylpyridine (4.329) is irradiated, the 3-ethoxycarbonyl-2,4,5-trimethylpyrrole (4.330) being obtained together with the corresponding 1,4-dihydropyridine (4.331)[280]. When a methyl group is absent on position 4, no ring contraction occurs; only a 1,4-dihydropyridine is yielded. Since the 1,4-dihydropyridines (4.331) are quite

(4.329)   (4.330)   (4.331)

R = CO$_2$CH$_3$

(4.332)   (4.333)

stable under these photolytic conditions, they cannot act as precursors of the pyrrole (for the thermolytic conversion of 1,4-dihydropyridines into pyrroles see section I.C.1.b). When (4.329) was irradiated in CH$_3$OD, the 1,4-dihydro compound was found to contain >90 per cent deuterium on position 4—i.e. (4.332). It is thought that the pyridine nitrogen removes the carbon hydrogen from the ethanol and that the resulting α-hydroxyradical transfers the hydrogen (deuterium) which is bound to the oxygen, to the 4-position, yielding (4.332), or to the 3-position, affording (4.333). Decomposition of (4.333) leads to the formation of the pyrrole (4.330).

Very interesting ring contractions were found to occur when 3-amino-2-bromopyridine and 3-amino-2-bromoquinoline were treated with potassium amide in liquid ammonia at −33°C. From 3-amino-2-bromopyridine (4.334) 3-cyanopyrrole (4.335) was obtained[281] in high yield (80 per cent), while from 3-amino-2-bromoquinoline (4.336) the 3-cyanoindole (4.338) was formed[282]. In this last reaction the yield on (4.338) was, however, rather low (about 20 per cent), since as the main product a compound was isolated which was proved by physical and chemical methods to be the cyanomethylisocyanide (4.337). It was found that the compound (4.337) could be quantitatively converted into (4.338) by treatment with potassium amide in liquid ammonia under conditions similar to those used for the conversion of (4.336) into (4.338), thus showing that (4.337) is a precursor of 3-cyanoindole. This finding is of considerable importance because of its mechanistic implications.

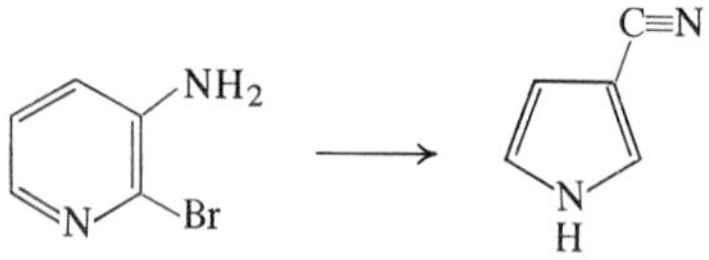

(4.334)   (4.335)

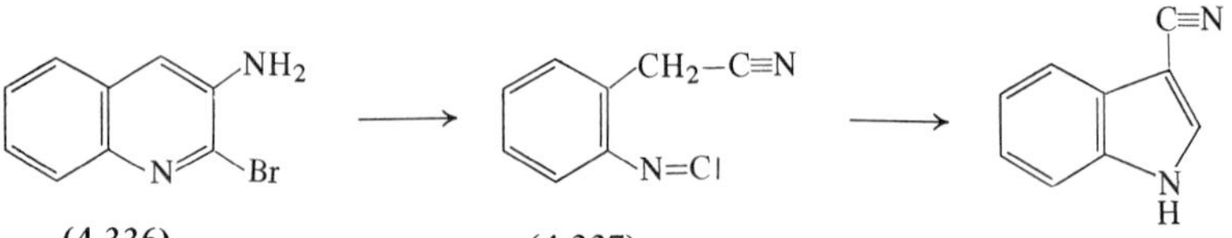

(4.336)   (4.337)   (4.338)

From the structure of the products obtained it is evident that in both reactions C_2-C_3 bond fission has occurred. This type of bond breaking is therefore of great interest, since a process leading to carbon-carbon bond fission in the pyridine ring is, unlike carbon-nitrogen bond fission, a very unusual phenomena which previously had rarely been observed (for other examples of carbon-carbon fission in pyridines and pyrimidines see sections I.C.2.a and II.C.1.a). The mechanism of these ring contraction reactions can be depicted as follows:

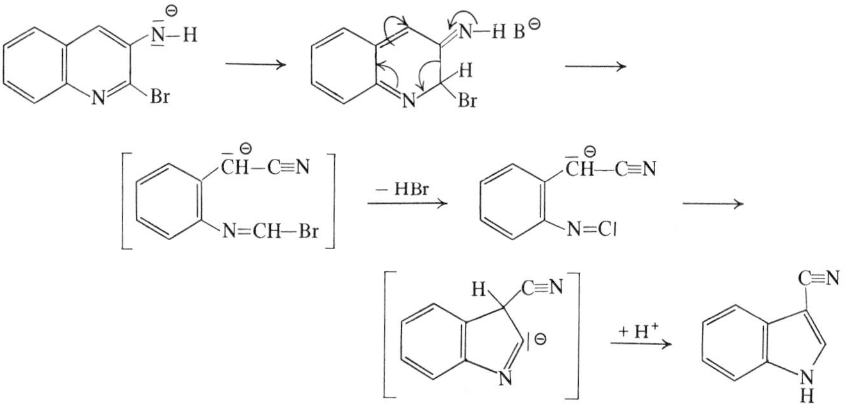

It has recently been found[282a] that the conversion of (4.334) into (4.335) can also be performed with lithium piperidide in piperidine. The following reactions are interesting to note though difficult to explain: when 3-amino-2-chloropyridine is treated with potassium amide in liquid ammonia it is not converted into 3-cyanopyrrole, but with lithium piperidide in piperidine ring contraction into (4.335) readily takes place[282b].

Remarkably, treatment of 3-hydroxy-2-bromopyridine (4.339) (R = H) and 3-hydroxy-2,6-dibromopyridine (4.339) (R = Br) with an excess of potassium

amide in liquid ammonia at $-33°C$ does not lead to the formation of a
3-substituted pyrrole (as with the 3-amino compound (4.334)), but yields a
2-substituted pyrrole, the pyrrole-2-carbonamide (4.340)[283] (R = H and R = Br
respectively). From the structure of these products it is clear that in these
reactions not a C_2-C_3 but a C_3-C_4 bond fission is involved; again, it should
be noted that this is quite an unusual phenomenon in pyridine chemistry.

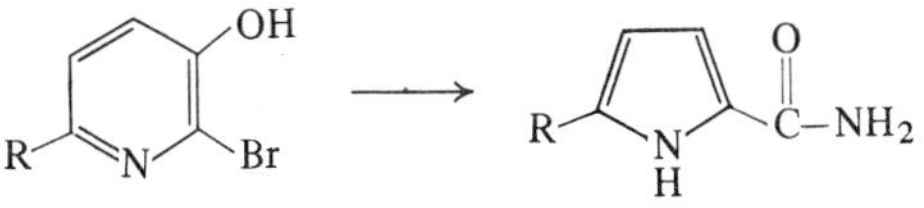

(4.339)　　　　　　　　(4.340)

R = H; Br

An interesting and uncommon feature in these ring contractions is that
(4.339) (R = H), when treated with lithium piperidide in piperidine, does not
yield a 2-substituted but a 3-substituted pyrrole derivative i.e. pyrrole-3-
carbopiperidide (4.340a). That the nature of the substituent has a strong
influence on the course of the ring contraction has been amply demonstrated[283a]. Thus the pyridine derivatives (4.340b) give, on treatment with
potassium amide in liquid ammonia, a mixture of both 3-cyano and 2-cyano
compounds (4.340c) and (4.340d) respectively, indicating that the substituents
on position 6 activate both C_2-C_3 and C_3-C_4 bond fission. With the
substituted 3-hydroxypyridines (4.340e) *only* the pyrrole-2-carbonamides
(4.340f) have been obtained[283b].

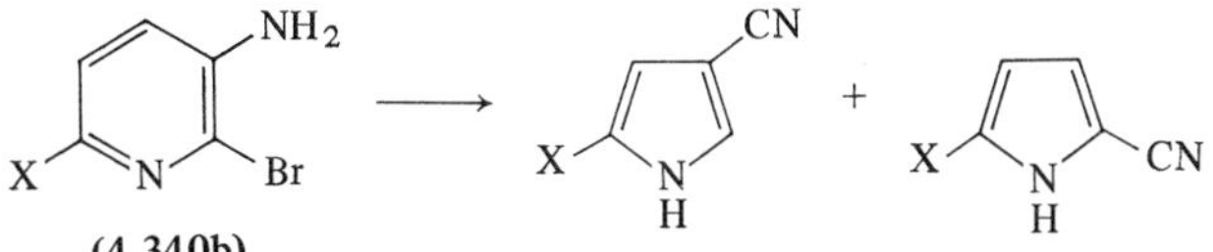

(4.340b)　　　　　　　(4.340c)　　　　　(4.340d)

X = Cl; Br; OC$_2$H$_5$

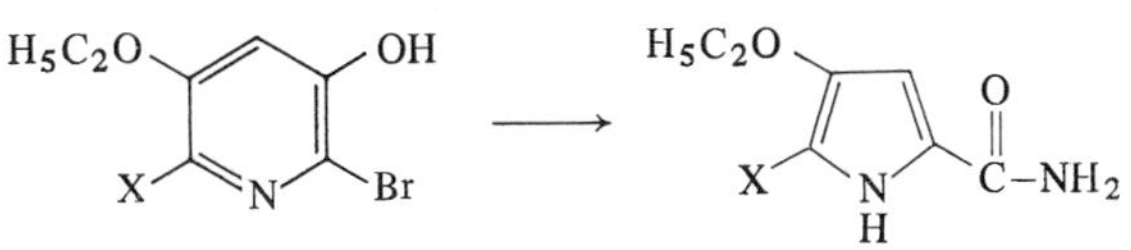

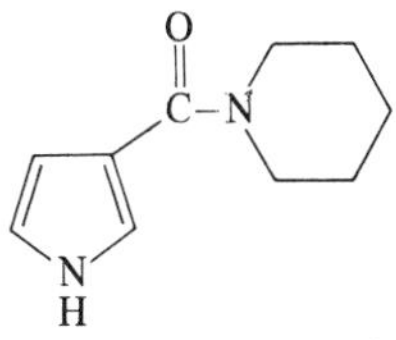

(4.340e)　　　　　　　(4.340f)　　　　　　(4.340a)

X = H, Br

It was recently observed that 4-bromo-3-hydroxyquinoline (4.340g), when
treated with potassium amide in liquid ammonia rather surprisingly yields
oxindole (4.340k)[284]. It is very likely that in this strong basic medium

4-bromo-3-hydroxyquinoline is converted via an anion into (4.340h) in which the ready occurrence of a nucleophilic attack at the azomethine bond can be envisaged; ring opening with concomitant loss of a bromide ion yields the keteneformamidinobenzene (4.340i) which gives ring closure into (4.340j) and a subsequent HCN fragmentation at the ring nitrogen[283b].

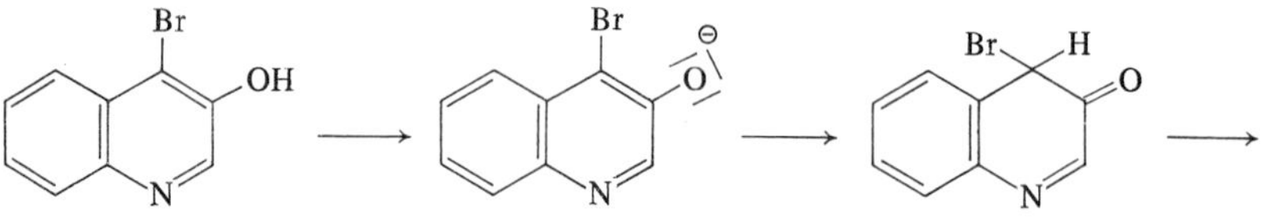

(4.340g)   (4.340h)

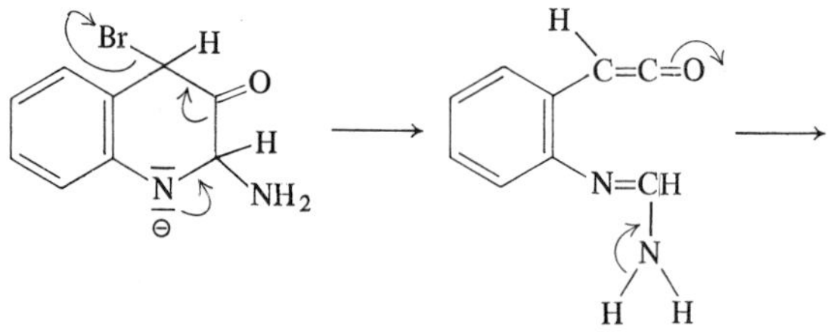

(4.340i)

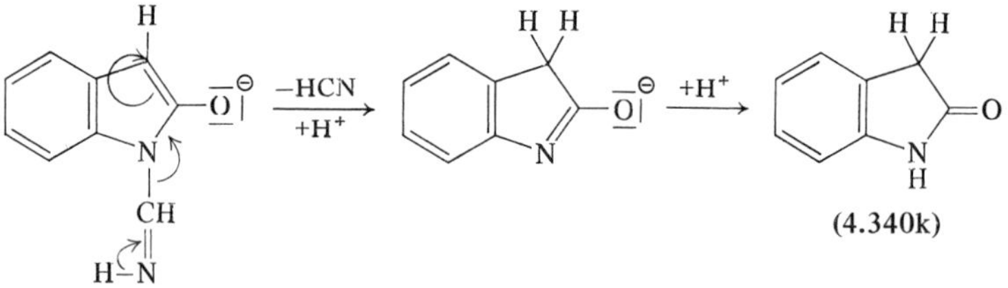

(4.340j)

Ring contraction has been found to occur when appropriately substituted quinolines or isoquinolines are oxidized with potassium permanganate. It has been established that whereas 3-phenylquinoline is converted by a cold dilute acidic solution of permanganate into N-benzoylanthranilic acid, the 2-methyl-3-phenylquinoline and 2-ethyl-3-phenylquinoline gave, amongst other compounds, the ring contracted product 1-alkanoyl-2-hydroxy-2-phenyl-3-indolinone (4.344). Apparently in all these reactions a 1,2-phenyl migration must take place[285,286]. Since it has been found that 4-hydroxy, 2-hydroxy and 2,4-dihydroxy derivatives of 3-phenylquinoline are not oxidized into N-benzoyl-anthranilic acid, it must be concluded[285] that these compounds are not

R = CH$_3$; C$_2$H$_5$ (4.341)

(4.342)   (4.343)   (4.344)

intermediates during the oxidation. The rearrangement mechanism fulfilling the fundamental requirement of placing the phenyl group on a carbon atom adjacent to the nitrogen can tentatively be suggested to occur to go via an initial formation of the pseudobase (4.341) and oxidation into the cyclic triol (4.342), followed by ring opening into o-acylaminobenzil (4.343) and finally cyclization into the N-acyl-3-indolinone (4.344).

Another example of an oxidative ring contraction is the rather high-yield conversion of 1-phenylisoquinoline (4.345) into 3-hydroxy-3-phenylphthal-imidine (4.347) by the action of potassium permanganate[287]. The mechanism is quite uncertain; it has been argued that in the intermediary arylphenylimine (4.346) the electrophilic carboxylate group attached to the nitrogen promotes the addition of water to the azomethine group, after which loss of potassium carbonate gives cyclization into (4.347). An equally attractive suggestion closely related to the one discussed above, is that oxidative formation of the intermediate (4.348) occurs which, by a sequence of steps involving a benzilic acid-type rearrangement, leads to the formation of (4.347).

The formation of phthalimide by oxidation of isoquinoline[288] could be explained by postulating the same mechanism.

Degassed Raney nickel is also found to perform ring contraction of the pyridine ring. When pyridine is boiled with degassed Raney nickel, besides large amounts of 2,2'-bipyridyl, a non-electrolytic complex C$_{28}$H$_{22}$N$_6$Ni (Λ = 3 Ω^{-1} in pyridine) is obtained containing one mole of 2,2'-bipyridyl and two moles of 2-(2-pyrrolyl)pyridine[289,290]; this method is of importance for the preparation of 2,2'-bipyridyl. It has been reported that ring contraction of quinolines into small amounts of indoles and carbazole can be achieved by this method[291]. So far little has been done to develop these ring contraction reactions as preparative methods for obtaining pyrroles or indoles, but they certainly deserve con-sideration to study the scope of the reaction and to unravel the mechanism.

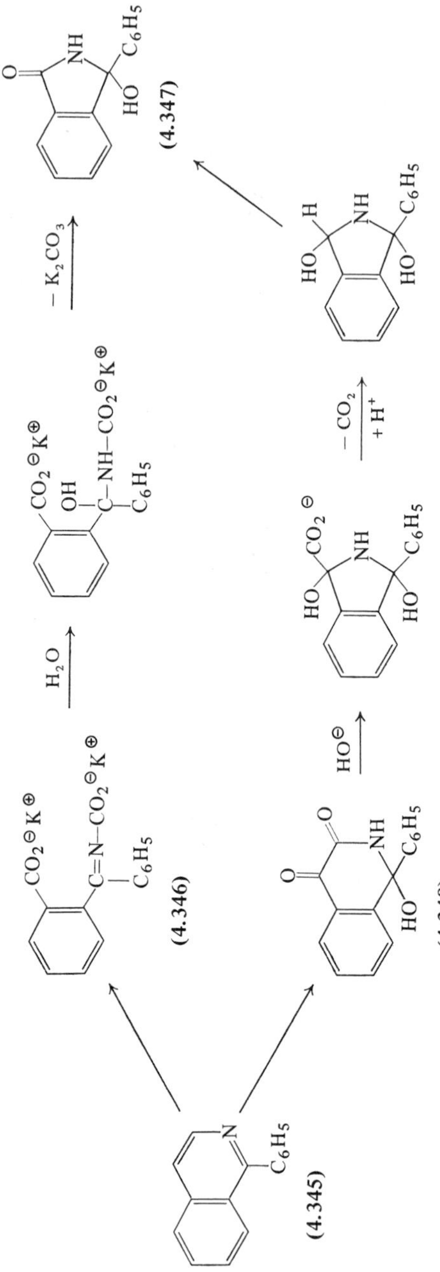
(4.345)
(4.346)
H_2O
$-K_2CO_3$
(4.347)
(4.348)
$HO^{\ominus}$
$-CO_2$
$+H^+$

d. *Tetrahydroquinolines into indoles, oxindoles and isatines; tetrahydroiso-quinolines into phthalimides*
The ring contraction of oxo derivatives of quinoline into oxo derivatives of indoles has been known for some time. The first reaction was reported in 1884, when the conversion by treatment with alkali of quinisatine (4.349) into 3-hydroxyoxindole (4.350), isatine (4.351) and 3,4-dihydroxycarbostyril (4.352) was described[292]. It can reasonably be assumed that (4.351) and

(4.349)

(4.350)   (4.351)   (4.352)

(4.352) are formed in a redox-type reaction between (4.349) and (4.350). Other examples of this type of ring transformation are the acid-catalysed conversion of 3-nitroso derivatives of 7-methoxy-2,4-dioxo-1,2,3,4-tetrahydroquinoline and of 7-methoxy-1-methylquinolone-2 into the corresponding isatine derivatives with loss of hydroxylamine. This reaction has proved to be a convenient method for the preparation of methoxyisatines which can only be prepared with great difficulty[293,294] by other methods. Another interesting method for the preparation of isatine by ring contraction of quinolines involves the oxidation of 3-amino-2,4-dioxo-1,2,3,4-tetrahydroquinoline with a $FeCl_3$ solution at $80°C$[295].

Other base-catalysed ring contractions with tetrahydroquinolines have been reported to take place with 3,3-dichloro-2,4-dioxo-1,2,3,4-tetrahydro-quinoline[296], its 1,8-trimethylene derivative (4.353)[297], its benzyl deriva-tive[298], and 3,3,6-tribromo-2,4-dioxo-1,2,3,4-tetrahydroquinoline[299]. This method has been developed into a new and reliable means for preparing substituted indigos[300].

The question of which carbon atom is ejected during these ring contraction reactions is important in order to get insight into the mechanism. Recent work on the ring conversion of [3-^{14}C]-labelled quinisatine (4.354) has revealed that only the carbon dioxide formed during the reaction is radioactive, indicating that the atom expelled during the reaction is carbon atom 3[301]. This result can readily be explained by the occurrence of a benzilic acid-type

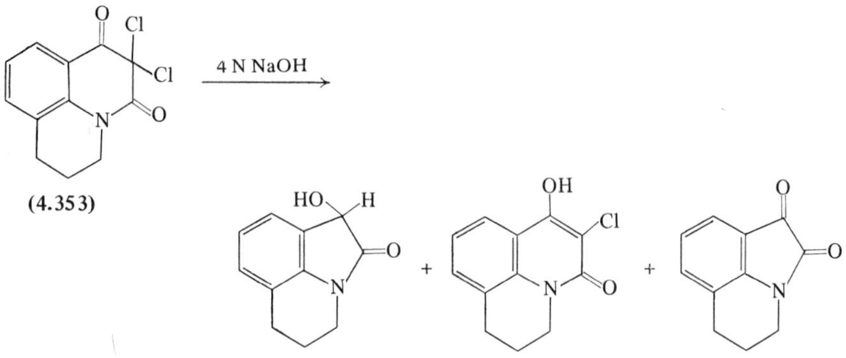

(4.353)

rearrangement in (4.355) involving migration of the carbonamide C-atom (= C_2) to carbon atom 4 and subsequent decarboxylation at C_4 in (4.356).

A somewhat analogous case is the conversion of 2H,4H-isoquinoline-1-thione-3-one (4.357) into phthalimide (4.361) by treatment with an excess of hydrogen peroxide in alkali[302]. This reaction must occur via the homophthalimide

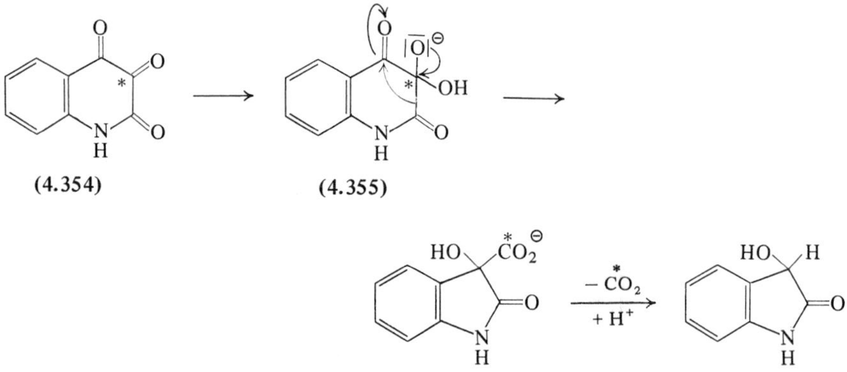

(4.354)   (4.355)

(4.356)   (4.350)

(4.358) since, in the presence of 4 equivalents of hydrogen peroxide instead of a large excess, (4.358) can be isolated. The homophthalimide (4.358) is presumably oxidized to the phthalonimide (4.359) (the isomer of quinisatine (4.349)), which, however, cannot be isolated owing to a rapid benzilic acid rearrangement in which the amido nitrogen migrates to the C4-atom yielding (4.360). On gentle heating carbon dioxide is evolved and further oxidation yields the phthalimide (4.361). That it is indeed the carbonyl group at C_3 in (4.357) which is eliminated—the overall reaction in effect deletes a methylene part—has been proved by studying the oxidation of [$3^{14}C$]-labelled (4.357)[302]. The carbon dioxide evolved is very highly radioactive. The occurrence of a Favorskii

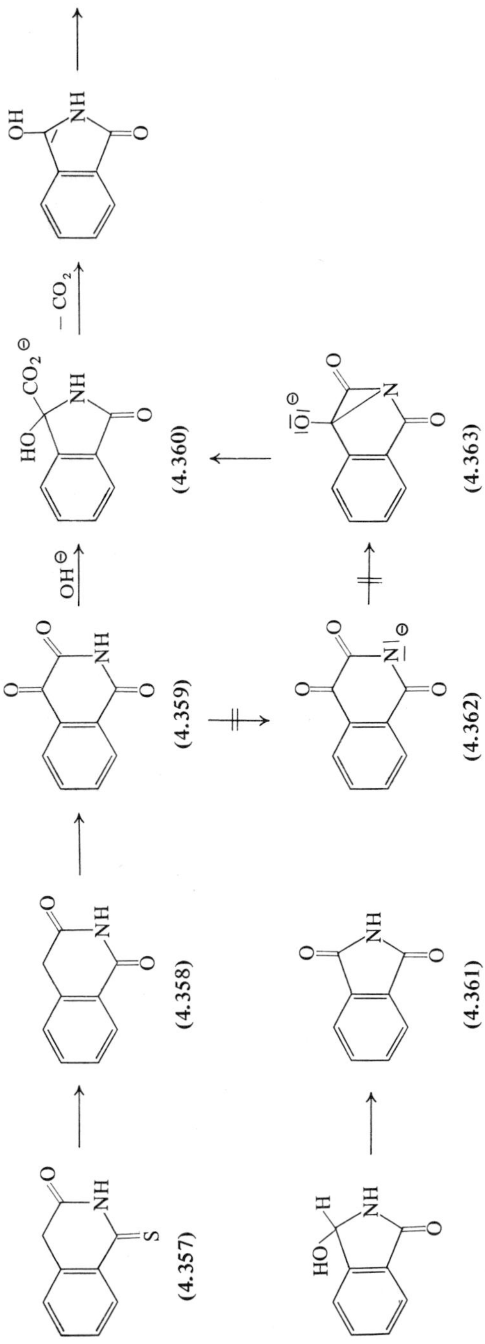
(4.357)
(4.358)
(4.359)
(4.360)
(4.361)
(4.362)
(4.363)
OH⊖
− CO2

rearrangement as an alternative path[303], involving the phthalonimide anion (4.362) and the three-membered aziridine ring (4.363), can be rejected by reason of the experimental finding that the N-methyl derivative of (4.358) is also found to give ring contraction into the corresponding N-methylphthalimide[302] without difficulty.

It has recently been found that the treatment with acid of the 3,4-dihydro-carbostyrils (4.364) which contains an acyl substituent (or its ketal) in position 4 gives a ring contraction rearrangement into 3-indolylacetic acid or ester (4.366)[304,305,306], the formation of which depends on the solvent (water or ethanol) used in these reactions. It is clear from the structure of the products obtained that the introductory step must be solvolysis of the N–CO bond which yields the phenyl derivatives (4.365).

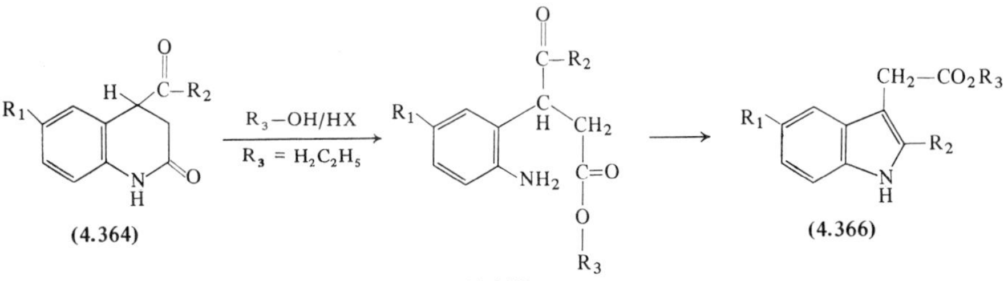

R_1	CH$_3$	NH$_2$	Cl	H	H	OCH$_3$	NO$_2$	OCH$_3$	OCH$_3$	H	CH$_3$	C$_6$H$_5$
R_2	C$_6$H$_5$	CH$_3$	CH$_3$	C$_6$H$_5$	CH$_3$	CH$_3$	CH$_3$	C$_6$H$_5$	Et-ϕ*	Et-ϕ*	Et-ϕ*	Et-ϕ*

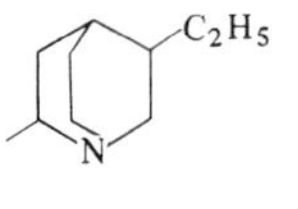

e. *Pyridines into 1,2,3-triazoles; quinolines into phenylpyrazoles*
Ring transformation of the pyridine ring into a 1,2,3-triazole ring is reported to occur when the N-alkyl-3-aminopyridinium bromides (4.367) are diazotized at low temperature (3-8°C) in an aqueous acidic medium[307,308]. The β-(1-alkyl-1,2,3-triazolyl-4)acroleine (4.368) is obtained; it is formed by the sequence of reactions shown opposite.

This ring transformation does not occur with (4.367) (R = H). A quite similar ring contraction has been reported to take place when the 1-aminoquinolizinium salt (4.369) reacts with aqueous nitrous acid giving the *cis-v*-triazolo[1,5-*a*]-pyridylacraldehydes (4.371)[309,310]. In both reactions the attack of the nucleophile water on the position *ortho* to the quaternary nitrogen and *para* to the diazonium group can be considered as an initial step in the ring opening.

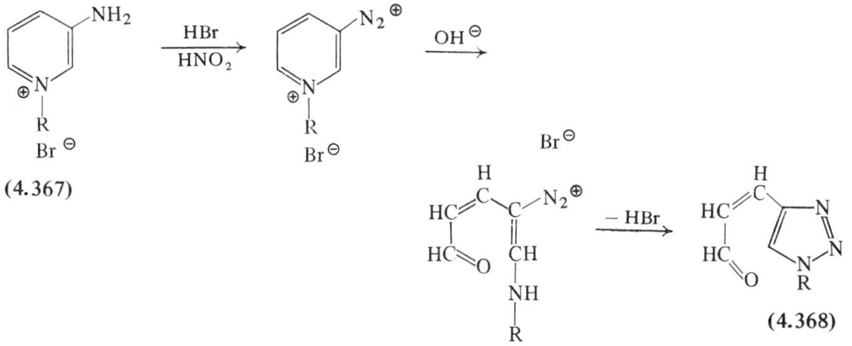

(4.367)

(4.368)

A point of further interest in this reaction is the fact that rotation around the C_1-C_{9a} bond in (4.370) occurs with retention of configuration, because the *cis*-aldehyde is always obtained first under conditions which do not promote *cis trans* isomerization.

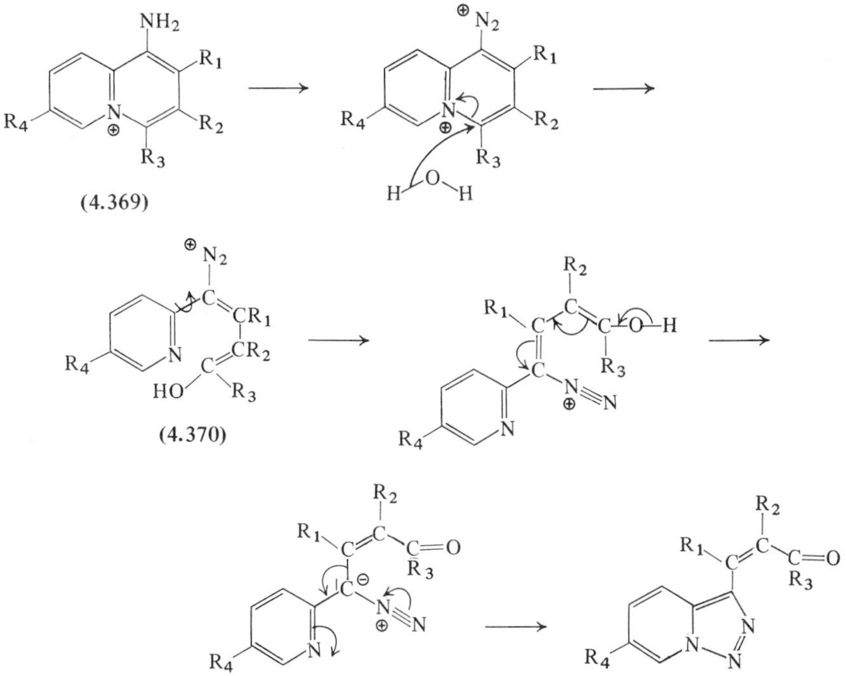

(4.369)

(4.370)

(4.371)

R_1 = H; D; Br. R_2 = H; CH_3. R_3 = H; D; $CO_2C_2H_5$. R_4 = H; CH_3

In analogy with the reported conversions of chromone derivatives into 3-(o-hydroxyphenyl)pyrazoles by treatment with hydrazine (see section I.A.1.b), 4-chloroquinoline (4.372) (R = H, X = Cl) was found to give on heating with hydrazine, the 3-(o-aminophenyl)pyrazole (4.374) (R = H)[311,312] and not 3,4-diaminoquinoline[313] as had previously been suggested. The same transformation was observed[311,314] with the 4-hydroxyquinoline and 4-hydrazinoquinoline. The reaction can be described to occur by an initial attack of the hydrazino group on position 2 of the quinoline ring, followed by ring opening into a o-iminophenylacylhydrazine (4.373) which, on ring closure, gives the pyrazole (4.374). The reaction has recently been extended to the 4-chloro-1,8-naphthyridines (4.375) which, on being allowed to react with hydrazine or substituted hydrazines at 150°C, gave 3-[2-amino-6-methyl-pyridyl-3]pyrazoles (4.376)[315].

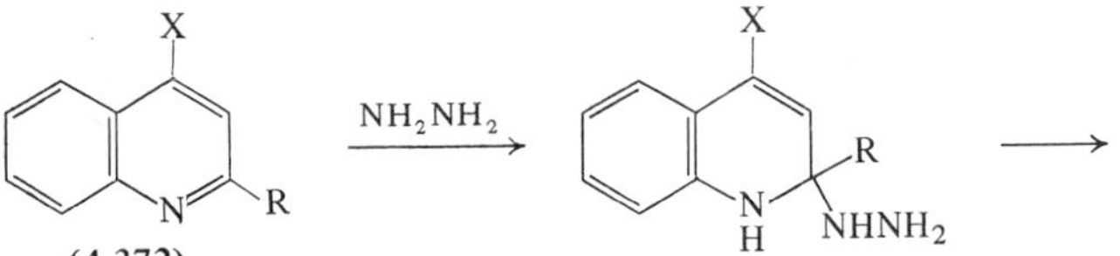

(4.372)

X	Cl	Cl	OH	NHNH$_2$
R	H	CH$_3$	H	H

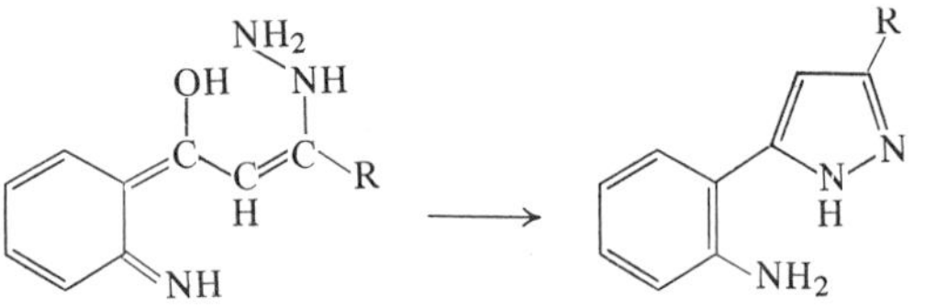

(4.373)   (4.374)

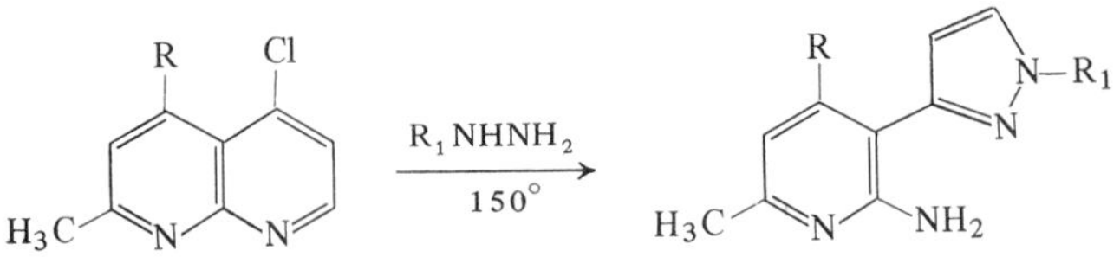

(4.375)   (4.376)

R = H; CH$_3$ R$_1$ = H; C$_6$H$_5$

2. RING TRANSFORMATIONS INTO OTHER SIX-MEMBERED HETEROCYCLES

a. *Pyridines into pyrimidines and pyridazines; quinolines into quinazolines; tetrahydroisoquinolines into isochromenes*

An intriguing interconversion of a tetrahydroisoquinoline into an isochromene derivative has recently been observed during treatment of the chlorohydrin

(4.376a) with triethylamine and aqueous sodium hydroxide in dioxane[315a]. A mixture of the isochromenes (4.376c) and (4.376d) is obtained; in the triethylamine reaction the major product is (4.376c), but the yield of (4.376d) increases when sodium hydroxide is used. As intermediate the open-chain compound (4.376b) is advanced.

(4.376a)

(4.376b)   (4.376c)   (4.376d)

An extensive investigation of the ocurrence of didehydropyridines during the reaction of monohalogeno- and dihalogenopyridines with potassium amide in liquid ammonia at $-33°C$[316,317], has revealed that 2,6-dibromopyridine (4.377) (X = Br) is converted under these conditions into 6-amino-2-methyl-pyrimidine (4.378) (yield 25 per cent)[318]. 2,6-Dichloropyridine (4.377) (X = Cl) was shown to react similarly, although more slowly; 2,6-difluoropyridine did not form a ring transformed product but gave in a high yield 2-amino-6-fluoro-pyridine. During a study of the influence of different substituents on position 2 of the pyridine ring on the occurrence or non-occurrence of this ring transform-ation[319,320], a great number of 6-substituted 2-bromopyridines were synthe-sized and subjected to the action of potassium amide in liquid ammonia under standard conditions. It appears that of all the 6-bromopyridine derivatives investigated, i.e. (4.379) (R = 2-NH$_2$; 2-NHC$_6$H$_5$; 2-CONH$_2$; 2-NCH$_3$C$_6$H$_5$; 2-CH$_3$; 2-C$_6$H$_5$; 2-OC$_2$H$_5$; 2-OC$_6$H$_5$; 2-p-OC$_2$H$_5$OC$_6$H$_4$; 2-m-OC$_2$H$_5$OC$_6$H$_4$; 2-p-FC$_6$H$_4$ and 2-NO$_2$) only 2-phenoxy- and 2-substituted phenoxy-6-bromopyridine were found to give substituted pyrimidines (4.380) in reasonable yields (40-50 per cent). However, the fact that apart from ring transformations other types of reactions have been found to occur (for instance: addition elimination reactions and cine-substitutions) makes it difficult to understand how the course of these ring transformation reactions is influenced by the substituents on position 2.

(4.377)   (4.378)

X = Cl; Br

(4.379)   (4.380)

$R = C_6H_5$; OC_6H_4-p-OCH_3; OC_6H_4-m-OC_2H_5; OC_6H_4-p-F; α-C_5H_4N

The bicyclic azacompound 2-bromoquinoline (4.381) has been found to give the same type of ring transformation on treatment with potassium amide in liquid ammonia; 2-methylquinazoline (4.382) is formed, besides 2-aminoquinoline (4.383)[318,321]. Although it has not been conclusively proved, it is very probable that when 2-bromo-1,5-naphthyridine is treated with potassium amide in liquid ammonia an amino-methyl derivative of pyrido[3,2-d]pyrimidine is obtained[322]. This result awaits further confirmation. As well as being important

(4.381)   (4.382)   (4.383)

as preparative methods, these ring transformations are also of considerable interest because it was evident from the earliest reported examples that C_3-C_4 bond fission occurred; this phenomenon had rarely before been encountered in pyridine chemistry. In order to rationalize the results, it was proposed that the probable initial step in these reactions was addition of the amide ion to position 4, yielding (4.384). The next step might then involve ring opening by fission of the C_3-C_4 bond, yielding the ketenimine (4.386) which has a strong tendency to ring closure. From the results of work on a similar ring transformation of deuterated 4-chloropyrimidines into s-triazines[323] (see section II.C.2.b) it is also possible that an acetylenic intermediate (4.387), formed via (4.385) or from the ketenimine (4.386), is involved in the base-catalysed conversion of 2-substituted 6-bromopyridines into 6-aminopyrimidine derivatives.

Only a few examples of ring transformation of a pyridine into a pyridazine derivative have been reported. Thus, it appears that the pyridinium hydroxide inner salt (4.386a, R = C_6H_5), on treatment with t-butylhydroperoxide, gives the corresponding (4.386b) (R = C_6H_5), while from (4.386a) (R = CH_3) a

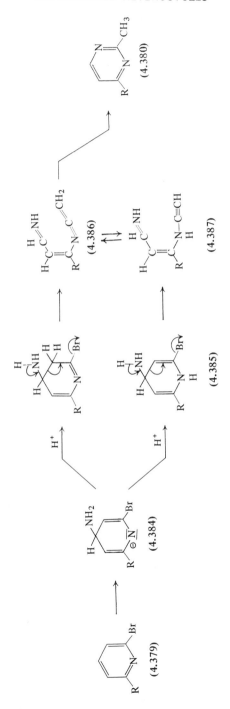

mixture of both 1,6-dihydropyridazines (4.386b) (R = CH$_3$) and (4.386c) is obtained[323a].

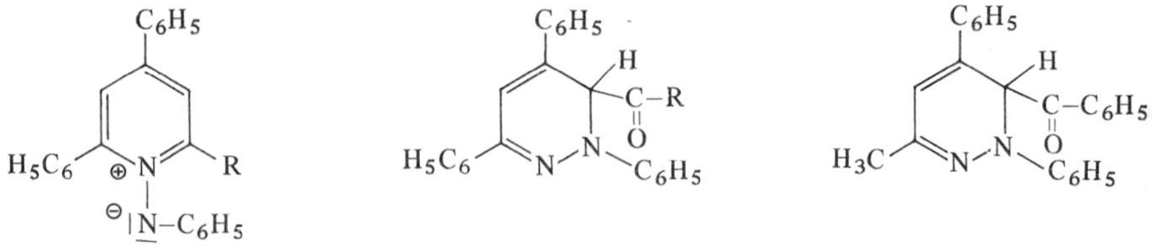

(4.386a)   (4.386b)   (4.386c)

A further example is the oxidation of tetraphenyl-N-aminopyridone (4.388) with lead tetraacetate in methylene chloride at room temperature, which reaction gave a ~60 per cent yield of the pyridazine (4.389) as well as small amounts of the pyridone (4.390) and 2,3,4,5-tetraphenylcyclopentadienone[324]. This oxidation process proceeds via a nitrene intermediate since oxidation of (4.388) in the presence of cyclohexene gives the nitrene adduct (4.391).

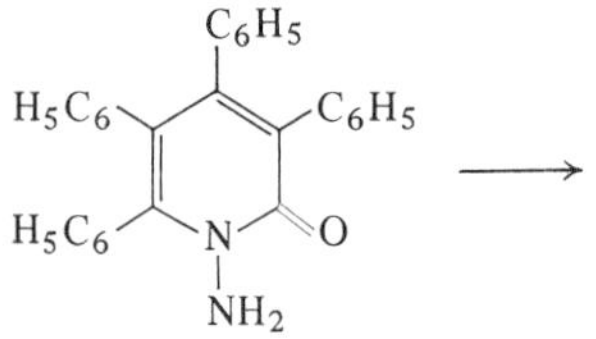

(4.388)

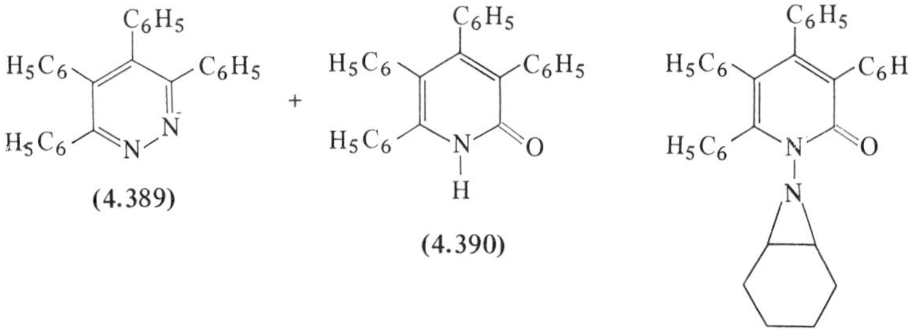

(4.389)   (4.390)   (4.391)

3. RING EXPANSIONS INTO SEVEN-MEMBERED COMPOUNDS

a. *Ring expansions of pyridines, dihydro- and tetrahydropyridines and piperidines into azepines*

The methods by which this ring enlargement are performed can be roughly divided into three groups: (1) those which involve an exocyclic carbonium ion inserted into the ring; (2) those occurring via Stevens rearrangement; (3) those which can be performed by nucleophilic substitutions with 1,4-dihydropyridines which contain a chloromethyl group on position 4.

Ring expansion of monocyclic and polycyclic heterocyclic ring systems by method (1), involving intermediary exocyclic carbonium ions, has been found to take place when an exocylic alcohol is treated with acid (in correspondence to reactions in the 1,4-dihydrobenzene series[325]), when diazomethane is used, or when an exocyclic aminomethyl group reacts with nitrous acid. Examples of reactions are the pinacol rearrangement of carbinol (4.392) with concentrated sulphuric acid into the hexahydroazepin-4-one (4.393)[326], the dehydration of 9-hydroxymethyl-9,10-dihydroacridine (4.394) by polyphosphoric acid at 160°C[327] or phosphorus pentoxide yielding the dibenzo[b,f]azepine (4.395) (R = H)[328], the conversion of N-methylacridinium salts (4.396) into N-methyl dibenzoazepine (4.395) (R = CH₃) by using diazomethane[329], and the ring expansion of N-benzylpiperidone-4 into N-benzylhexahydroazepin-4-one also by treatment with diazomethane[330].

(4.392)   (4.393)   (4.394)   (4.395)

(4.397)   (4.398)   +   (4.399)   (4.396)

Carbon insertion via a Tiffenau–Demjanov rearrangement was observed when the 4-(aminomethyl)-4-piperidinol (4.397) reacts with nitrous acid, yielding the hexahydroazepine (4.398)[331] and the spiro compound (4.399). The same rearrangement occurred when 1-amino-3-aminomethylpiperidine was treated with nitrous acid at low temperature. Together with compounds containing the unchanged piperidine ring, a mixture of N-methyl-4,5,6,7-tetrahydro- and 2,5,6,7-tetrahydroazepine together with 3-hydroxy- and 4-hydroxyhexahydro-azepines was obtained in low yield (8 per cent)[332].

An interesting variation of this type of ring transformation is the solvolysis of aziridinium salts such as (4.401) which are easily prepared from the 3,4,5,6-tetrahydropyridine (4.400) with diazomethane. Methanolysis of the bicyclic aziridinium salt (4.401) at reflux temperature, followed by treatment of the reaction product with a base, gave a compound whose structure was conclusively proved to be 3-ethyl-3-methoxy-N-methylhexahydroazepine

(4.402)[333]. Using an aqueous solution of base, the same ring expansion occurred, yielding the 3-hydroxy instead of the 3-methoxy compound. In this ring expansion the direction of opening of the aziridine ring follows the general pattern, that is cleavage of the bond between nitrogen and the tertiary carbon atom[334].

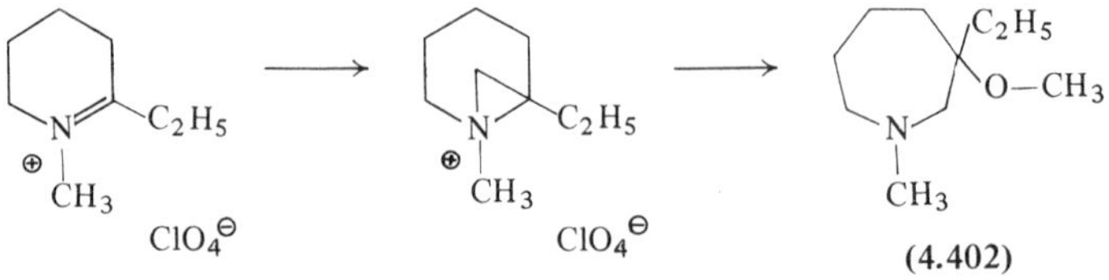

(4.400)   (4.401)   (4.402)

Analogously, the conversion of the tricyclic aziridinium salt (4.403) into the bicyclic aza[5.4.0]undecane (4.404) has been reported[333]. It is important to note that the occurrence of this type of ring expansion is dependent on the stereochemical position of the tertiary carbon atom. Thus methanolysis of the tetracyclic aziridinium salts (4.405) gives a product in which the aziridine ring has been opened, yielding the methoxymethyl derivative (4.406); none of the rings was expanded. Due to steric interaction of the attacking nucleophile with the tetracyclic constrained molecule the tertiary carbon-nitrogen bond is not readily broken; hence solvolysis of the primary carbon atom in the aziridine ring occurs[334].

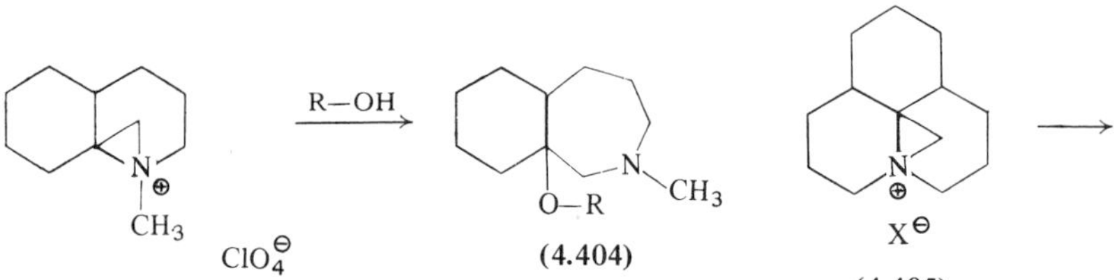

(4.403)

R—OH

(4.404)

R = CH₃; C₂H₅

(4.405)

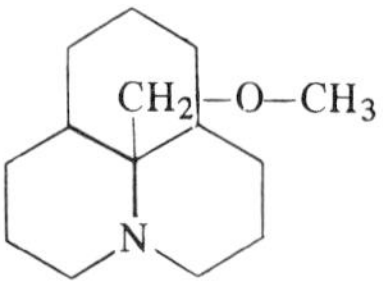

(4.406)

During a very extensive study of the fragmentation reaction of carbonium ions containing amino groups in the γ-position[335] (carbimonium salts and olefins are generated), it was found that solvolysis of 4-tosyloxymethyl-quinuclidine (4.407) leads to ring expansion into the methylene hexahydro-azepine (4.411). The main factor in this ring expansion is that the primarily

(4.407)   (4.408)

(4.409)   (4.410)   (4.411)

(4.412)   (4.413)

formed carbonium ion (4.408) isomerizes into the bicyclic tertiary carbonium ion (4.409)[336], which, because it is unable to lose a proton during formation of the bicyclic system (4.412) (violation of Bredt's rule), gives fragmentation via (4.410) into the hexahydroazepine (4.411). Since the reaction takes place at about a quarter of the rate of the corresponding reaction of its homocyclic counterpart 1-tosyloxymethyl bicyclic[2.2.2]octane (4.413), it is concluded that the ring expansion of the quinuclidine cannot occur synchronously but that it takes place in separate steps.

Method (2), Stevens rearrangement of N-ylides, has proved to be excellent for ring expansion of certain spiropiperidinium salts. An extensive study on the occurrence of Sommelet and/or Stevens rearrangements during treatment of N,N-aralkylisoindolinium salts with phenyllithium[337], revealed that 2,2-pentamethylene-5,6-benzoisoindolinium bromide (4.414) is converted by this reagent into the 1,2-pentamethylene-5,6-benzoindoline (4.415). In this reaction the ylid (4.416) or possibly the recently advocated formulation (4.417)[338] is very probably the intermediary stage in this reaction.

Considering method (3), interesting ring expansions are observed when 4-chloromethyl-1,4-dihydropyridines are subjected to comparatively mild treatment with nucleophiles such as CN^-, $AlkO^-$ or H_2O. One of the first examples of this ring enlargement was reported[339] when diethyl 2,6-dimethyl-4-chloromethyl-1,4-dihydropyridine-3,5-dicarboxylate (4.418) (R = C_2H_5) was treated with potassium cyanide in boiling ethanol and the 4,5-dihydroazepine (4.419) (R = C_2H_5) was formed. In addition to (4.419), a pyrrole derivative was

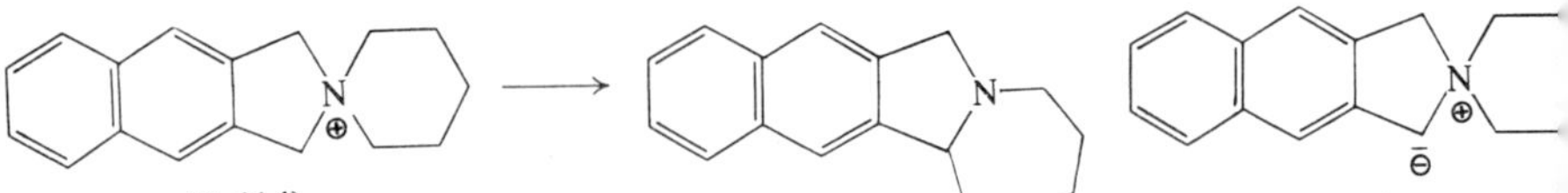

(4.414)   (4.415)   (4.416)

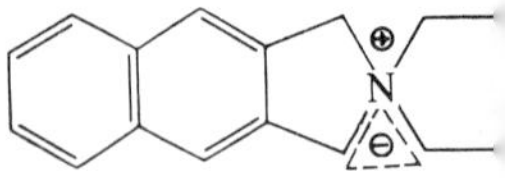

(4.417)

formed[339,340] which appeared to be the compound (4.420) (R = C_2H_5) and not, as originally supposed, ethyl 2-methyl-4-cyanomethylpyrrole-5-carboxylate[340].

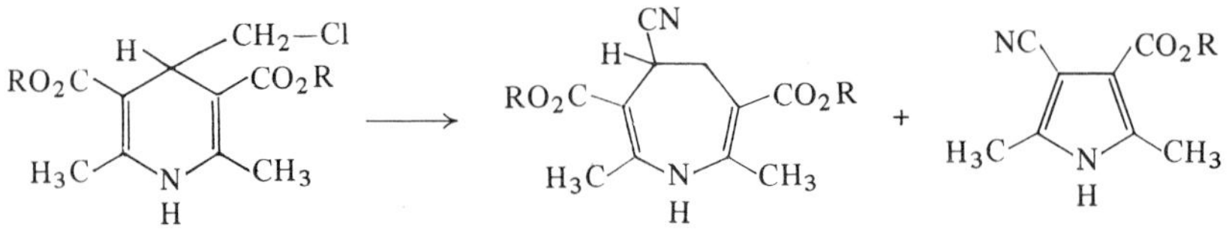

(4.418)   (4.419)   (4.420)

No trace of a pyrrole derivative was formed when the reaction was carried out at room temperature, although there was nearly quantitative replacement of the chlorine by the cyano group. A study of the kinetics of the ring enlargement revealed[341] that the reaction was a generally base-catalysed second order reaction. A mechanism has been suggested which involves as rate-determining step the formation of the 1,4-dihydropyridine anion (4.421), a fast rearrangement into the 4H-azepine (4.422), which by an 1,4-addition of hydrogen cyanide, yields compound (4.419). In accordance with the suggestion that the first step is

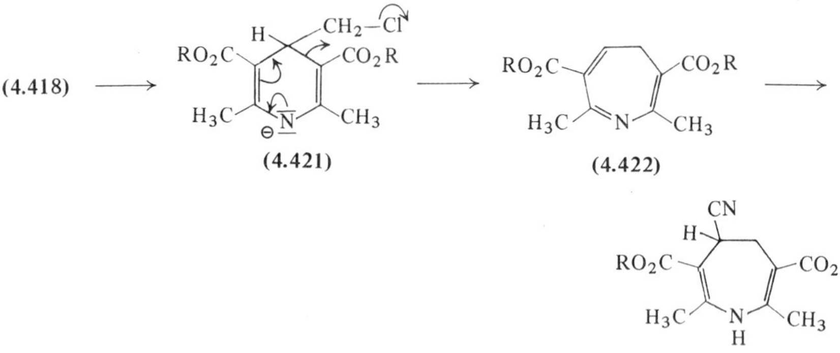

(4.418) ⟶ (4.421)   (4.422) ⟶

(4.419)

base-catalysed, is the finding that when (4.418) is treated with the more powerfully nucleophilic but non-basic iodide ion it remains unchanged[341]. Further it has been established[341] that (4.419) is formed by a fast reaction between hydrogen cyanide and the 4H-azepine (4.422), prepared by an independent method. It was further established that the pyrrole derivative (4.420) is obtained by ring contraction of the 4H-azepine (4.419)[343] and not of the 1,4-dihydropyridine. The resonance-stabilized carbanion (4.423) at C_4 is formed in the alkaline solution; this carbanion is converted by a transannular reaction (a well-known phenomenon in the seven-membered ring compounds) into the bicyclic compound (4.424). The pyrrole (4.420) is yielded by splitting off the acrylic ester.

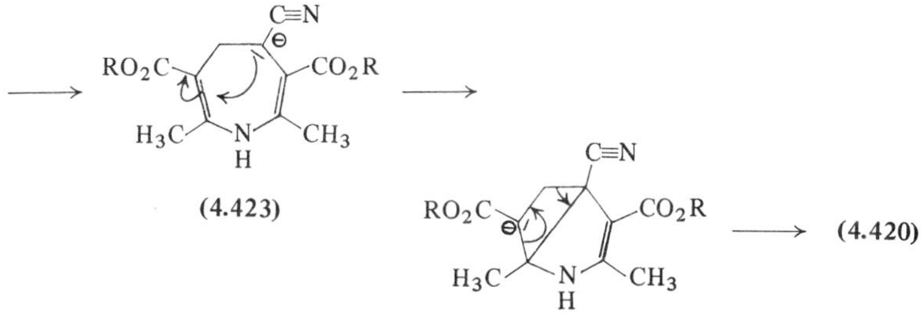

In correlation, the 3,5-diacetyl-1,4-dihydropyridine (4.425) has recently been found to undergo the same type of ring expansion during treatment with an aqueous alcoholic potassium cyanide solution, leading to the formation of 4-cyano-4,5-dihydroazepine (4.426) (R = CN)[344]. In the reaction of (4.425)

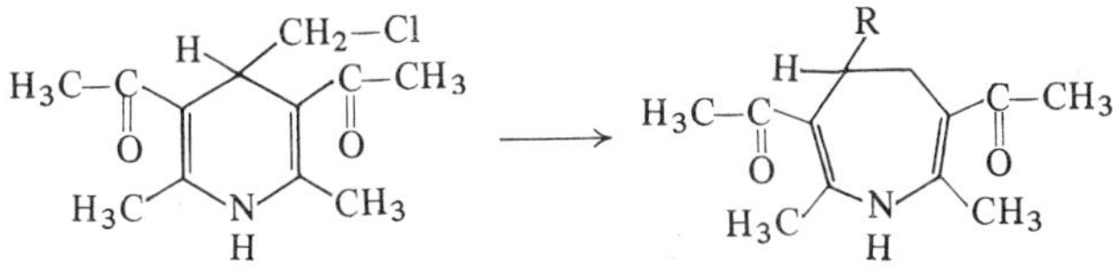

with the nucleophile water, in addition to a 1,4-dihydropyridine derivative, it is quite unexpectedly found that the 2,3-dihydrofuran (4.429) is obtained[344]. It is postulated that initially the 4,5-dihydroazepine (4.427) is formed and that this is hydrolysed into the open-chain intermediate (4.428) and further dehydrated into (4.429).

The reaction of (4.418) with sodium ethoxide is found to take place in quite a complicated manner, since it appears that the structure of the seven-membered compound which is formed depends strongly on the reaction conditions applied.

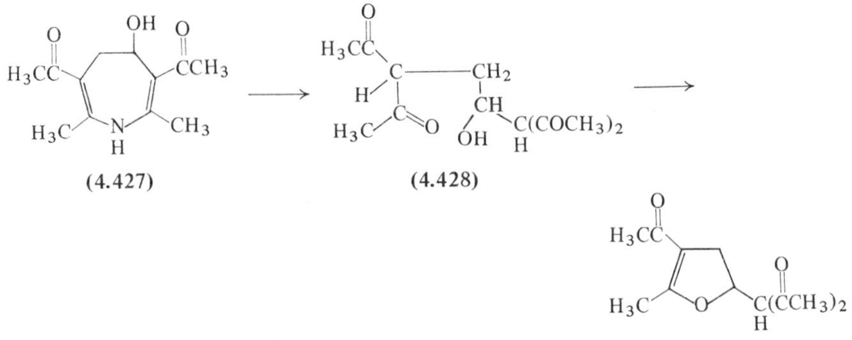

(4.427)　　　　　　(4.428)

(4.429)

On treatment with sodium ethoxide in the cold, the 4-ethoxy-4,5-dihydroazepine (4.430) is formed[342], with sodium ethoxide in boiling ether the 4H-azepine (4.431) is obtained, while by boiling with a sodium ethoxide solution the 3H-azepine (4.432) is yielded[343]. This last compound is formed by isomerization of the 4H-azepine, probably via a transannular interaction.

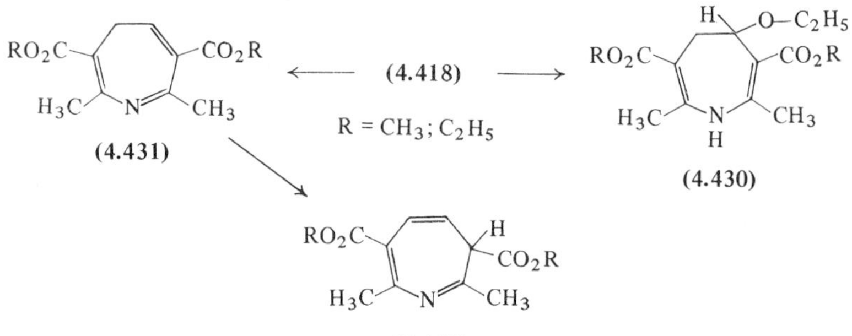

(4.431)

(4.418)

R = CH$_3$; C$_2$H$_5$

(4.430)

(4.432)

Ring expansion and ring contraction have been reported to occur simultaneously when (4.433) reacts with liquid ammonia. Together with methyl 2-methylpyrrole-3-carboxylate (4.435), the 4,5-dihydroazepine-3,6-dicarboxylate (4.434) is formed[250]. The presence of the strange side-chain attached to the seven-membered ring indicates that a quite complicated reaction has occurred. Presumably, the first step in this reaction is the rearrangement of the conjugated base of (4.433) into the 2,7-dimethyl-4H-azepine-3,6-dicarboxylate (4.422) (R = CH$_3$). After addition of ammonia to give (4.436), bridging between the C$_2$ and C$_4$ position takes place, leading to the formation of the bicyclic compound (4.437). Base-induced loss of the NH-bridging proton, followed by ring opening, gives the dihydropyrrole (4.438). Through aromatization of the ring by removal of the methyl 3-aminocrotonate (4.439) the pyrrole (4.435) is formed. Addition of (4.439) to the 4*H*-azepine (4.432) (R = CH$_3$), followed by hydrolysis during work-up, gives (4.434). Support for the occurrence of a bridged-type intermediate such as (4.437) comes from the fact that, with the primary amines,

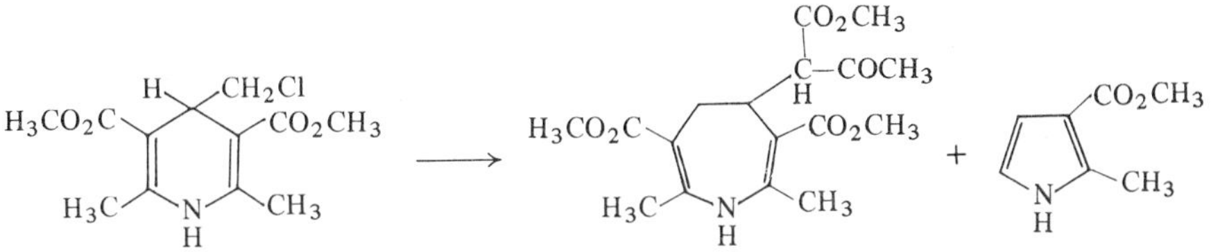

(4.433)   (4.434)   (4.435)

(4.422, R = CH$_3$) $\longrightarrow$

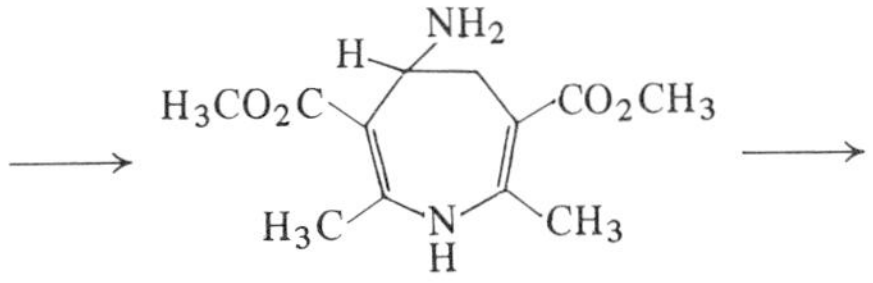

(4.436)

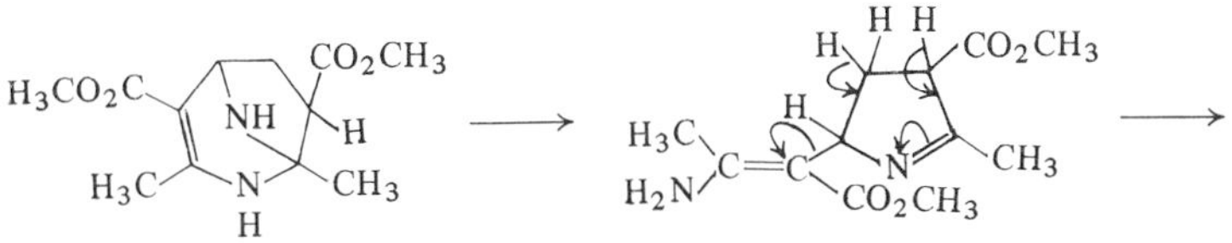

(4.437)   (4.438)

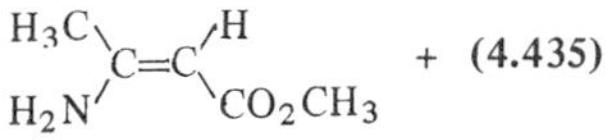

(4.439) 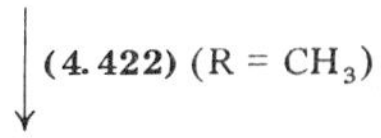   + (4.435)

(4.422) (R = CH$_3$)

(4.434)

methylamine or benzylamine, the bridged bicyclic compounds (4.437) (CH$_3$ or C$_6$H$_5$CH$_2$ instead of H) can be isolated because further rearrangement cannot occur as a NH bridging proton is not available[250]. With aqueous ethanolic hydrogen selenide the selenium-bridged 4,5-dihydroazepine (4.440) is obtained[250]. In the analogous case of sulphur the formation of (4.441) (R = CO$_2$CH$_3$ or COCH$_3$) has been reported[250,345].

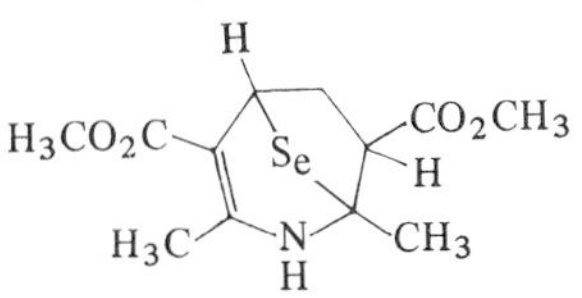

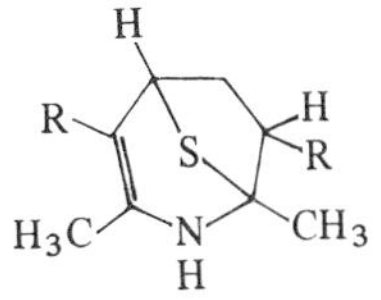

(4.440)   (4.441)

Ring enlargement is also observed in 1,4-dihydropyridines containing a substituent on the N_1-position[346] when they undergo base treatment Examples are the rearrangement of the N-phenyl-2,4,6-trimethyl-4-chloromethyl-1,4-dihydropyridine (4.442) into the 1-phenyl-1H-azepine (4.443)[346], and the conversion of the 1,2,6-trimethyl-4-chloromethyl-1,4-dihydropyridine (4.444) into the 1,2,7-trimethyl-1H-azepine (4.445)[251] under the influence of potassium t-butoxide. As no hydrogen is available on the N_1-position, these conversions must take place by a mechanism different from that described above for the conversion of (4.418) into (4.419) for instance. Some indications of the course of the reaction have been obtained by studying the conversion under carefully controlled conditions. Two precursors of (4.443) have been isolated,

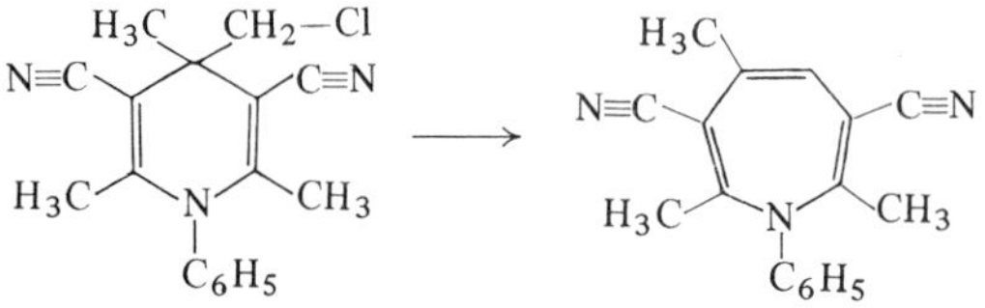

(4.442)   (4.443) 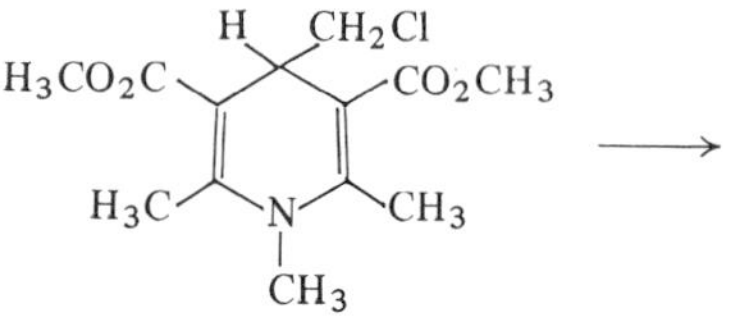   (4.444)

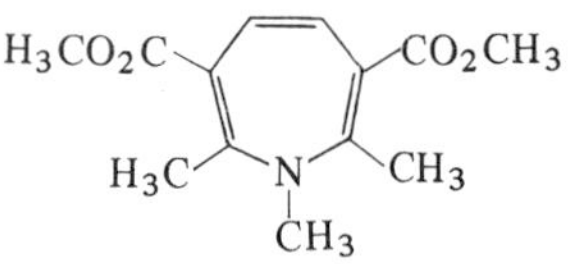

(4.445)

the bicyclic compound (4.446) and the 4-methylene-1H-azepine (4.448); another precursor, the 2-methylene-2,5-dihydroazepine (4.447) was shown by uv-spectroscopy also to be present in solution[346]. It resisted however, isolation, because of its extreme lability.

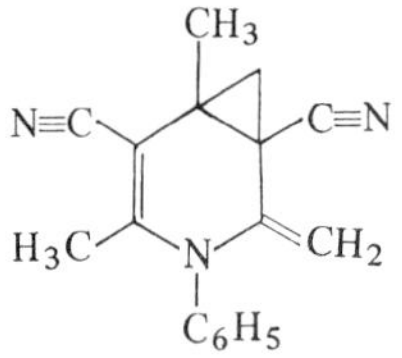

(4.446) 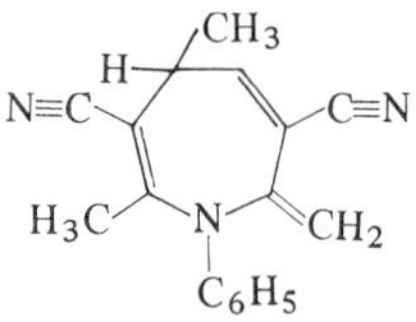   (4.447) 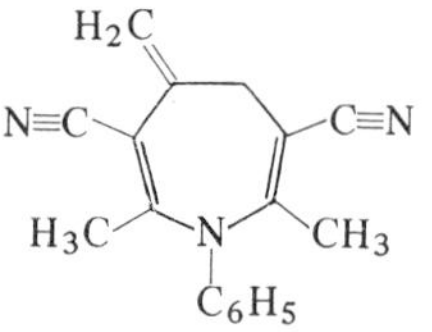   (4.448)

From these product studies it is apparent that the primary process is the conversion of (4.442) into (4.446). From the results of the deuterium exchange reactions, found to occur in the methyl groups on position C_2 and C_6, and studies of the kinetics of the reaction of the N-phenyl compounds containing different substituents in various positions on the phenyl ring with bases of different strengths (HO⁻, ⁻OtBu, ⁻OC$_6$H$_5$), it has been established that an unimolecular ionization of the C–Cl bond or a carbenoid mechanism can be discounted. From the results so far obtained it seems most likely that primary formation of the ion (4.449) occurs and that a subsequent electron shift (as shown) leads to the formation of the bicyclic methylenecyclopropane (4.446).

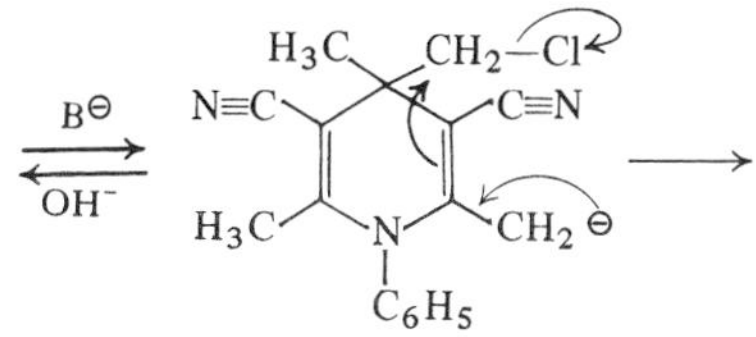

(4.449)

The simultaneous occurrence of ring expansion and ring contraction in a 1,2-dihydropyridine ring system has been observed when the 9a-vinyl-9aH-quinolizine-1,2,3,4-tetracarboxylate (4.450) is irradiated[347]. The pyrrolo-[1,2-a]azepine (4.452) is formed; it is postulated that this compound is obtained through ring opening into the 10-membered ring system (4.451), followed by cyclization and proton shift. The same ring expansion–ring contraction reaction is observed with the 6,9a-dimethyl- and 9aH-6-trans-styryl derivative of the parent system[347].

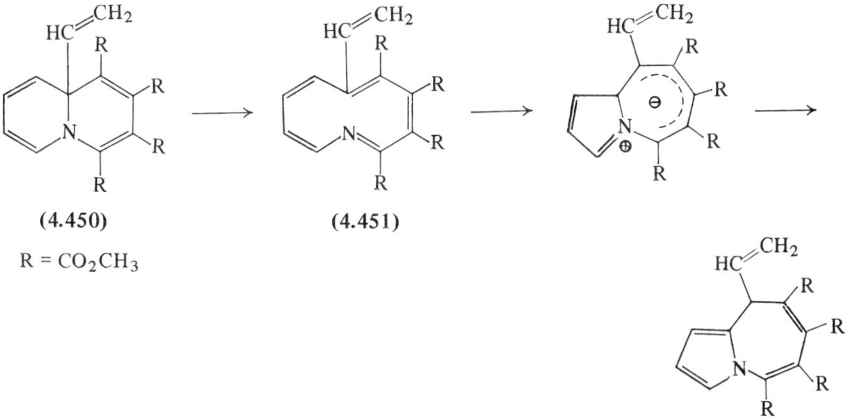

(4.450)   (4.451)

R = CO₂CH₃

(4.452)

Ring expansion of compounds containing a tetrahydropyridine ring has recently been demonstrated with the dibromocyclopropane derivative (4.453) which is converted into the 1H-benzazepine derivative (4.454) when heated under refluxing conditions in pyridine[348].

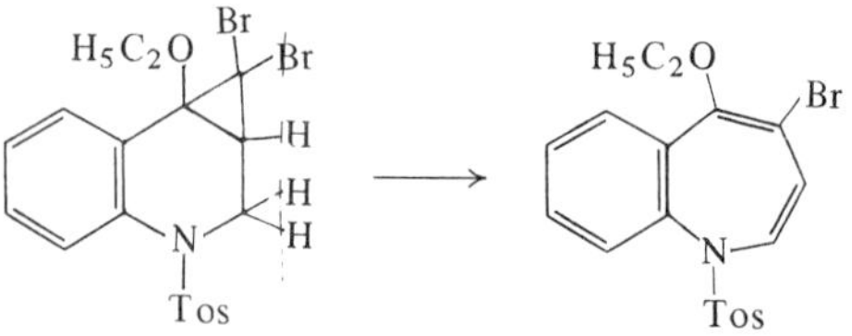

(4.453)   (4.454)

b. *Ring expansions of piperidines and pyridines into diazepines; piperidines into oxazepines; quinolines and isoquinolines into benzoxazepines and benzodiazepines*

The well-known Schmidt reaction[349] and Beckmann rearrangement[350] have been found to be useful methods for ring enlargement of six-membered nitrogen heterocycles. When the 4-piperidones (4.455) react with hydrazoic acid, hexahydro-1,4-diazepinones (4.456) (= 5-homopiperazinone) are obtained in good yield[351]. The same type of ring expansion has been reported to occur[351] by Beckmann rearrangement of the ketoxime of (4.455) ($R_2 = R_3 = R_4 = R_5 = CH_3$; $R_1 = H$), which leads to the formation of (4.456) ($R_2 = R_3 = R_4 = R_5 = CH_3$; $R_1 = H$). With the non-symmetrically substituted 3-ethoxycarbonyl piperidones-4 (4.457), in the Schmidt reaction both isomeric products, (4.458)

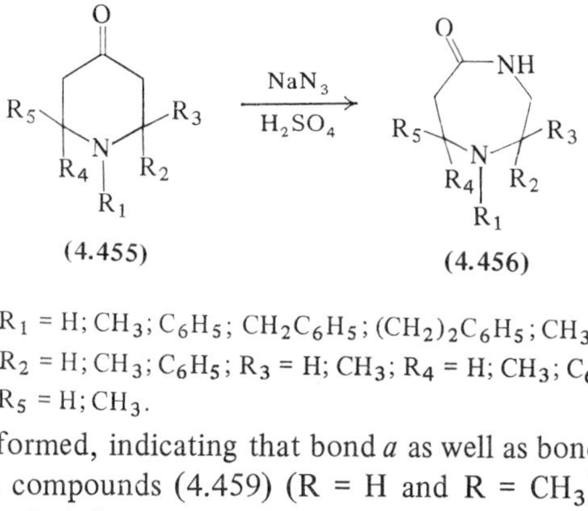

(4.455)   (4.456)

$R_1 = H; CH_3; C_6H_5; CH_2C_6H_5; (CH_2)_2C_6H_5; CH_3$.
$R_2 = H; CH_3; C_6H_5; R_3 = H; CH_3; R_4 = H; CH_3; C_6H_5$.
$R_5 = H; CH_3$.

and (4.459), are formed, indicating that bond *a* as well as bond *b* migrate in this reaction[351]. The compounds (4.459) (R = H and R = CH_3) are, in fact, not really isolated but their formation is deduced from the structure of the products

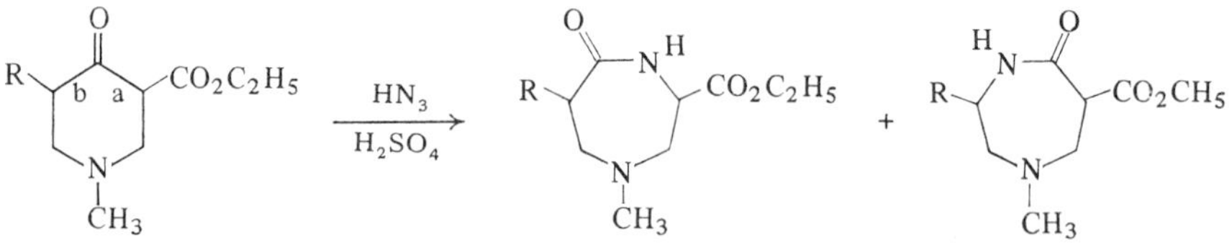

(4.457)   (4.458)   (4.459)

R = H; CH_3

obtained by acid hydrolysis of the reaction mixture obtained in the Schmidt reaction. It has been suggested that the steric effect of the group R on the 5-position strongly determines[351,352] the course of the rearrangement. It seems, however, not to apply in the reaction of 2,6,6-trimethylpiperidone-4 (4.460) which gives (4.461) only and does not yield even a trace of its isomer (4.462)[353]. It has been claimed[354] that when the N-substituted 1,2,3,4-tetrahydro-4-oxoquinoline (4.463) is treated with sodium azide and sulphuric acid ring expansion occurs leading to the formation of the 1,5-benzodiazepinone (4.464). The structure assignment has, however, been arrived at without rigorous proof. The suggestion has been made that the 1,4-benzodiazepinone (4.464a) was obtained[354b] and not the 1,5-benzodiazepine (4.464). Quite similarly, when 4-oxo-1,2,3,4-tetrahydroisoquinolines were treated with sodium azide and sulphuric acid the 1,3,4,5-tetrahydro-2H-1,4-benzodiazepin-2-ones were obtained[354a]. This ring expansion has also achieved with the corresponding ketoximes using polyphosphoric acid.

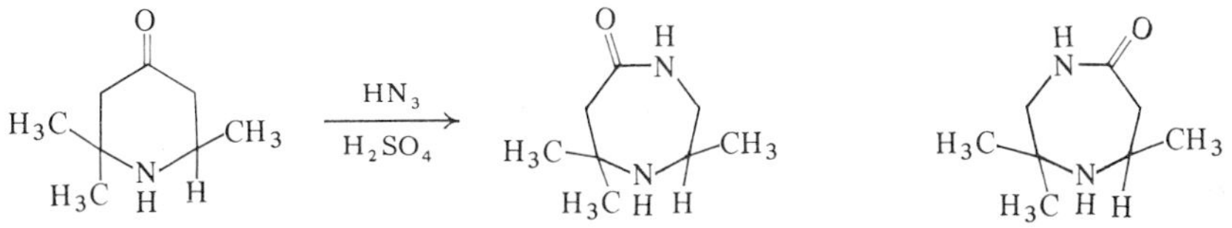

(4.460)   (4.461)   (4.462)

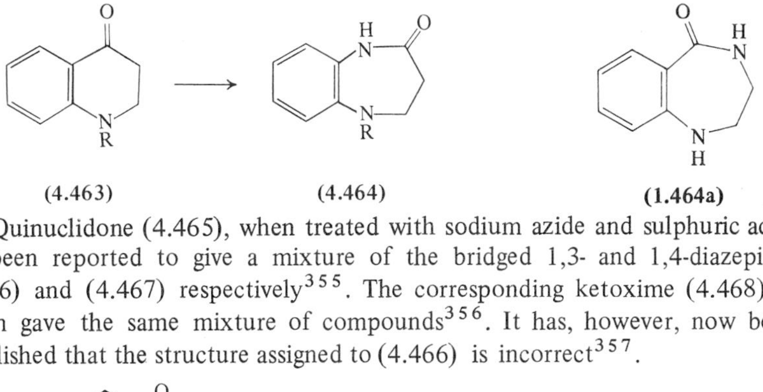

(4.463)   (4.464)   (1.464a)

3-Quinuclidone (4.465), when treated with sodium azide and sulphuric acid, has been reported to give a mixture of the bridged 1,3- and 1,4-diazepines (4.466) and (4.467) respectively[355]. The corresponding ketoxime (4.468) in oleum gave the same mixture of compounds[356]. It has, however, now been established that the structure assigned to (4.466) is incorrect[357].

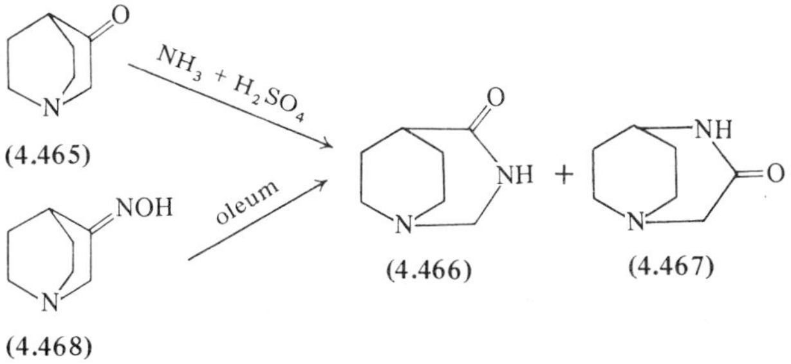

(4.465)

(4.468)   (4.466)   (4.467)

Schmidt ring expansion has been observed to lead in the bicyclic amino-
ketoximes 1-aza-bicyclo[3,3,0]octane (4.469) and 1-aza-bicyclo[4,3,0]nonane
(4.471) to the lactams (4.470) and (4.472) respectively. Thus in both cases the
migration of bond *b* (the C—C bond most distant from the nitrogen atom in the
ring) takes place[358]. The fact that bond *b* migrates and bond *a* does not is

(4.469)   (4.470)   (4.471)   (4.472)

evident from a consideration of the structure of the intermediate (4.473)
involved in the Schmidt reaction. The protonated nitrogen reduces the aptitude
of the neighbouring carbon-carbon σ-bond (bond *a*) to migrate to the electron
deficient azide nitrogen, therefore causing preferential rearrangement of bond *b*.

(4.473)

From this mechanistic consideration it can be expected—and indeed experi-
mentally verified—that in a bicyclo compound, in which the keto group is part
of a six-membered ring instead of a five-membered ring, the additional
methylene group diminishes the inductive effect and therefore makes possible a
less directive specificity: thus with hydrazoic acid 1-azabicyclo[4.4.0]decan-4-
one (4.474) gives both (4.475) and (4.476)[358].

Treatment of the stable radical of the piperidine nitroxide (4.479) with
hydroxylamine hydrochloride leads to a spontaneous Beckmann-type rearrangement into the 2,2,7,7-tetramethylhexahydro-1,4-diazepin nitrogen oxide radical
(4.480)[359]. Beckmann rearrangement of the ketoxime (4.477) with acetic
anhydride and sulphuric acid gave a product which was believed to be the
pyrido[1,2]1,4-diazepinium bromide (4.478)[360]. However, it was shown by an
independent synthesis of (4.478) that this structure assignment was
incorrect[361].

A remarkable ring expansion occurs during the lithium aluminium hydride
reduction of the ketoxime of the phenothiazine derivative (4.481). The product

(4.474)   (4.475)   (4.476)

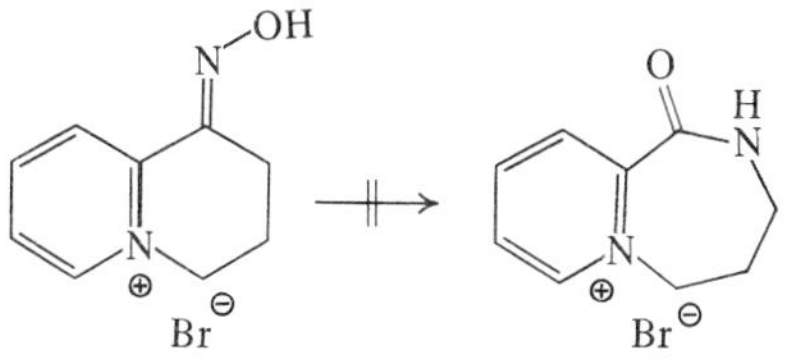

(4.477)   (4.478)

(4.479)   (4.480)

obtained has been shown to be the compound (4.482)[362] and not, as originally suggested, the unrearranged tetracyclic amine (4.483)[363], although it was logical to expect the formation of this latter compound. It is somewhat surprising that an exclusive migration of the phenyl group takes place during reduction rather

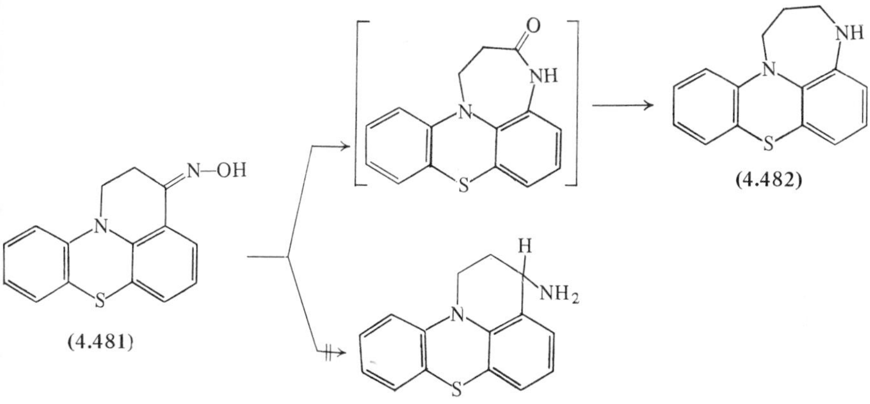

(4.481)   (4.483)   (4.482)

than the formation of an unrearranged amino product because in the lithium aluminium hydride reduction of a number of phenyl-substituted derivatives of the ketoxime of acetophenones both rearranged and unrearranged products have been found to be formed[364].

Quite a number of interesting papers have been published dealing with the formation of compounds with a seven-membered ring and products with a five-membered ring from the azaaromatics quinoline and isoquinoline-N-oxides during irradiation. When a solution of 2-phenylquinoline-N-oxide

(4.484) $(R = C_6H_5)$ is irradiated with a high pressure mercury lamp in an aprotic solvent[365-367] the main product yielded is the benz[d] 1,3-oxazepine (4.488) $(R = C_6H_5)$[365], but 3-phenylcarbostyril (4.485) $(R = C_6H_5)$ and the N-benzoyl 2-hydroxy-2,3-dihydroindole (for this ring contraction reaction see section I.C.1.c) are also formed.

The contention that (4.488) is an oxazepine derivative and does not have the oxaziridine structure (4.486) (as originally proposed[366,368]) is based on the following facts: (1) the compound under investigation does not liberate iodine from a solution of potassium iodide (oxaziridines, being in general strong oxidizing reagents, are able to do so); (2) lithium aluminium hydride reduction gives N-benzoylindole (oxaziridines under similar conditions are reduced to imines); (3) the compound has great stability on prolonged heating in benzene solution in the presence of water; (4) the compound showed a 1650-1680 cm^{-1} absorption frequency in the infrared spectrum. Recently this benzo-1,3-oxazepine structure has been confirmed[367,369].

In all the uv-induced ring transformations of the azaaromatic N-oxides it is suggested that the unstable oxaziridine derivative, 1-aH-1-a-phenyloxaziridino [2,3a]-quinoline (4.486), is primarily formed. Fast isomerization of the 1,2-epoxide (4.486) into the 2,3-epoxide (4.487) and subsequent ring expansion yields the desired seven-membered heterocycle (4.488). It seems very attractive to formulate this reaction as a symmetry-allowed thermal [1,5]-sigmatropic shift[375], although zwitterionic or free radical intermediates cannot be excluded[375a]. It is evident[365] that the 2,3-dihydroindole results from solvolysis of the 1,3-

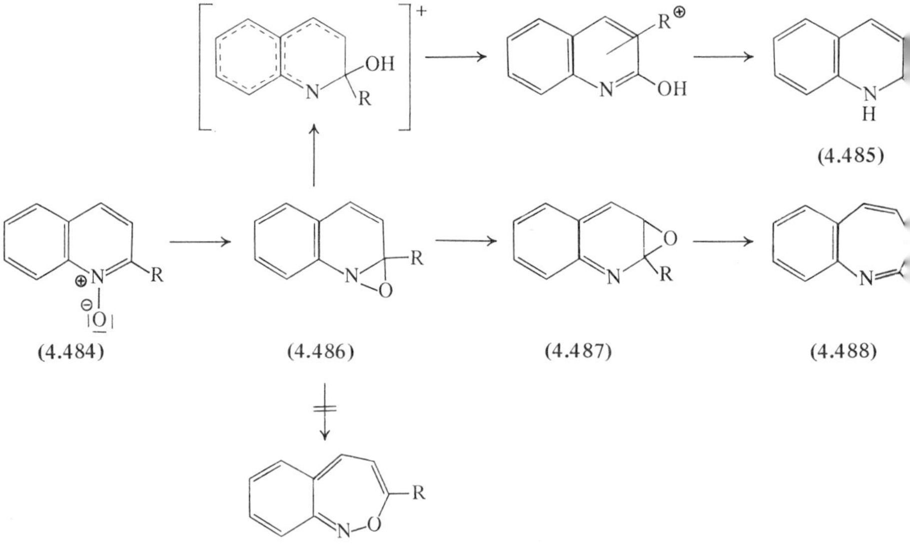

(4.485)

(4.484)   (4.486)   (4.487)   (4.488)

(4.489)

oxazepine (4.488). It does not appear that 3-phenylcarbostyril is formed from the benzo[d] 1,3-oxazepine (4.488) either photochemically or thermally[365]; however it has been suggested that a carbonium ion rearrangement of the phenyl group is involved[263,272]. The formation of benz[c] 1,2-oxazepine (4.489) has not been detected, probably because the stable benzenoid system is lost in the valence tautomerization[370]. Irradiation of the corresponding 3-methyl, 4-methyl and 6-methyl derivative of (4.484) (R = C_6H_5) has confirmed the results discussed above[365]. Irradiation of the 2-cyanoquinoline-1-oxides

(4.490)   (4.491)   (4.492)

R_1	CN	CN	CN	CN
R_2	H	CH_3	Cl	OCH_3

(4.490)[366,367] gave identical results: the benzo[d] 1,3-oxazepines (4.491) were isolated and not the 1aH-oxazirino[2,3a] quinoline derivatives (4.492), originally proposed[366,368].

The photochemical behaviour of the 1-phenyl- and 1-cyanoisoquinoline-N-oxides (4.493) shows a striking similarity to that of the 2-phenyl- and 2-cyanoquinoline-N-oxides. From spectroscopic data and observation of their chemical behaviour the products were assigned the benzo[f] 1,3-oxazepine structure (4.495)[367,371], instead of the oxaziridine structure (4.494) as previously suggested[366,372]. This benzo[f] 1,3-oxazepine structure assignment has been conclusively proved by a recent X-ray crystallographic study of the product, which is obtained by photolysis of 4-bromo-1-cyanoisoquinoline-N-oxide in acetone, i.e. (4.496)[373]. Since the infrared and ultraviolet spectra of (4.496) are very similar to the products (4.495), the structure of these compounds can therefore be ascertained. Photolysis of the isoquinoline-N-oxides (4.493) (R_1 = CN, R_2 = CH_3 or R_1 = CN, R_2 = H) in toluene with a medium pressure mercury lamp gave the corresponding 2a, 7a-dihydrobenzofuro[2,3b]-azete derivatives (4.497) by transannular bridging between C_2 and C_5 in benz[f] 1,3-oxazepine derivatives (4.495)[374].

Irradiation (>330 mμ) of tetrahydroacridine-N-oxide (4.498) in an aprotic solvent yields the ring-expanded product 2,7-tetramethylenebenz[d] 1,3-oxazepine (4.501) as the main product in a 70 per cent yield; in addition the ring-contracted compound 5,6,7,8,9,10-hexahydrocyclohept[b]indol-10(5H)-one (4.502) is obtained in a 10 per cent yield[369,376]. Both products originate

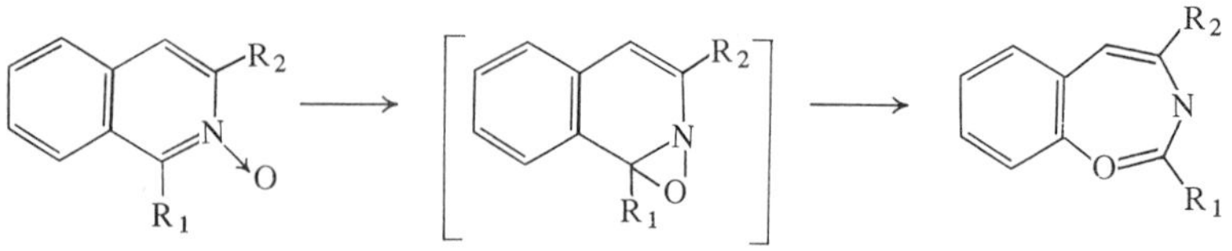

(4.493)   (4.494)   (4.495)

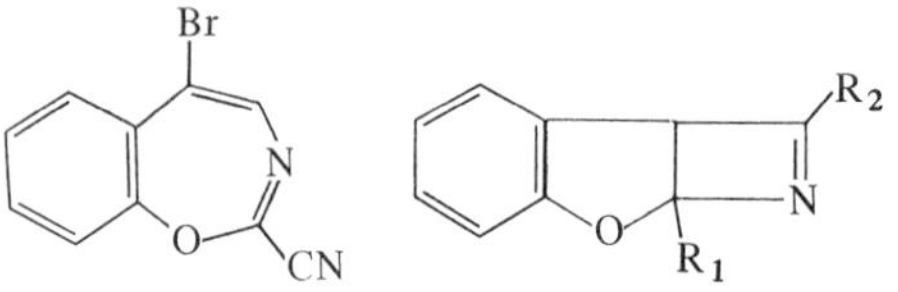

(4.496)   (4.497)

R₁	C₆H₅	C₆H₅	CN	CN
R₂	H	CH₃	H	CH₃

from the common 2,3-epoxy compound (4.500) which is formed by isomerization from (4.499).

Attempts to prepare the 3,4-benzo-1,6-oxido[10]2-azaannulene (4.504) by ultraviolet irradiation of acridine-N-oxide (4.503) have failed[376]; only (4.505) has been obtained. This failure may be due either to a high rate ratio of route b

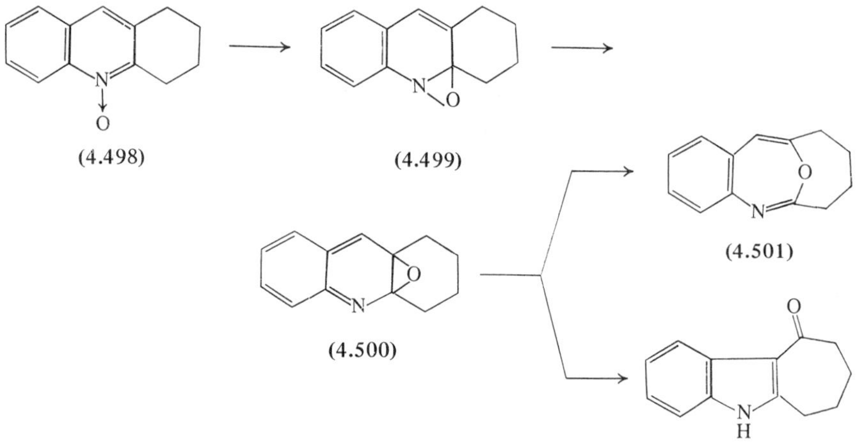

(4.498)   (4.499)

(4.500)

(4.501)

(4.502)

to route a or, alternatively, to a facile photoisomerization of (4.504) into (4.505). This theory is supported by the finding that the benz[d] 1,3-oxazepine-2-carbonitrile (4.506) is easily photoisomerized into the 3-methyl-2-cyanoindole (4.507)[377].

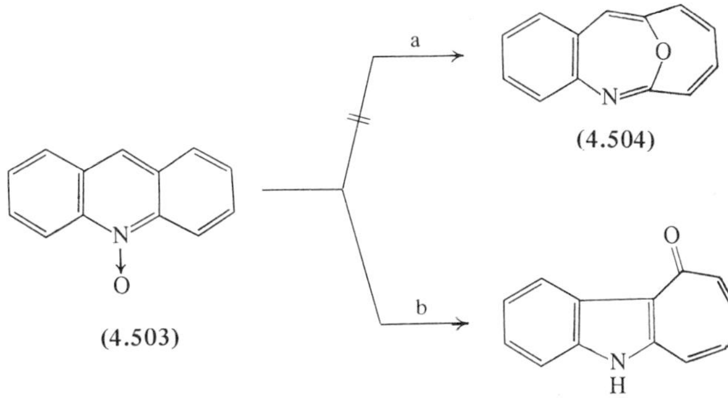

(4.504)

(4.503)

(4.505)

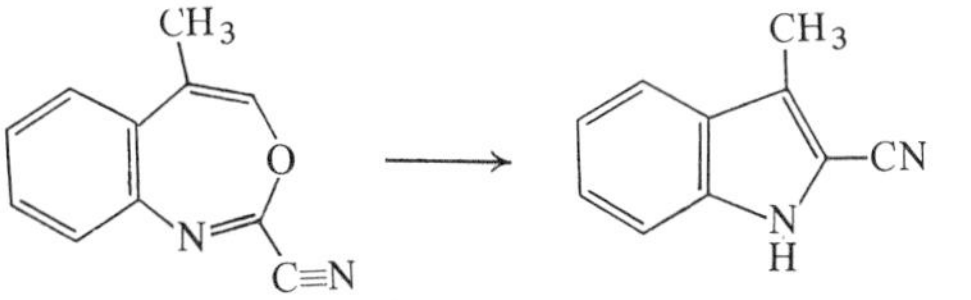

(4.506)   (4.507)

It has recently been found that the meso-ionic 1,3-dipolar *N*-ethoxycarbonyl-iminopyridinium ylides (4.508) give on irradiation 1-ethoxycarbonyl-1*H*-1,2-diazepines (4.510)[378-384]; this structure has been established by X-ray analysis of its $Fe(CO)_3$-complex[378]. In this photochemically induced ring expansion, it is postulated that, in analogy to the mechanism advanced for the pyridine *N*-oxides or pyridinium dicyanomethylide (see section I.C.1.c), the diazirine (4.509) is a possible intermediate; by valence tautomerization it is converted into (4.510). Evidence based mainly on the results of chemical reactions indicates that the diazirine (4.509) is in equilibrium with the 1,2-diazepine (4.510) which is the

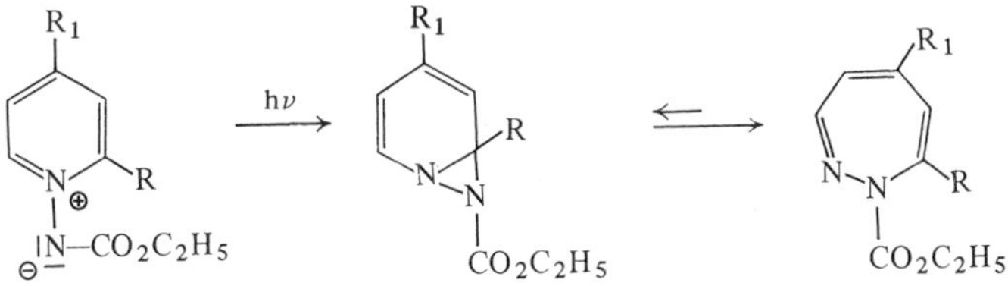

(4.508)   (4.509)   (4.510)

R	H	CH_3	H	CH_3
R_1	H	H	CH_3	CH_3

predominant isomer[379]. Photochemical rearrangement leading to 1H-1,2-diazepines has also been found to occur with the N-alkoxysulphonylimino-pyridinium   ylides[379,383]   and   with   the N-acyliminopyridinium ylides[379,383,384].

Expansion of a piperidine ring into a seven-membered heterocycle takes place during pyrolysis under reduced pressure of 1-methylanabasine-1-oxide (4.511), a mixture of 1-methylanabasine and the hexahydro-1,2-oxazepine (4.512) being obtained[385]. Analogous ring enlargements were earlier reported with five-membered heterocycles; an example is the conversion of nicotine-1-oxide into 6-(3-pyridyl)tetrahydro-2H-1,2-oxazine (see chapter 3, section I.C.5.c). Mechanistically, this ring enlargement is closely related to the Meisenheimer rearrangement[386] of N-allyl- and N-benzylamine-oxides into o-allyl- and o-benzylhydroxylamines of which a quite intensive study has been made. In line with this view, it has been found[385] that neither 1-methylpiperidine oxide nor its 2-methyl derivative show ring expansion. With 2-arylpiperidine-N-oxides the same type of rearrangement into a hexahydro-1,2-oxazepine has been reported[387].

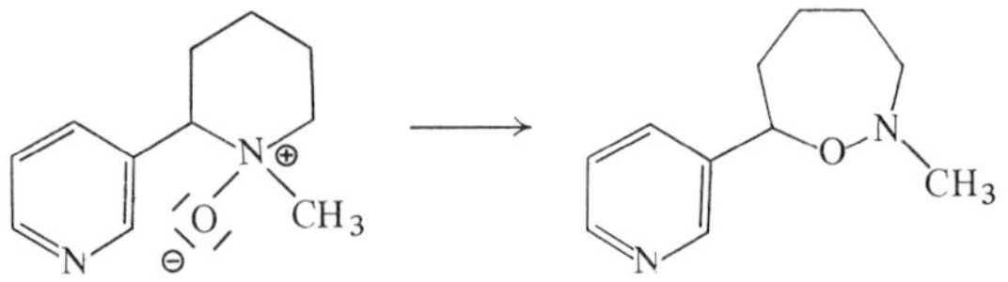

(4.511)   (4.512)

Thermal degradation of the isoquinolinotetrazine derivative (4.513) gives a compound to which the structure of the dihydrobenzodiazepine (4.514) has been assigned[388]. Sufficient evidence has been presented to prove that (4.515) is not the product formed in this thermal decomposition.

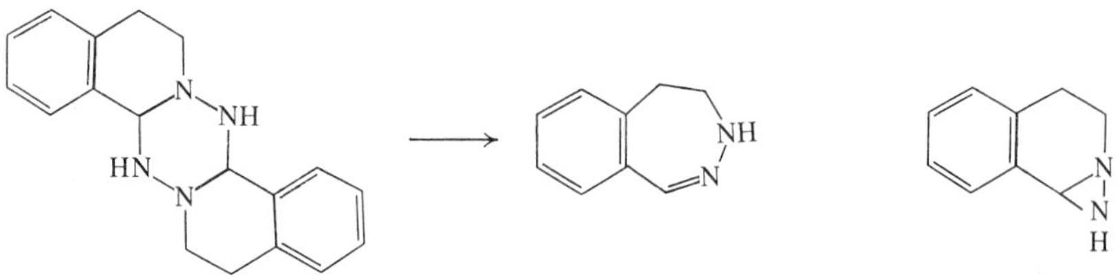

(4.513)   (4.514)   (4.515)

4. RING EXPANSIONS INTO HETEROCYCLES CONTAINING MORE THAN SEVEN ATOMS

Only a few ring enlargements of nitrogen heterocycles have been shown to lead to a new heterocycle containing a ring with more than seven atoms. The Sommelett–Hauser rearrangement of the piperidinium salt (4.516) into the nine-membered compound (4.519) by means of sodamide in liquid ammonia is an example of a rearrangement of this type[389]. It proceeds by an internal nucleophilic attack of the carbanion of the ylid (4.517) on position 2 of the phenyl ring, giving (4.518), which, in a subsequent allylic rearrangement, gives (4.519). Attempts to obtain (4.521) from the N-methyl derivative (4.520) by a

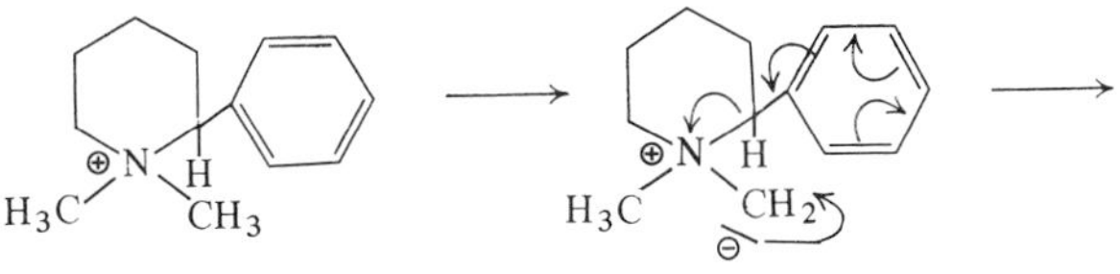

(4.516)   (4.517)

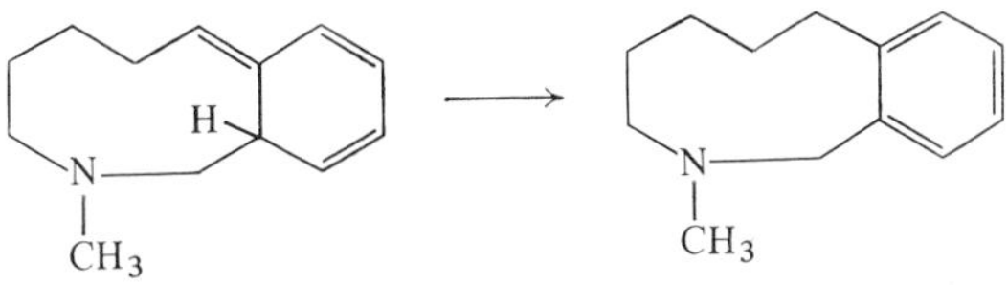

(4.518)   4.519)

renewed sequence of reactions failed. This was apparently due to the fact that the approach of the carbanion to the aromatic ring produced a transition state with a considerably strained conformation.

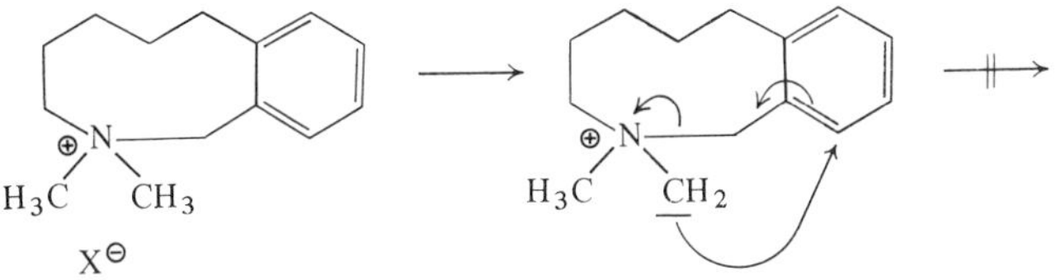

(4.520)

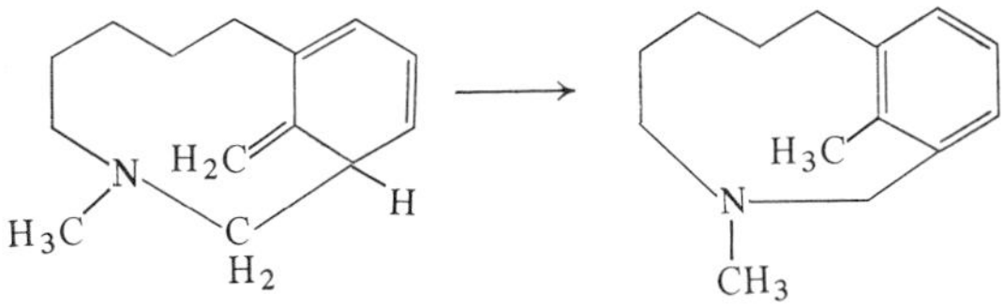

(4.521)

Hydrogenation of the piperidone-2 (4.522), which is thus substituted on the nitrogen with a β-hydroxypropionyl group, using palladium as catalyst and tetrahydrofuran as a solvent, has been found to give ring expansion into the ten-membered ring compound (4.525)[390]. By reductive loss of toluene the N-hydroxypropionyl group is formed (4.523) which reacts further, very probably via the semi-acetale (4.524), into (4.525). An indication for the transient existence of (4.524) is provided by the experimental finding that the compound (4.526) yields (4.527), the five-membered stable analogue of (4.524), on cyclization[391]. The fact that further conversion of (4.527) into the nine-membered compound (4.528) does not take place is probably due to steric hindrance in the resulting ring, if such a ring is formed.

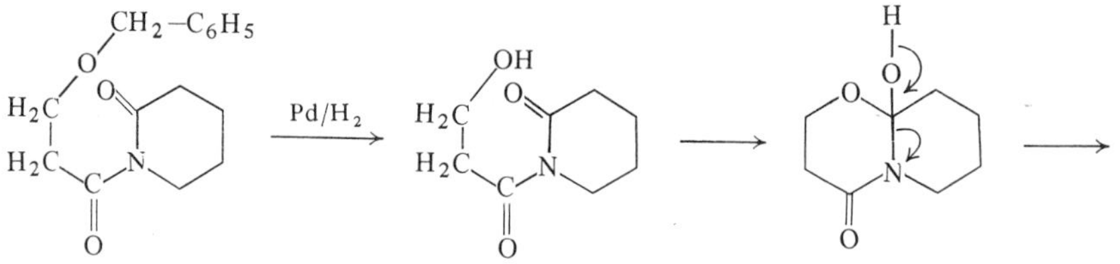

(4.522)   (4.523)   (4.524)

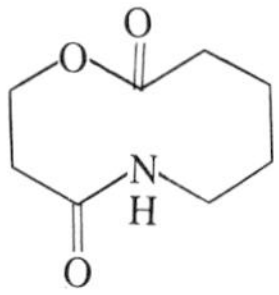

(4.525)

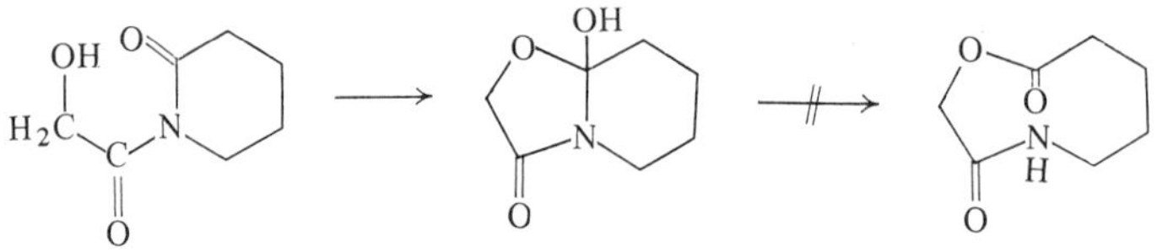

(4.526)   (4.527)   (4.528)

II. RING TRANSFORMATIONS OF SIX-MEMBERED HETEROCYCLES CONTAINING TWO IDENTICAL HETEROATOMS

A. Ring transformations of six-membered heterocycles containing two oxygen atoms

1. 1,4-DIOXANES INTO 1,3-DIOXOLANES AND PIPERAZINES; 1,3-DIOXANES INTO 1,3-OXAZINES

Relatively few ring transformation have been reported with six-membered heterocycles containing two oxygen atoms in the ring. Ring contraction of the

1,4-dioxane system has been found to occur[392] (although it was not immediately recognized as such[393]) when 2,3-dichloro-1,4-dioxane (4.529) is treated with glycol; bis 1,3-dioxolane (4.530) is yielded, the structure of which has been firmly established by crystallographic techniques.

An 8 per cent conversion of 1,4-dioxane (4.531) in piperazine (4.532) has been achieved by gas-phase reactions at 370°C at 11.5 atm in the presence of ammonia, using active alumina as catalyst[394].

(4.529)   (4.530)   (4.531)   (4.532)

In the 1,2-dioxane series (cyclic peroxides), the reductive ring contraction of phthalylperoxide (4.533) and of ascaridol (4.535) into phthalic anhydride (4.534) and the endoxyde (4.536) respectively has been reported[395,396]. Both reactions occur through the action of triphenylphosphine. The mechanism of

(4.533)   (4.534)   (4.535)   (4.536)

these reductive ring contractions, which is strongly related to that reported for the extensively studied reaction of organic peroxides with triphenyl-phosphine[397], is assumed to involve the bimolecular formation of a transient quaternary phosphonium adduct of the cyclic peroxide with the phosphorus compound. The phosphonium salt decomposes spontaneously into triphenyl-phosphine oxide and the compound without oxygen.

In the 1,3-dioxane series, the ring transformation of the carbonates of 1,3-diols (4.537) into the 1,3-oxazine derivatives (4.538) has been reported to occur during the lithium chloride catalysed reaction of (4.537) with phenyliso-cyanate at about 220°C[398]. This reaction is strongly related to the conversion of 1,3-dioxolan-2-one into oxazolidines (see vol. 1, chapter 3, section II.A.1). In analogy to the proposal made for the occurrence of a non-octet stabilized 1,3-dipolar intermediate in the thermolysis of oxiranes and 1,3-dioxolan-2-ones in the presence of isocyanates, it may be advanced that the conversion of the 1,3-dioxane into an 1,3-oxazine involves the 1,4-dipolar compound (4.539).

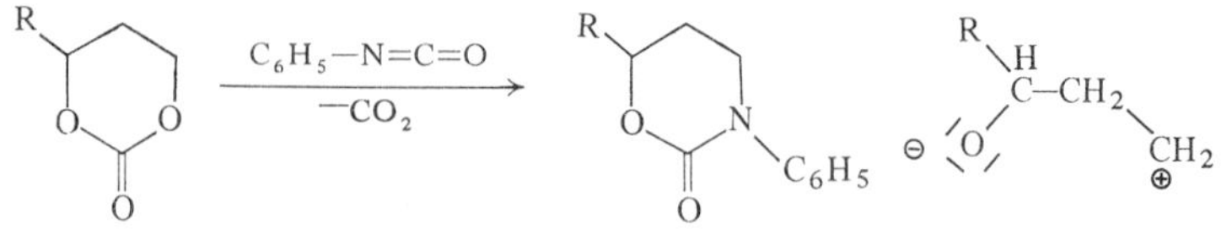

(4.537)   (4.538)   (4.539)

$R = H; CH_3$

B. Ring transformations of six-membered heterocycles containing two sulphur atoms

1. RING CONTRACTIONS OF SIX-MEMBERED HETEROCYCLES CONTAINING TWO SULPHUR ATOMS

a. *1,4-Dithiins into thiophenes*

This ring contraction which has a broad application, can be achieved by several methods. One is the pyrolytic method. If 2,5-diphenyl-1,4-dithiin (4.540) (R = H) is heated at 190°C in an inert atmosphere there is loss of sulphur and formation of 2,4-diphenylthiophene (4.541) (R = H)[399]. The 3-nitro and 3-bromo derivative of 1,4-dithiin can similarly be converted into 2-nitro- and 2-bromo-2,4-diphenylthiophene[400,401] respectively. The following conversions

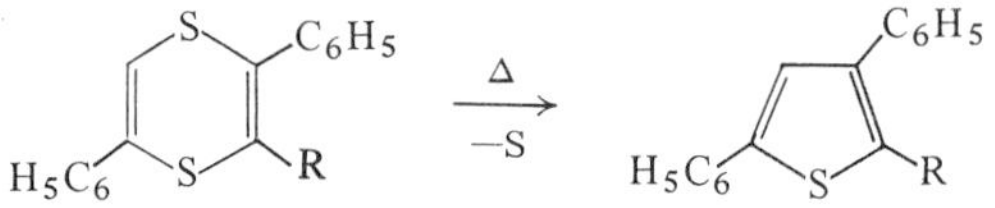

(4.540)   (4.541)

$R = H; NO_2; Br$

have similarly been carried out: tetracyano-1,4-dithiin (4.542) ($R_1 = R_2 = CN$) into tetracyanothiophene[402]; 2,5-dimethyl-1,4-dithiin (4.542) ($R_1 = CH_3$, $R_2 = H$) into 2,4-dimethylthiophene[403]; tetraalkoxycarbonyl-1,4-dithiin (4.542) ($R_1 = R_2 = CO_2R$) into tetraalkoxycarbonylthiophene[404]; tetraphenyl-1,4-dithiin (4.542) ($R_1 = R_2 = C_6H_5$) into tetraphenylthiophene[405]; and 2,5-diphenyl-3,6-dimethoxycarbonyl-1,4-dithiin[406] (4.542) ($R_1 = C_6H_5$, $R_2 = CO_2CH_3$) into the corresponding thiophene (4.543).

More recently it has been demonstrated that there is an analogous ring contraction of the tetrahydrotetracyanotetrathiaanthraquinone (4.544) into the corresponding benzo(1,2b:4,5b')dithiophene (4.545)[407].

Surprisingly it has been found[399] that (4.540) (R = H) can be converted into the 5-thenaldehyde (4.541) (R = OCH) by treatment with the Vilsmeyer reagent, dimethylformamide and phosphoryl chloride. It can be assumed that primarily

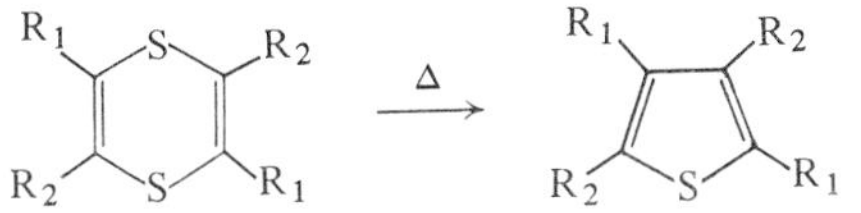

(4.542)   (4.543)

R_1	CN	CH_3	CO_2R	C_6H_5	C_6H_5
R_2	CN	H	CO_2R	C_6H_5	CO_2CH_3

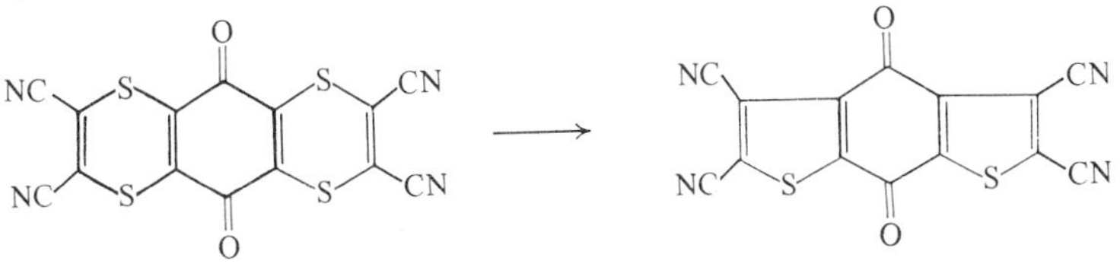

(4.544)   (4.545)

formylation occurs into position 3, yielding (4.540) (R = OCH) which subsequently rearranges into (4.541) (R = OCH). The fact that in all cases 2-substituted thiophene derivatives are obtained from 3-substituted 2,5-diphenyl-1,4-dithiins—not the 4-substituted ones—clearly indicates that it is the sulphur atom at position 1 which is eliminated during the ring contraction. The thermal process requires a $(4n + 2)$ disrotatory electrocyclic reaction leading to a cis fused resonance—stabilized thiirane type intermediate (4.546a) $\leftrightarrow$ (4.546b)[400,408]. The bond distance between position 3 and 5 makes such interaction possible, since 1,4-dithiin possesses a boat configuration[409]. In the case of the 3-substituted derivatives (R = NO_2, Br, OCH), the contribution of (4.546a) is, however, more important since the presence of these electron-attracting substituents makes further delocalization of the negative charge possible.

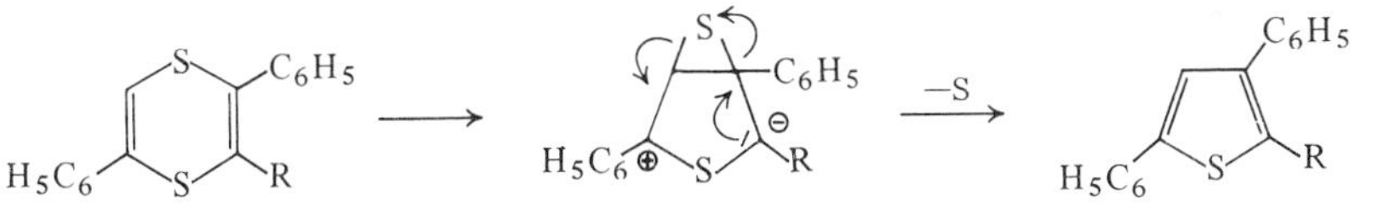

(4.540)   (4.546a)   (4.541)

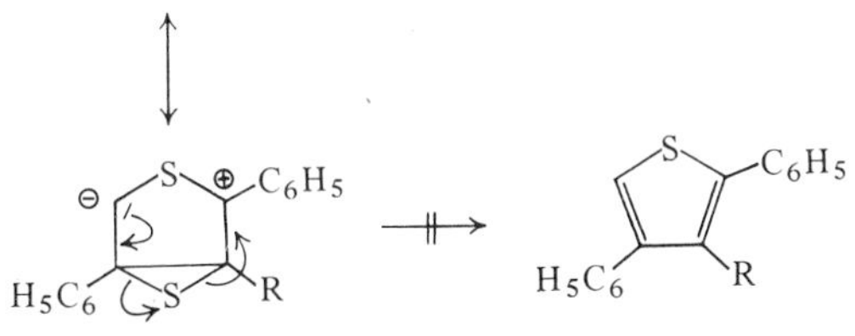

(4.546b)

Removal of sulphur is also achieved during pyrolysis of thianthrene (4.547) with copper at high temperature[410] (with the formation of dibenzothiophene (4.548)) and during treatment of its 5-oxide (4.549) with *n*-butyllithium[411].

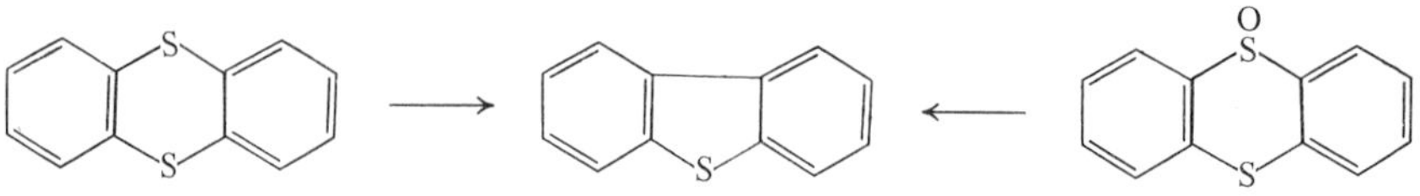

(4.547)   (4.548)   (4.549)

Hydrogen peroxide in acetic acid is also an effective reagent in the achievement of ring contraction of 1,4-dithiins into thiophenes. In contrast to the pyrolytic reactions, however, a mixture of both isomeric thiophenes, the 2-nitro- and 4-nitro-3,5-diphenylthiophene, is obtained[400] from (4.540) (R = NO$_2$). 3,6-Disubstituted 2,5-diphenyl-1,4-dithiins (4.550) are converted into 2,4-disubstituted 3,5-diphenylthiophene (4.551) in reasonable yields[401] and the 5,8-dioxo-5,8-dihydro-1,4-benzodithiin (4.552) is converted into the benzo-thiophene 4,7-quinone (4.553)[412]. It has been suggested[399,400] that during

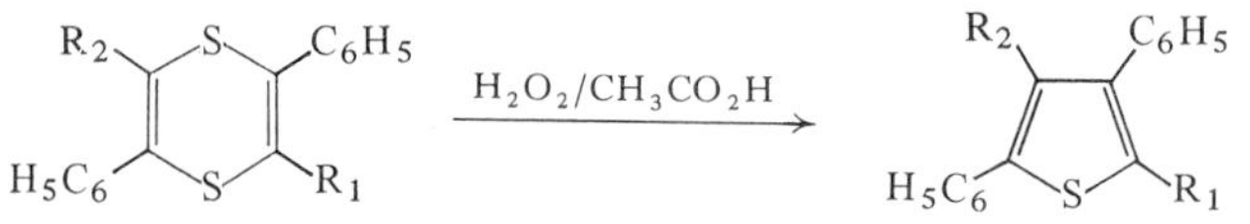

(4.550)   (4.551)

R$_1$	Br	NO$_2$	Br
R$_2$	Br	NO$_2$	NO$_2$

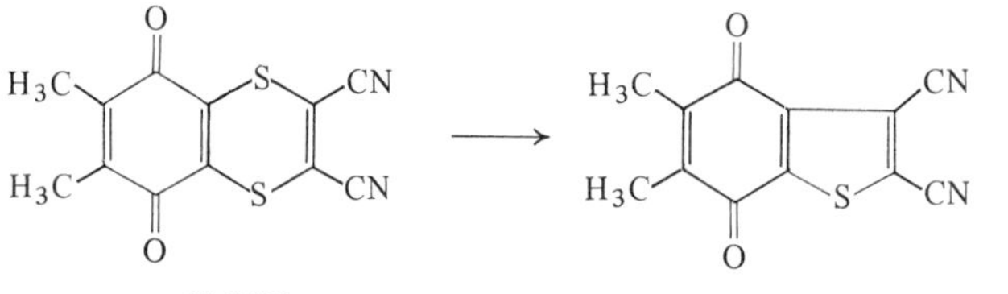

(4.552)   (4.553)

this oxidative ring contraction the 1,4-dithiin is first converted into a monosulphone, which thereupon, through loss of sulphur dioxide, is converted into a thiophene derivative. The intermediary occurrence of a monosulphone, however, has been questioned[413,414] since the monosulphone of (4.540) (R = H and R = Br) is found to be stable under these oxidation conditions. The sulphoxides rather than the monosulphones are considered to be the inter-mediates in these oxidative ring contractions. Preparation of the sulphoxides of

(4.540) showed that they were very unstable and decomposed easily into thiophenes. Recently the intermediary occurrence of a sulphoxide during these ring contractions has been demonstrated[414a]; reaction of 5,8-dihydroxy-6,7-dimethyl-2,3-dicyano-1,4-benzodithiin with 40 per cent cold peracetic acid gave the corresponding S-oxide of this 1,4-benzodithiin which was apparently stable enough to be isolated and was found to readily undergo ring contraction into 5,6-dimethyl-2,3-dicyano-1-benzothiophene-4,7-quinone.

b. *1,3-Dithiins into thiolium salts; 1,3-dithianes and 1,4-dithianes into 1,3-dithiolanes*

1,3-Dithiins also show interesting ring contraction reactions. When the 1,3-dithiin (4.554) is treated with sulphuryl chloride in ether–acetic acid and perchloric acid is added to the reaction mixture, dehydrogenation takes place yielding the 1,2-dithiolium salts (4.555)[415]. The carbon present at position 2 in (4.554) was discovered to be found after working up into the formed aldehyde; it was identified as the hydrazone of this compound. This dehydrogenation could also be carried out with bromine in acetic acid[415].

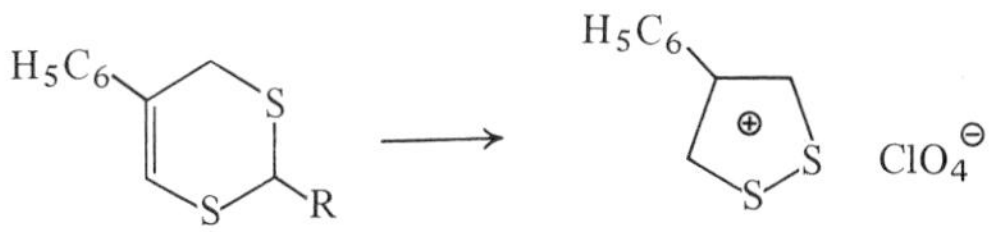

(4.554)   (4.555)

R = H; C_6H_5; p-OCH$_3$C$_6$H$_4$

2-Phenyl-1,3-dithiane-5-ol (4.556) has been converted into the 1,3-dithiolane (4.557) by treatment with phosphoryl chloride in pyridine[416,417]. In this ring contraction a considerable degree of stereospecificity is maintained since the *trans*-2-phenyl-4-chloromethyl-1,3-dithiolane (4.557) was obtained from the *trans* derivative (4.556) and the corresponding *cis*-1,3-dithiolane was formed from a *cis* derivative. Assuming that the breakdown of the initially formed

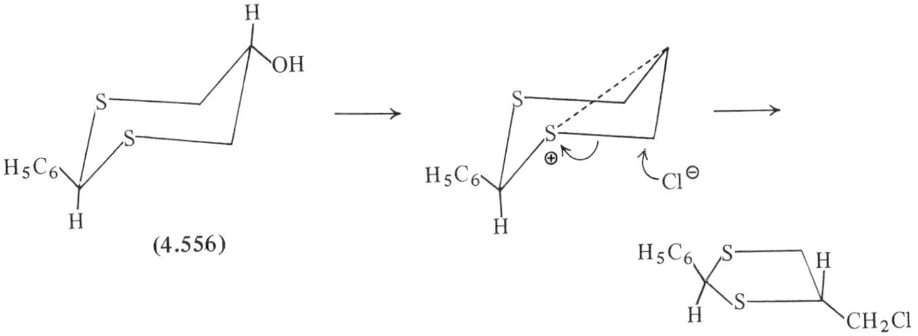

(4.556)

(4.557)

phosphorus-containing intermediate involves participation of the neighbouring sulphur atom, different bridged sulphonium ions are obtained from the isomeric *cis* and *trans* alcohols; the configuration of these strongly determines the geometry of the 1,3-dithiolane formed in the subsequent reaction with the chloride ion. It has been assumed that a similar bridged sulphonium ion is intermediate in the fluorination of 1,4-dithiane (4.557a) over potassium tetrafluorocobaltate (III) which leads to the formation of a mixture of five different polyfluoro derivatives of 2-methyl-1,3-thiolane (4.557d) and seven different polyfluoro derivatives of (4.557a)[417a]. Since it is known that in general a carbonium ion is stabilized by a sulphur atom at the β-position as well as at the α-position[417b], it is evident that if the reaction proceeds via the carbonium ion (4.557b) participation of the β-sulphur leads to the bridged sulphonium ion (4.557c) from which rearrangement into the 2-methyl-1,3-dithiolane is clearly feasible.

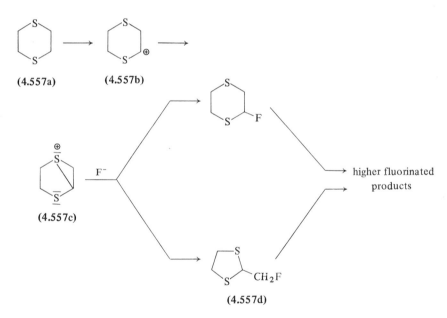

(4.557a)   (4.557b)

(4.557c)

higher fluorinated products

(4.557d)

c. *1,2-Dithiins into thiophenes and pyridazines*

The preparation of 1,2-dithiin (4.559) (R = H) has recently been described[418,419]. From uv data (λ_{max} = 470 – 480 nm) and nmr data (τ = 3.87 and τ = 4.03) it was shown to have the structure of the cyclic disulphide (4.559) rather than the open-chain thioketene (4.558). It was found to lose sulphur readily on heating or on uv irradiation, yielding thiophene (4.560) (R = H)[419].

Similarly, from the 3,6-disubstituted 1,2-dithiins (4.559) 2,5-disubstituted thiophenes (4.560) are obtained[418]. This ring contraction is of special interest from a biochemical point of view, since there is a possibility that the naturally occurring thiophene derivatives[420] may originate from 1,2-dithiins since these compounds are known to be present in plants.

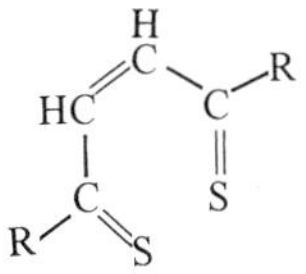

(4.558) 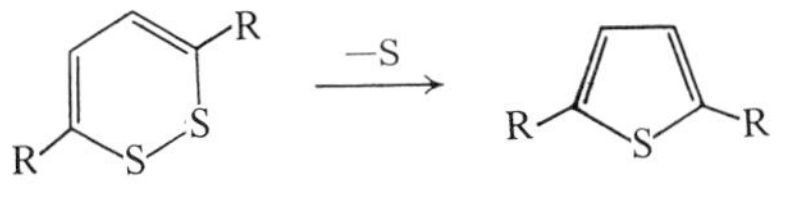 (4.559) (4.560)

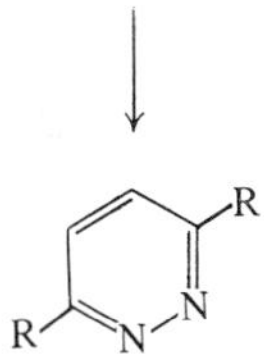

(4.561)

R = H; C_6H_5; p-$CH_3C_6H_4$; p-$OCH_3C_6H_4$; 2-thienyl

Desulphurization of the dibenzo[c,e]dithiin (4.562) takes place when this compound is pyrolysed with copper bronze at high temperature: the dibenzothiophene (4.563) is yielded[421]. Loss of sulphur dioxide during the pyrolysis of the corresponding S-dioxide leads to the formation of the corresponding benzothiophene[422].

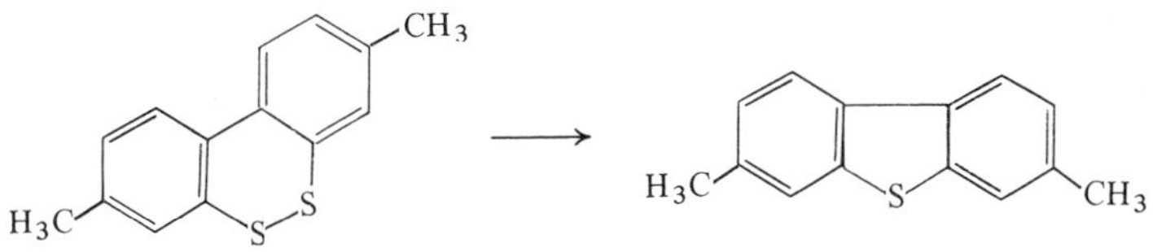

(4.562) (4.563)

Ring transformation of 1,2-dithiins into 3,6-disubstituted pyridazines (4.561) has been reported to occur when (4.559) is treated with hydrazine in pyridine or in dimethylformamide.

C. Ring transformations of six-membered heterocycles containing two nitrogen atoms (pyrimidines and quinazolines)

1. RING CONTRACTIONS OF PYRIMIDINES AND QUINAZOLINES

a. *Pyrimidines into pyrroles, imidazoles, pyrazoles, isoxazoles and triazoles; tetrahydropyrimidines into oxazolidines and imidazolidines; quinazolines into indazoles*

In correlation with the reported conversion of 3-amino-2-bromopyridine into 3-cyanopyrrole (see section I.C.1.c), it has been established that 5-amino-4-chloro-2-phenylpyrimidine (4.564) reacts with potassium amide in liquid ammonia to form 4(5)cyano-2-phenylimidazole (4.565)[423,424]. 4-Chloro-5-methyl-2-phenylpyrimidine (4.566) is also converted into an imidazole, that is 4(5)-ethynyl-2-phenylimidazole (4.567)[423,424], which is surprising since a similar reaction with 2-bromo-3-methylpyridine does not give a pyrrole derivative under these reaction conditions.

(4.564)   (4.565)

(4.566)   (4.567)

Whereas in the formation of 3-cyanopyrrole from 3-amino-2-bromopyridine it is clear that carbon-carbon bond breakage must occur between positions 2 and 3 in the pyridine ring, in the conversion of (4.564) → (4.565) and (4.566) → (4.567) it is *a priori* not evident whether the carbon-carbon bond fission has taken place between position 4 and 5 or between position 5 and 6 in the pyrimidine ring. In order to clarify which of these two routes is involved, the reaction of the ^{14}C-labelled pyrimidine derivative (4.568) with potassium amide was investigated. It was established that in the imidazole derivative formed, the ethynyl group was attached to the radioactive carbon atom (4.569)[425]. This evidence was based on the fact that oxidation of (4.570), obtained by hydrogenation of (4.569), with hydrogen peroxide in basic solution, yields a 1,2,4-oxadiazole (4.571)[425,426] which holds its radioactivity on C_5, that is on

(4.568)   (4.569)   (4.570)

(4.571)

the carbon, attached to the ethyl group. This leads to the unequivocal conclusion that it is the carbon-carbon bond between position 5 and 6 of the pyrimidine ring which is broken. The mechanism (see p. 118) can be tentatively advanced. In the first step, loss of a proton takes place from the methyl group in this strong basic media, thus yielding the resonance-stabilized anion (4.572). After ring fission at the C_5-C_6 position, the anionic ethynyl open-chain intermediate (4.573) is formed which is in equilibrium with the ionic secondary propargylic halide (4.574). From the results of base-promoted solvolyses of propargylic bromides[427], it is possible that the ring closure into (4.576) will proceed through the carbon (or carbenoid) species (4.575). However, a direct ring closure of (4.573) into (4.577) cannot be excluded, since in the analogous ring contraction of the 5-aminopyrimidine (4.564) into 4(5)-cyanoimidazole (4.565) it is the open-chain cyano compound (4.578), which very probably gives direct ring closure.

The conversion of pyrimidines into imidazoles has also been reported to occur during pyrolysis of the 4-azidopyrimidine (4.579), giving the imidazole 1-carbonitrile (4.582)[428] in high yield (88 per cent). The formation of the N-cyano compound is consistent with the production of a singlet nitrene (4.580), which, through the bicyclic intermediate (4.581), gives ring contraction into (4.582). The fact that the isomeric cyano compound (4.583) is not obtained can be explained in terms of stabilization of the intermediate (4.581) which is due to the orbital overlap of the bridgehead nitrogen with the orbitals containing the nitrogen and carbon lone pairs; this would lead to collapse to (4.582) rather than to (4.583).

For many years there has been considerable interest in the ring trans-formations caused by the action of hydrazine and substituted hydrazines on pyrimidine and its derivatives. This interest arose through the finding that mutation occurs in microorganisms when they are subjected to the action of hydrazine and methylhydrazine. Therefore the hydrazinolysis of the naturally occurring pyrimidine bases[429-431], the nucleotides and nucleo-

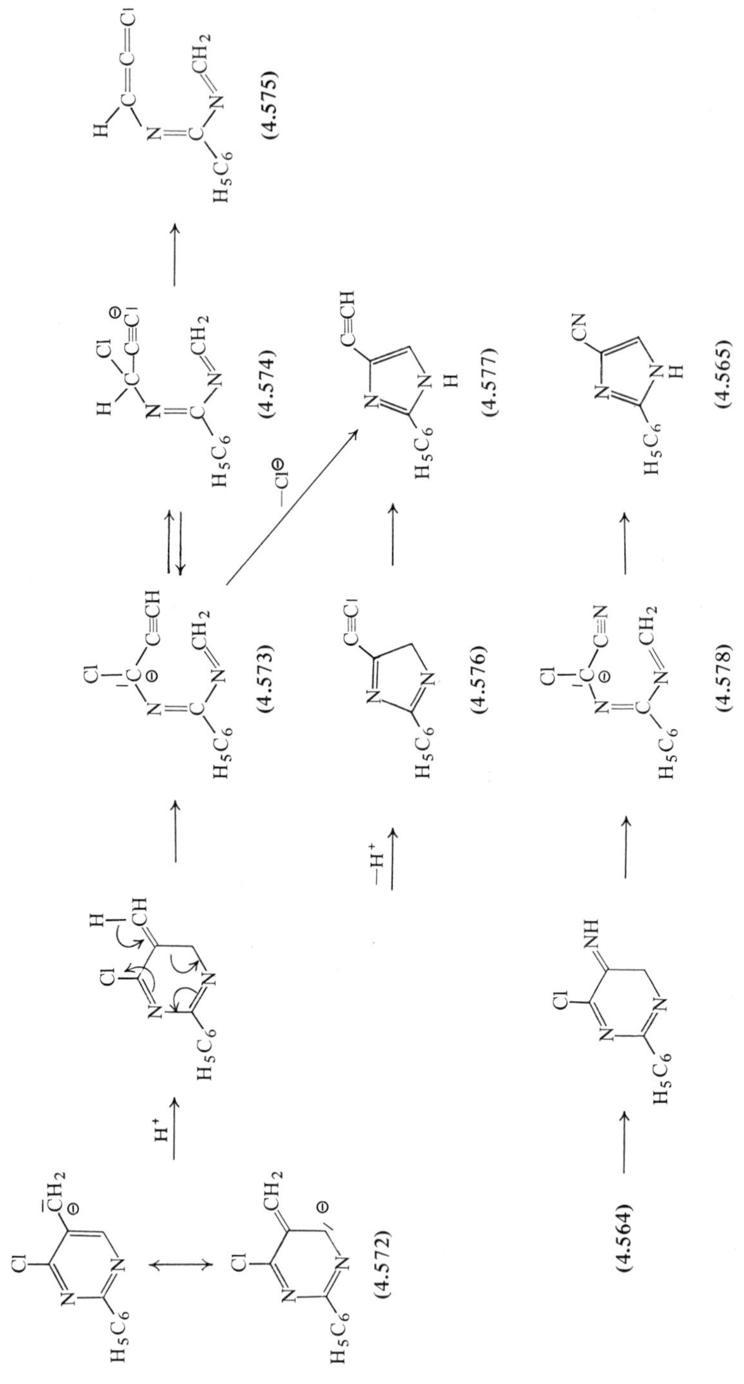
(4.572)
(4.573)
(4.574)
(4.575)
(4.576)
(4.577)
(4.564)
(4.578)
(4.565)

(4.579) (4.580) (4.582)

(4.581) (4.583)

sides[429,432-436], and the ribo and desoxyribonucleic acids[437-442], has been the subject of investigations. A primary example of ring contraction of this type is the conversion of uracil (4.584) (R = H) into pyrazolon-5 (4.585) and urea by reaction with hydrazine[429]. Formally, the N–C–N part of the pyrimidine ring is divided from the C_4–C_5–C_6 fragment, which recycles with hydrazine into the corresponding pyrazole. Extension of this reaction with methylhydrazine and sym-dimethylhydrazine has led to the isolation of the 1-methyl-3-ureido-pyrazolid-5-one (4.586) and its 1,2-dimethyl derivative (4.588)[430]. Hydrolysis with acid or base gives the 1-methyl- or 1,3-dimethylpyrazolon-5 (4.587) and (4.589). The ring contraction of uracil into pyrazolone-5 probably takes place by an initial attack of hydrazine either at position 4 or at position 6; which position is attacked has not been established[431] (see scheme on p. 121).

Analogously, from thymine (4.584) (R = CH_3) with hydrazine, methyl-hydrazine and dimethylhydrazine, the corresponding 4-methyl derivatives of (4.585), (4.586) and (4.588) are formed; however, the rate at which the reactions take place is lower than with uracil[430,431]. The ring contraction of cytosine (4.590) into pyrazole derivatives has been found to take place only when the reaction is carried out with hydrazine hydrate at 90°C; the yield is low. The products are 3-aminopyrazole (4.591) and a compound tentatively postulated to be N,N-di(3-pyrazolyl)hydrazine (4.592). The formation of (4.591) and (4.592) is found to be concurrent rather than sequential. In neutral and weak acidic solutions only replacement of the amino group by a hydrazino group takes place. The reaction of the pyrimidine bases with methylhydrazine yielding ureidopyrazolidones can be best described to occur by attack of methylhydrazine at position 4. Hydrazinolysis of the nucleoside uridine also

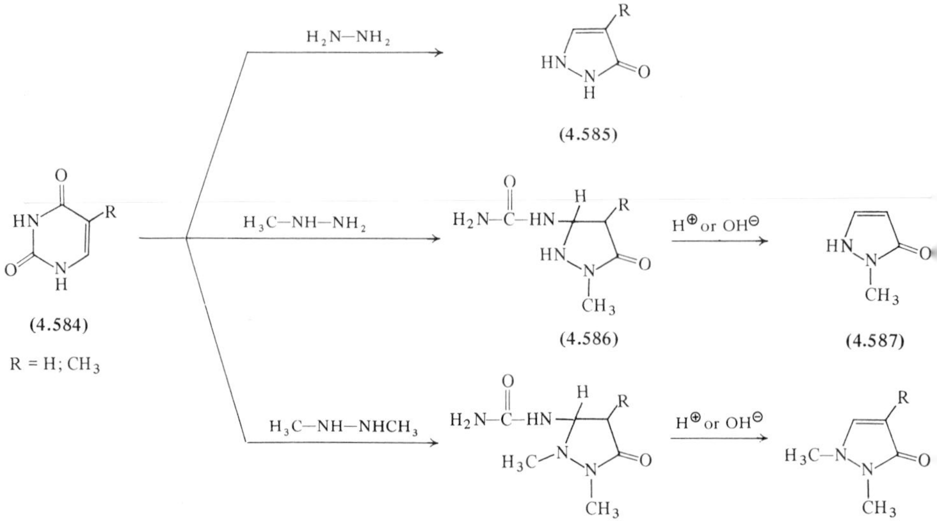

(4.584)

R = H; CH$_3$

(4.585)

(4.586)

(4.587)

(4.588)

(4.589)

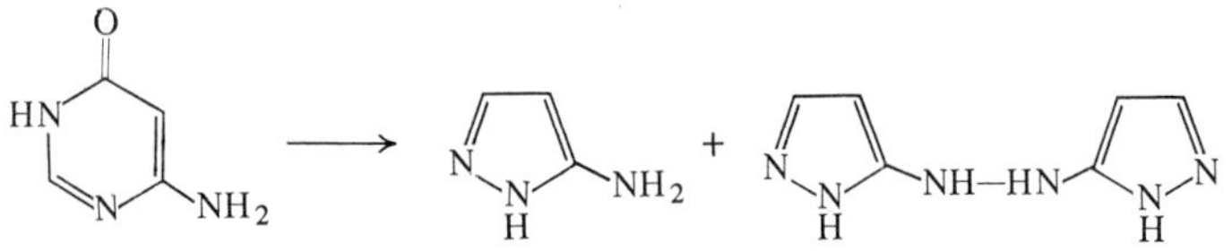

(4.590)   (4.591)   (4.592)

yields a pyrazolone, together with a compound which is probably ribosylurea. By this method it appears to be possible to establish the position of the phosphate group in the cytidylic acids a and b[443,440].

Uracil-5-carboxaldehyde (4.593) is found to be converted into the 4-alkyl-ideneureidopyrazolone-5 (4.594) on treatment with hydrazine or methyl-hydrazine[443]. With phenylhydrazine and hydroxylamine, normal condensation products were obtained[444]. The fact that this rearrangement was not observed with the less basic hydrazines (such as phenylhydrazine) or N,N-dimethyl-hydrazine suggests the reaction sequence presented on p. 122.

Simple pyrimidines, not directly connected with the naturally occurring pyrimidine bases, were also the subject of investigations with hydrazines and substituted hydrazines. When an aqueous solution of pyrimidine (4.595) ($R_1 = R_2 = H$), 4-methyl- (4.595) ($R_1 = H; R_2 = CH_3$) or 4,6-dimethylpyrimidine (4.595) ($R_1 = R_2 = CH_3$) is allowed to react with aqueous hydrazine at 190°C, pyrazole (4.596) ($R_1 = R_2 = H$), 3(5)-methylpyrazole (4.596) ($R_1 = H; R_2 = CH_3$) or 3,5-dimethylpyrazole (4.596) ($R_1 = R_2 = CH_3$) respectively are

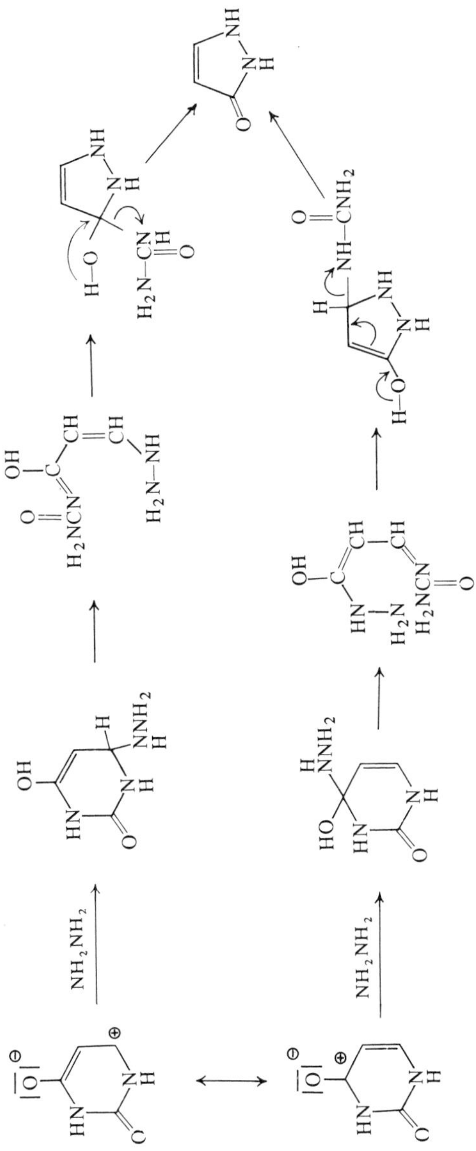

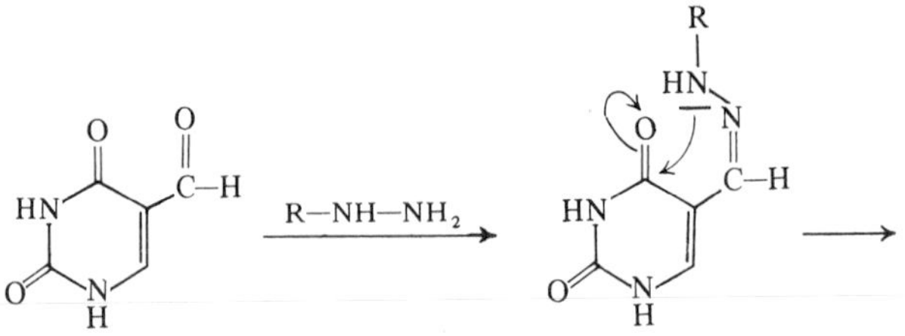

(4.593)

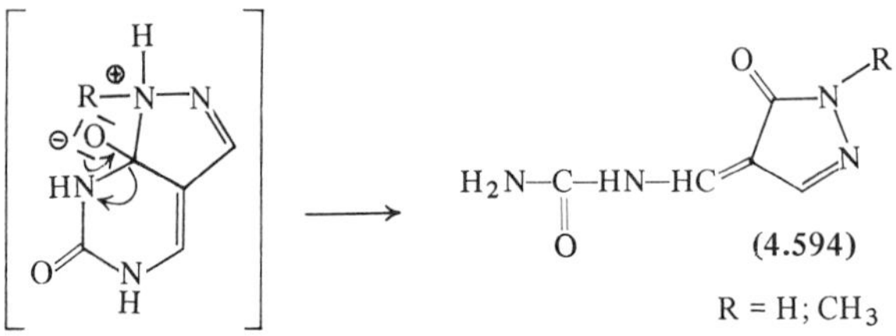

(4.594)

R = H; CH₃

obtained in excellent yields[445,446]. The three pyrimidines, when treated with methylhydrazine sulphate, yield 1-methylpyrazole, an isomeric mixture of 1,3-dimethyl- and 1,5-dimethylpyrazole or 1,3,5-trimethylpyrazole. It is observed that the methyl groups in the 4- and/or 4,6-positions of the pyrimidine ring strongly retard the rate of the ring contraction but that quaternization of the ring by attachement of a methyl group to the nitrogen increases the rate dramatically. Thus 1,4,6-trimethylpyrimidinium iodide (4.597) ($R_1 = R_2 = CH_3$, X = I) when treated with hydrazine, gives, even at room temperature, formation of 3,5-dimethylpyrazole (4.596) ($R_1 = R_2 = CH_3$)[446].

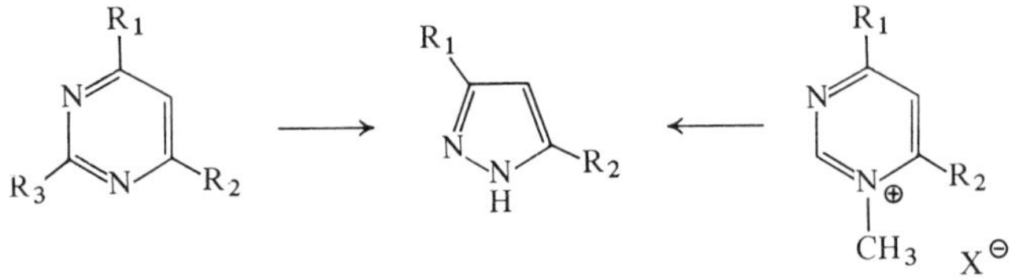

(4.595)   (4.596)   (4.597)

R_1	H	CH₃	CH₃
R_2	H	CH₃	H
R_3	H	H	H

Further, it was found that 2-phenyl-6-methylpyrimidone-4 is converted into 3-methylpyrazolone-5, and that 2-methylmercapto-6-methylpyrimidone-4, via the 2-hydrazino compound, is also converted into the same compound[447]. In all these ring contractions, an initial attack of the hydrazine on the polarized

N_1-C_6 linkage of the pyrimidine ring can be considered to take place, yielding the dihydro compound (4.598). Fission of the N_1-C_6 bond gives the open-chain hydrazino intermediate (4.599), which, by internal addition across the C=N bond, leads to ring closure into the dihydropyrazole (4.600). Subsequently, aromatization by loss of formamidine (or formamide) yields (4.596). It has definitely been proved that it is carbon atom 2 which is expelled during these ring contractions[446,447]. From this mechanism, the rate-retarding effect of the C-methyl groups and the rate enhancement of the N-methyl groups can easily be understood. It is proposed[446] that in the reactions with 4-methylpyrimidine primary addition takes place both on the N_1-C_6 and on the N_3-C_4 linkage.

The formation of pyrazole has also been reported when 1,2-dihydro-2-imino-1-methylpyrimidine is treated with hydrazine[448]. In this case an initial attack at the 6-position has been postulated.

(4.598) (4.599)

(4.600) (4.596)

Very recently it was found that when 4,6-diethoxy-, 4,6-dichloro- and 4,6-dithiomethylpyrimidine are treated with aqueous hydrazine at ~200°C, 3-methyl-1,2,4-triazole (4.602) is formed[445,449]. It has been proved that in all these reactions 4,6-dihydrazinopyrimidine (4.601) is the key intermediate. The fact that 3-ethyl-1,2,4-triazole is obtained from 5-methyl-4,6-diethoxy-pyrimidine, and 3,5-dimethyl-1,2,4-triazole from 2-methyl-4,6-diethoxy-pyrimidine indicates that it is the $C_2-N_3-C_4-C_5$ fragment which is divided from the N_1-C_6 part of the pyrimidine ring and that is instrumental in the formation of the 1,2,4-triazole. It has been found that 4,6-dihydrazino-pyrimidine can be converted into the 1,2,4-triazole (4.602) simply by heating in water; this indicates that the nucleophile water is also able to bring about the conversion of (4.601) into (4.602). The mechanism suggested for this ring contraction is quite different from that in the pyrazole formations previously described. It involves addition of the nucleophile in position 2, ring opening into the open-chain hydrazino compound (4.603), followed by ring closure into the dihydro-1,2,4-triazole (4.604). Electron shift via the cyclic transition state (4.605), with subsequent prototropy, leads to the formation of the product (4.602).

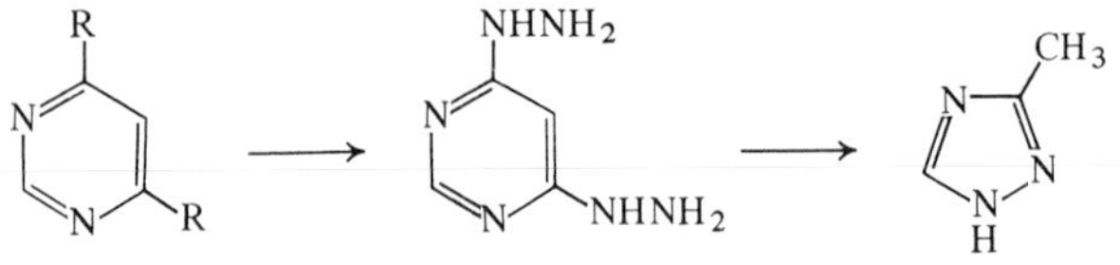

(4.601)   (4.602)

R = Cl; OC$_2$H$_5$; SCH$_3$

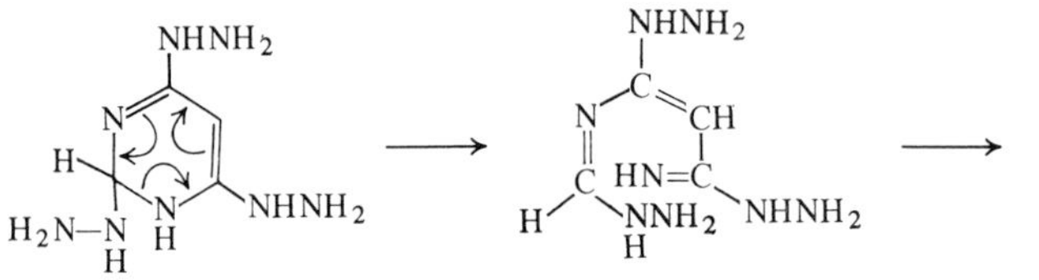

(4.603)

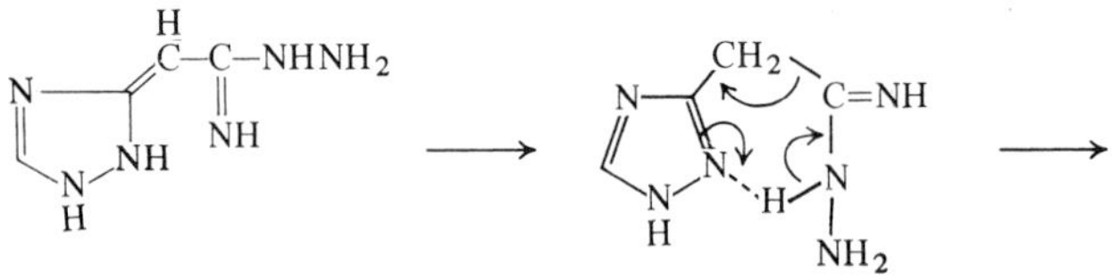

(4.604)   (4.605)

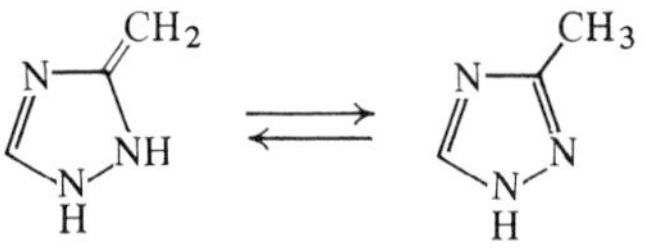

(4.602)

Another type of pyrimidine ring fragmentation into a N_3–C_4–C_5–C_6 and N_1–C_2 part has recently been discovered; it occurs during the hydrazinolysis of 4-ethoxy- and 4-chloropyrimidine into 3-aminopyrazole (4.607) (R_2 = H)[450] and 4-ethoxy-5-nitropyrimidine into 3-amino-4-nitropyrazole (4.607) (R_2 = NO$_2$)[451]. In these reactions also the corresponding 4-hydrazinopyrimidine (4.606) is the compound which is primarily formed and which undergoes the ring contraction. Since it has been found[450] that 4-ethoxy-6-phenylpyrimidine (4.608) is converted into 3-amino-5-phenylpyrazole (4.609) and the 4-ethoxy-2-

phenylpyrimidine (4.610) is converted into 3-aminopyrazole (4.611), it is evident (if we assume that a phenyl group largely blocks the attack of a nucleophile on that position) that in these compounds and probably in 4-hydrazinopyrimidine (4.606) (R_2 = H) and in 4-hydrazino-5-nitropyrimidine (4.606) (R_2 = NO_2) also, both the 2- and 6-position are vulnerable to nucleophilic attack. From the data so far available the mechanism of these ring contractions can be considered to start by a primary attack of the nucleophile on either of the positions 4 and/or 6 and to follow the route indicated on p. 126.

The high susceptibility of the pyrimidine ring to nucleophilic attack can further be demonstrated by the recent finding that hydroxylamine hydrochloride can convert 4,6-dimethylpyrimidine (4.612) (R_1 = R_2 = CH_3) and 4-methyl-6-phenylpyrimidine (4.612) (R_1 = CH_3, R_2 = C_6H_5) into 3,5-dimethylisoxazole (4.613) (R_1 = R_2 = CH_3) and 3-methyl-5-phenylisoxazole (4.613) (R_1 = CH_3, R_2 = C_6H_5) respectively[452].

Interestingly, it has been found that refluxing of the pyrimidine-N-oxide (4.614) with aqueous acid also gives hydrolytic ring contraction into the isoxazole derivatives (4.613)[453]. This reaction is described as an attack by water on the 2-position of the conjugate acid of the N-oxide. It has further been reported that arylpyrimidines by use of zinc and aqueous acetic acid are converted into arylpyrroles[453a]. In the mechanism of this ring interconversion the site of initial reduction attack by zinc is believed to be at the 1,6- rather than the 2,3-bond.

There have been a number of publications providing a rather broad coverage of the ring contractions observed during the reaction of derivatives of the cyclic amides, dialuric acid, alloxan and 5-methyl-, 5-bromo- and 5-aminobarbituric acid with base. Dialuric acid (= 5-hydroxybarbituric acid) (4.615) (R = H), on treatment with an aqueous base[454-456] yields oxazolidine-2,4-dione (4.616) (R = H) and not as originally proposed, the four-membered tartronimides[457]. The mono N-alkyl (aryl) derivatives of (4.615) give the 5-substituted oxazolidine-2,4 diones (4.616) (R = alkyl, aryl) while N,N unsymmetrically disubstituted derivatives (4.617) can yield a mixture of both N-substituted compounds (4.618) and (4.619)[456,458]. Thus, for example, 5-benzyl-1-methyl-3-phenyl-dialuric acid gives a mixture of 5-benzyl-3-methyl- (15 per cent) and 5-benzyl-3-phenyloxazolidine-2,4-dione (55 per cent)[459]. If water which is just boiling is used instead of an alkaline solution for the performance of the ring contraction of the N,N-dialkyl (aryl) dialuric acids, the yields are generally high, since the N-alkyloxazolidine-2,4-diones formed are more stable in water than in an alkaline medium. In all the base-catalysed ring contractions of dialuric acids into oxazolidine-2,4-diones, a carbamoyl derivative (4.620) and/or (4.621) is postulated as an intermediate formed by an initial attack of the 5-hydroxy group on the 2-position with subsequent ring fission of the 1,2-bond of the dialuric acid (route a or b). Support for this intermediate is the fact that in the reaction of

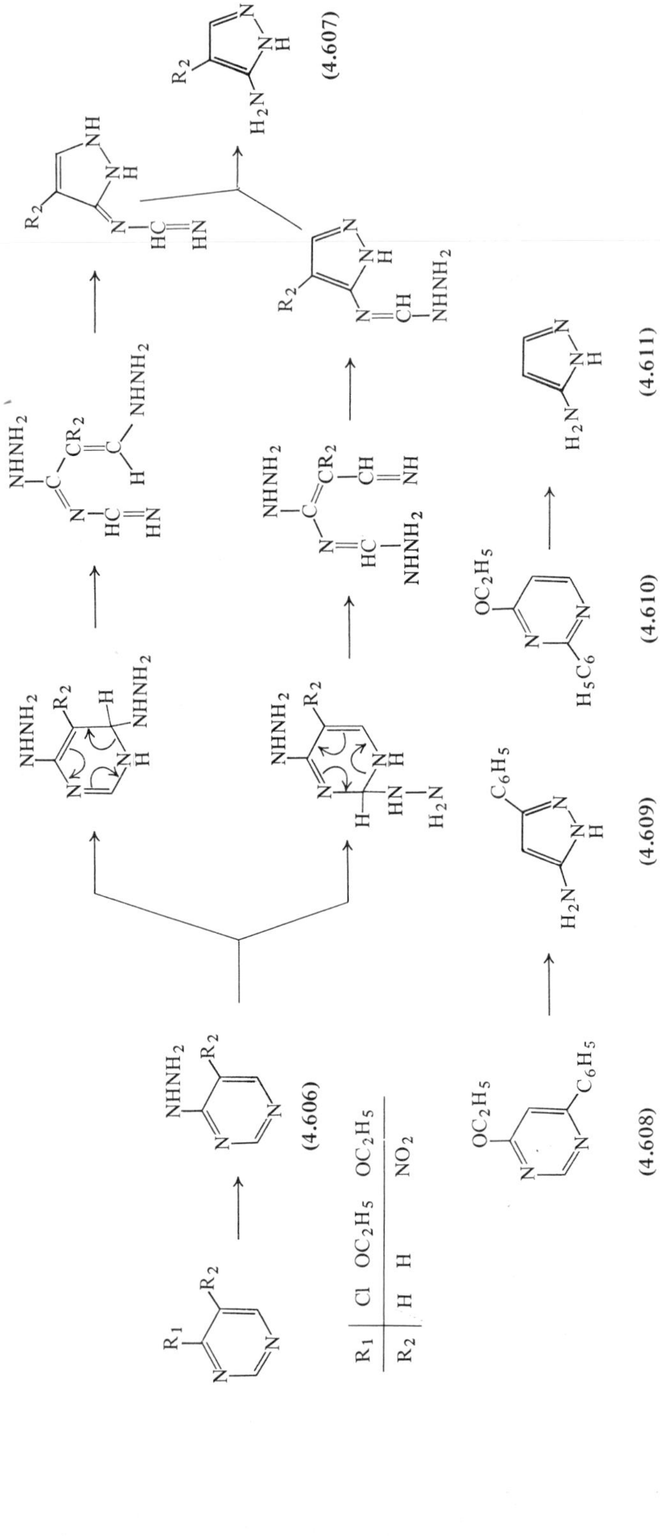

(4.607)

(4.606)

(4.608)

(4.609)

(4.610)

(4.611)

R₁	Cl	OC₂H₅	OC₂H₅
R₂	H	H	NO₂

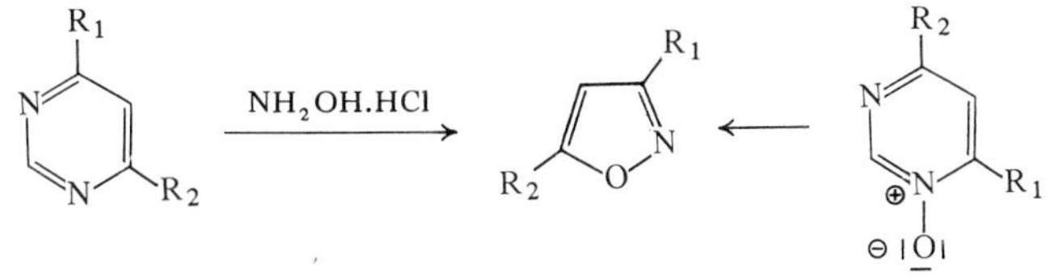

(4.612)　　　　　　　(4.613)　　　　　　(4.614)

R_1	CH_3	CH_3
R_2	CH_3	C_6H_5

R_1	H	CH_3	CH_3
R_2	C_6H_5	C_6H_5	CH_3

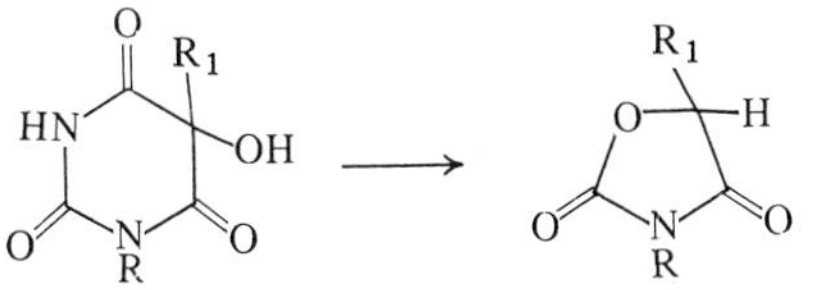

(4.615)　　　　　　　(4.616)

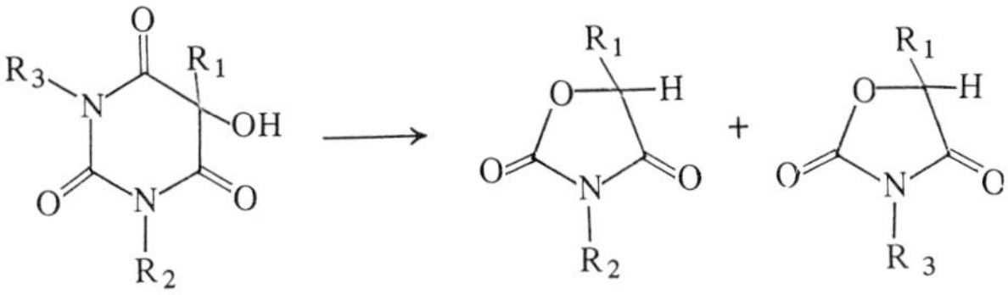

(4.617)　　　　　(4.618)　　　　　(4.619)

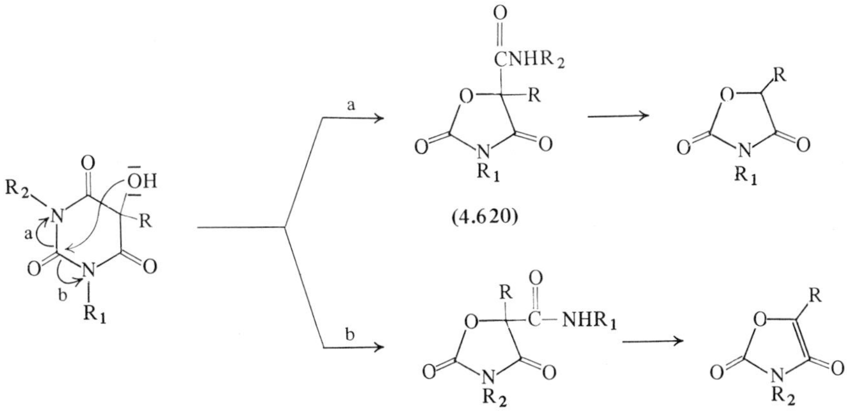

(4.620)

(4.621)

5-benzyl-1-methyl-3-phenyldialuric acid with boiling water, a similar carbamoyl derivative has indeed been isolated[459].

Carbamoyl derivatives of oxazolidine-2,4-dione are also formed if an ethanolic solution of the methylamine salts of 1,3,5-trimethylbarbituric acid (4.622) is boiled with air passing through[462]. This salt probably undergoes an initial oxidation into the 5-hydroxy compound (4.623) which is converted into (4.624) by the mechanism discussed above.

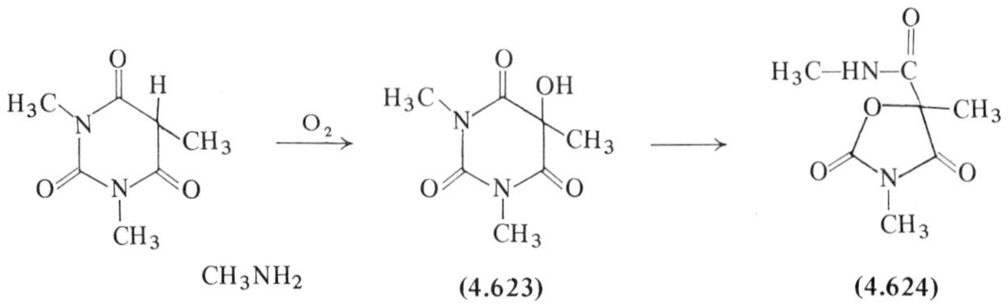

(4.622)   (4.623)   (4.624)

Treatment of alloxan or its hydrate (4.625) with alkali gives ring contraction into the salt of alloxanic acid (4.628)[460-464]. The reaction can most conveniently be performed with aqueous barium hydroxide. There has been some dispute regarding the mechanism of this ring contraction reaction. It is proposed that either a primary attack of the nucleophile takes place on carbon atom 4, yielding (4.626), which subsequently rearranges by a carbon-nitrogen bond migration (route a), or that a base-catalysed proton abstraction from the hydroxy group on position 5 occurs giving (4.627), which, in a benzilic acid rearrangement involving a carbon-carbon shift, is transformed into (4.628)

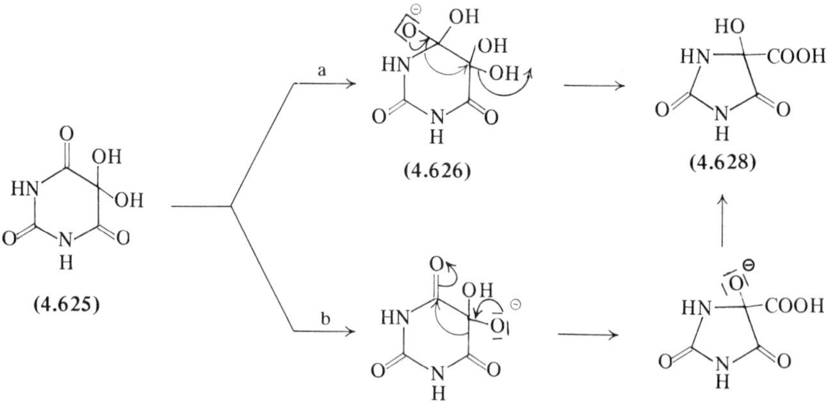

(4.625)   (4.626)   (4.628)

(4.627)

(route *b*). A ^{14}C-labelling experiment has unambiguously proved[462] that route *a* is the preferred route in the pH region 7-13: from 5-^{14}C alloxan hydrate (4.629) an alloxanic acid is obtained in which the ring carbon attached to the carboxyl group is radioactive. It is clear that route *b* would give a labelled alloxanic acid with its radioactivity in the carboxyl group. The results of a study of the kinetics of this ring contraction also support the occurrence of this carbon-nitrogen shift[465].

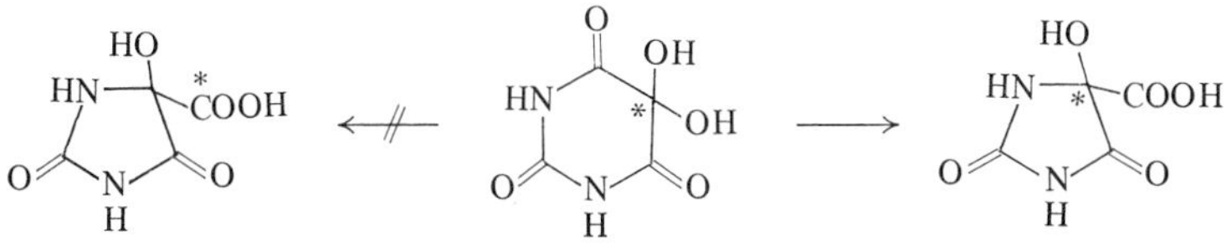

(4.629)

In good correlation with the results discussed above, it has been found that this ring contraction also occurs[463,464] in the presence of secondary amines (morpholine, piperidine, pyrrolidine, dimethylamine). The products obtained, however, are the salts (4.631) and not, as originally proposed, the monohydrate

(4.630)   (4.631)

of alloxan acid amides (4.630). Treatment of anhydrous alloxan with sodium ethoxide in ethanol, and acidification of the reaction mixture with hydrochloric acid, gives ring contraction into ethyl alloxanate[464]. It has been proposed that the benzilic-acid-type rearrangement (route *b*) occurs in this reaction.

5-Bromobarbituric acid and many of its derivatives give, when treated with aqueous or alcoholic alkali, a mixture of derivatives of both iminooxazolidin-4-one and hydantoin[454-456,466,467]. The N-alkyl(aryl)5-bromobarbituric acids show similar behaviour. For example, 5-bromo-1,5-diphenylbarbituric acid (4.632) (R = C_6H_5) gives a mixture of 5-phenyl-2-(phenylimino)oxazolidin-4-one (4.633) (R = C_6H_5) (45 per cent) and 1,5-diphenylhydantoin (4.634) (R = C_6H_5) (21 per cent). A bromoacylurea has been isolated as a side-product and it

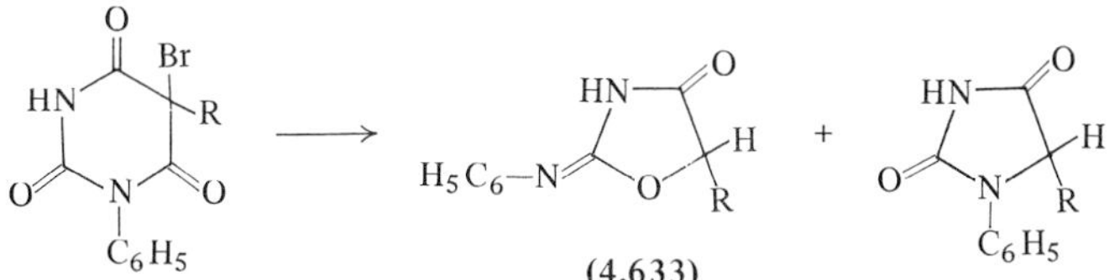

(4.632)   (4.633)   (4.634)

is proposed that this is the probable intermediate in these reactions. Similar reactions have been reported to occur with 5-benzyl, 5-ethyl and 5-methyl derivative of 5-bromo-1-phenylbarbituric acid, yielding mixtures of (4.633) and (4.634)[456]. With the *N,N*-dialkyl 5-bromobarbituric acids hydantoins are obtained exclusively; when the 1- and 3-substituents are different, a mixture of both isomeric hydantoins is formed. An interesting example of a ring contraction of a barbituric acid derivative into a hydantoin derivative by alkali is reported[468] with the 5-(arylamino) derivative 1,2,3,4-tetrahydro-4-methylquinoxalino-2-spiro-5-(hexahydro-2,4,6-triketopyrimidine).

Related ring contractions which deserve attention are the conversions of 1,3,10-trimethylflavinium perchlorate (4.636) when treated with nucleophilic

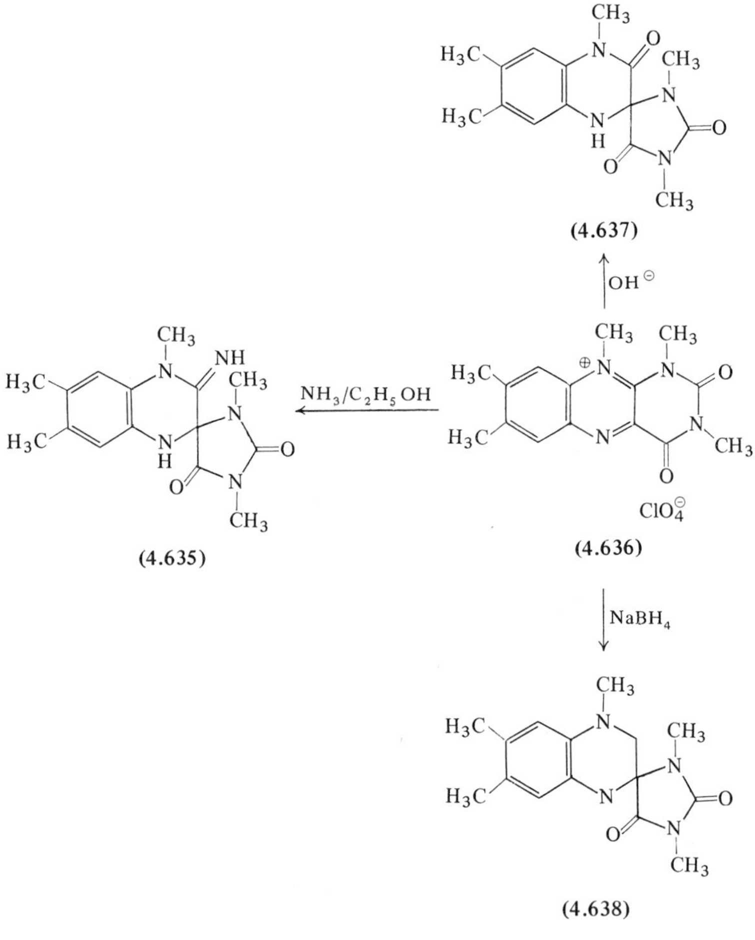

(4.637)

(4.635)

(4.636)

(4.638)

reagents (ammonia in ethanol, alkali or sodium borohydride) into the spiro-hydantoins (4.635), (4.637) and (4.638) respectively[469]. In these reactions, the reaction pathway can be described to occur by an initial nucleophilic (for example OH⁻) attack on the highly activated azomethine bond at the positive nitrogen (4.639). Ring opening gives the quinoxalone (4.640) and subsequent ring closure yields (4.637). The assumption of a quinazolone intermediate like (4.640) does not seem unreasonable since it is known that a compound like (4.641) can certainly be recyclized in alkaline solution into the spiro compound (4.642).

(4.636) ⟶

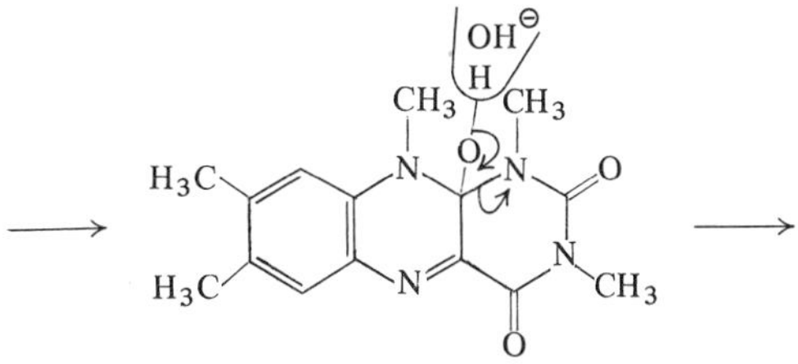

(4.639)

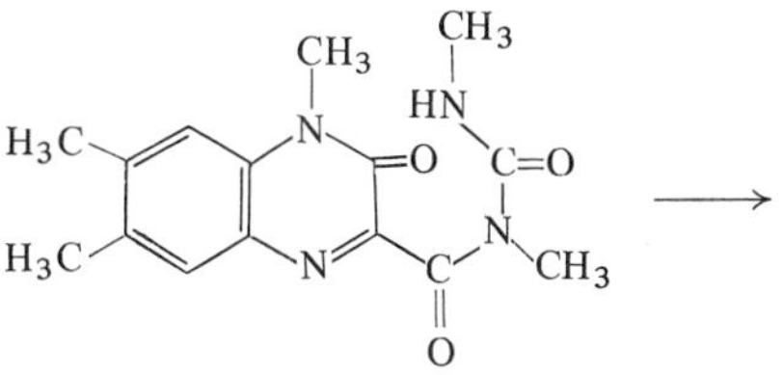

⟶

(4.637)

(4.640)

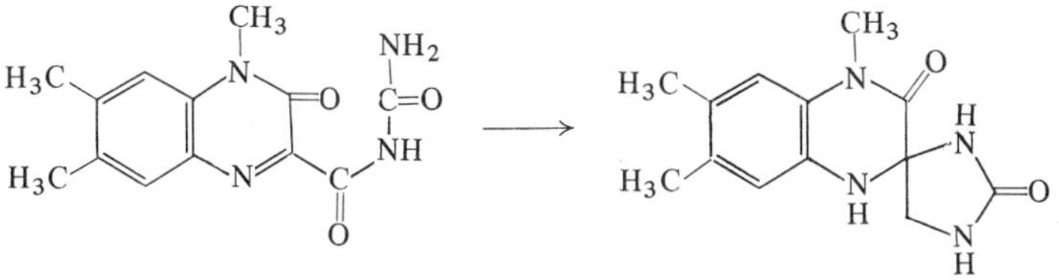

(4.641) ⟶ (4.642)

Another interesting example of a hitherto unreported ring contraction in the nucleoside area is the conversion of D-arabinofuranosyl-5-bromouracil (4.643) with 0.1 N sodium hydroxide solution into 1-(β-D-arabinofuranosyl)-2-oxo-Δ⁴-imidazoline-4-carboxylic acid (4.644)[470]. The same compound (4.644) is also obtained from the 5-fluoro compound (4.647) by prolonged treatment with alkali. When this reaction was carried out for a short duration, the open-chain compound (4.646) was obtained (the fluorine ion is removed more slowly than the bromide ion) indicating that in these ring contractions the transient intermediates are probably the open-chain compounds of the type (4.646) and the dihydroimidazole (4.645).

Ring contraction of the 1,4,5,6-tetrahydropyrimidine derivative (4.648)–formed by the reaction of 2-sulphanylamidopyrimidine with bromine in

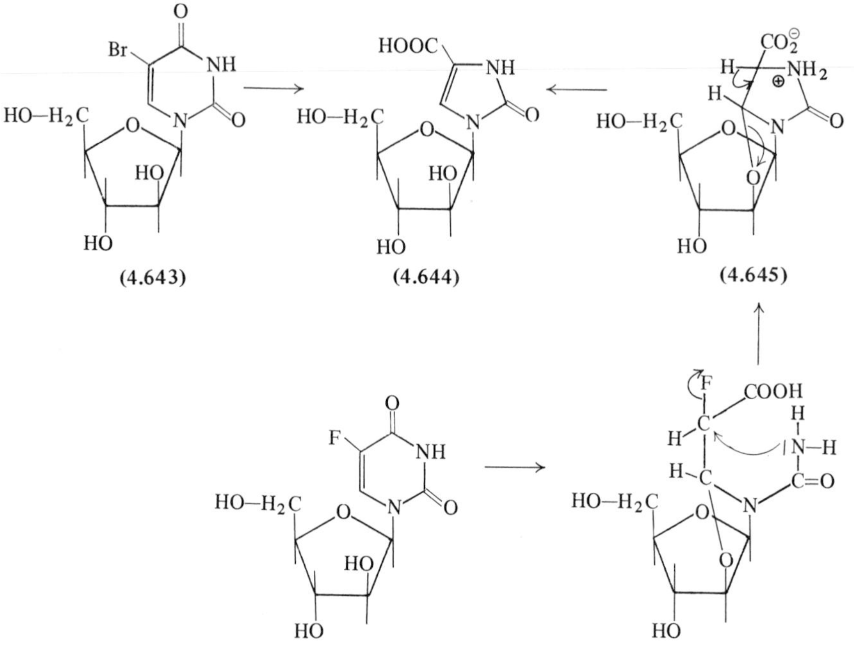

(4.643)   (4.644)   (4.645)

(4.647)   (4.646)

methanol–into the imidazole (4.649) is reported to occur during treatment with alkali[471,472]. Very little is known about the mechanism. But one thing is certain: 4-formylimidazole (4.650) is not an intermediate in these reactions. Corresponding ring contractions were also reported with the hexahydro-pyrimidines (4.650a); they are converted into the dihydroimidazoles (4.650b) on treatment with sodium methoxide[472a,472b].

Very recently, it was found that irradiation of the quinazoline-1-oxide (4.651) gives the N_1-acetylindazole (4.653)[473]. This has been explained by the

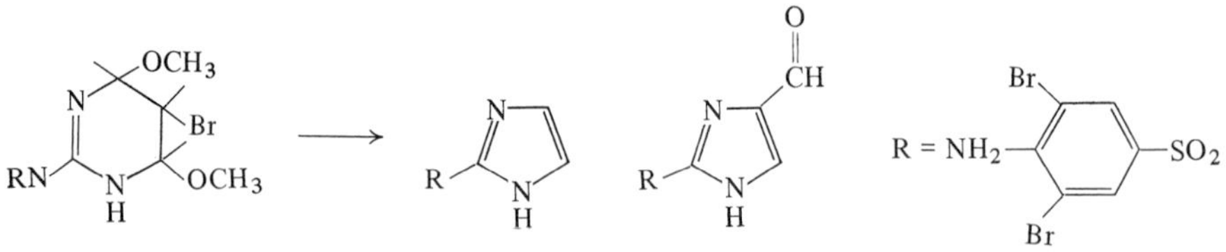

(4.648)   (4.649)   (4.650)

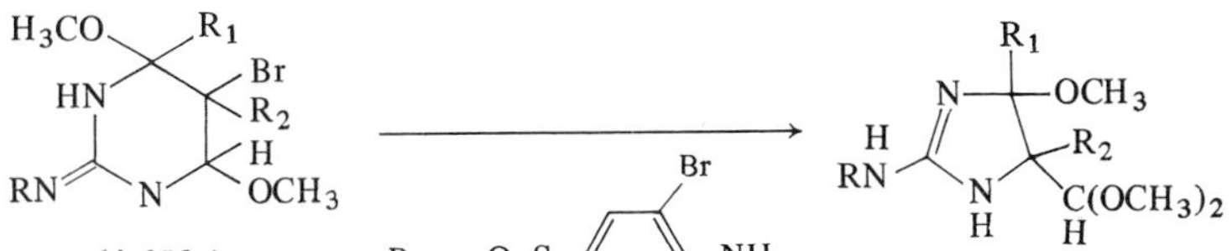

(4.650a)

R = — O₂S—⟨ ⟩—NH₂

(4.650b)

R₁	CH₂Br	H
R₂	H	CH₃

R₁	CH₂OCH₃	H
R₂	H	CH₃

intermediate formation of the seven-membered compound (4.652); this hypothesis is supported by the fact that the compound (4.654) indeed gives, on irradiation, an indazole (4.655). For similar conversions in the quinoline-N-oxide field, we refer to section I.C.1.c where the mechanism of these related conversions is discussed in detail. Irradiation of 5-methylpyrimidine-N-oxide (4.656) in benzene at 280 mμ does not lead to a ring-contracted product but to the appearance of the photo isomer, the conjugated nitrile (4.660)[474].

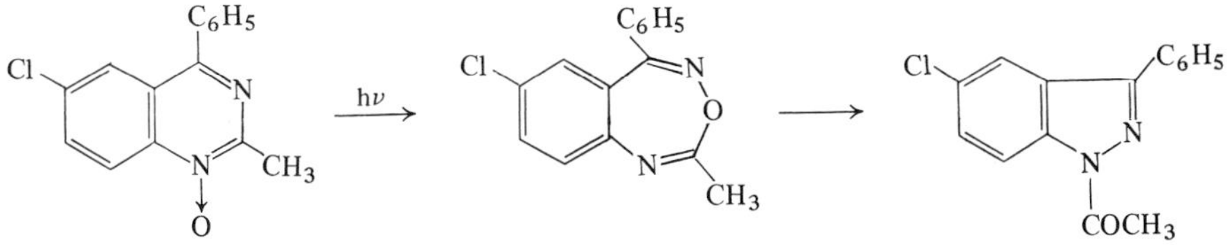

(4.651)　　　　　　(4.652)　　　　　　(4.653)

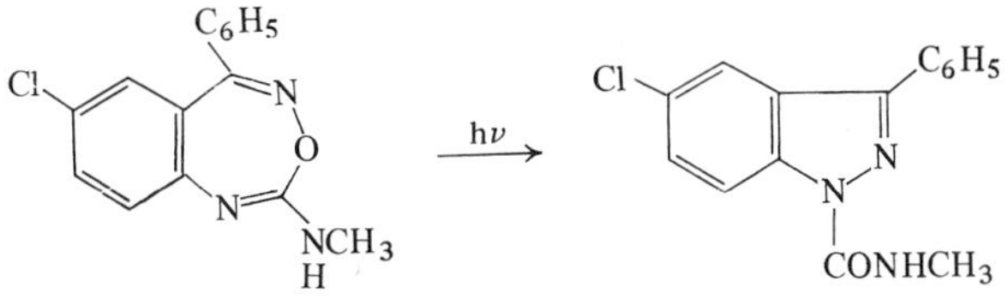

(4.654)　　　　　　(4.655)

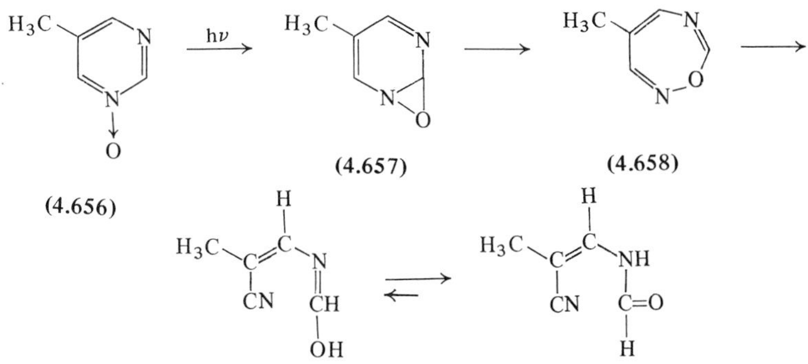

(4.656)　　(4.657)　　(4.658)

(4.659)　　(4.660)

The same series of steps is involved; a 1,3-photocyclization into the oxaziridino compound (4.657), valence tautomerization to form the oxadiazepine (4.658), which shows N—O rupture and a concomitant hydrogen shift into (4.659), which is in keto-enol tautomeric equilibrium with (4.660). Very recently, the photo-induced ring contraction of pyrimidine-N-oxides into imidazoles has been found [464a].

2. RING TRANSFORMATIONS OF PYRIMIDINES INTO OTHER SIX-MEMBERED HETEROCYCLES

a. *Pyrimidines into pyridines*

In section I.C.2.a the conversion of pyridines into pyrimidines has been extensively discussed. Until very recently, however, there were no examples of the reverse reaction, that is of pyrimidines into pyridines. Now it has surprisingly been found[475] that when the 5-aryl-2-methoxypyrimidine (4.661) is heated with ethanolic ammonia the 3,5-diarylpyridines (4.662) are formed (as well as the corresponding 2-aminopyrimidines (4.663)). Methyl-aminolysis of (4.661) gave no ring conversion into pyridine derivatives. A quite similar reaction was found to occur when pyrimidine was heated with methylamine and water at about 180-200°C, yielding 2-methyl-5-ethylpyridine (4.664)[476]! The reaction mechanism is obscure in both cases; it seems, however, reasonable to assume that in both ring conversions, a degradation of the pyrimidine ring into a C_2-fragment = (aryl)acetaldehyde (4.665) takes place and this reacts with ammonia or methylamine (probably in an aldol-type reaction) to form a pyridine derivative[477,478].

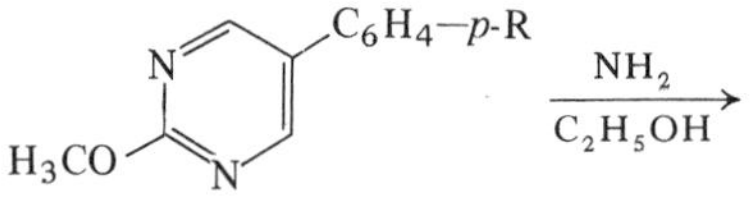

(4.661)

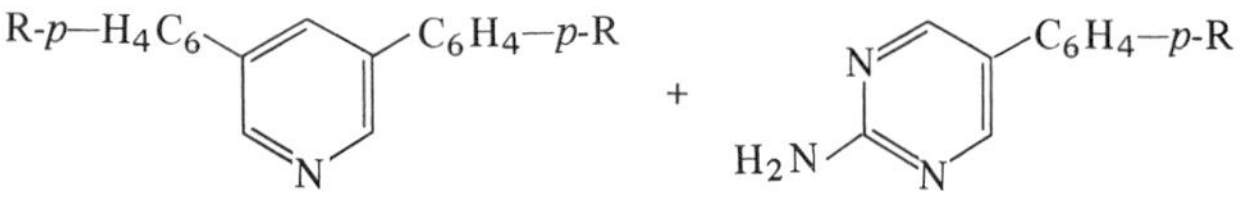

(4.662)   +   (4.663)

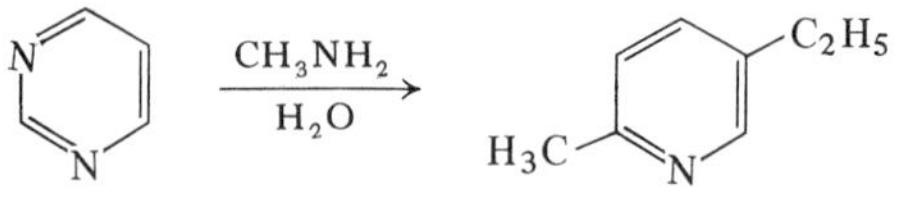

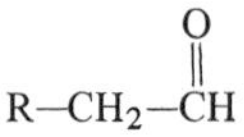

(4.664)   (4.665)

$(R = H; C_6H_4-p-R)$

b. *Pyrimidines into s-triazines*

An extensive investigation of the formation of didehydropyrimidines from 4-
and 5-halogenopyrimidines by treatment with potassium amide in liquid
ammonia at low temperatures[317,479-481] has revealed that these compounds
frequently undergo carbon skeleton rearrangements[482,483] in this reaction
medium. When the 2-substituted 4-chloropyrimidines (4.666) are treated with
potassium amide in liquid ammonia at −33°C, it appears that only small
amounts of the corresponding 4-amino compounds (4.668) are
obtained[482,483]. The main products of the reaction are 2-substituted 4-methyl
s-triazines (4.667). These ring transformations have been found to be of a very
general nature; they are also of importance from a preparative point of view
since unsymmetrically substituted s-triazines, which are difficult to produce by
other methods, can readily be obtained.

(4.666)   (4.667)   (4.668)

$R = CH_3; \ C_2H_5; \ C_6H_5; \ N(CH_3)_2;$ $; \ N-(C_6H_5)$
                                                            $|$
                                                            CH_3

When an alkyl or aryl substituent is present in position 2, the yields are up
to about 40 per cent; with a dialkylamino, piperidino or morpholino
substituent, the yield increases to about 80-90 per cent. The reaction with the
corresponding 4-bromo- and 4-iodopyrimidine gives almost identical results. The
amount of the corresponding 4-aminopyrimidines is increased in the series J >
Br > Cl. Further, the 2,5-dimethyl-4-chloropyrimidine (4.669) was found to be
convertible into the s-triazine derivative (4.670), which can be also obtained
from 4-chloro-2-ethylpyrimidine (4.671). Much work has been devoted to
elucidating the mechanism of these reactions. Because (4.670) is formed from

(4.669)   (4.670)   (4.671)

(4.669), it is clear[482] that in this ring transformation no hetaryne is involved,
since the presence of a methyl group on position 5 prevented the formation of
4,5-didehydropyrimidine as intermediate. These results further show that it is

the C_5 carbon atom of the pyrimidine ring which takes part in the formation of the methyl group (or ethyl group) attached to the s-triazine ring. An answer to the important question of which carbon-carbon bond is broken during the ring transformation that is the C_5-C_6 or the C_4-C_5 bond, has been provided by a study of the reaction with ^{14}C-labelled pyrimidines[484,485]. When 4-^{14}C-4-chloro-2-phenylpyrimidine (4.672) was treated with potassium amide in liquid ammonia at $-33°$C all the radioactivity in the radioactive 4-methyl-2-phenyl-s-triazine (4.673) formed was contained in the carbon atom of the s-triazine nucleus attached to the methyl group. This result proves unequivocally that the ring transformation takes place by fission of the C_5-C_6 bond (fission a) in the pyrimidine ring and not of the C_4-C_5 bond (fission b).

(4.672)   (4.673)

A study of the reaction of potassium amide in liquid ammonia with 4-chloro-5-deuterio-2-phenylpyrimidine (4.674) has revealed[486] that in the s-triazine (4.675) formed no deuterium is present. Since under these conditions no D/H exchange was observed in the starting substance (4.674) and in 4-deuteriomethyl-2-phenyl-s-triazine (4.676) (which was independently prepared by treatment of 4-methyl-2-phenyl-s-triazine with sodium methoxide in deuterated methanol) there was only about a 20 per cent exchange in the side-chain, it appears that an intermediate in which all deuterium is rapidly exchanged must be present during the ring transformation. An acetylenic intermediate (4.678) is proposed because it is in equilibrium with the ketenimine (4.679) or with its anionic form (4.680) and/or (4.681).

(4.674)   (4.675)   (4.676)

All the results obtained so far seem feasible when it is postulated that the initial attack of the amide ion takes place at position 6, that is at the position not occupied by the halogen atom[482,484,486]. (Such an unusual nucleophilic attack is also thought to occur in the conversion of 2,6-dibromopyridine into 4-amino-2-methylpyrimidine (see section I.C.2.a)). The resonance-stabilized ion (4.677) can either be protonated and subsequently converted into the

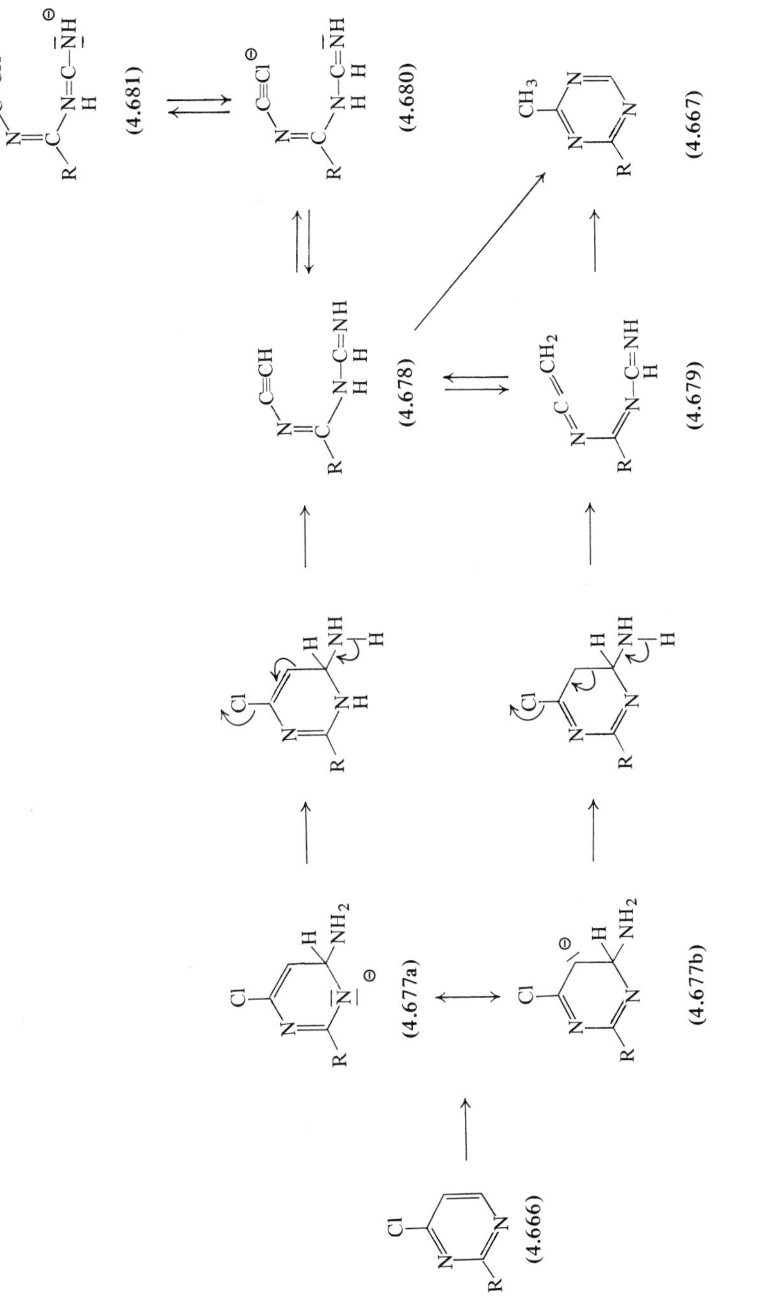
(4.666)
(4.677a)
(4.677b)
(4.678)
(4.679)
(4.680)
(4.681)
(4.667)

ketenimine (4.679) or it can be transformed immediately into the acetylene (4.678). Just as in the conversion of 2,6-dibromopyridine and 2-bromoquinoline into 4-amino-2-methylpyrimidine and 2-methylquinazoline respectively, no open-chain intermediates can be isolated in the conversion of (4.666) into (4.667). However, indication for the occurrence of open-chain intermediates may be taken from the fact that the yield of the conversion of 4-chloro-2-methylpyrimidine into 2,4-dimethyl-s-triazine is nearly doubled when the reaction is carried out in very dilute solutions[482]: the intermediates formed in these dilute solutions are less able to polymerize with other particles and therefore more inclined to react in a different manner, that is, an intramolecular ring closure.

Investigation of the behaviour of the pyrimidine ring towards potassium amide, when groups of $-I$ and $-M$ character ($CO_2C_2H_5$, $CONH_2$, $COOH$, CN) are introduced into position 5 in 4-chloro-2-phenylpyrimidine (4.682) has revealed that complex reaction mixtures are formed from which in all cases 4-(methylsubstituted)-2-phenyl-s-triazines (4.683) can be isolated together with the corresponding 4-aminopyrimidines (4.684)[487]. In the case of R = $CO_2C_2H_5$, $CONH_2$ and CO_2H, it was found that, in addition to (4.683) and (4.684), 2-phenyl-4-methyl-s-triazine (4.685) and the 4-aminopyrimidine (4.686) are formed; the ratio of both these products is in all cases about (5 to 6):1. Since this ratio is about the same as reported for the products (4.667) ($R = C_6H_5$) and

(4.682)	(4.683)	(4.684)	(4.685)	(4.686)
R = $CO_2C_2H_5$	39%	12%*	~ 4%	~0.5-1%
R = $CONH_2$	40%	10%	16%	~ 3%
R = CO_2H	~1.5%	68%	6%	~ 1%
R = CN	15%	10%	—	—

* This 4-amino product consists of a mixture of 4% ($R = CO_2C_2H_5$) + 8% ($R = CONH_2$).

(4.668) ($R = C_6H_5$) obtained from 4-chloro-2-phenylpyrimidine (4.666) ($R = C_6H_5$), it is suggested that in this potassium amide–liquid ammonia system the compounds (4.682) ($R = CO_2H$, $CONH_2$, $CO_2C_2H_5$) lose the substituent from position 5, yielding 4-chloro-2-phenylpyrimidine. A mechanism according to which this removal can take place is indicated below; it is similar to the potassium amide catalysed removal of a benzoyl group from o-halobenzophenone[487a].

It is a striking fact that treatment with potassium amide of the compounds 5-ethoxy-, 5-methoxy-, 5-phenyl- or 5-thiomethyl-4-chloro-2-phenylpyrimidine (4.687) (all the groups on position 5 are characterized by a +*M*, −*I* effect), does not lead to the formation of substituted *s*-triazines, but gives the open-chain conjugated 1,3-diazapenta-2,4-dienes (4.688)[488,489]. When appropriately labelled [14]-C-pyrimidines have been used, and the position of ^{14}C is established in this open-chain intermediate, it has been proved that the formation of these compounds has taken place by an initial attack of the amide ion on position 6[489].

(4.687)   (4.688)

R = OCH$_3$; OC$_2$H$_5$; SC$_2$H$_5$; C$_6$H$_5$

A group of pyrimidines, which are also found to be easily converted into *s*-triazines, is the 4-amino-5-nitrosopyrimidines (4.689). On treatment with acetic anhydride, phosphoryl chloride, or benzenesulphochloride, these compounds are converted into the cyano-*s*-triazines (4.694)[490,491]. The yields vary, depending on the character of group R, from 30 per cent to 70 per cent: groups with a strong electron-donating character promote the reaction considerably and give the highest yields. The presence of a nitroso group on position 5 and an amino group on position 6 makes possible a tautomeric equilibrium, (4.689) ⇄ (4.690), in which the oxime form (4.690) can readily react with acetic anhydride, phosphoryl chloride or benzenesulphochloride (the reaction with the acetyl derivative is given in the reaction scheme). The presence of the acetate group, which is a better leaving group than the hydroxyl group, makes possible a cisoid elimination via (4.691) or a transoid elimination via (4.692), leading ultimately to the intermediate (4.693), which can recyclize into the amino-cyano-*s*-triazines (4.694). It has been mentioned that the presence of a strong

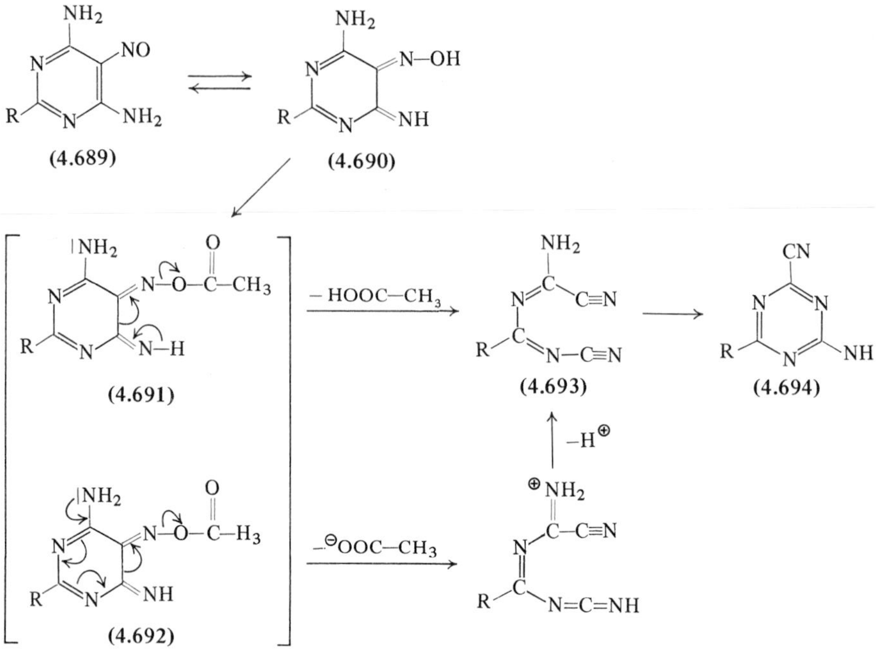

$$R = C_6H_5; p\text{-}OCH_3C_6H_4; \beta\text{-}C_5H_4N; SCH_3; N(CH_3)_2; NH_2$$

electron-donating group on position 2 promotes ring opening via a transoid elimination of the acetate ion. This transoid elimination is known to occur preferentially and at a higher rate than the cisoid elimination[490a].

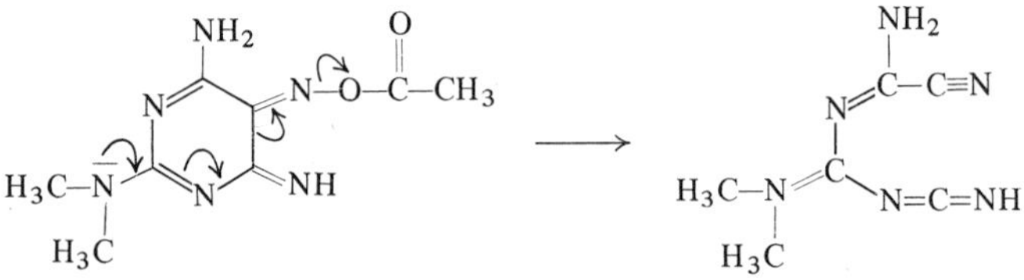

Oxidative ring interconversions of pyrimidines into s-triazines have been reported to occur when 6-substituted 2,4-diaminopyrimidines (4.694a) are treated with peracetic acid or trifluoroperacetic acid, the ammelide (4.694b) being obtained[491a]. No mechanism has been given. A thermally induced rearrangement of a pyrimidine derivative into s-triazine has been observed[491b] when 2-dimethylamino-4-acetamido-5-acetoxyimino-6-oxo-5,6-dihydro-pyrimidine (4.694c) is heated to just above its melting point, 2-dimethylamino-4-cyano-6-oxo-1,6-dihydro-s-triazine (4.694e) being formed. It has been suggested that in this last conversion, fragmentation occurs leading to the formation

of the acetate (4.694d) in which through a subsequent addition of the NH
group to the isocyanate, ring closure takes place yielding (4.694e).

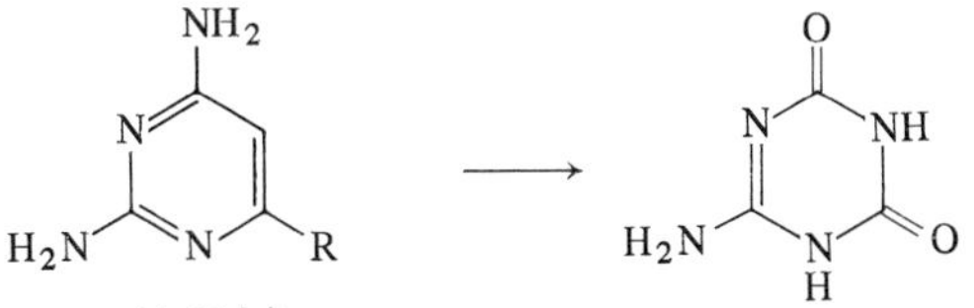

(4.694a)

R = Br, Cl; OH; NH$_2$; OC$_6$H$_5$

(4.694b)

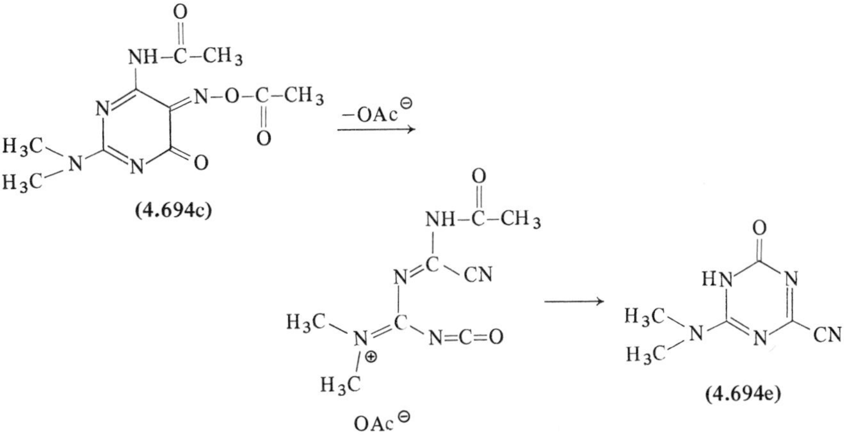

(4.694c)

(4.694d)

(4.694e)

3. RING EXPANSIONS OF QUINAZOLINES

a. *Quinazoline-3-oxides into 1,4-benzodiazepines and phenylbenz[f] 1,3,5-
oxadiazepines*

The ring expansions which have been reported to occur when the 2-chloro-
methyl- and 2,2-dichloromethylquinazoline-3-oxides are treated with various
nucleophilic reagents are important, since in these reactions the derivatives of
1,4-benzodiazepine-3-oxides—a class of compounds of great pharmacological
interest—are obtained in reasonable to good yields. 6-Chloro-2-chloromethyl-4-
phenylquinazoline-3-oxide (4.696) (X = Cl) or the quaternary salt (4.696) (X =
$^{\oplus}$N(CH$_3$)$_3$) reacts with ammonia or primary amines to form the ring-enlarged
product 2-amino- or 2-alkylamino-7-chloro-5-phenyl-3H-1,4-benzodiazepine-4-
oxide (4.697)[492-496] and not a corresponding quinazoline derivative.

$$\text{(4.695)} \qquad \xleftarrow{\text{HN}(C_2H_5)_2} \qquad \text{(4.696)} \qquad \xrightarrow{R-NH_2}$$

(4.695) structure: quinazoline-3-oxide with Cl, $CH_2-N(C_2H_5)_2$, C_6H_5

(4.696) structure: quinazoline-3-oxide with Cl, CH_2X, C_6H_5

(4.697) structure: 1,4-benzodiazepine with Cl, NH—R, H_5C_6

$$R = H;\ CH_3;\ n\text{-}C_4H_9;\ CH_2CH{=}CH_2;\ CH_2CH_2OH;\ CH_2CH_2OCH_3;\ CH_2CH_2NH_2$$

Because of the interesting pharmacological properties of these 1,4-benzo-diazepines, a thorough study has been made of the many aspects of these ring expansion reactions. It has been found that in the reaction of the 2-iodomethyl derivative (4.696) (X = I) with ammonia a normal replacement of the iodine by the amino group takes place almost quantitatively, whereas from the 2-bromo-methyl derivative (4.696) (X = Br) a quinazoline derivative as well as a 1,4-benzodiazepine derivative is formed[492]. Using secondary aliphatic amines normal replacement into (4.695) takes place; however with the secondary amines, dimethylamine and pyrrolidine, both products, a quinazoline- and a 1,4-benzodiazepine derivative, are obtained[497-499].

The effect of a great variety of substituents on the reaction course has been reported; these substituents include substituents in the phenyl ring[492,494,495], 2- or 4-pyridyl groups[500], 2-thienyl[495], cyclohexyl[495], substituents in the benzo ring (bromine[494], methyl[494,495], trifluoromethyl[501,502], nitro[503], methoxycarbonyl[504] and alkylthio[505]) and substituents in both the benzo- and phenyl rings. With electron-releasing groups in position 6 in (4.696), ring enlargement occurs concurrently with a normal replacement reaction; the presence of electron-releasing groups in both the 6- and 8-position, on the contrary, only gives formation of substituted quinazoline-3-oxides.

With alcoholic sodium hydroxide solution 6-chloro-2-chloromethyl-4-phenyl-quinazoline-3-oxide (4.696) (X = Cl) undergoes a similar ring expansion. The product obtained is the corresponding 7-chloro-1,2-dihydro-2-oxo-5-phenyl-3H-1,4-benzodiazepine-4-oxide (4.698)[506,507]. In a similar way, a large number of quinazolines containing different substituents in the benzo group[501,503,508], or being unsubstituted in position 4[509], have been converted into substituted 1,4-benzodiazepines by reaction with a sodium hydroxide solution.

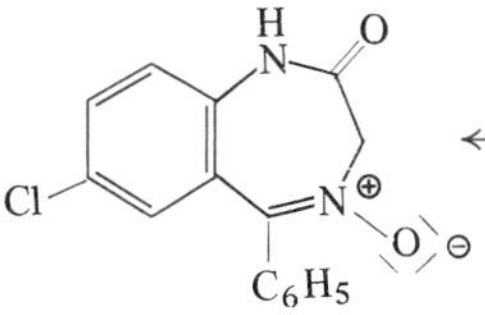

(4.698)   (4.696)   (4.699)

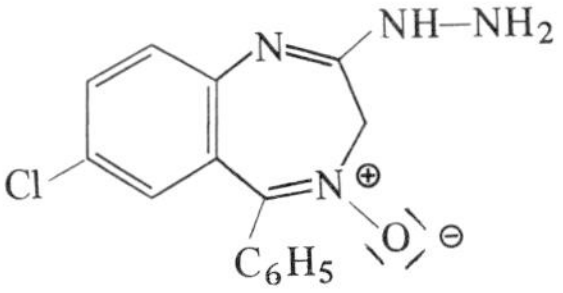

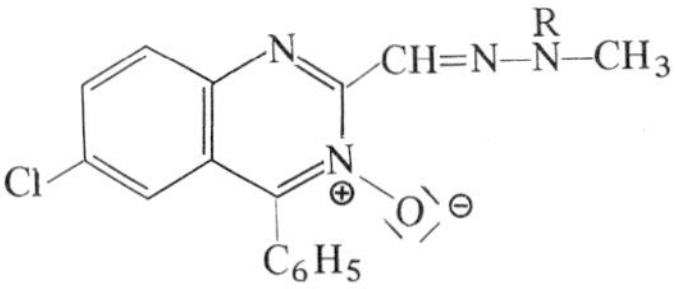

(4.700)

R = H; CH$_3$

With hydrazine, (4.696) undergoes ring expansion into the 2-hydrazino-1,4-benzodiazepine (4.699)[510]. Interestingly enough, with 1,1-dimethylhydrazine and methylhydrazine, no ring enlargement takes place, but an oxidation process indicated by the formation of the hydrazone (4.700)[510] occurs. A study of the mechanism of these ring expansion reactions has led to the proposal[511] that the initial attack of the nucleophile (illustrated, for example, by the OH$^-$ ion) takes place on position 2 of the quinazoline ring (4.701), for this position is known to be vulnerable to attack by nucleophilic reagents[507]. Ring opening leads to the chloroamidobenzene derivative (4.702) which is protonated to form the oxime (4.703). Ring closure yields, via (4.704), the 1,4-benzodiazepine (4.699). The fact that an independently prepared specimen of (4.703) readily cyclizes under identical conditions into the product (4.699)[507] clearly demonstrates that in

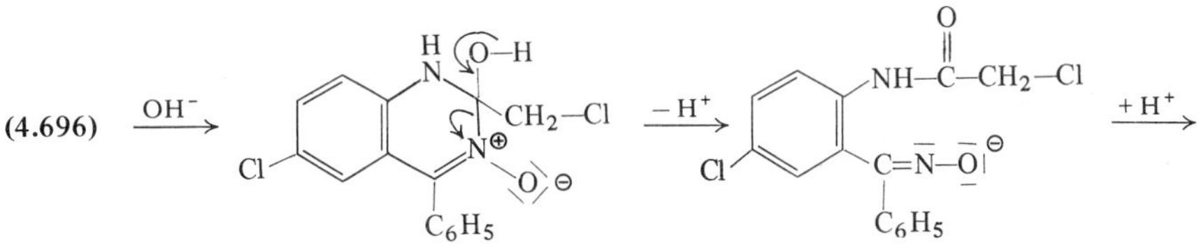

(4.696)   (4.701)   (4.702)

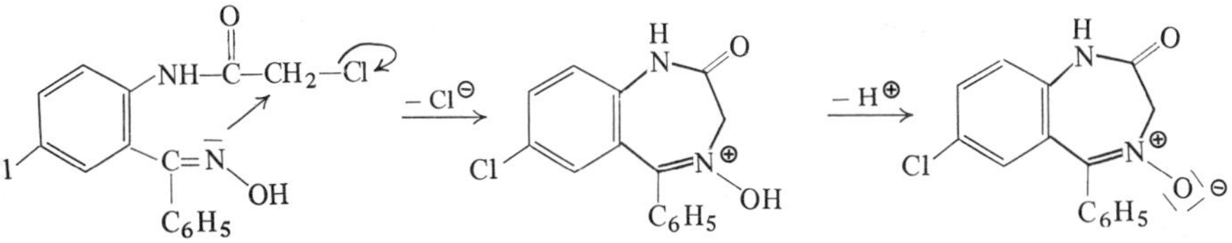

(4.703)   (4.704)   (4.699)

this sequence of reactions the *anti* oxime (4.703) is a probable intermediate. However, all attemps to isolate the *anti* oxime have failed: the rate of ring closure of the *anti* oxime apparently far exceeds that of the ring opening reaction[511] which occurs first. It was found that the 2,2-dichloromethyl-quinazoline-3-oxide (4.705) on treatment with alkali, also gives ring expansion[511]: with 2 equivalents of sodium hydroxide the seven-membered 1,4-benzodiazepine-4-oxide (4.706) was formed in nearly quantitative yield, while with 1 equivalent of sodium hydroxide a mixture of (4.706) (17 per cent) and an *anti* oxime (4.707) (48 per cent) was obtained. That (4.707) is inter-

mediate was clearly established from observations of the uv and ir spectra of the reaction mixture during the course of the ring expansion[511] as well as by the fact that further treatment of the *anti* oxime (4.707) with alkali gives (4.706)[511].

In correlation with uv induced ring expansions reported with quinoline- and isoquinoline-*N*-oxides, which carry a cyano or phenyl group (see section I.C.3.b), it has been found that irradiation of a solution of the 4-phenylquinazoline-3-oxide (4.708) in benzene also gives ring enlargement, yielding the 2-phenyl-benz[*f*] 1,3,5-oxadiazepines (4.710) (R = H, CH$_3$)[367,512,513]. It has been conclusively proved that a seven-membered compound and not the isomeric oxaziridine derivative (4.709) was obtained[512]: (1) by uv spectrometry (λ_{max} 330-340 mμ, in agreement in position and intensity with that reported for 2-phenylbenz[*f*] 1,3-oxazepine derivatives), (2) ir spectrometry (a strong absorption at 1640 cm^{-1} (two C = N functions)) and (3) chemical evidence based on the fact that solvolysis of the irradiation product obtained from (4.708) (R = R$_1$ = H) in boiling methanol easily gives formation of 2-phenylbenzoxazole (4.711), benzoxazole (4.712) and *o*-formamidophenol (4.713). Based on trapping experiments of the intermediary 1,2-epoxyquinolines by amines in the photolysis of 2-cyanoquinoline-*N*-oxides[271], and on the fact that very likely the photolysis of aromatic-*N*-oxides proceeds via oxaziridines, it seems that for the quinazoline-*N*-oxides similar intermediates (4.709) are probable. Irradiation of the quinazoline-1-oxide (4.714)[513] does not, as might be expected, give the

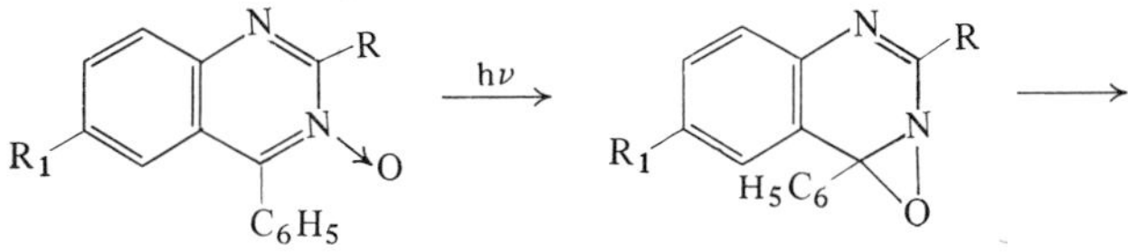

(4.708)   (4.709)

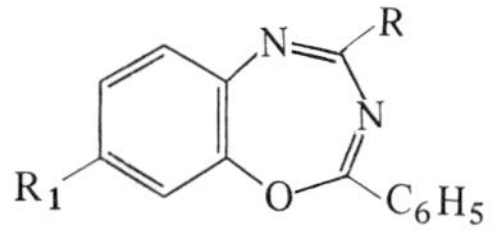

(4.711)

(4.710)

R	CH₃	C₆H₅	H
R₁	Cl	H	Cl

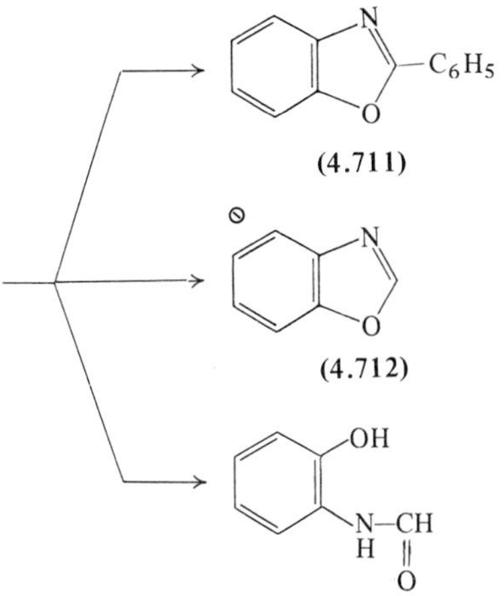

(4.712)

(4.713)

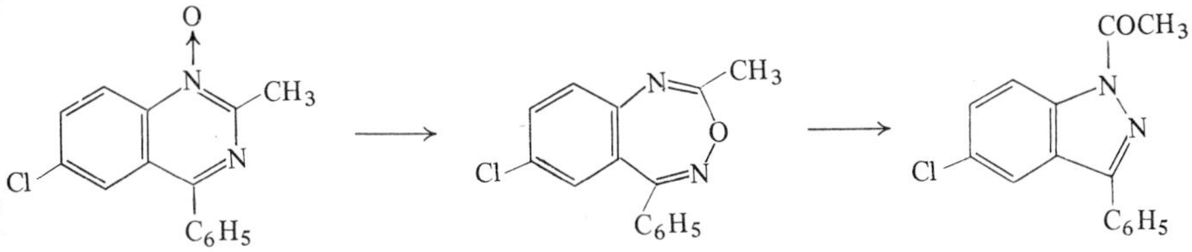

(4.714)   (4.715)   (4.716)

benzoxadiazepine (4.715), but the *N*-acetylindazole (4.716), which, as suggested, could be formed by photolytic decomposition of (4.715)[514]. It has been shown[513a] that, in diffuse daylight 2-(R-substituted)-6-chloro-4-phenyl-quinazoline-3-oxides (4.708) (R_1 = Cl; R = CH_2Cl, CH_2J, CH_3—OC—CH_2,

C_2H_5—S—CH_2, C_6H_5—S—CH_2, NC—S—CH_2, O⟨ ⟩N—CH_2, p-$CH_3C_6H_4$-

C_6H_5—O—CH_2, $NHCH_2$) easily decompose, as suggested by the partial disappearance of the long wavelength band and an increase of the extinction of the short wavelength. Attempts to isolate benzo-1,3,5-oxadiazepines as intermediates have failed, since these compounds are easily solvolysed, forming besides open-chain compounds. 5- and 6-membered heterocycles. In fact, in the case of

6-chloro-2-chloromethyl-4-phenylquinazoline-3-oxide   the   6-chloro-2-phenyl-benzoxazole was isolated.

In an attempt to obtain spectral evidence for the presence of stable, isolable intermediates in the photoisomerization of (4.708) (R = CH_3, R_1 = Cl) the uv spectra of the reaction mixture was measured at several intervals. Isobestic points were found at 289, 232 and 345 mμ, indicating the absence of stable intermediates[513].

D. Ring transformations of six-membered heterocycles containing two nitrogen atoms (pyridazines, pyrazines and their benzo derivatives)

1. RING CONTRACTIONS OF PYRIDAZINES, PYRAZINES, CINNOLINES, PHTHALAZINES AND QUINOXALINES

a. *Pyridazines into pyrazoles, pyrroles and triazoles*

When 1-phenyl-3,5-dichloro-6-(1H)-pyridazone (4.717) is treated with alkali, ring contraction into 3-hydroxy-1-phenylpyrazole-5-carboxylic acid (4.722) takes place[515,516]. It has been suggested that the chlorohydroxy compound (4.718) and the 3,5-dihydroxy compound (4.719) are intermediates in this ring contraction reaction[517,518]. By a series of reactions involving prototropy of (4.719) into the α-diketone (4.720), addition of a hydroxyl ion to the C=O at position 6 (4.721), ring contraction by the electron shift as indicated and dehydration, the desired product (4.722) is ultimately yielded. Although the mechanism is somewhat speculative, there is some evidence that this 3,5-dihydroxy derivative (4.719) is a key intermediate in this ring contraction. It appears that all 1-phenyl-6-(1H)-pyridazones bearing substituents in position 3 and 5, which are convertible either by alkali or by acid into the 3,5-dihydroxy-pyridazone, give under these conditions ring contraction into (4.722)[517]. Hence, when the 3,5-dialkoxy compound (4.723) is treated with acid (47 per cent hydrobromic acid is very useful), the 3-hydroxypyrazole (4.722) is obtained. The action of a base on this 3,5-dialkoxy compound (4.723) leads only to the formation of the 3-alkoxy-5-hydroxypyridazone (4.725). However, by subsequent treatment of (4.725) with acid the second alkoxy group is converted into a hydroxyl group forming the compound (4.719) which then by further action in acid similarly gives the 3-hydroxypyrazole-5-carboxylic acid (4.722). The same results were obtained with the compounds (4.724) and (4.726).

Further investigation of the scope of this ring contraction involving a study of the behaviour of 3,4-dihydroxy-, 3,4-dimethoxy-, 3,4-dichloro- and 3,4,5-tri-chloro-1-phenyl-6-(1H)-pyridazone revealed that these compounds were not inclined to undergo this ring contraction[517,518]. It is therefore reasonable to conclude that the formation of an α-diketone (4.720) is an essential requirement

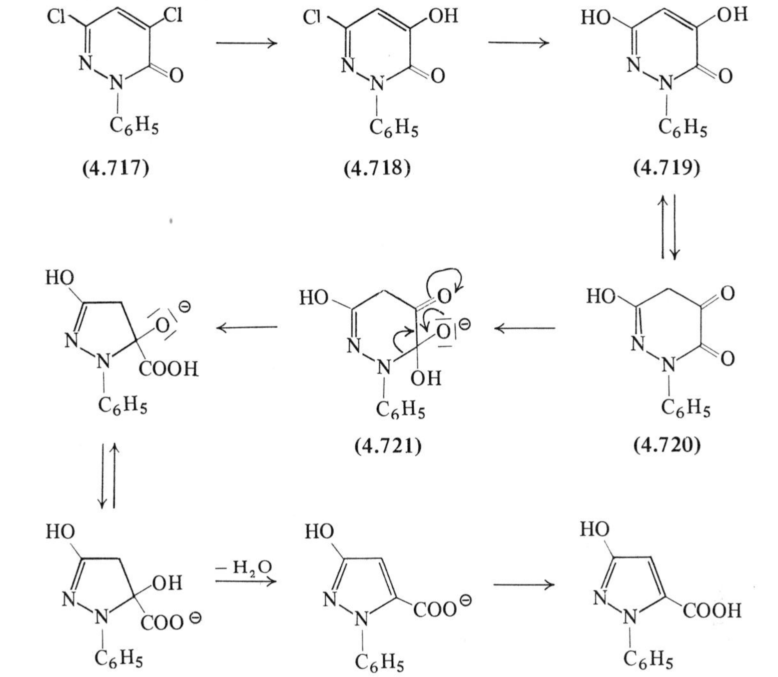

(4.717)   (4.718)   (4.719)

(4.721)   (4.720)

(4.722)

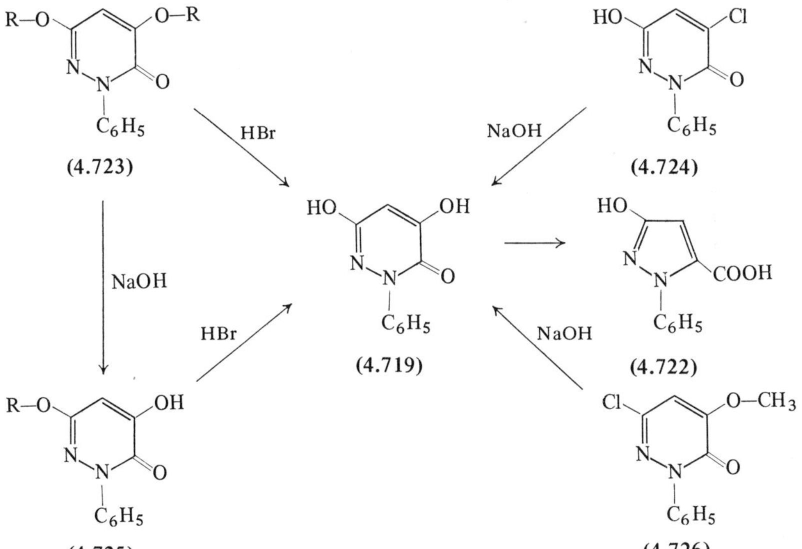

(4.723)   HBr   NaOH   (4.724)

NaOH   (4.719)   NaOH   (4.722)

HBr

(4.725)   (4.726)

R = CH₃; C₂H₅

for the occurrence of this ring contraction. This has been supported by the conversions of the 5-hydroxy- (4.727) (R = H) and 5-ethoxy-3-methyl-1-phenyl-6-(1H)-pyridazone (4.727) (R = C_2H_5) into the 1-phenyl-3-methylpyrazole-5-carboxylic acid (4.728) in an alkaline medium[519,520], and the recently reported rearrangement of the 4-nitropyridazone (4.729) into the corresponding 4-nitropyrazole (4.730)[520].

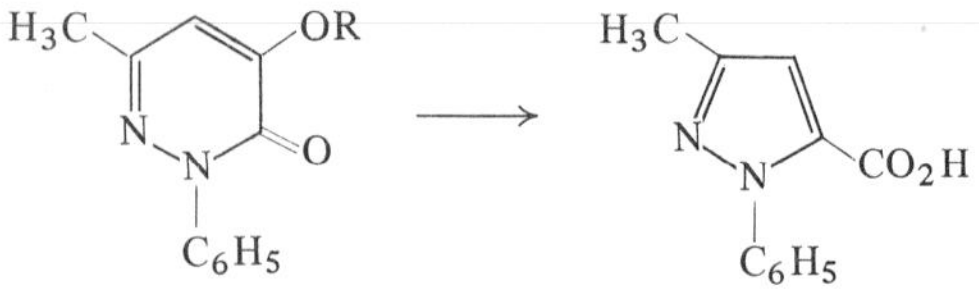

(4.727)   (4.728)

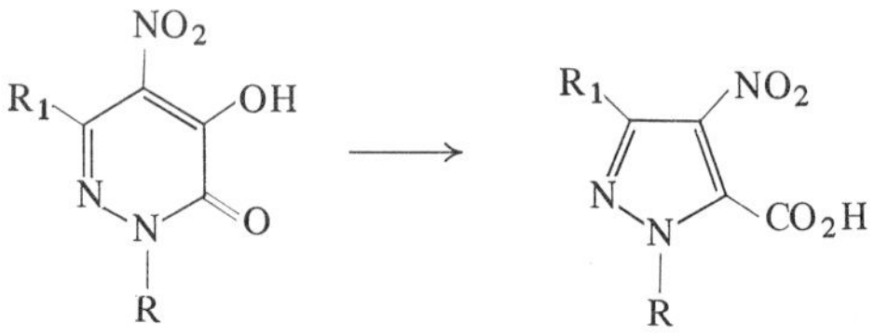

(4.729)   (4.730)

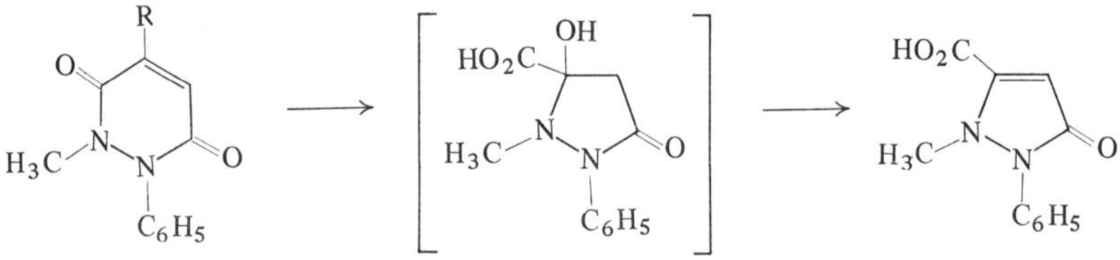

(4.731)   (4.732)   (4.733)

In a similar experiment[517,521,522] it was found that the 4-substituted 1-phenyl-2-methyltetrahydropyridazine-3,6-dione (4.731) is, on treatment with alkali, converted into the 5-pyrazolone-3-carboxylic acid (4.733). An example of a ring contraction which illustrates again, how unexpected and intriguing ring transformations can be, is the base-catalysed conversion of 4,5-dichloro-1-phenyl-6-(1H)-pyridazone (4.734) into the 3-hydroxypyrazole-5-carboxylic acid (4.722) and not the 4-hydroxy isomer (4.736) which would have been expected from the reaction course discussed above. The explanation put forward is that in this ring transformation both a *cine* substitution and a normal substitution reaction take place, yielding (4.719) which in the next step is converted into (4.722). This reaction also serves to support the hypothesis that a 3,5-dihydroxy compound operates as intermediate. That the occurrence of this ring contraction is very sensitive to the nature of the halogeno atoms can be demonstrated by the fact that base treatment of the 4,5-dibromo-1-phenyl-6-(1H)-pyridazone does

(4.734)   (4.735)   (4.736)   (4.719)   (4.722)

not give the 3-hydroxypyrazole (4.722), as in the case of the 4,5-dichloro compound, but gives the 4-hydroxypyrazole (4.736)[523]. This result throws considerable doubt on the original suggestion that this type of ring contraction only takes place when hydroxy groups are present on position 3 and 5 of the pyridazone. A careful and detailed study of the mechanism involved is required. A very recent report suggests[523a] that a pyridazyne intermediate is involved in the ring contraction of both (4.734) (X = Cl) and (4.717).

It is interesting that when 4-amino-3,6-dibromopyridazine (4.737) is treated with potassium amide in liquid ammonia, ring contraction occurs, yielding the compound 3-cyanomethyl-1,2,4-triazole (4.738)[524]. The reaction is in some ways analogous to the ring contraction of 3-amino-2-bromopyridine into 3-cyanopyrrole discussed in section I.C.1.c and the ring contraction of 5-amino-4-chloro-2-phenylpyrimidine into a 4(5)-cyanoimidazole discussed in section II.C.1.a. However, it differs from them in two respects: (a) the cyano group is not directly attached to the five-membered ring; and (b) one more nitrogen atom is present in the five-membered ring than in the starting substance. Although there are certain differences, basically the type of reaction mechanism discussed in sections I.C.1.c and II.C.1.a applies in the formation of the compound (4.738).

Similar to the conversion of pyridine-N-oxides into pyrroles on uv irradiation discussed in section I.C.1.c, it has been established that the photolysis of the pyridazine-N-oxides (4.739) yields 3-acetyl-, and 3-benzoylpyrazoles (4.740) respectively[525,526]. In the case of (4.739) (R_1 = Cl, R_2 = CH$_3$) and (4.739) (R_1 = OCH$_3$, R_2 = CH$_3$) the deoxygenated pyridazines and 4-hydroxymethyl-pyridazines are formed as by-products. The scope of this uv-induced ring contraction is, however, not very great since the pyridazine-N-oxides (4.739) (R_1 = R_2 = H; R_1 = R_2 = CH$_3$; R_1 = H, R_2 = CH$_3$) do not give pyrazoles[525]. An oxaziridino intermediate is proposed in this reaction[525,526] that has its

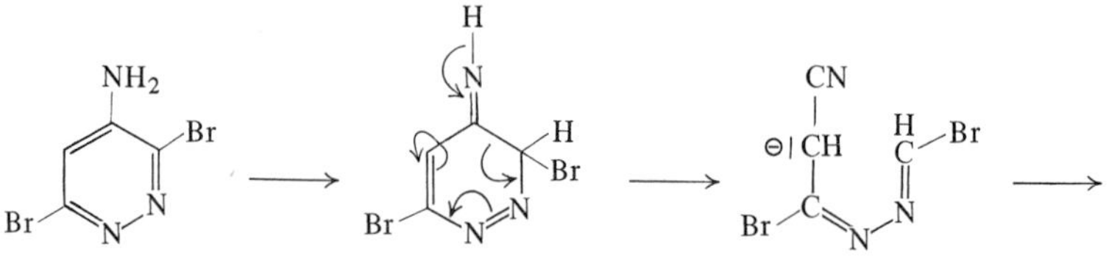

(4.737)

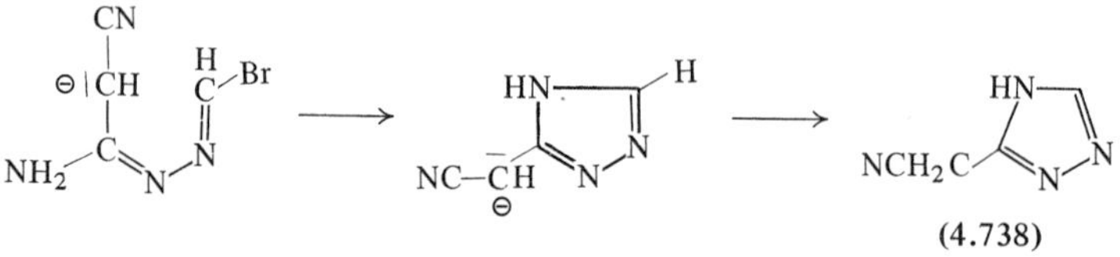

(4.738)

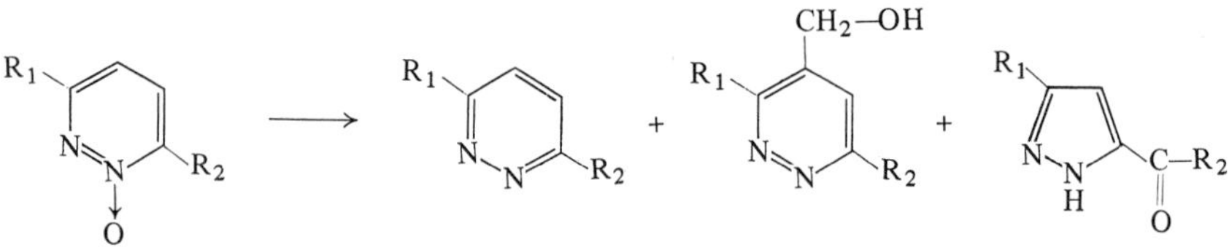

(4.739)   (4.740)

R_1	Cl	OCH_3	C_6H_5
R_2	CH_3	CH_3	C_6H_5

analogy in the photochemistry of pyridine-N-oxides and quinoline-N-oxides (see section I.C.1.c). Photochemically or thermally, this oxaziridino intermediate is converted into the diazo compound (4.739a) as suggested[526] (a) by the appearance during irradiation of (4.739) ($R_1 = R_2 = C_6H_5$) of a consistent yellow colour which fades in the dark, and (b) by the fact that immediately after irradiation an ir absorption at 2070 cm^{-1} is observed.

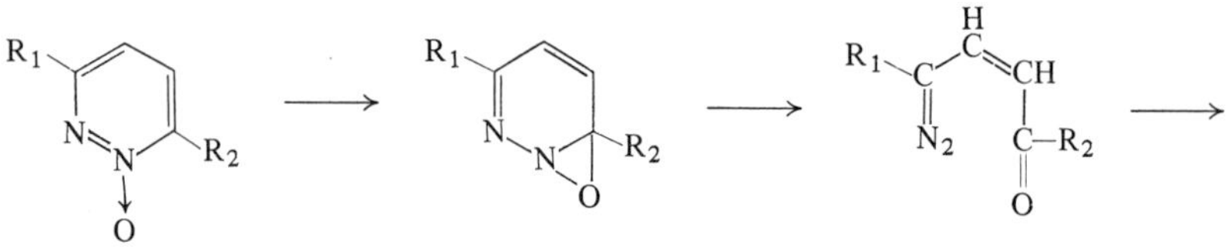

(4.739)   (4.739a)

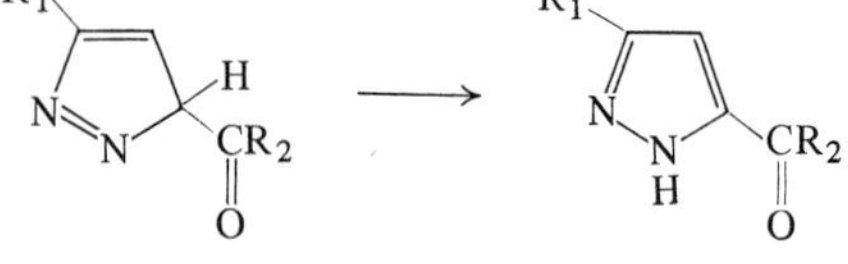

Irradiation of the 1,2-dihydropyridazine-1,2-dicarboxylate (4.741) leads to the formation of two photoisomers, the valence tautomer (4.745) (61 per cent) and the pyrrole (4.744) (14 per cent)[527]. The pyrrole (4.744) is envisaged as a further photoproduct of the open-chain compound (4.742) and (4.743)[527].

(4.741) (4.742) (4.743) (4.744) (4.745)

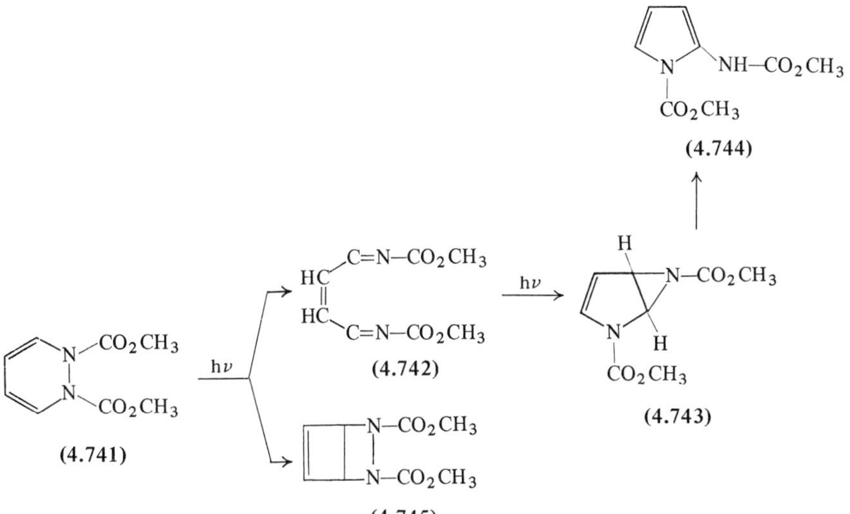

An interesting but rather unexplored type of ring contraction is found to occur during reduction of the 1,4-dihydropyridazines through the agency of refluxing acetic acid. By this method, 1-tosyl-1,4-dihydropyridazine (4.746) is converted into 1-tosylamidopyrrole (4.747)[528]. This ring contraction is probably initiated by an acid-catalysed addition of the ethanol to the olefinic bond prior to carbon-nitrogen bond rupture.

Reductive ring contraction has been found to occur when 4,5,6-triphenyl-3(2*H*)-pyridazinethione (4.748) and 3-anilinopyridazine (4.750) are treated with

(4.746) (4.747)

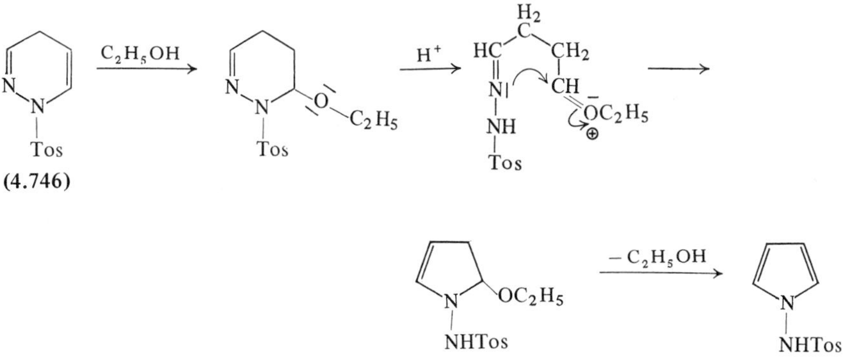

Raney nickel yielding, respectively, 2,3,4-triphenylpyrrole (4.749)[529] and 2-anilino-Δ^2-pyrroline (4.751)[530]. Which nitrogen atom in the ring is expelled during reduction has not been established. The ring contraction of 4-hydroxy-imino-3-methyl-5-phenyl-1(H)-dihydropyridazine into 2-methyl-5-phenylpyrrole by the action of zinc and acetic acid is somewhat similar[531].

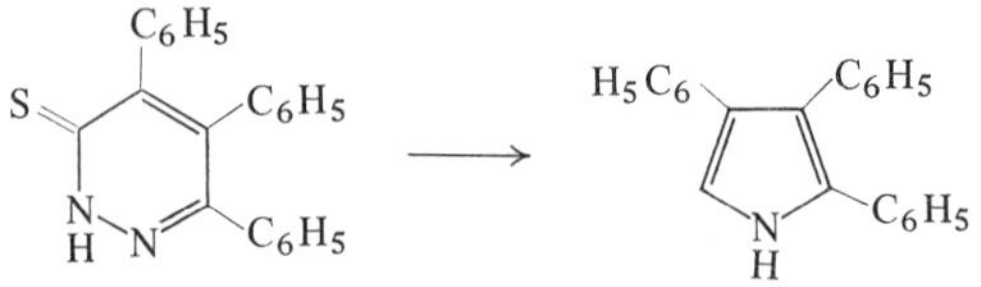

(4.748)   (4.749)

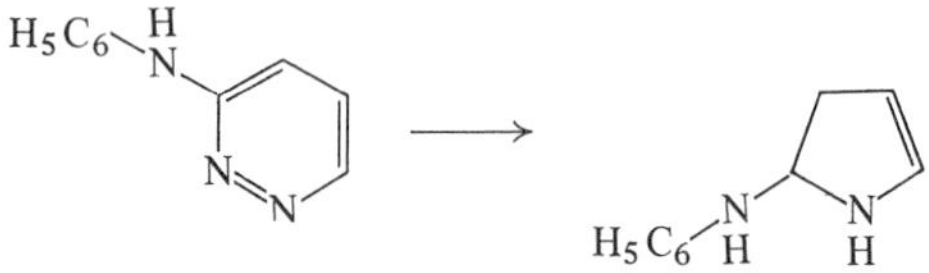

(4.750)   (4.751)

Electrochemical reduction of pyridazines has been observed to give ring contraction into pyrroles. Thus, it has been found[531a] that by this method 5-t-butyl-3,6-diphenyl-1-methylpyridazinium iodide (4.751a) is reduced in acid into the N-methylpyrrole (4.751d) and the pyrrole (4.751e). The formation of these products might be thought to involve primarily a two-electron reduction into the different tautomeric dihydropyridazinium ions (4.751b) and (4.751c), which, by hydrogenolysis of the nitrogen-nitrogen bond and subsequent ring closure as shown in the following scheme, are converted into both pyrroles.

b. *Pyrazines into imidazoles; dihydropyrazines into Δ^2-imidazolines and imidazoles*

Irradiation of pyrazine-N-oxides and of 2,3-dihydropyrazines gives photo-rearrangement into imidazoles. From 2,5-dimethylpyrazine-N-oxide (4.752) (R = CH_3) two products, 2-acetyl-4-methylimidazole (4.754) (R = CH_3) and 2,4-dimethylimidazole (4.756) (R = CH_3), are formed[532]. The same ring contraction was also observed with 2,5-diphenylpyrazine-N-oxide (4.752) (R = C_6H_5). From the results reported for the pyridine-, quinoline- and isoquinoline-N-oxide, it seems highly probable that ring contraction commences with the formation of an oxaziridine derivative. In the 2,5-disubstituted pyrazine-N-oxide (4.752), however, two different isomeric oxaziridine derivatives are possible i.e. (4.753) and (4.755). The imidazoles (4.754) and (4.756) are formed via the pathways *a* and *b* which are analogous to those discussed for photo-isomerizations in sections I.C.1.c and II.C.1.a.

The results obtained from studies on photoreactions of α-dicarbonyl systems, prompted an investigation of the photodecomposition of the structurally related

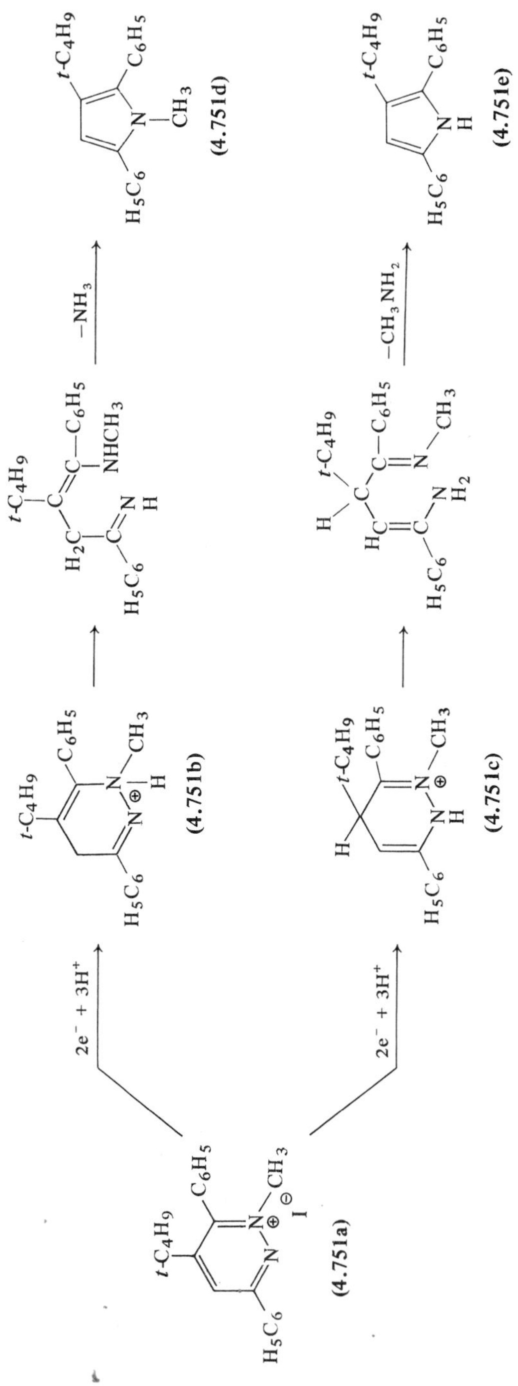
(4.751a)
(4.751b)
(4.751c)
(4.751d)
(4.751e)
2e⁻ + 3H⁺
−NH₃
−CH₃NH₂

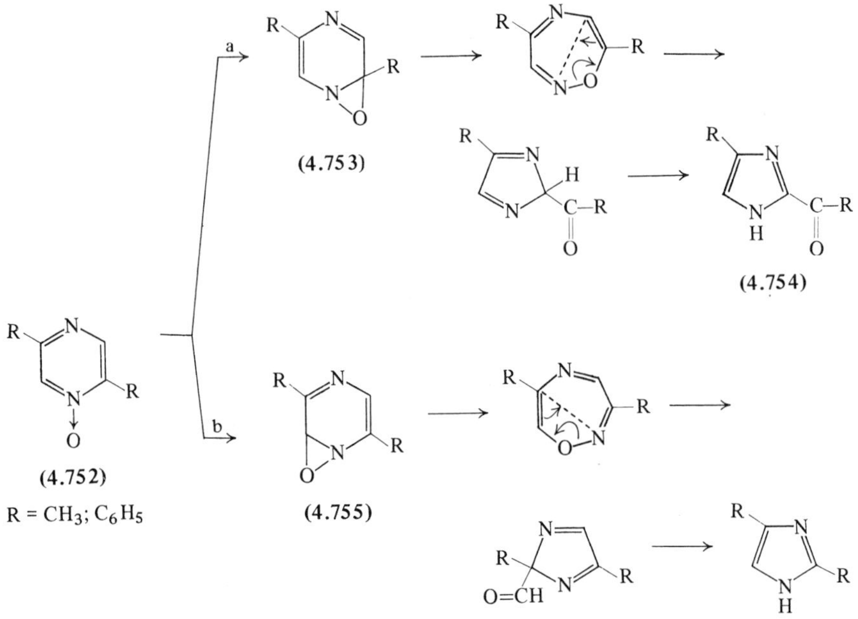

(4.753)

(4.754)

(4.752)

R = CH₃; C₆H₅

(4.755)

(4.756)

2,3-dihydropyrazine and its derivatives because of the practical and theoretical implications. A smooth rearrangement into a mixture of imidazoles was observed when an ethanolic solution of the 2,3-dihydro compound (4.757) was irradiated in a nitrogen atmosphere. It gives 75 per cent of 1-methyl-4,5-diphenylimidazole (4.758), together with the 1-ethoxymethyl-4,5-diphenylimidazole (4.759) (10 per cent)[532a]. It is reasonable to postulate that the reaction proceeds by a

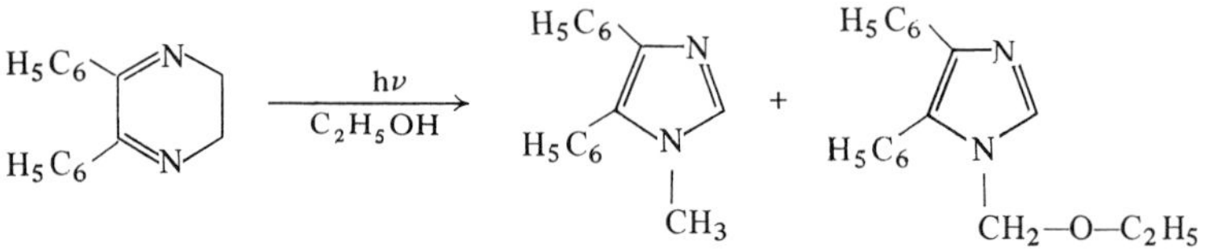

(4.757)   (4.758)   (4.759)

photolytic ring opening into the enediimine (4.760). Recyclization generates the resonance-stablized intermediate (4.761a) ↔ (4.761b) which is converted either by proton transfer into the N-methylimidazole (4.758) or into the imidazoline (4.762) by addition of ethanol. This imidazoline can be easily oxidized into (4.759) by the oxygen which remains in the sweeping nitrogen.

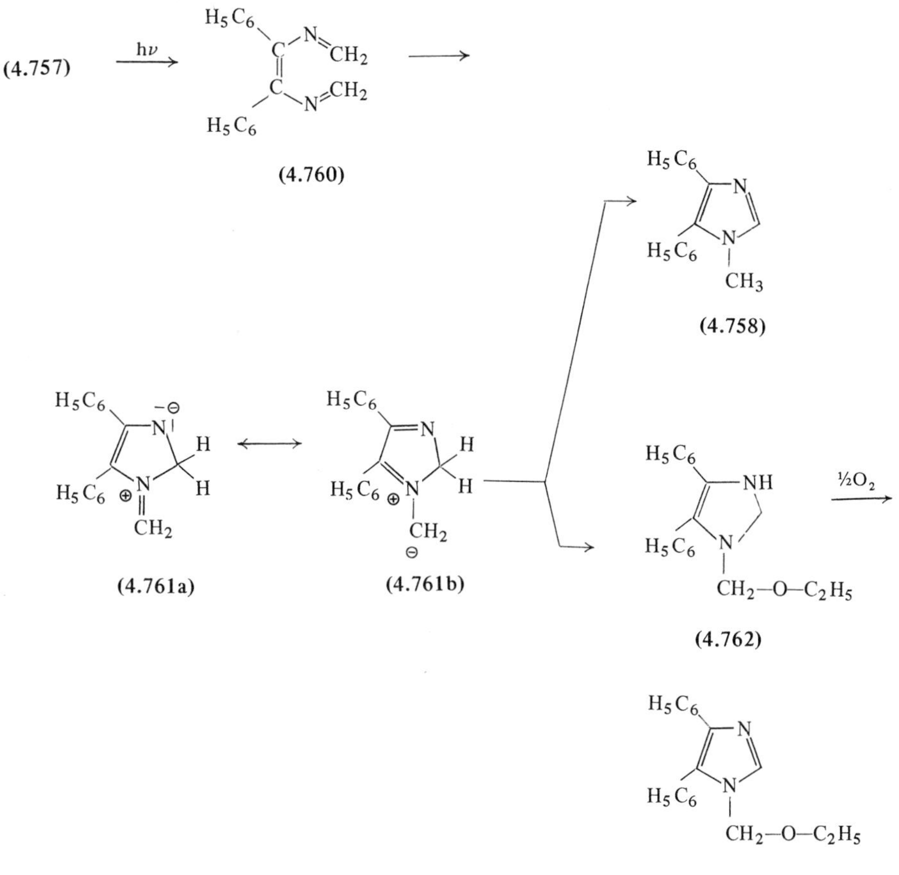

(4.757) $\xrightarrow{h\nu}$

(4.760)

(4.758)

(4.761a) ⟷ (4.761b)

(4.762)

(4.759)

The mechanism by which this enediimine (4.760) is formed from the 2,3-dihydropyrazine (4.757) is not reversible[532a]. This is elegantly demonstrated by the fact that photolysis of the optically active 2-isobutyl derivative (4.763) gives a reaction mixture which contains the starting material in which less than 5 per cent of racemization has occurred.

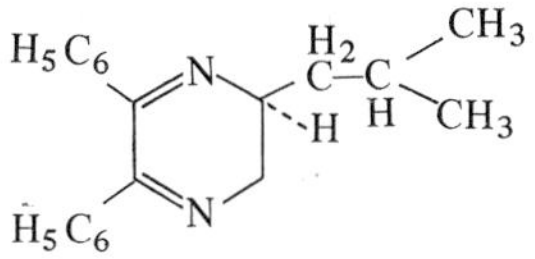

(4.763)

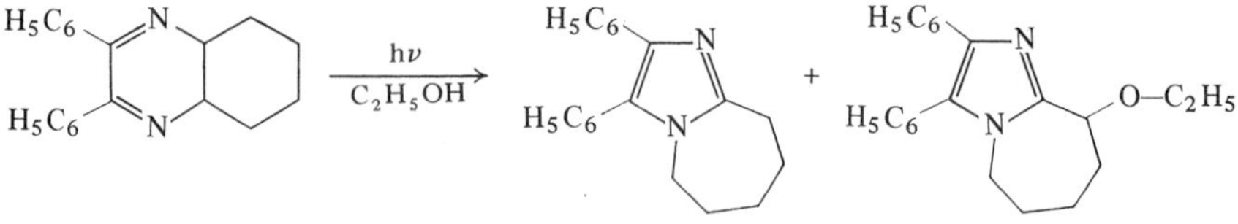

(4.764)   (4.765)   (4.766)

This photoreaction is important from a preparative point of view, since a number of 5,6-dialkyl-(aryl)-2,3-dihydropyrazines have been shown to undergo ring contraction into imidazoles which are otherwise difficult to prepare. The conversion of the tetramethylene 2,3-dihydropyrazine (4.764) into the bicyclic imidazoles (4.765) and (4.766) serves as a good illustration.

The reaction of hydrogen peroxide on (4.757) has been found[532b] to be of considerable interest, since it might be considered as a model for some of the spontaneous reactions exhibited by partially reduced pyrazines on exposure to air. The products formed were N,N-dibenzoylethylenediamine, N-benzoylethylenediamine and 2-phenyl-Δ^2-imidazoline; no 5,6-diphenyl-2,3-dihydropyrazine-1,4-di-N-oxide was formed. The ring contraction reaction is of special interest since the principles of a new synthesis of Δ^2-imidazolines are involved. The compound 5,6-diphenyl-1,2,3,6-tetrahydropyrazine-6-hydroperoxide has been suggested as the primary product of oxidation; 2-phenyl-Δ^2-imidazoline is formed by a ring expansion–ring contraction sequence.

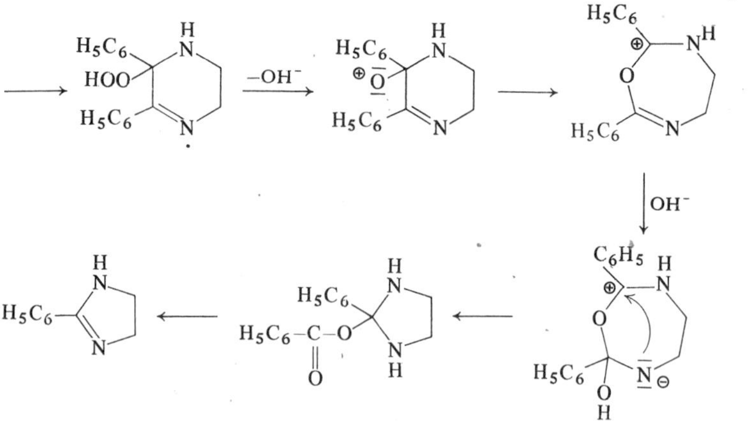

A recent report states that 2-chloropyrazine (4.767), on treatment with potassium amide in liquid ammonia at 75°C, is converted into 2-cyanoimidazole (4.768), which is normally difficult to obtain, in a yield of about 35 per cent together with the imidazole (4.769) (13-15 per cent) and 2-aminopyrazine[533] (4.770). In order to throw some light on the mechanism of these independently operating ring contractions (2-cyanoimidazole is not a precursor of imidazole), 2-chloro-2-$[^{14}C]$ pyrazine, 2-chloro-2-$[^{15}N]$ pyrazine and 2-chloro-4-$[^{15}N]$ pyrazine were synthesized and the distribution of the isotopic atoms in the 2-cyanoimidazole and imidazole was established[533a]. It was proved conclusively that both ring contractions start by a nucleophilic attack at position 3, i.e. (4.771), and that the imidazole is formed by fission of the C_2-C_3 bond (route a). The results of the experiments with the labelled compounds also lead to the conclusion that the reaction did not take place via a pathway involving the seven-membered triazepine (4.772) (see route b) as intermediate. The formation of the cyanoimidazole can reasonably be assumed to occur by a mechanism involving a C_3-C_4 bond fission (route c). At present it is not clear at which stage of the reaction the dehydrogenation takes place. It is suggested that a Cannizzaro type disproportionation reaction occurs with iminomethyl-imidazole, a precursor of (4.768). This compensated oxidation-reduction process should lead to the formation of 2-aminomethylimidazole; however this compound has not so far been isolated.

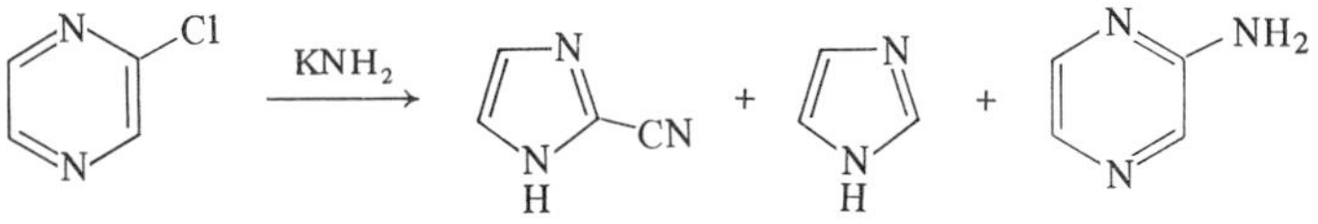

(4.767)   (4.768)   (4.769)   (4.770)

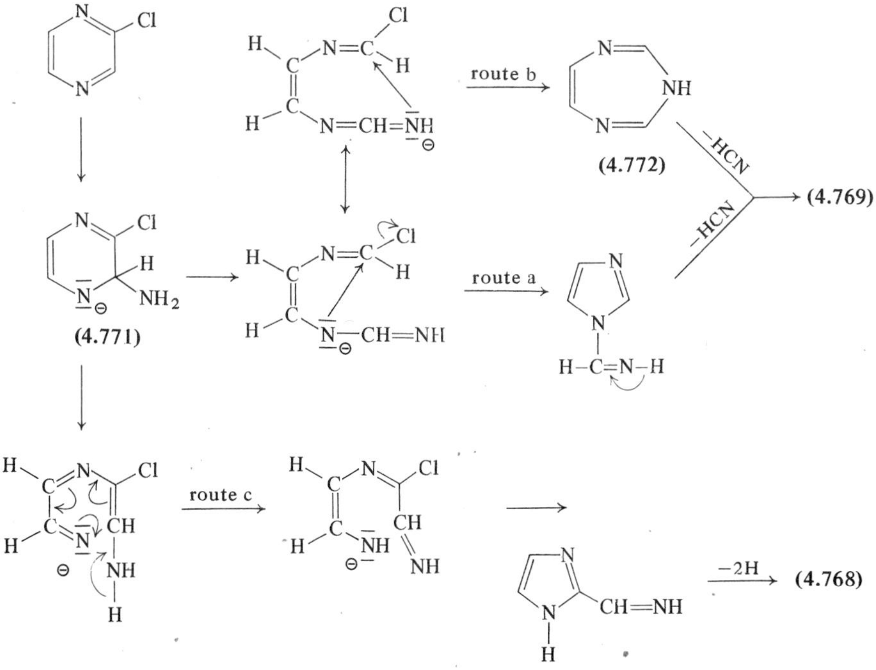

c. *Cinnolines into indoles and indazoles; phthalazines into isoindoles and isoindolines; phthalazinones into phthalimidines and isobenzofurans*

In analogy to the reported reductive ring contraction of 1,4-dihydropyridazines into pyrroles (see section II.D.1.a) it has been observed that 4-alkyl(aryl)-cinnolines (4.773), on treatment with amalgamated zinc in aqueous acetic acid under reflux, give the 3-substituted indoles (4.775)[534-536]. Reduction of 3-cinnolone (4.776) with zinc and sulphuric acid gives 1-aminooxindole (4.777)[537]. In the conversion of (4.773) → (4.775), the 1,4-dihydrocinnolines (4.774) are proposed as intermediary products. In order to establish which nitrogen atom is eliminated during this ring contraction, the reduction of 4-phenylcinnoline-2-^{15}N was investigated[534]. It appears that the 3-phenylindole formed does not contain ^{15}N, clearly indicating that it is the nitrogen at position 2 of the cinnoline ring which is eliminated during the reduction. The 1,4-dihydrocinnoline (4.778) can give either N—N bond fission into the amino-imino compound (4.779), which then cyclizes into (4.780) and aromatizes by a subsequent loss of ammonia into (4.782), or may be converted by ring contraction into N-aminoindole (4.781) which is further converted into (4.782). However, the fact that in aqueous acid the 1,4-dihydrocinnoline (4.778) does not react in the absence of amalgamated zinc, but in its presence is converted into 3-phenylindole (4.782) in a nearly quantitative yield supports the idea that the reductive ring contraction must take place via the amino-imine (4.779)[534].

The conversion of 1,4-dihydrocinnolines into indoles by electrochemical reduction has also been reported[534a]. Although it has been shown that in an acidic medium 1,4-dihydrocinnolines are in equilibrium with N-amino-indoles[534b], this latter compound cannot be intermediate in this electro-chemically induced ring contraction since it has been found that under the applied conditions this compound cannot be further reduced. The route is thought to involve a two-electron reduction process leading to an iminoaniline.

(4.773) (4.774) (4.775) (4.776) (4.777)

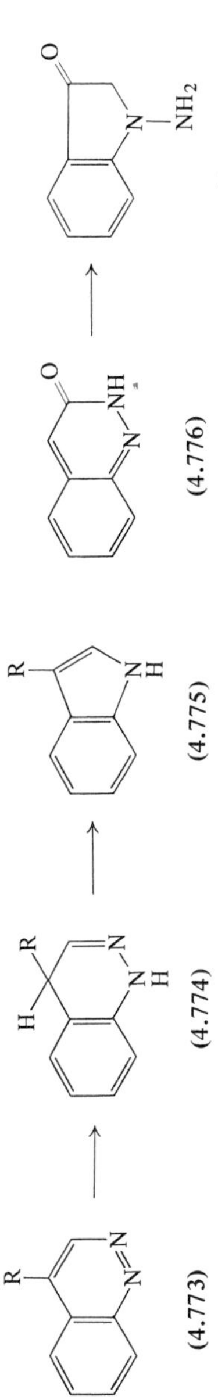

R = H; CH_3; C_6H_5; 4-OCH_3—C_6H_4; 4-OH—C_6H_4; 2,5-$(OH)_2$—C_6H_3; 3,4$(OH)_2$—C_6H_3; 2,3-$(OH)_2C_6H_3$; 2,5-$(OCH_3)_2$—C_6H_3;
3,4-$(OCH_3)_2$—C_6H_3; 2,3-$(OCH_3)_2$—C_6H_3

(4.773, R = C_6H_5) (7.778) (4.779) (4.780) (4.781) (4.782)

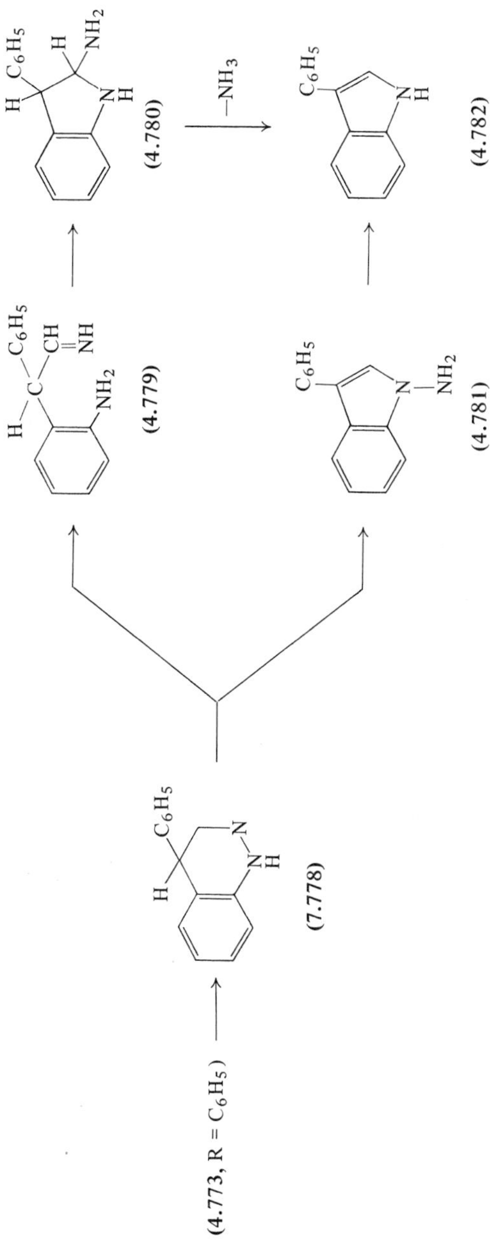

Results obtained by polarography of 1,4-dihydrocinnolines showing the occurrence of a wave, which could be due to the formation of this compound, confirm the suggestion.

It has been reported[534c] that reductive ring contraction also occurs when 4-methyl- and 3,4-dimethyldihydrocinnolines are heated with formic acid and formamide leading to the formation of the corresponding 1-formamidoindole. It cannot be excluded that the occurrence of the equilibrium 1,4-dihydrocinnoline $\rightleftarrows$ 1-aminoindole is favoured in the formic acid–formamide system.

Catalytic dehydrogenation of 3-phenyl-5,6,7,8-tetrahydrocinnoline does not give 3-phenylcinnoline but leads to the formation of 2-phenylindole[538].

Reduction of 1,3-dimethylcinnolinium iodide (4.783) by the Clemmensen method gives 1,2-dimethylindole (4.784) and 1,2-dimethylindoline (4.785), while the 2-methylcinnolinium salts (4.786) give the indole (4.787) and the indoline (4.788), which thus do not contain a methyl group on the nitrogen atom[539]. It seems that in these reductions it is always the nitrogen atom on position 2 which is eliminated; this finding has been advanced as a useful method of determining the site of quaternization of cinnolines.

(4.783)   (4.784)   (4.785)

(4.786)   (4.787)   (4.788)

R	CH$_3$	H	H
R$_1$	H	CH$_3$	C$_6$H$_5$

In correlation with the reported conversion of 8-nitroquinoline into 7-nitrooxindole by oxidation (see section I.C.1.c) is the conversion of 5- and 8-nitrocinnoline on warming with hydrogen peroxide in acetic acid into 4- and 7-nitroindazoles[540].

Ring contraction with N–N bond fission was reported to occur during reduction of 4-chlorophthalazine (4.789) (R = H) and its 1-methyl derivative (4.789) (R = CH$_3$) on treatment with zinc and hydrochloric acid[541-543] and also on treatment with tin and hydrochloric acid[541]. The dihydroisoindole (4.790) was obtained[541] and not, as originally suggested[544] in the case of 1-methyl-4-chlorophthalazine, 1-methyldihydropseudoisoindole. No information is available which nitrogen atom is eliminated during this ring contraction. It seems reasonable to assume that the reaction mechanism is somewhat similar to that reported for the reduction of cinnolines into indoles. Similarly,

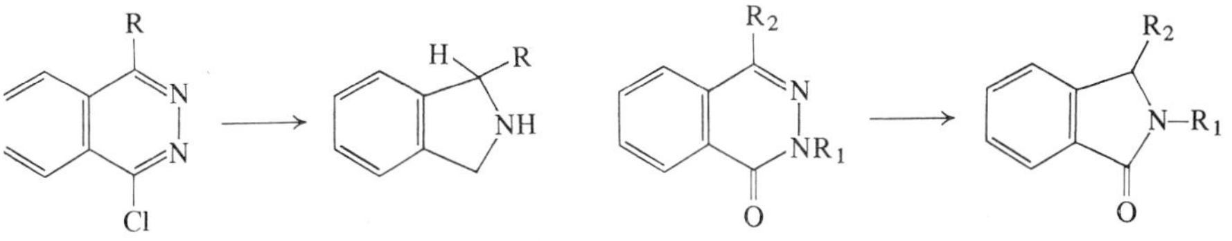

(4.789)   (4.790)   (4.791)   (4.792)

R = H; CH$_3$

phthalazones (4.791), on reduction with zinc and hydrochloric acid were found to give phthalimidines (4.792)[545,546]. An interesting modification of this reaction is the use of quaternary phthalazone salts as substrates in these reactions[547]. In the phthalazine-1,4-dione series, it has been reported that the compound (4.793), when heated for a short time with a sodium hydroxide solution, gives ring contraction leading to the formation of derivatives of N-aminophthalimide, i.e. (4.794)[548]. Similarly, condensation of the 1,4-dione (4.795) with aromatic aldehydes gives, via the adduct (4.796), formation of the phthalimide (4.797)[549].

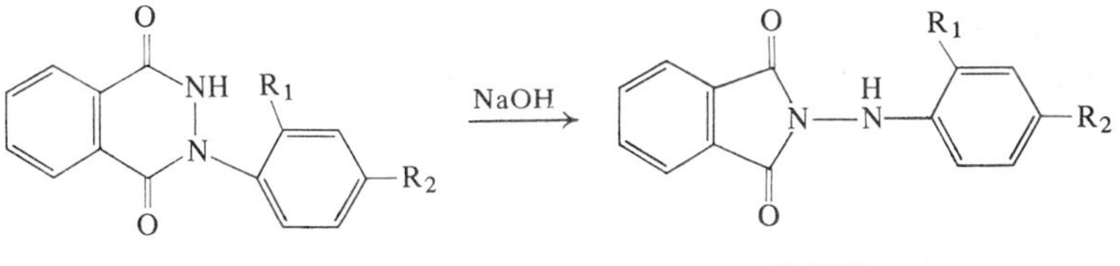

(4.793)   (4.794)

R$_1$	H	NO$_2$
R$_2$	NO$_2$	NO$_2$

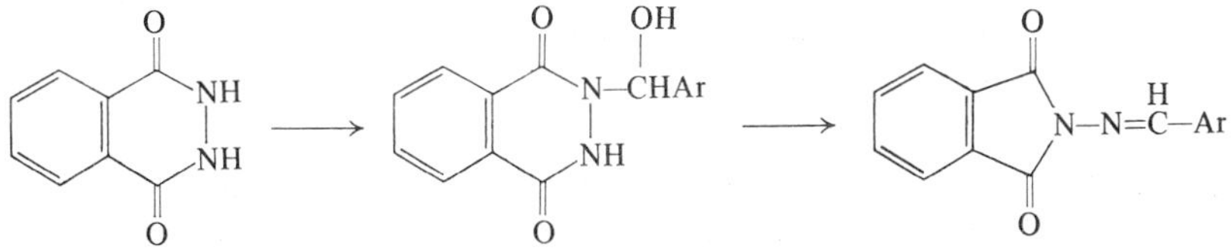

(4.795)   (4.796)   (4.797)

Interestingly, ring contraction of phthalazine and its derivatives has also been conducted by electrochemical reduction[549a]. Thus, phthalazine in acid solution is converted into isoindoline, o-xylene-α,α'-diamine also being yielded; the ratio in which these compounds are produced is dependent primarily on the pH. 1-Methylphthalazine (4.797a) is, in acid solution, found to be converted with a four-electron reduction into 1-methylisoindole (4.797b); at a more negative potential a further two-electron reduction yields the 1-methylisoindolinium ion (4.797c). The following scheme has been suggested:

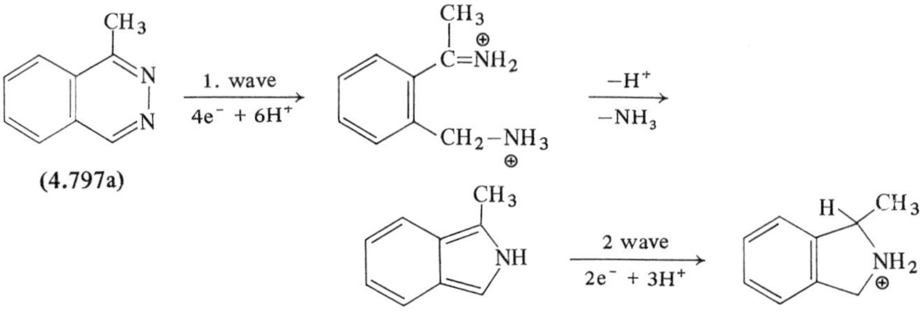

$$\begin{array}{c}\text{(4.797a)}\quad\xrightarrow[\;4e^- + 6H^+\;]{\text{1. wave}}\quad\xrightarrow[\;-NH_3\;]{-H^+}\quad\text{(4.797b)}\quad\xrightarrow[\;2e^- + 3H^+\;]{\text{2 wave}}\quad\text{(4.797c)}\end{array}$$

A corresponding ring contraction into phthalimidines has been found to occur during the electrochemical reduction of methoxy- and amino-phthalazines[549b] and phthalazinones[549c]. In the latter compounds, the carbon-nitrogen double bond has been shown to be saturated before hydrogenolysis of the nitrogen-nitrogen bond. Thus, the 4-methylphthalazinone (4.797d) gives, in a four-electron reduction, 1-methylphthalimidine (4.797f) via (4.797e). The equivalent ring contraction of 2,3-dihydro-2,3-dimethyl-1,4-phthalazinedione and the corresponding dihydrophthalazinone (4.797g) into 2-methylphthalimidine (4.797h) at low pH has also been reported[549c,d]. Detailed polarographic data are used to establish which intermediates are involved[549c] in these ring contractions.

Ring contraction under photolytic conditions has also been reported with dihydrophthalazines and phthalazine-N-oxides. Irradiation of the 1,2-dihydro-phthalazine (4.798) in benzene in a Pyrex glass resulted in the photochemical generation of the azomethine-imine (4.799)[550]. This conversion is therefore of importance since the imine obtained is claimed to be the first stable compound of this type. Whether in this reaction a diaziridine (4.800) or azo compound (4.801) is intermediate has not been established.

The photolysis of phthalazine N-oxide (4.802) in acetone under evolution of nitrogen, yields a reaction mixture which contains the deoxygenated products dibenzoylbenzene (4.808) and, surprisingly, 1,3-diphenylisobenzofuran (4.807)[551]. Compare the photolysis of 3,6-diphenylpyridazine-N-oxide in which a pyrazole derivative is formed but *no* furan derivative is yielded (see section

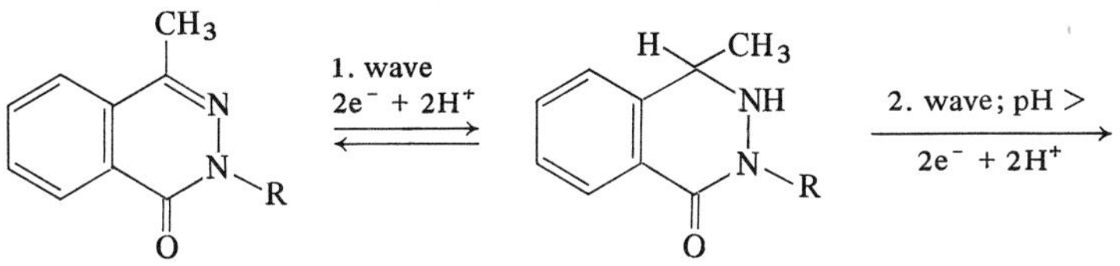

(4.797d)   (4.797e)

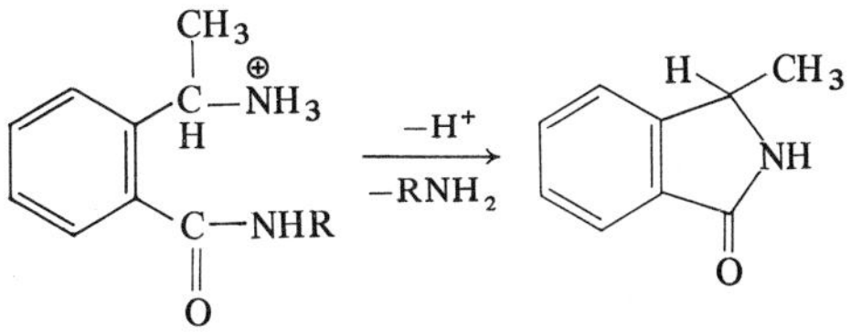

(4.797f)

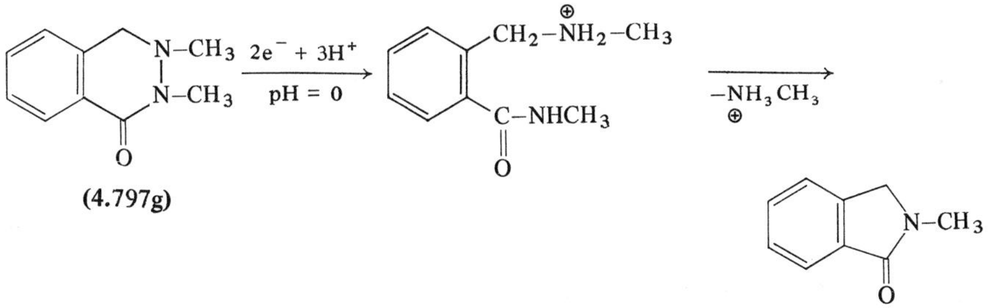

(4.797g)

(4.797h)

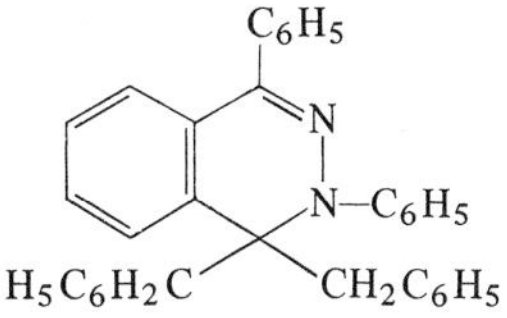

(4.798) 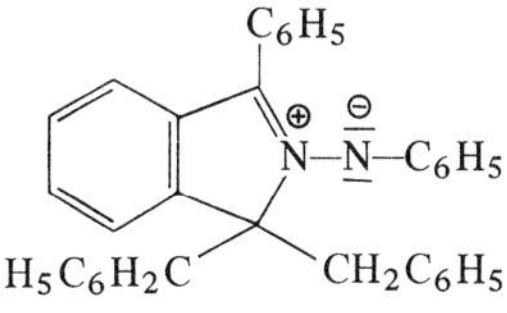   (4.799)

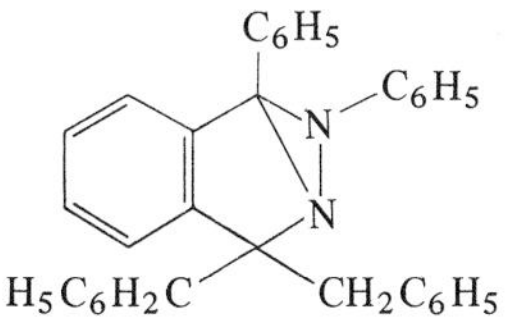

(4.800) 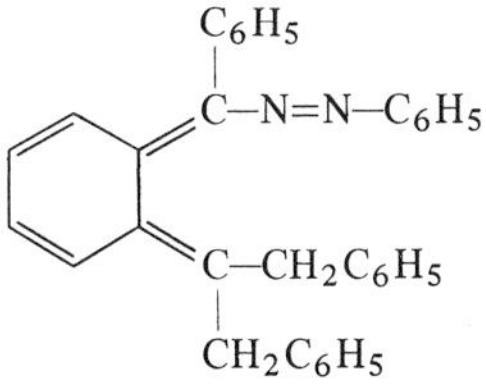   (4.801)

II.D.1.a). The formation of (4.807) is remarkable, since so far no further examples are available of the conversion of aromatic *N*-oxides into heterocyclic compounds in which all nitrogen atoms have been removed from the ring. The formation of (4.807) can be explained by an initial formation of the oxaziridine intermediate (4.803), which photochemically or thermally rearranges via (4.804) into the diazo compound (4.805). Since it is known that diazo compounds decompose on photolysis it is probable that the diazo compound decomposes via the carbene (4.806) into the isobenzofuran (4.807). The isobenzofuran is converted into dibenzoylbenzene (4.808) by a thermal or ultraviolet light induced process in the presence of oxygen.

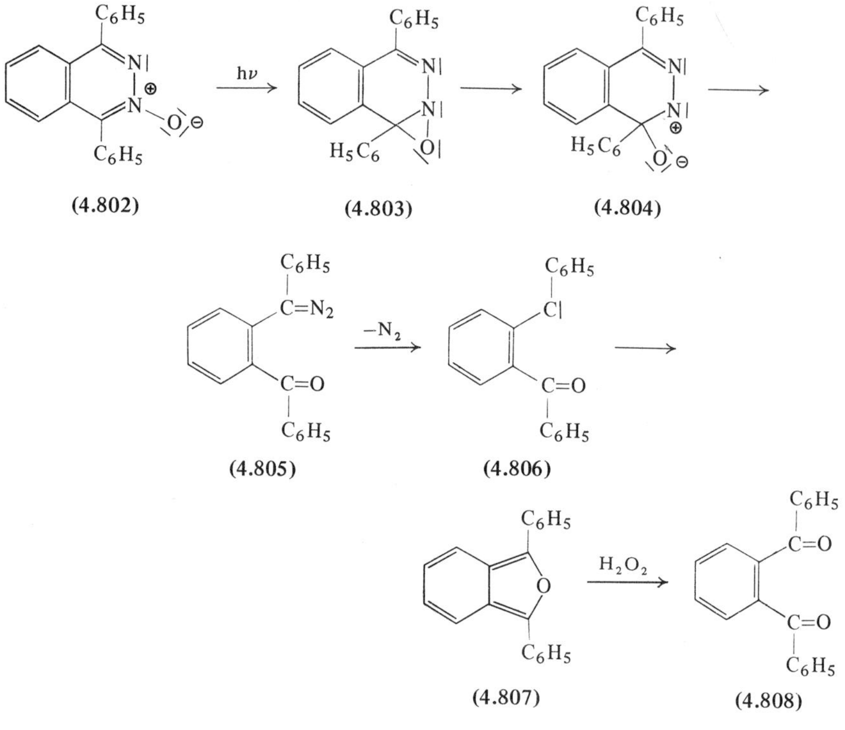

(4.802)   (4.803)   (4.804)

(4.805)   (4.806)

(4.807)   (4.808)

d. *Quinoxalines into benzimidazoles*

Until recently only a relatively few ring contractions of quinoxalines were known. The pseudobase (4.809), when treated with ethanolic hydrochloride, gives the 1,2-diphenylbenzimidazole (4.812)[552]. This reaction is suggested to take place via the primarily formed 2,3-dihydro-2-benzoylbenzimidazole (4.810) which, with loss of ethylbenzoate, is converted into the dihydrobenzimidazole

(4.811). A subsequent oxidation which, it is assumed, is performed by the nitro group present, yields (4.812). The presence of the 1-phenyl group is also found to be an essential requirement for the occurrence of this reaction, since the

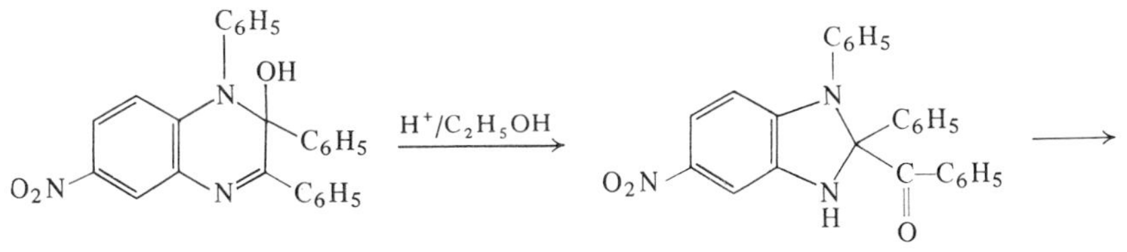

(4.809)   (4.810)

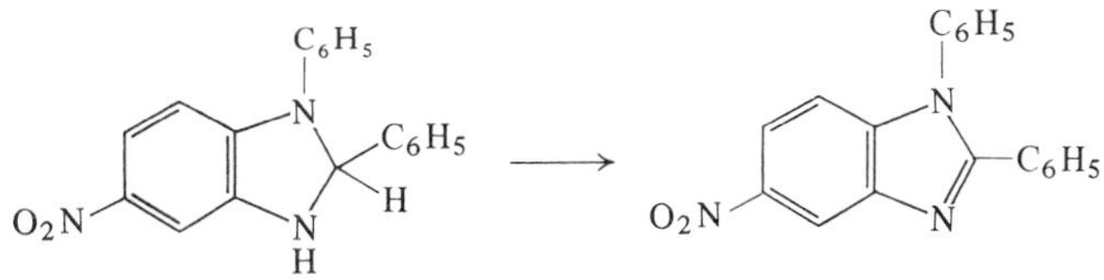

(4.811)   (4.812)

1-methyl derivative does not show this ring contraction. The quinoxazoline-2,3-dione (4.813), which is formed by a reaction between equimolar amounts of o-phenylenediamine and oxamide, reacts with o-phenylenediamine (not with its 4-methyl or 4-nitro derivative) to yield the bis benzimidazole (4.814)[553].

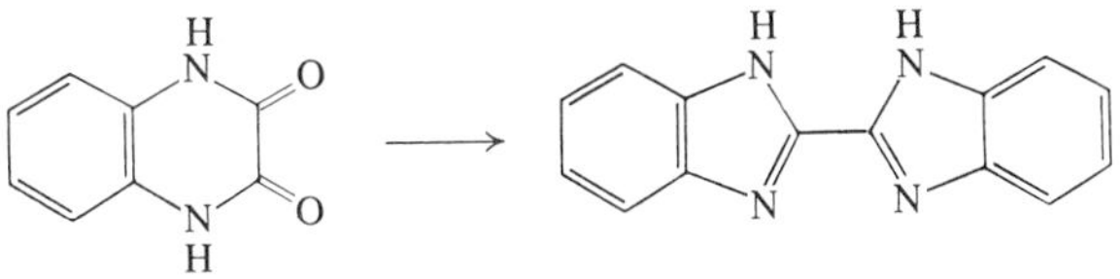

(4.813)   (4.814)

Several ring contractions into benzimidazoles have been found to occur on treatment of substituted quinoxalines with potassium amide in liquid ammonia. Heating 2,3-diphenylquinoxaline (4.815) with potassium amide in liquid ammonia at 140°C produces a 30 per cent yield of 2-phenylbenzimidazole (4.821)[554]. It can be assumed to take place via (4.816), (4.818) and (4.820). Phenylmethylimine (4.819) has, however, not been isolated. This ring contraction cannot be affected by other bases such as potassium hydroxide in water or in ethanol, sodium methoxide in methanol or sodium hydride in toluene. No indications have been obtained for the occurrence of a benzilic acid-type rearrangement in the intermediate (4.816) which would lead to phenyl migration and would yield the phenyldihydroquinoxaline (4.817).

Almost identical results were obtained when the 2-chloroquinoxaline (4.822) (R = H, CH$_3$, C$_6$H$_5$) was treated with potassium amide in liquid ammonia at low

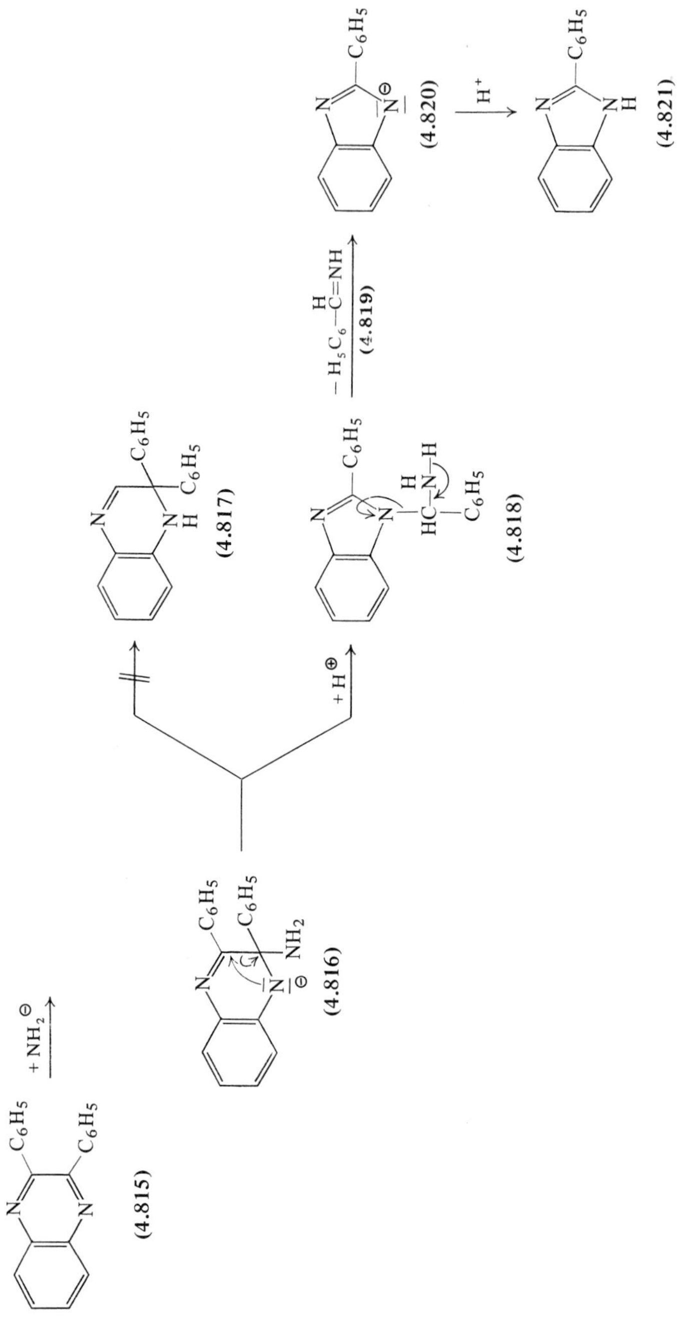
(4.815)
$+NH_2^{\ominus}$
(4.816)
(4.817)
$+H^{\oplus}$
(4.818)
$-H_5C_6-\overset{H}{C}=NH$
(4.819)
(4.820)
H^+
(4.821)

temperature[555]. In all cases benzimidazole (4.823) is formed as well as the corresponding 2-aminoquinoxalines. This indicates that ring contraction is not prevented by the presence of a substituent on position 3 (methyl or phenyl), and that it is carbon atom 3 in the quinoxaline ring which is eliminated. Considering

(4.822)   (4.823)

R = H; CH$_3$; C$_6$H$_5$

the mechanism of this ring contraction it seems likely that an initial attack on the amide ion at position 3 takes place, leading to the intermediate (4.824)[555]. Whether in this reaction course an isonitrile (4.825) (route a) or iminochloride (4.826) (route b) is involved, is at the moment an open question. It is tentatively suggested that this ring contraction takes the course shown on p. 168. The formation of a nitrile has been confirmed by the fact that with (4.822) (R = C$_6$H$_5$) it is possible to isolate benzamidine, which is probably formed by the addition of potassium amide to benzonitrile. See for comparison of the corresponding reaction of 2-chloropyrazine section II.D.1.b.

It has recently been found[556] that hydrazinolysis of 2-oxo-1,2-dihydro-quinoxaline (4.828) in boiling 50 per cent aqueous hydrazine leads to the formation of 2-methylbenzimidazole (4.831). This interesting reaction again demonstrates the strong tendency of hydrazine to bring about ring trans-formation; it is thought to occur by an initial addition of hydrazine at the C=N bond (4.829), thus a 3,4-addition analogous to addition reactions in 6- and 7-oxopteridines with nucleophilic reagents[556a]. Ring contraction into (4.830), as indicated, and the reductive conversion of the −HC=N−NH$_2$ into a methyl group by a Wolff–Kishner-type process, together with dehydration, gives (4.831).

(4.828)   (4.829)

(4.830)   (4.831)

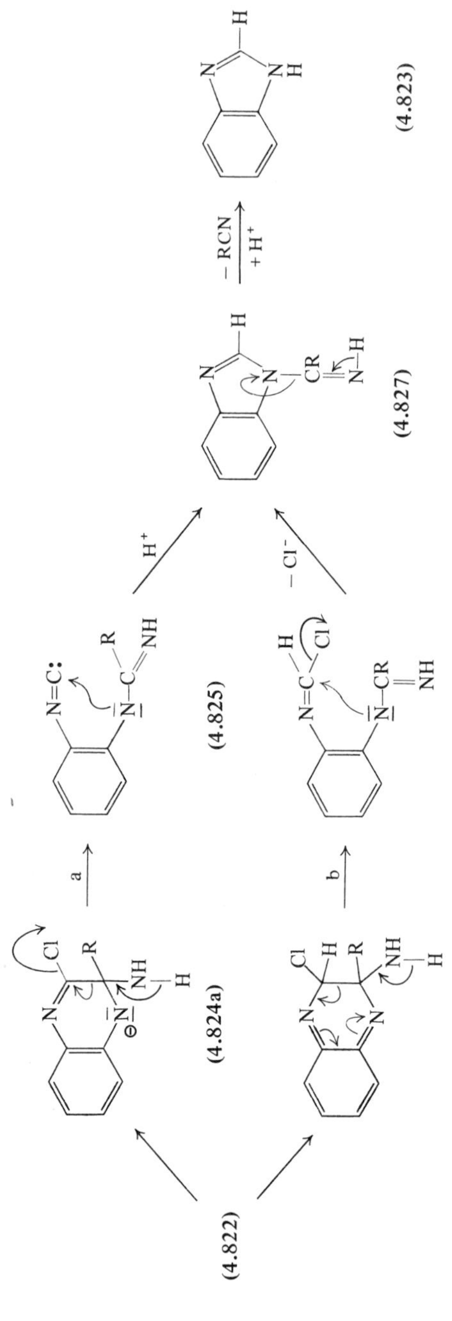
(4.822)
(4.824a)
(4.824b)
a
b
(4.825)
(4.826)
H⁺
− Cl⁻
(4.827)
− RCN
+ H⁺
(4.823)

Quinoxaline-1-oxides (4.832), bearing a substituent at C_2 (aryl- or cyano group), a carbonyl group at C_3 and hydrogen at N_4 do not give a Boekelheide rearrangement when heated with acetic anhydride as usually observed with aromatic N-oxides, but are converted into the benzimidazole derivatives 1-acyl-3-acetyl-2-benzimidazolinones (4.835) and/or 1,3-diacetyl-2-benzimida-zolinones (4.836)[557]. Just as in many photodecompositions of aza aromatic-N-oxides, an oxaziridine derivative (4.833) is postulated as intermediate in this conversion.

Likewise, 2-phenylquinoxaline-1,4-dioxide, on treatment with acetic anhydride, gives the benzimidazolone (4.835) ($R_1 = C_6H_5$, $R_2 = R_3 = H$). A previous rearrangement of this di-N-oxide into 3-oxo-2-phenyl-3,4-dihydro-

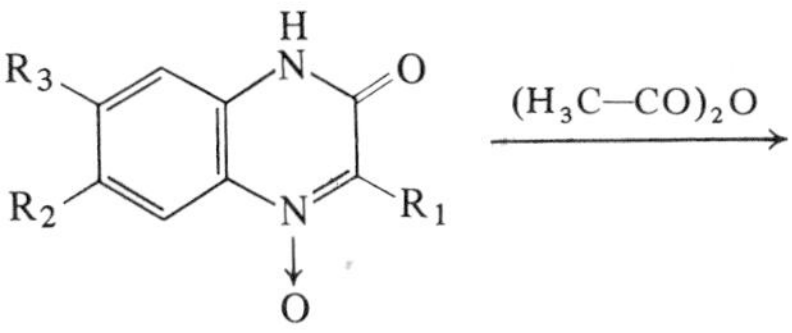

(4.832)

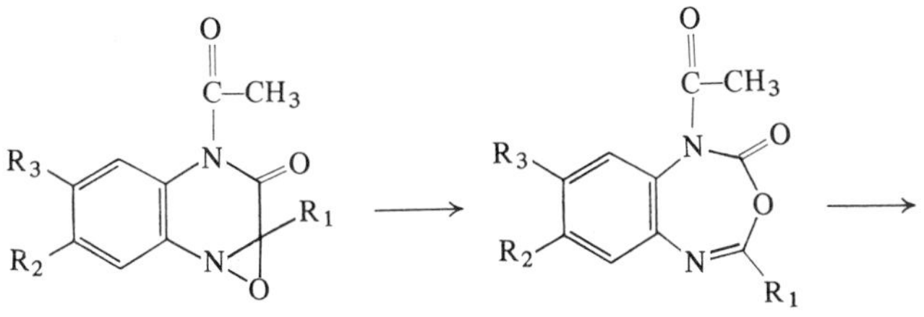

(4.833)   (4.834)

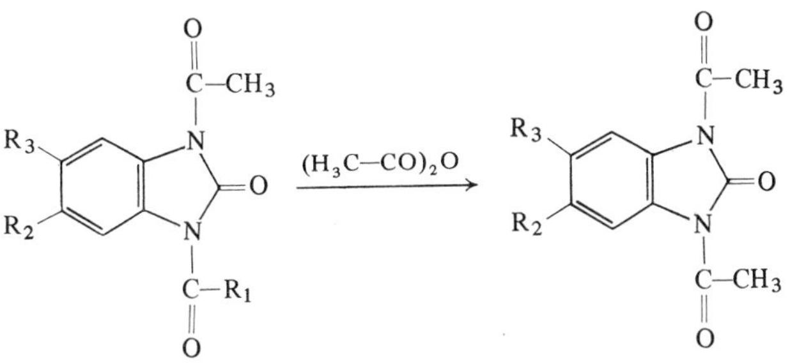

(4.835)   (4.836)

$R_1 = C_6H_5; p\text{-}NO_2C_6H_4; p\text{-}ClC_6H_4; CN. \ R_2 = H; OC_2H_5; OCH_3; Cl.$
$R_3 = H; Cl$

quinoxaline-1-oxide (4.832) (R_1 = C_6H_5, R_2 = R_3 = H) apparently takes place at an early stage in the reaction[557]. It has been speculated that the benzimidazolones are formed via the benzo[d] 1,3,6-oxadiazepine (4.834). The intermediates (4.833) and (4.834) have, however, not been isolated, although ultraviolet spectroscopy has revealed their presence as intermediates. Nevertheless, ring contractions of 1,3,5-benzoxadiazepines into N-aroylbenz-imidazoles have been reported[559], and, moreover, the ring expansion of quinoxaline-1-oxides (4.837) into the seven-membered oxadiazepine derivatives (4.839), probably via (4.838), has been demonstrated by uv irradiation of a solution of the quinoxaline-1-oxides[559].

(4.837)   (4.838)   (4.839)

R_1	C_6H_5	C_6H_5	H
R_2	C_6H_5	H	C_6H_5

2,4-Dioxotetrahydropteridine-5-oxides (4.840), on heating with acetic anhydride, show the same type of ring contraction, the 3-substituted 1-methyluric acids (4.844) being obtained[558]. Since [7-^{14}C]labelled (4.840) (R = CH_3) gave radioactive 1,3-dimethyluric acid and from [6-^{14}C]labelled (4.840) (R = CH_3) no radioactive 1,3-dimethyluric acid was obtained a mechanism must be proposed that essentially involves the elimination of the carbon atom at position 7. It is suggested that isomerization into (4.843) occurs via the 7-acetoxy compound (4.841) (because nucleophilic attack of the N-oxide oxygen on the neighbouring C_6 atom is now facilitated) and via the oxaziridino derivative (4.842); the compound (4.843) is then further hydrolysed into the uric acid (4.844).

2. RING TRANSFORMATIONS OF PYRIDAZINES AND PYRAZINES INTO OTHER SIX-MEMBERED HETEROCYCLES

a. *Pyridazines and pyrazines into pyrimidines; pyridazines into pyrazines; cinnolines into quinazolines; tetrahydropyridazines into oxadiazines*

Ring transformations of little preparative significance but of great theoretical interest are the photochemically induced conversions of pyrazine and its methyl derivatives in the vapour phase at 2537 Å and at low pressure. Pyrimidine is formed from pyrazine; a mixture of both 4- and 5-methylpyrimidine and

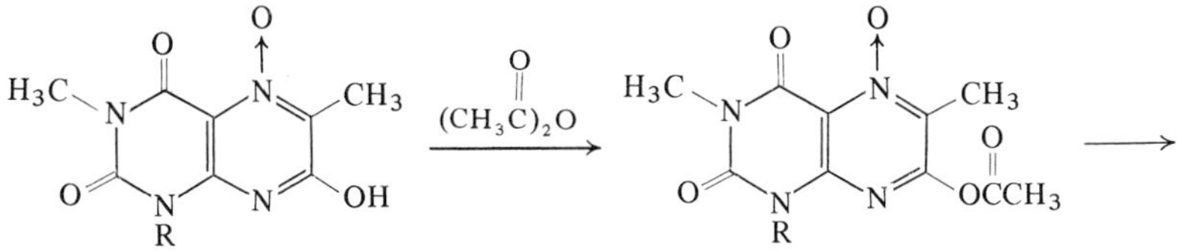

(4.840)   (4.841)

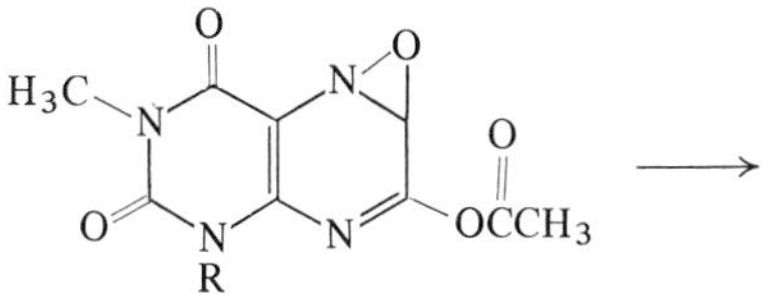

(4.842)

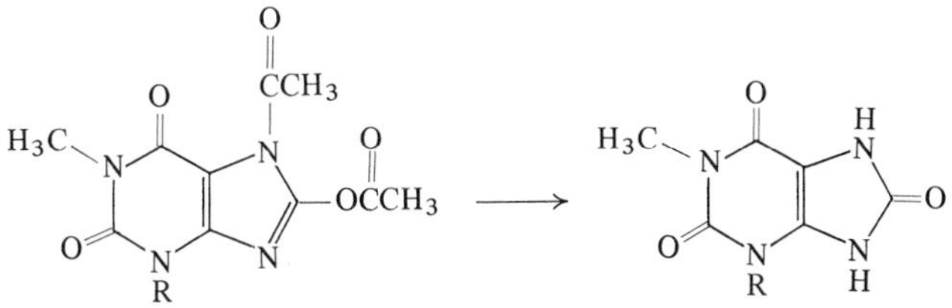

(4.843)   (4.844)

$$R = CH_3; C_6H_5$$

probably 2-methylpyrimidine is obtained from 2-methylpyrazine; 4,5-dimethyl-pyrimidine is yielded from 2,6-dimethylpyrazine, and a mixture of 2,5- and 4,6-dimethylpyrimidine[560] is obtained from 2,5-dimethylpyrazine. The results of the photo decomposition of the dimethylpyrazines are of special importance, since they make it possible for us to decide whether diaza derivatives of benzvalene, bicyclohexadiene and/or prismane are the intermediates in this reaction[547]. Surveying the results of the photo irradiation of 2,5-dimethylpyrazine (4.845) (for the other dimethylpyrazines similar reasoning can be used), it is clear that the exclusive formation of 2,5-dimethyl- (4.849) and 4,6-dimethylpyrimidine (4.847) and the fact that 2,4-dimethylpyrimidine (4.854) is not present, can only be explained by a pathway in which a diaza derivative of a benzvalene-type intermediate i.e. (4.846) and (4.848), occurs (see scheme on p. 172). Derivatives of diazabicyclohexadiene (4.850), (4.852) and (4.853) and diazaprismane (4.851) are therefore rather unlikely. The fact that these isomerizations occur with light of 2537 Å and not with light of wavelength 3130 Å has been interpreted as an indication that the excited state involves the singlet π-π^* transition and that triplet states do not play an important role.

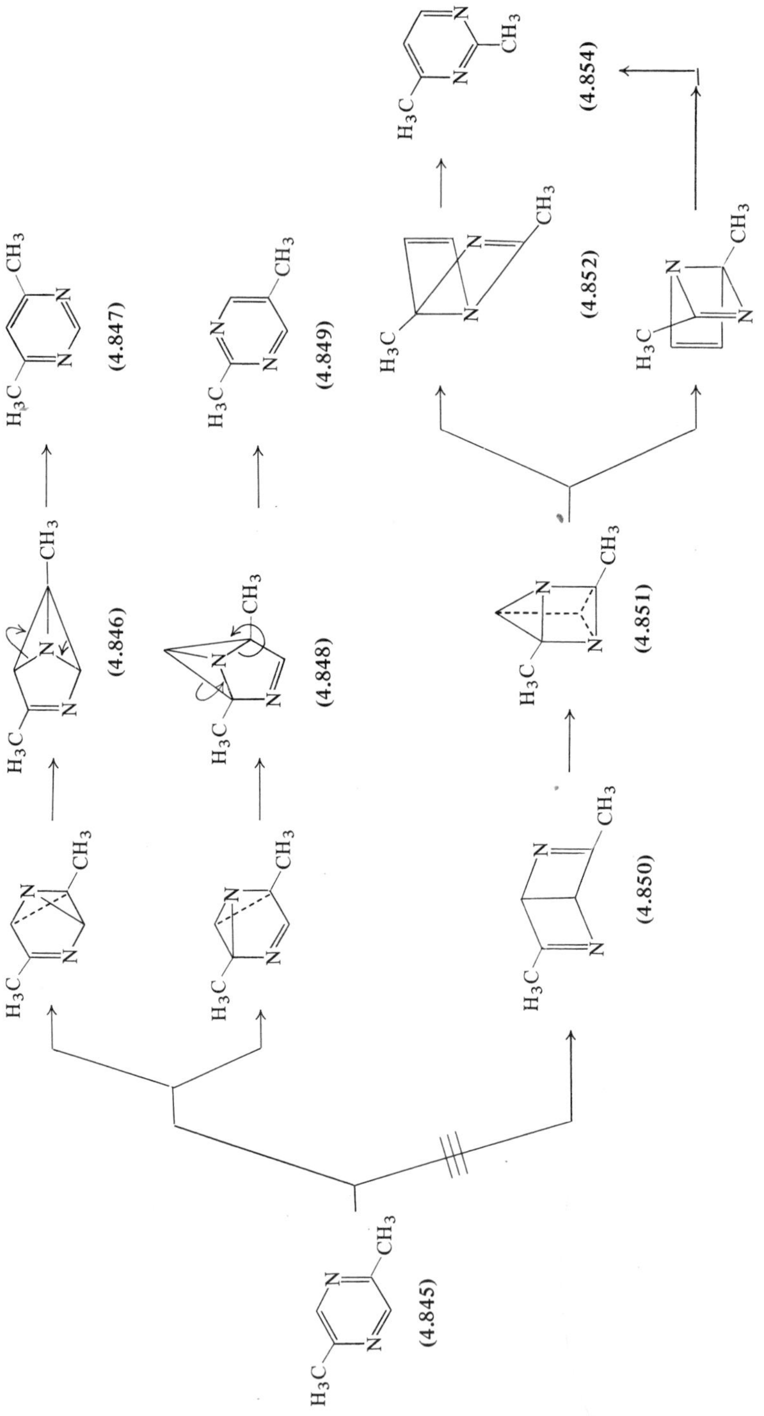
(4.845)
(4.846)
(4.847)
(4.848)
(4.849)
(4.850)
(4.851)
(4.852)
(4.853)
(4.854)

Irradiation of pyrazine in organic solvents with uv light of wavelength 2537 Å
leads to the formation of pyrimidine (1 per cent); no pyridazine has been
found[561]. Thermal decomposition of pyrazine gives isomerization into
pyrimidine (3 per cent), acetylene and hydrogen cyanide (36 per cent)[562].

A pyrazine-pyridazine photoisomerization has been found to occur when
2,5-difluoro-3,6-dichloropyrazine (4.854a) is irradiated with light of wavelength
254 nm, 3,6-difluoro-4,5-dichloropyridazine (4.854b) being obtained[562a]. It has
been established that the rearrangement starts from the n-π* singlet state; the
occurrence of a diazaprismane-type intermediate is unlikely since it would not
explain the formation of (4.854b).

(4.854a)   (4.854b)

Recently it has been reported that pyrolysis of perfluoropyridazine (4.855)
gives perfluoropyrimidine (4.857) (60 per cent) with about 3 per cent of
perfluoropyrazine (4.860), while photo irradiation of (4.855) gives exclusively
(4.860)[563]. The explanation for this is probably that different intermediate
stages in the reaction are involved. It has been proposed that the thermal
isomerization would occur via tetrafluorodiazabenzvalene (4.856) and that in
the photoreaction tetrafluorodiazaprismane (4.859) is formed via the diaza
Dewar benzene (4.858). Quite recently, the photochemical isomerization of
hexafluorocinnoline (9.860a) into hexafluoroquinazoline (4.860c) has been
reported[563a]; the benzodiazabenzvalene (4.860b) is suggested as intermediate.

When the 4-nitroso-1,4-dihydropyrazines (4.861) are treated with phosphorus
pentachloride, ring opening into the 1-amino-1,4-dihydropyrimidines (4.863)
occurs; this reaction also is important as a method of preparation[564]. The
seven-membered 1,2,4-triazepine (4.862) has been postulated as intermediate.
This reaction is closely analogous to the conversions of nitrosopyrroles into
pyrimidones-4 (chapter 3, section I.C.5.a) and of nitrosopyrazoles into 1,2,4-
triazines (chapter 3, section II.C.2.a).

The Diels–Alder adduct of azodibenzoyl and cyclopentadiene (4.864) gives
on thermolysis a quantitative isomerization into the *cis* bicyclic oxadiazine
(4.866)[565]. The reaction is found to be first order in a range of solvents (energy
of activation: 23.2 kcal mol^{-1}; entropy of activation: −9 e.u. (350 K)).
Spectrometric analysis of the reaction mixture during rearrangement showed one
or more isobestic points. These data, combined with the fact that addition of
dienophiles did not divert the reaction pathway strongly, point to a cyclic
mechanism, i.e. (4.864) → (4.865) → (4.866), which can formally be interpreted
as a [3,3]-sigmatropic shift.

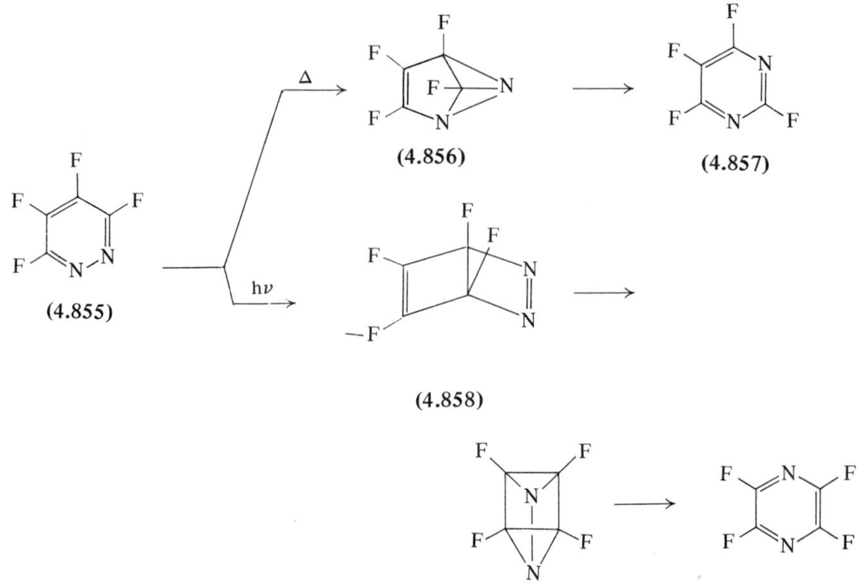

(4.855)

(4.856)

(4.857)

(4.858)

(4.859)

(4.860)

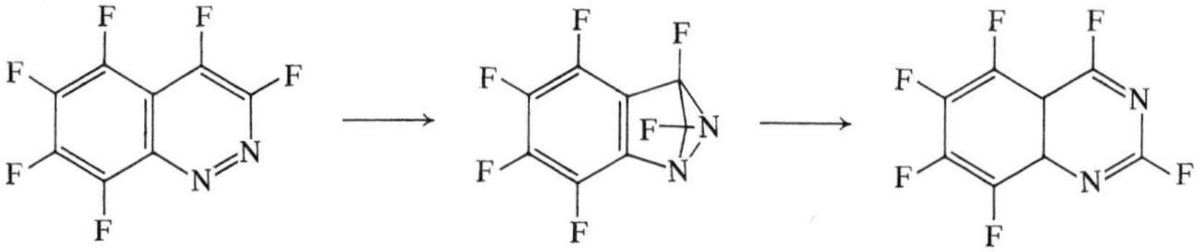

(4.860a)

(4.860b)

(4.860c)

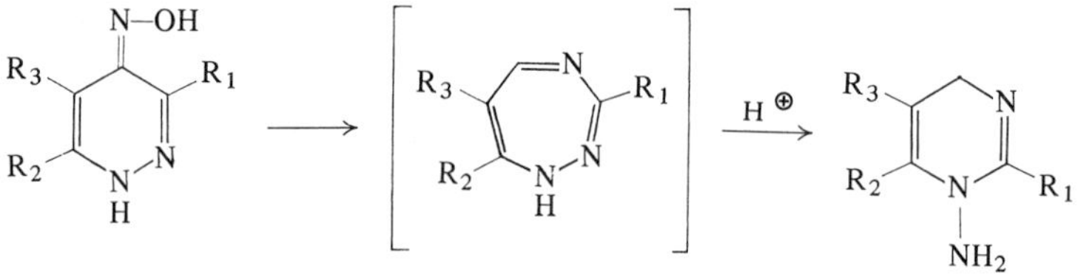

(4.861)

(4.862)

(4.863)

R_1	C_6H_5	C_6H_5	CH_3
R_2	C_6H_5	C_6H_5	C_6H_5
R_3	H	C_6H_5	H

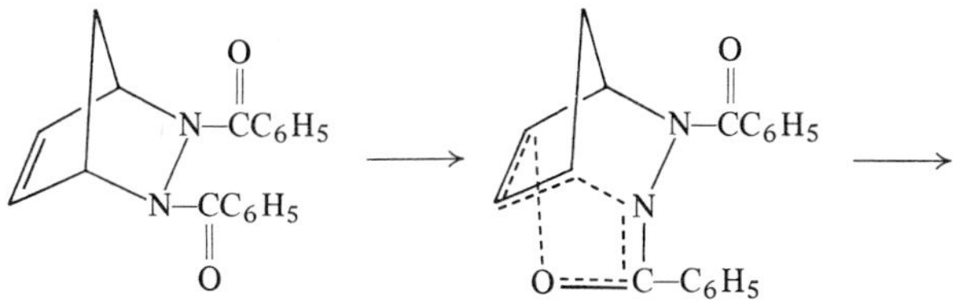

(4.864)   (4.865)

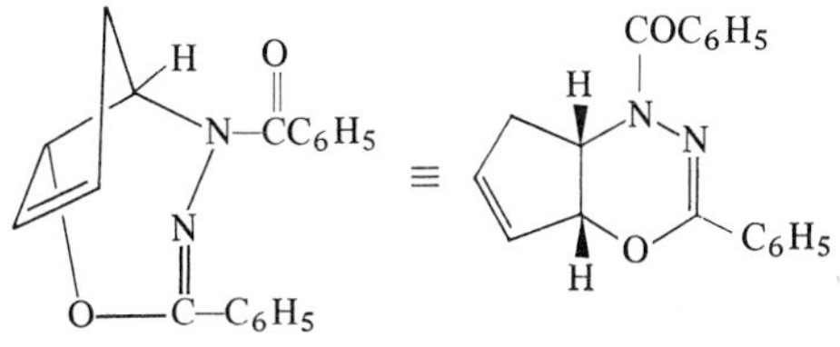

(4.866)   (4.866)

3. RING EXPANSIONS OF PYRIDAZINES AND QUINAZOLINES INTO SEVEN-MEMBERED HETEROCYCLES

a. Pyridazines into 1,2-diazepines; quinoxalines into oxadiazepines

The 1,3-dipolar adduct of diazopropane and the pyridazone (4.867), i.e. (4.868), gave, on thermal decomposition a mixture of the compounds (4.869) (46 per cent), (4.870) and (4.871)[566]. In the photochemical decomposition of (4.868) ($\lambda > 300$ nm) in benzene, the yield of the 1,2-diazepinone (4.869) is lower (12 per cent). In a photo-sensitized decomposition (benzophenone) 35 per cent of (4.869) is yielded. The formation of (4.869) was rationalized by a radical mechanism involving (4.872) in which, by a 3,4-bond cleavage, a compound with two relatively stable radicals (4.873) is formed; recombination gives (4.869). For the ring expansion of quinoxalines into oxadiazepines, see section II.D.1d.

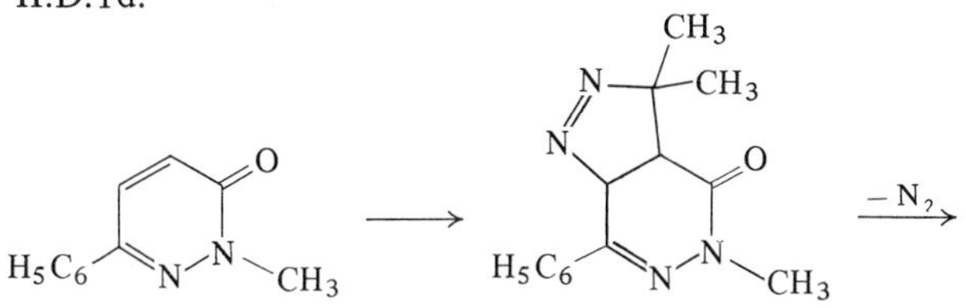

(4.867)   (4.868)

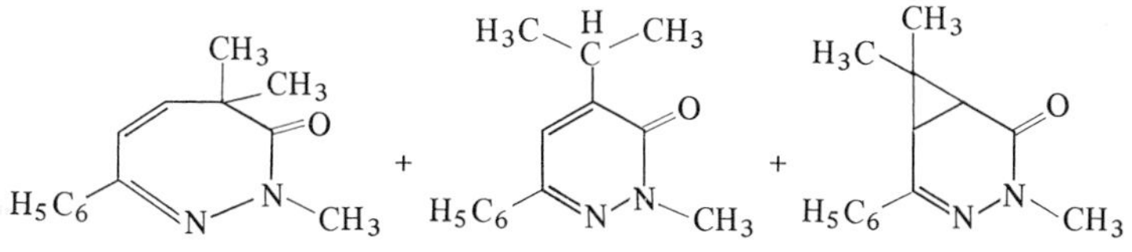

(4.869)   (4.870)   (4.871)

(4.868) $\longrightarrow$ (4.872) $\longrightarrow$ (4.873) $\longrightarrow$ (4.869)

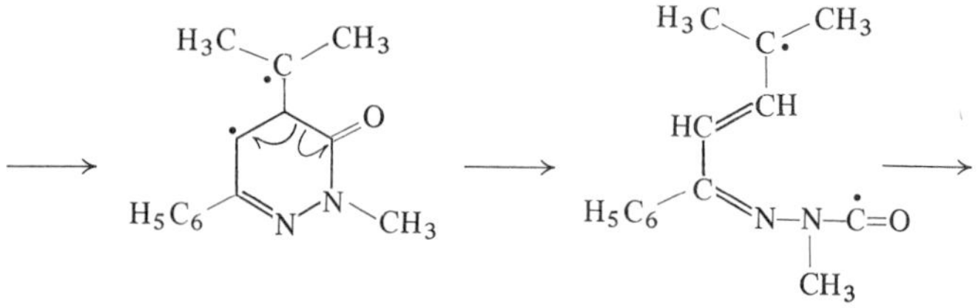

III. RING TRANSFORMATIONS OF SIX-MEMBERED HETEROCYCLES CONTAINING TWO DIFFERENT HETEROATOMS

A. Ring transformations of six-membered heterocycles containing one oxygen and one sulphur atom

1. RING CONTRACTION INTO FIVE-MEMBERED HETEROCYCLES

a. *Oxathiins into furans*

A number of alkyl- and/or aryl-substituted furans are conveniently prepared by heating 1,2-oxathiin-2,2-dioxides[567,568]. This reaction is important as a method of preparation since the 1,2-oxathiin-2,2-dioxides (4.874) (δ sultones), are easily prepared (from α,β- or β,γ-unsaturated ketones by treatment with a cold mixture of concentrated sulphuric acid and acetic anhydride[567,569]) and the ring contraction can readily be accomplished when an intimate mixture of the compounds (4.874) with calcium oxide and quinoline is heated at 230°C. A large number of 2,4-disubstituted, 2,3,5- and 2,4,5-trisubstituted and 2,3,4,5-tetrasubstituted furans (4.875) can be prepared by this method without difficulty. Not much information is available on the mechanism of these reactions.

Thermolysis of phenoxathiin (4.876), a derivative of 1,4-oxathiin, at 200°C in the presence of copper has been reported to give sulphur extrusion, yielding the dibenzofuran (4.877)[570]. However, these results could not be confirmed[571,572], even when (4.876) was heated (with reduced copper bronze) at temperatures of 285 to 290°C.

2. RING TRANSFORMATIONS INTO OTHER SIX-MEMBERED HETEROCYCLES

a. *Tetrahydro-1,4-oxathiin into tetrahydro-1,4-dithiin*

It has been reported that 2,2,3-trichlorotetrahydro-1,4-oxathiin (4.878) and the isomeric 2,3,3-trichlorotetrahydro-1,4-oxathiin (4.879) are hydrolytically con-

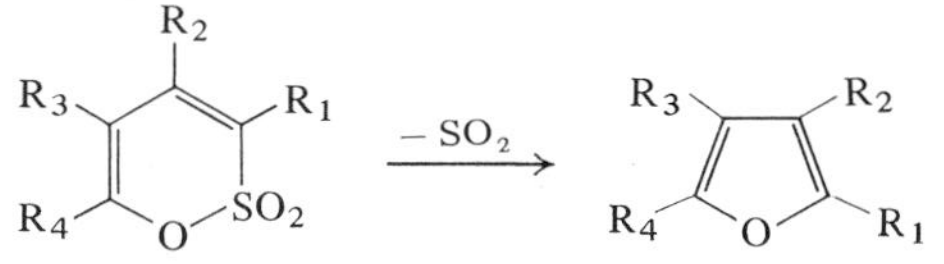

(4.874) $\xrightarrow{\;-SO_2\;}$ (4.875)

R1	CH3	n-C3H7	t-C4H9	C6H5	CH3	C2H5	CH3	CH3	CH3	CH3	CH3	CH3
R2	H	H	H	H	CH3	CH3	C6H5	H	(CH2)4	(CH2)4	CH3	(CH2)4
R3	CH3	CH3	CH3	CH3	C6H5	CH3	H	(CH2)3	(CH2)4	CH3	(CH2)4	(CH2)4
R4	H	H	H	H	H	CH3	CH3	H	CH3	CH3	(CH2)4	(CH2)4

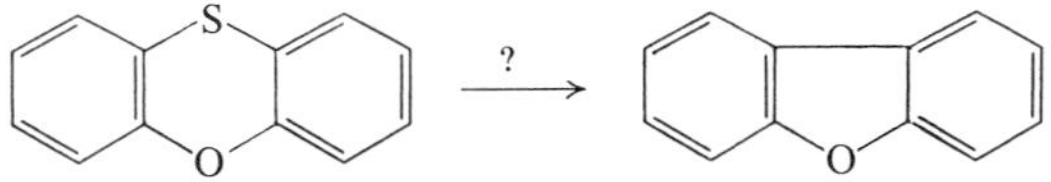

(4.876) $\xrightarrow{\;?\;}$ (4.877)

verted into tetrahydro-1,4-dithiin (4.881)[573]. Mercaptoethanol (4.880) is the intermediate primarily formed; on dehydration it yields (4.881).

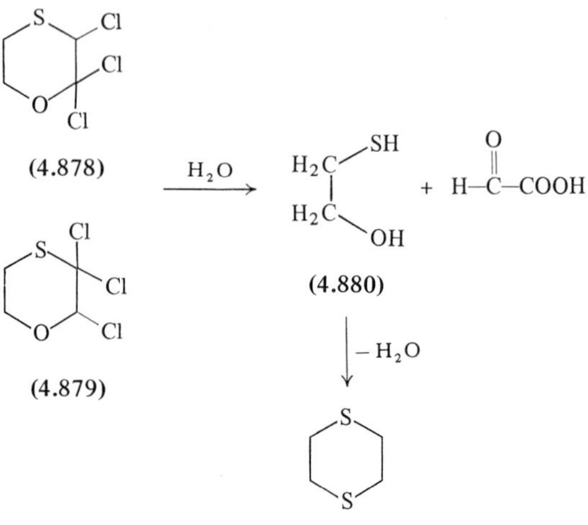

3. RING EXPANSION REACTIONS

a. *3,1-Benzoxathians into dibenzodithiocins*

Two reports on the ring expansion of 3,1-oxathians deal with the photolysis of the 3,1-benzoxathian-4-one (4.881a) in chloroform or hydrocarbon as a solvent. It results in the formation of 6*H*, 12*H*-dibenzo[*b,f*] [1.5] dithiocin (4.881f)[573b] and not of the isomeric 11*H*, 12*H*-dibenzo[*c,g*] [1.2] dithiocin (4.881e) as originally proposed[573a]. When the photolysis was carried out at 77° K in an infrared cell, a species with a v(C=O) frequency at 1803 cm^{-1} was obtained. This absorption band disappeared on heating to −40°C while simultaneously a 900 cm^{-1} absorption band which is characteristic for (4.881f) appeared. The 1803 cm^{-1} species is thus clearly a precursor of (4.881f). It has been postulated that this species is 2-thiobenzpropiolactone (4.881g). It is proposed that in the formation of (4.881g) a photo-induced C–S bond fission as well as a C–O bond fission, leading to the intermediates (4.881b) or (4.881c) respectively, may occur, although there are insufficient data to distinguish between these two possible intermediates. It has been found that these bond fissions are thermally reversed. The resonance-stabilized intermediate (4.881d) is formed by thermal expulsion of an aldehyde or ketone.

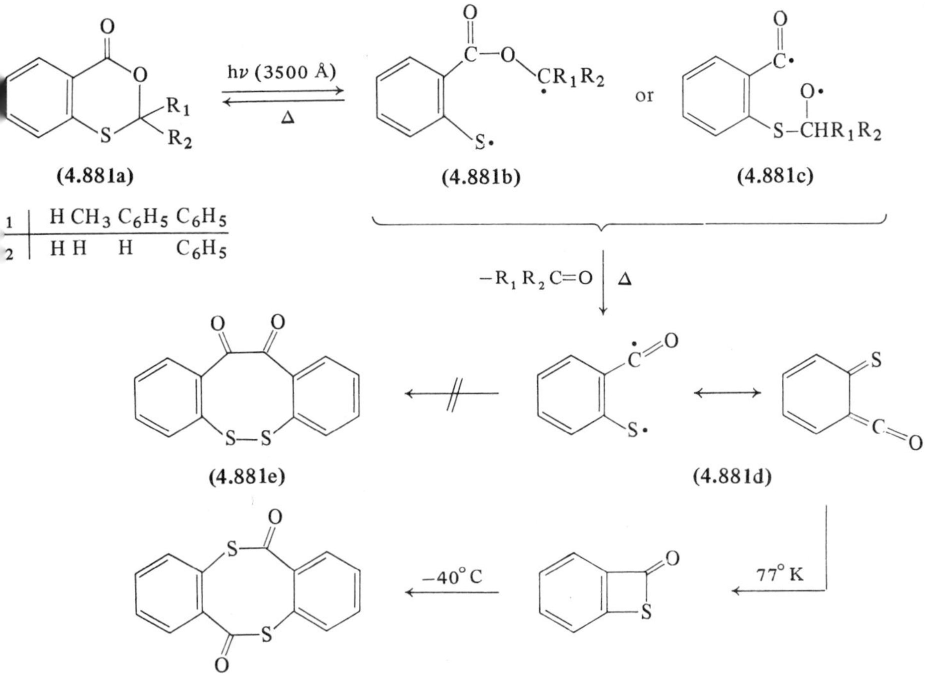

(4.881a)   (4.881b)   (4.881c)

| 1 | H | CH$_3$ | C$_6$H$_5$ | C$_6$H$_5$ |
| 2 | H | H | H | C$_6$H$_5$ |

$-R_1R_2C=O$ Δ

(4.881e)   (4.881d)

(4.881f)   (4.881g)

$-40°$ C $77°$ K

B. Ring transformations of six-membered heterocycles containing one oxygen and one nitrogen atom

1. RING TRANSFORMATIONS OF 1,2-OXAZINES

a. 3,6-Dihydro-1,2-oxazines into Δ³-pyrrolines, pyrroles, pyridines and α-pyrones

3,6-Dihydro-1,2-oxazines (4.882), compounds which are readily synthesized by treating nitroso compounds with substituted 1,3-butadienes, are found to be excellent starting materials for the preparation of N-heterocycles, mainly Δ³-pyrrolines and pyrroles. Which of these compounds is formed depends on the character of the substituents present in the dihydro-1,2-oxazines[574]. Several methods are available for bringing about this ring contraction; which one should be used depends on the structure of the compound. A study has been made of the reductive ring contraction of (4.882) into Δ³-pyrrolines (4.884) with zinc and acetic acid at enhanced temperatures. It is of interest that by using these reagents, the fission of the N–O bond occurs with retention of the double bond[575,576]. When the same reaction is carried out at low temperature the

4-amino-but-2-ene-1-ol (4.883) is obtained; it is obviously an intermediate during the conversion of (4.882) → (4.884).

(4.882)   (4.883)   (4.884)

R_1	H	H	H	CH_3	C_6H_5
R_2	C_6H_5	p-$CH_3C_6H_4$	C_2H_5	C_2H_5	H
R_3	C_6H_5	p-$CH_3C_6H_4$	C_2H_5	C_2H_5	CH_3

A reductive ring contraction, leading to a pyrrole derivative instead of a pyrroline derivate, has been found to occur with the adduct of p-chloronitroso-benzene with 1-acetoxybutadiene, i.e. (4.885): besides reductive N—O fission, the acetyl group is removed, leading to γ-aminobutenal (4.886) which cyclizes into the N-arylpyrrole (4.887)[577]. This ring closure must be very rapid as in the presence of an excess of phenylhydrazine no hydrazone can be isolated.

(4.885)   (4.886)   (4.887)

It is interesting that the adducts of nitroso compounds with 1,3-butadiene-1-carboxylate, i.e. (4.888) react during reduction with zinc and acetic acid not into the corresponding pyrroline derivatives (4.890) as expected but into 3-hydroxy-3,6-dihydropyridone-2 (4.891), which, when boiled with acid, can be converted into the N-substituted pyridone-2 (4.892)[577]. Apparently in the case of 4-aminobutenol (4.889) loss of methanol occurs showing that dehydration into (4.890) is not the favoured pathway.

Ring contraction of 3,6-dihydro-1,2-oxazines into Δ^3-pyrrolines can also be achieved by treatment with acids. The compound (4.893), when heated with 33 per cent phosphoric acid at 60°C, gives the N-(p-hydroxyphenyl)-Δ^3-pyrroline (4.896)[578]. It is possible that this conversion occurs via the resonance stabilized phenyl derivative (4.894) and the compound p-(4-hydroxy-2-butenyl-amino)phenol (4.895) as intermediates. The last compound can be isolated when

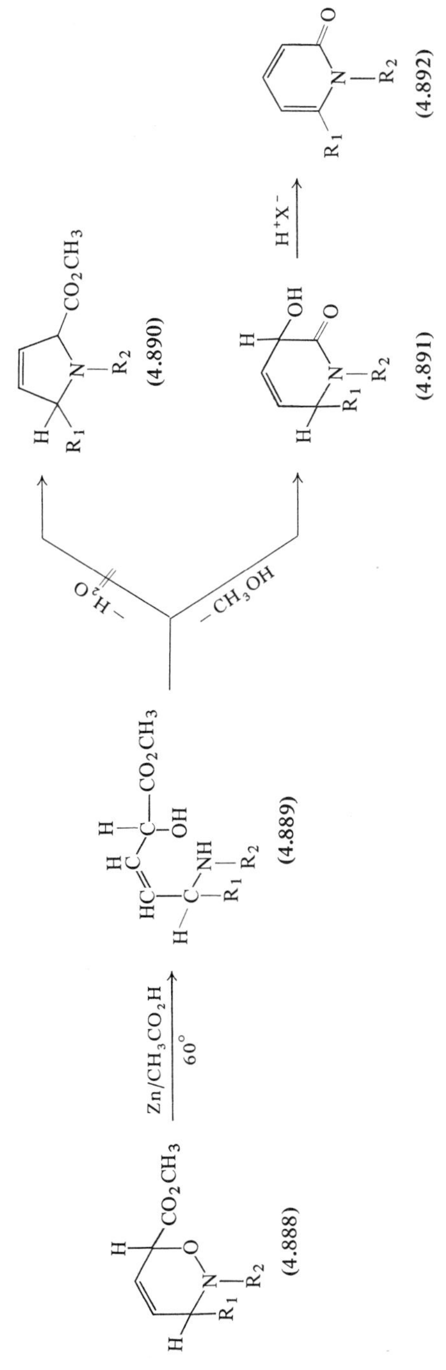

(4.888)

$\xrightarrow[60°]{Zn/CH_3CO_2H}$

(4.889)

$-H_2O$

$-CH_3OH$

(4.890)

(4.891)

$\xrightarrow{H^+X^-}$

(4.892)

the reaction is carried out with somewhat milder conditions (room temperature, 20 per cent phosphoric acid) and it can easily be dehydrated into (4.896). The ring contraction can thus be postulated to take place by the following series of reactions:

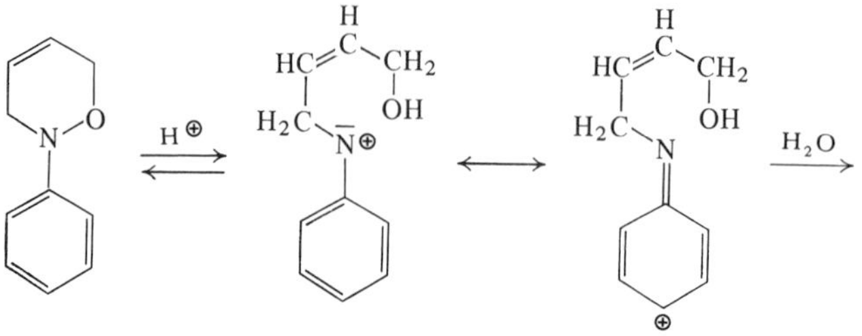

(4.893)   (4.894)   (4.894)

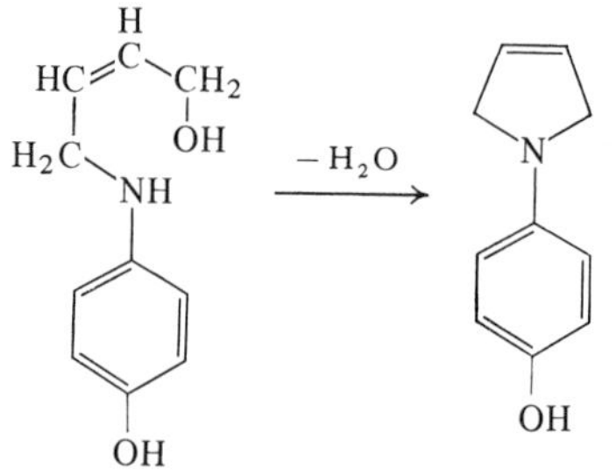

(4.895)   (4.896)

    The same reaction also occurs with hydrogen chloride or hydrogen bromide, yielding N-(p-halogenophenyl)Δ^3-pyrroline[579]. The adducts of nitroso compounds with 1,3-butadiene-1-carboxylates, i.e. (4.888), on treatment with triethylamine in methanol or, more effectively, by chromatography with a 10 to 20 fold excess of aluminium oxide in a non-polar solvent, are very smoothly converted into pyrrole-2-carboxylates (4.897). The fact that the ring contraction of the 3,6-dihydro-1,2-oxazines containing a carbonyl group on position 6 readily occurs can be shown in the reaction of nitrosobenzene with ketones of the type (4.898); at room temperature there is no formation of the adduct (4.899), but rearrangement of the pyrrole ketone (4.900) subsequently occurs by ring contraction of this adduct[580]. In the absence of a carbonyl function at position 6, higher temperatures are necessary for the reaction to occur. For example, 3,6-dihydro-1,2-oxazine itself needs a much higher temperature (about 170-180°C) to be converted into pyrrole; but nevertheless the yield is excellent[581].

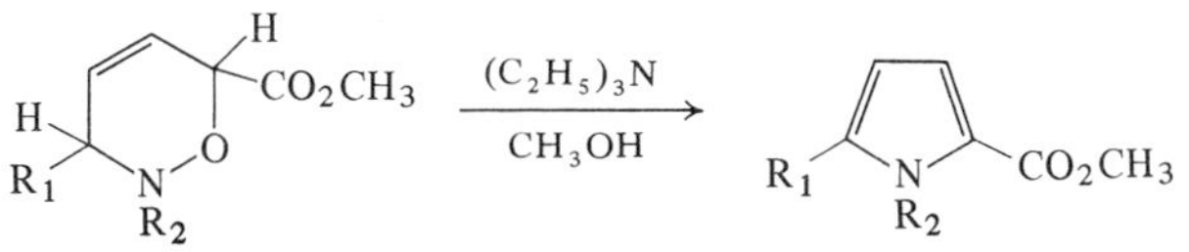

R₁ = H; CH₃; C₆H₅; CO₂CH₃. R₂ = C₆H₅; p-Cl C₆H₄; p-Tosyl

(4.888) → [(C₂H₅)₃N / CH₃OH] → (4.897)

$R_1 = H; CH_3; C_6H_5; CO_2CH_3. R_2 = C_6H_5; p\text{-}Cl\,C_6H_4; p\text{-}Tosyl$

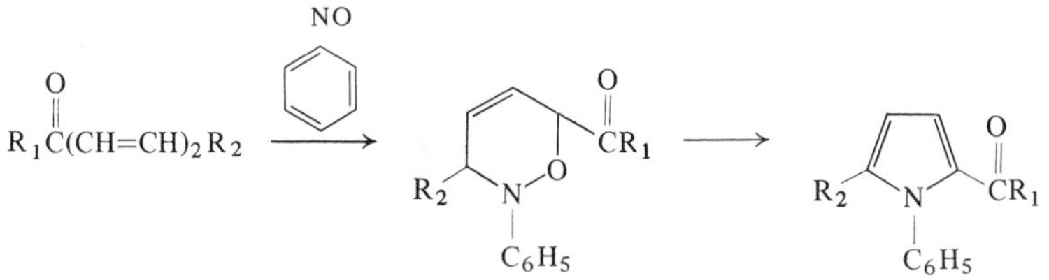

$R_1\overset{O}{\overset{\|}{C}}(CH=CH)_2R_2$ (4.898) → (4.899) → (4.900)

R₁ = C₆H₅; p-C₆H₅C₆H₄; p-NO₂C₆H₄; m-NO₂C₆H₄; p-OCH₃C₆H₄; 2-thienyl; C—(CH₃)₃.

R₂ = C₆H₅; 2-furyl.

The important step in the ring contraction reaction of (4.888) into (4.897) is again the fission of the nitrogen-oxygen bond of the 1,2-oxazine ring which yields the α-keto-δ-amino esters (4.902). This fission is initiated by the formation of the anionic intermediate (4.901)[577,582], which process of decomposition is similar to that occurring in the known conversion of N-alkoxyammonium salt into aldehydes and tertiary amines by treatment with base[583]. Compounds of type (4.902) are known[584] to cyclize readily into pyrrolines; the ring closure involved is favoured by the *cis* orientation of the double bond. Subsequently the 2-hydroxy-Δ³-pyrrolines (4.903) aromatize by dehydration with great ease.

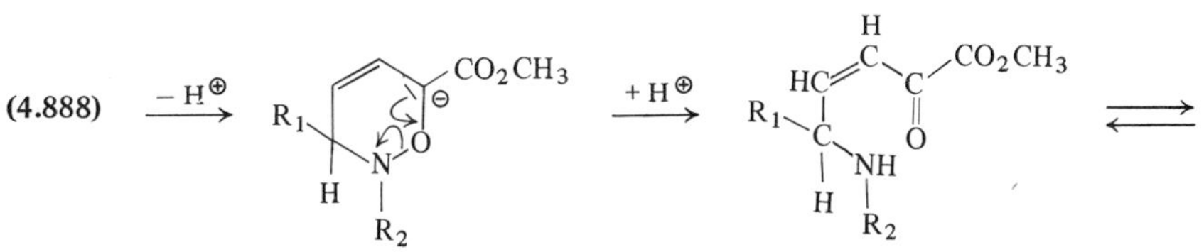

(4.888) —$-H^{\oplus}$→ (4.901) —$+H^{\oplus}$→ (4.902) ⇌

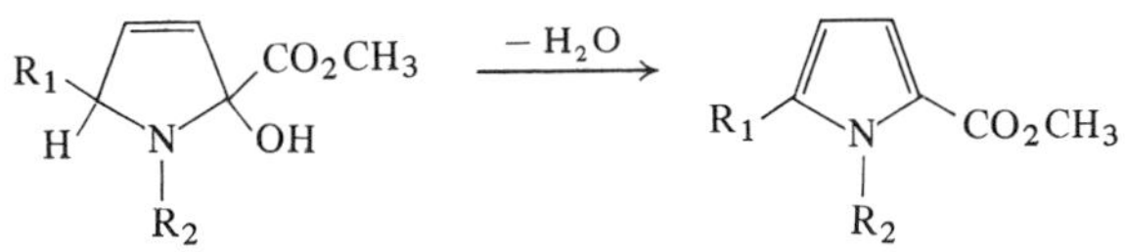

(4.903) —$-H_2O$→ (4.897)

Let us now examine more closely the various steps in the mechanism shown on p. 183. As indicated in this scheme, the conversion is base catalysed. This reaction thus requires a hydrogen in position 6 which is sufficiently acidic to be ionized under these conditions. It is to be expected that groups at position 6, which stabilize the anion formed, i.e. alkoxycarbonyl groups, aldehyde groups, ketones, cyano- or nitro groups, and π-deficient heterocycles, will promote the pyrrole formation[574,577]. The following example will serve to illustrate this point: it appears that the compounds (4.904) (R = α- or γ-pyridyl) give ring contraction into (4.906) in a weak basic medium, while (4.904) (R = β-pyridyl) under identical conditions does not react at all; the carbanion (4.905), which has α- and γ-pyridyl in position 6, is evidently more resonance-stabilized. From these considerations it is clear that the reaction can also be carried out in an acidic medium: protonation of the pyridyl group enhances its electron-accepting property and therefore promotes the reaction[582].

(4.904)   (4.905)   (4.906)

R = α-, β- or γ-pyridyl

The rate of the base-catalysed step, (4.888) → (4.901), appears to be much slower than that of the conversion (4.901) → (4.902). This conclusion is based on the experimental finding that when the reaction of (4.888) (R_1 = CH_3, R_2 = p-ClC_6H_4) was carried out with triethylamine and D_2O in pyridine, no H/D exchange in (4.888) was observed. Moreover, when the reaction mixture was investigated by nmr spectroscopy at several intervals, a continuous formation of the pyrrole (4.897) (R_1 = CH_3, R_2 = p-ClC_6H_4) was observed[577].

The intermediary formation of a 2-hydroxy-Δ^3-pyrroline (4.903) in the ring contraction is very likely because of the fact that it is possible to isolate from the reaction mixture, obtained when (4.888) (R_1 = CH_3; R_2 = p-tosyl) reacts with aluminium oxide, two stereoisomeric Δ^3-pyrrolines, the compounds (4.907) and (4.908), which can both be converted into the same pyrrole, i.e.

(4.907)   (4.908)

(4.897) (R_1 = CH_3, R_2 = p-tosyl)[577]. In the sequence of reactions, the ring closure of the γ-aminoketone into the 2-hydroxy-Δ^3-pyrroline is found to be reversible: when the conversion of (4.909) → (4.912) is carried out with potassium hydroxide in methanol instead of triethylamine in methanol, the concentration of the keto-amino compound (4.910) is found to be increased. This is evident from the fact that in the reaction of (4.909) with potassium hydroxide in methanol, besides the N-substituted pyrrole (4.912), the hydroxypyridone-2 (4.913) is obtained[577]; it is formed from (4.910) by loss of methanol. Apparently the dehydration of (4.910) → (4.911) is considerably retarded and this enables (4.910) to enter into side reactions.

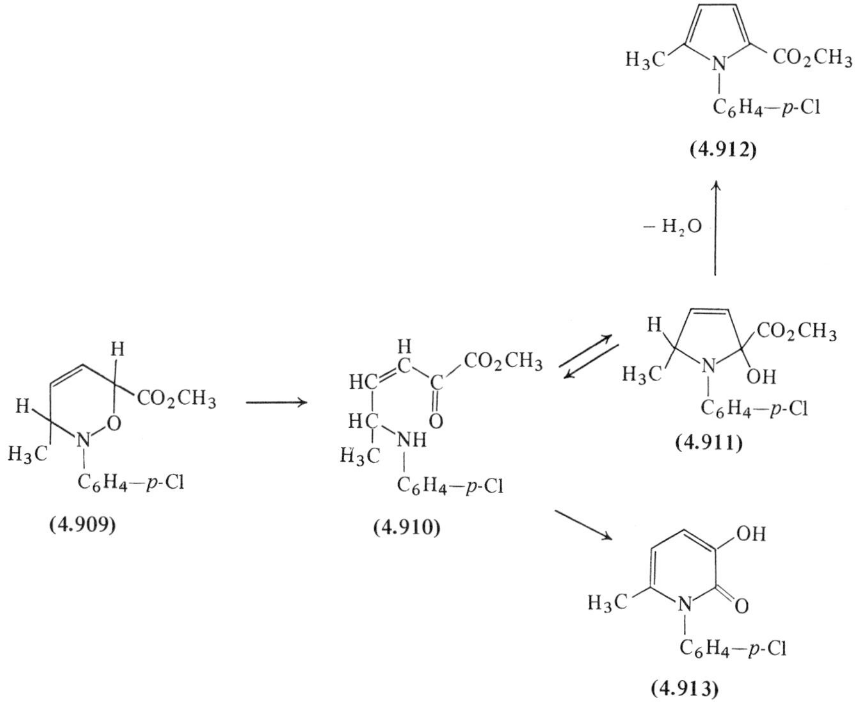

(4.912)

$-H_2O$

(4.911)

(4.909)

(4.910)

(4.913)

In contrast to the 6-methoxycarbonyl derivative (4.909), 3-methoxycarbonyl-3,6-dihydro-1,2-oxazine (4.914) gives with aluminium oxide the 3-amino-α-pyrone (4.917)[577]. The mechanism is analogous to that given for the formation of pyrroles from the 6-alkoxycarbonyl compound and can be described to occur as an isomerization of (4.914) via (4.915) into (4.916). It would be expected from the results discussed above—and indeed has been verified—that the 3,6-bis(methoxycarbonyl)-3,6-dihydro-1,2-oxazine (4.919) shows the characteristic formation of both a pyrrole, i.e. (4.918), and the α-pyrone (4.920)[577].

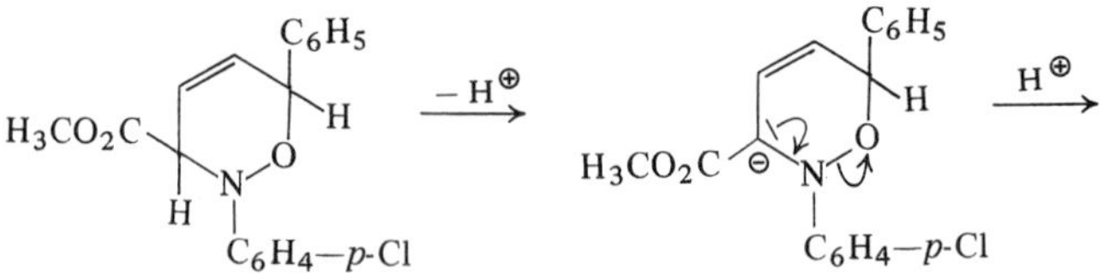

(4.914) (4.915)

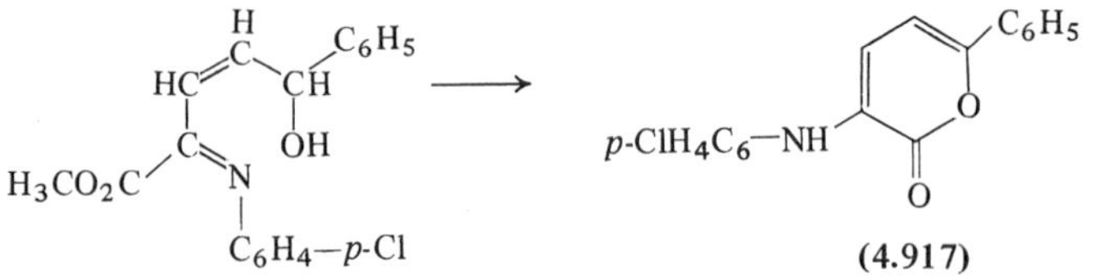

(4.916) (4.917)

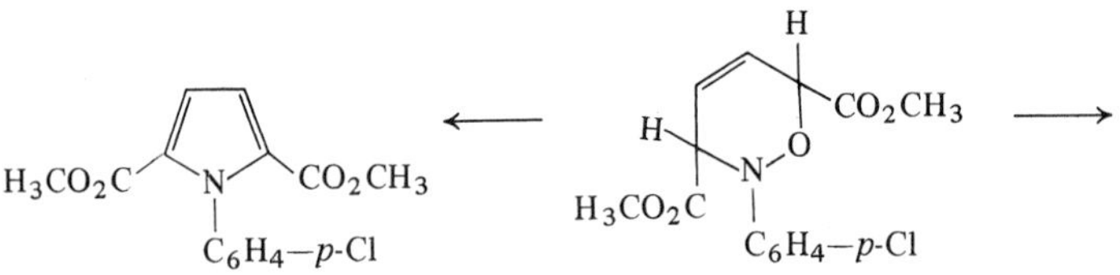

(4.918) (4.919)

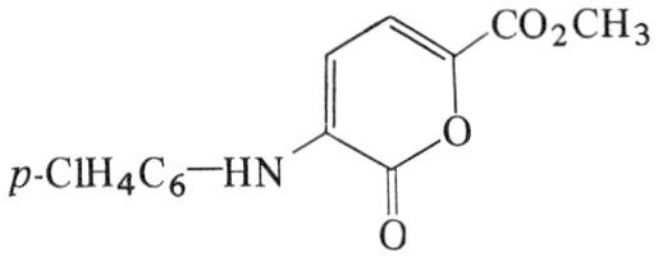

(4.920)

Electrolysis of appropriately substituted 1,2-oxazines can also lead to ring contraction. It has been found that the phthalimidine (4.920d) is formed when the 4-(4'-methoxyphenyl)-2,3-benzoxazin-1-one (4.920a) is electrolysed in acid[584a]. The reaction of (4.920a), which compound contains a cyclic oxime structure, involves an initial two-electron hydrogenolysis of the protonated nitrogen-oxygen bond into the ketimine (4.920b) which is found to be stable and in fact can be isolated. A further two-electron reduction gives, via (4.920c), the compound (4.920d).

Recently the photolysis of the 3,6-dihydro-2-phenyl-1,2-oxazines (4.921) into the N-phenylpyrroles (4.925) has been reported[585]. Also this reaction provides us with a convenient method for the preparation of substituted pyrroles. The reaction very probably involves the formation of the cis-γ-amino-α,β-unsaturated ketone (4.923) which is formed by an initial homolysis of the nitrogen-oxygen bond, i.e. (4.922), followed by intramolecular allylic hydrogen abstraction. Ring closure to the Δ^3-pyrroline (4.924) and subsequent dehydration yields (4.925).

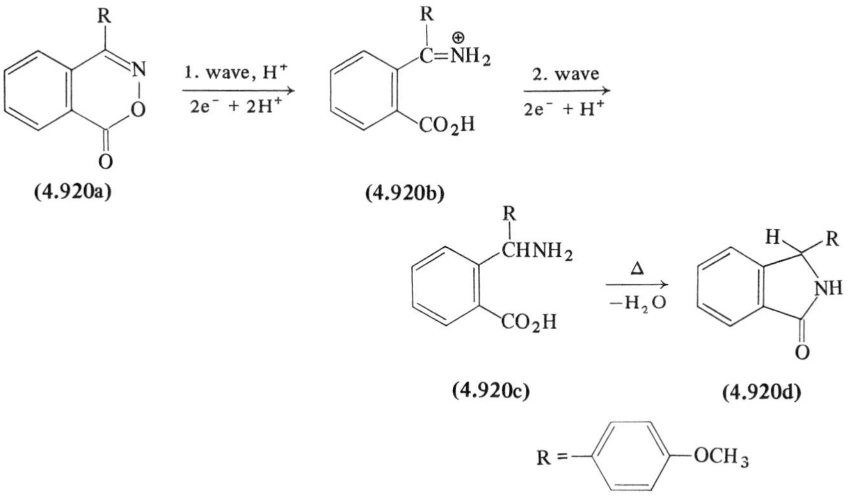

(4.920a)   (4.920b)

(4.920c)   (4.920d)

$$R = \text{---}\langle\text{benzene ring}\rangle\text{---OCH}_3$$

Direct evidence for a γ-aminoketone is provided by the fact that photolysis of 3,6-dimethyl-2-phenyl-3,6-dihydro-1,2-oxazine at a temperature of $-180°C$ (liquid nitrogen) gave the appearance of a peak at 1695 cm^{-1} which was assigned to a carbonyl frequency for a *s-cis* orientation in an α,β-unsaturated ketone[586].

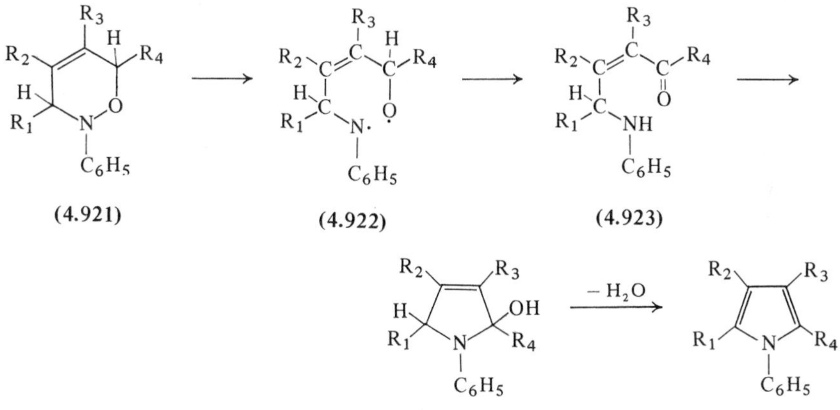

(4.921)   (4.922)   (4.923)

(4.924)   (4.925)

R_1	H	CH$_3$	H	C$_6$H$_5$	H
R_2	H	H	CH$_3$	H	H
R_3	H	H	CH$_3$	H	H
R_4	H	CH$_3$	H	C$_6$H$_5$	C$_6$H$_5$

Irradiation of a solution of 1,1,4-triphenyl-2,3-benzoxazine (4.927) in benzene or in acetonitrile with Pyrex-filtered light gives, very interestingly, the photo product (4.929); this is also formed by uv irradiation of the nitrone (4.926)[587]. With regard to the mechanism of the photo isomerization of the benzoxazine (4.927) into the oxazirane (4.929), a direct transformation appears likely, although intervention of the nitrone (4.926) or the nitroso compound (4.928) cannot be excluded.

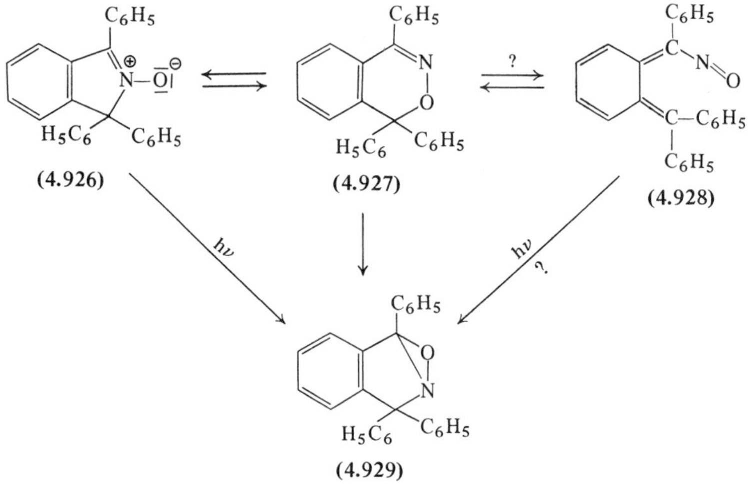

Thermolysis of the perfluorotetrahydro-2-methyl-1,2-oxazine (4.930) at 350°C in the presence of iron gives perfluoro 1-methyl-2-pyrrolidone (4.932) as main product. In the presence of platinum at 480°C the perfluoro-1-methylazetidine (4.931) is the major product; (4.932) is only obtained in traces[588]. The effect of the different behaviour exhibited by (4.930) towards platinum and iron upon defluorination[589] can be depicted as opposite.

2. RING TRANSFORMATIONS OF 1,3-OXAZINES

a. *1,3-Oxazines into pyrimidines; 3,1-benzoxazines into quinazolines and benzimidazoles*

The ring transformation of 1,3-oxazines and their benzo derivatives into pyrimidine and quinazoline derivatives respectively is brought about by treatment with ammonia or appropriate primary amines: 2,4,5-triphenyl-1,3-oxazin-6-one (4.933) is converted into the 4-pyrimidones (4.934)[590], and the dihydro-2,4-dioxo-1,3-oxazines (4.935) are converted into the uracils (4.936)[591]. In the bicyclic series, a great number of differently substituted 3,1-benzoxazones (4.937) have been treated with ammonia or primary alifatic or aromatic amines

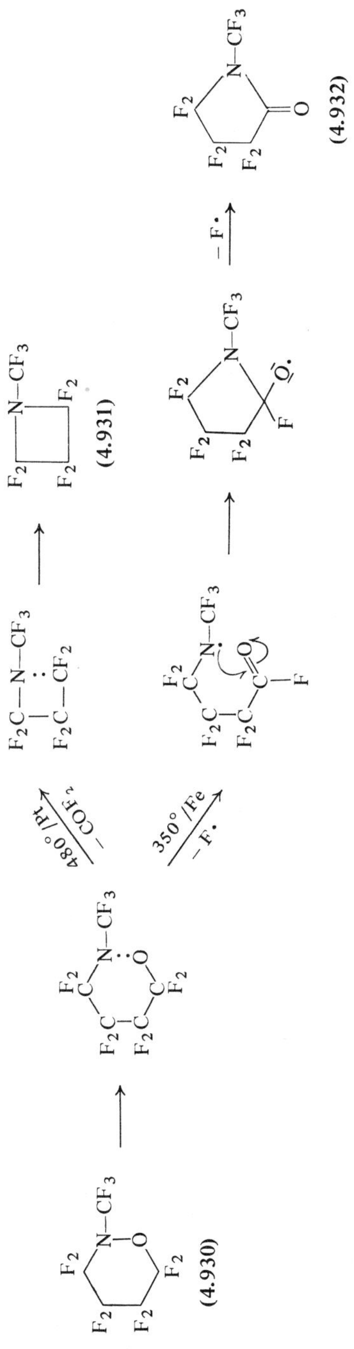

(4.930) (4.931) (4.932)

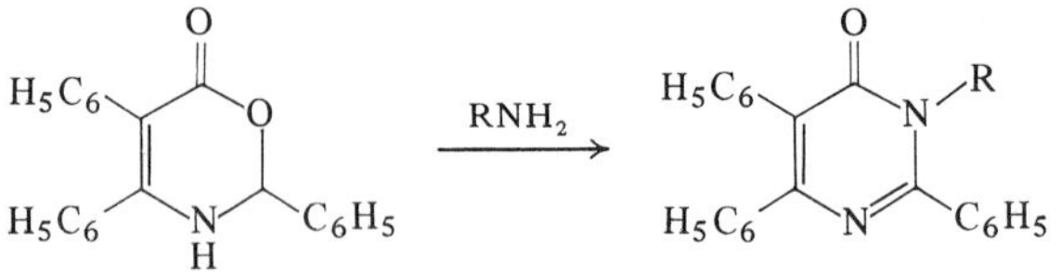

(4.933)   (4.934)

R = CH$_3$; C$_2$H$_5$; CH$_2$CH$_2$N(C$_2$H$_5$)$_2$;
CH$_2$CH$_2$NH$_2$; CH$_2$CH$_2$OH

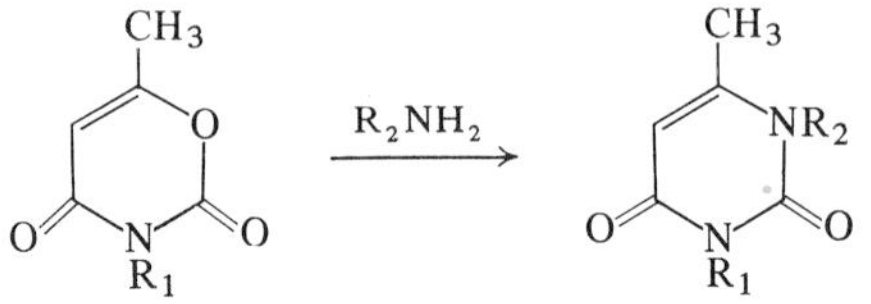

(4.935)   (4.936)

R$_1$	C$_2$H$_5$	C$_2$H$_5$	C$_6$H$_5$
R$_2$	H	C$_2$H$_5$	CH$_3$

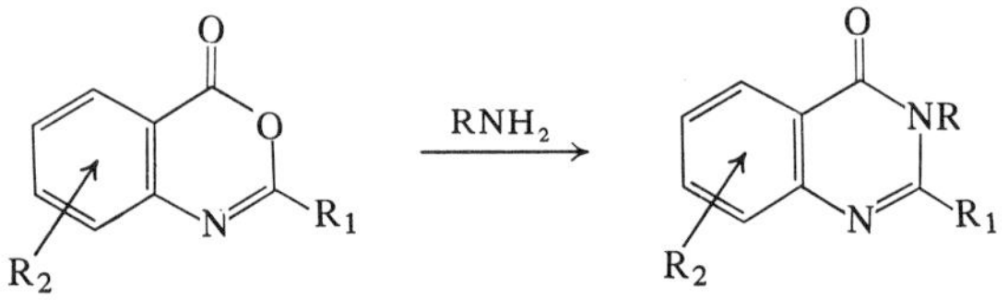

(4.937)   (4.938)

to give, in an identical manner, 4-oxo-3,4-dihydroquinazolines (4.938)[592,593].
With dibasic amines the same conversion into (4.938) has also been
reported[596,597]. For a review on these conversions and a survey of the
compounds obtained by this method, we refer to the reference cited in reference
592. The reaction of benzoxazones with amines was shown to occur in two
steps: the first involved ring opening by fission of the C$_4$–O bond into an
o-amidobenzamide (4.939), and the second step was the ring closure of this
amide[594]. By this mechanism it is possible to explain the influence of the
nucleophilic nature of the amines and the steric influence of groups of different

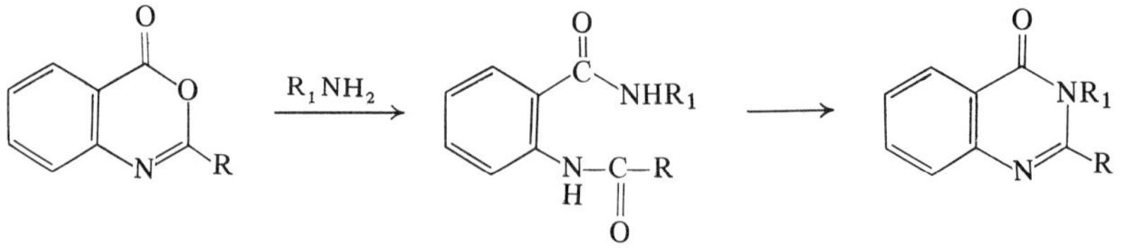

(4.939)

size and electronic character present in the amine and in the benzoxazones, upon the course of the reaction. From steric considerations it is clear that amines having a primary amino group directly attached to a tertiary carbon atom cannot react with benzoxazones[595].

Recently, the reaction of the benzoxazone (4.940) with aminoguanidine has been reported[598]. It appears that, together with the s-triazolo[1,5-c] quinazoline (4.943), the 3-guanidino-2-methyl-4-quinazolone (4.942) is obtained[598]. The reaction has been studied with a great variety of substituents in both the benzene ring and in position 2 of the heterocyclic ring. Since an open-chain compound of type (4.941) (R = Cl) was isolated from the 7-chloro derivatives of (4.940), the reaction may be considered to occur in the following way:

(4.940)   (4.941)

(4.942)   (4.943)

The action of nucleophiles on isatoic anhydride is well documented[599,600] and has been employed for synthesizing quinazolines, particularly of 4-oxo-3,4-dihydro- and 2,4-dioxo-1,2,3,4-tetrahydroquinazolines[601]. Recently reported examples are the reaction of isatoic anhydride with compounds of type $R-N = CHR'$[602] and those with thioamides, leading to a number of 2-substituted and/or 2,3-disubstituted 4-quinazolones[603-605]. For an extensive survey, we refer to reference 592. Interesting applications of these ring transformation reactions of isatoic anhydride (4.945) are the formation of (4.946) from (4.945) and pyridone-2, and of (4.944) from (4.945) and 3-hydroxyisothiazole. In both cases the primary step involves N-acylation[606]. In the reaction of isatoic anhydride (4.945) with equimolar amounts of N,N-diarylformamidine at elevated temperature, with liberation of an equivalent amount of carbon dioxide and arylamine, 3-aryl-4-oxo-3,4-dihydroquinazolines have been obtained[607]. In this reaction the o-aminobenzoyldiarylamidine (4.947) is probably the intermediate.

(4.944)   (4.945)   (4.946)

(4.945)   (4.947)   Ar = C_6H_5; m-$CH_3C_6H_4$; p-$CH_3C_6H_4$

A reaction of general application and wide scope is the formation of 2,4-dioxo-1,2,3,4-tetrahydroquinazolines (4.950) from substituted isatoic anhydrides (4.948) and amines. The reaction appears to be strongly dependent upon the position and character of the substituents in the benzene ring and the amines used[592]. Thus 5,7-dibromoisatoic anhydride and aniline gives 2-amino-3,5-dibromobenzanilide with merely a trace of 6,8-dibromo-3-phenyl-2,4-dioxo-1,2,3,4-tetrahydroquinazoline, while with 5,7-dibromoisatoic anhydride and ethylamine a good yield of the 6,8-dibromo-3-ethyl-2,4-dioxo-1,2,3,4-tetrahydroquinazoline is obtained[608]. A detailed study covering a large number of substituted amines have revealed that in these reactions fission at bond a, leading to o-ureidobenzoic acid (4.949), can take place as well as fission at bond b yielding, under CO_2 evolution, the o-aminobenzenecarbonamide (4.951).

Isatoic anhydride (4.945) reacts with urea or thiourea when melted or when refluxed in DMF to form derivatives of 2,4-dioxo-[609] or 4-oxo-2-thio-tetrahydroquinazolines (4.953)[610]. According to the authors, acylation of the urea or thiourea takes place first, leading to (4.952) which gives ring closure by loss of an amine and carbon dioxide. Similar conversions were also reported for 1-methylisatoic anhydride[610].

The conversion of benzoxazones into 2,4-dioxo-tetrahydroquinazolines can also be carried out with acids, provided that a nitrogen atom is present in the substituent attached to position 2. Thus the 2-phenylimino-1,2-dihydro-benzoxazone (4.954) when treated with phosphoric acid and phosphorus pentoxide at 150°C, gives the 2,4-dioxo-1,2,3,4-tetrahydroquinazoline (4.957)[611]. The reaction can be shown to occur as an initial attack of the acid on the carbonyl group yielding (4.955), after which ring opening occurs into the anion of the mixed anhydride of a o-(substituted aminophenyl) carboxylic acid and phosphoric acid (4.956). The anion of the ureido nitrogen attacks the carbonyl group to cleave to the phosphoric acid moiety and leads to (4.957).

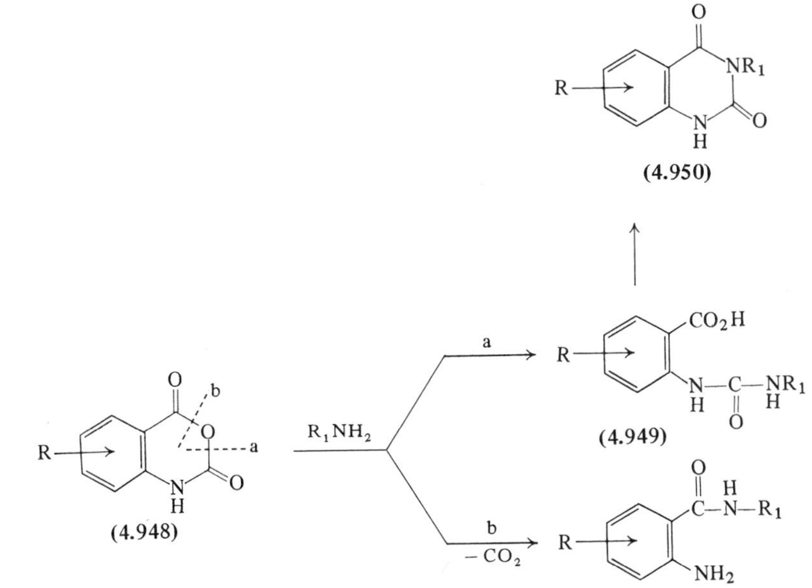

(4.950)

(4.948)

R_1NH_2

a

b

$-CO_2$

(4.949)

(4.951)

(4.945)

$X = C\diagdown^{NHR}_{NHR}$

$-CO_2$

(4.952)

$-RNH_2$

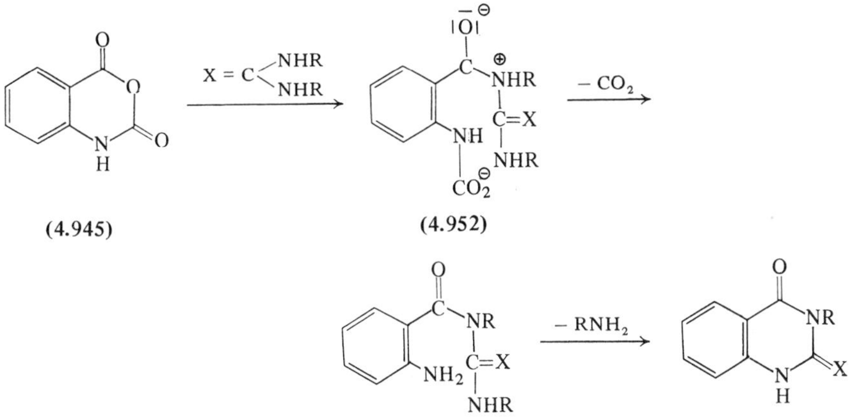

(4.953)

X = O, S

By a similar reaction the 2-(phenylimino)- and 6-(alkylamino)-1,3-oxazines (4.958) and (4.960) respectively rapidly rearrange into the uracils (4.959) and (4.961)[612,613].

3,1-Benzoxazones (4.937), when azidolysed with hydrazoic acid, behave in a manner quite different from that expected. Instead of N-acylamino acid derivatives being obtained, there is ring contraction into both 1-acylbenzimidazolone (4.962) and the o-carboxyphenyltetrazole (4.963)[614]. This result is

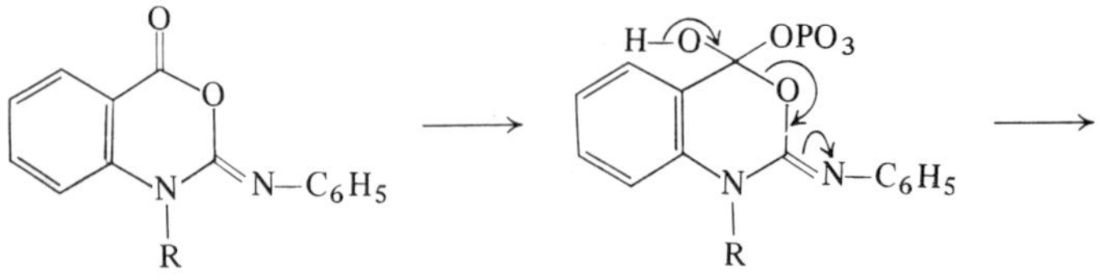

(4.954)

(4.955)

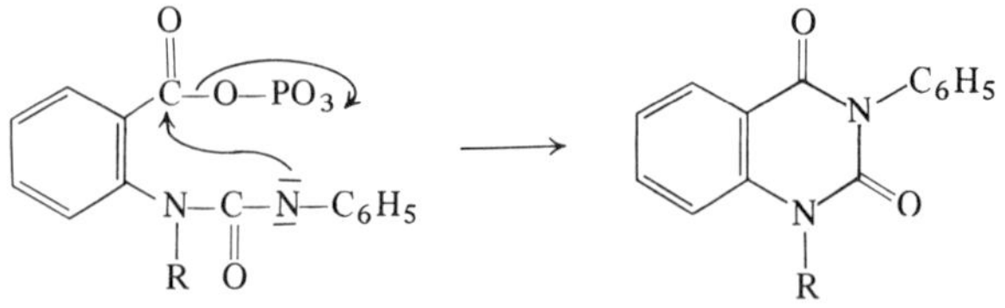

(4.956)

(4.957)

R = H; CH$_3$

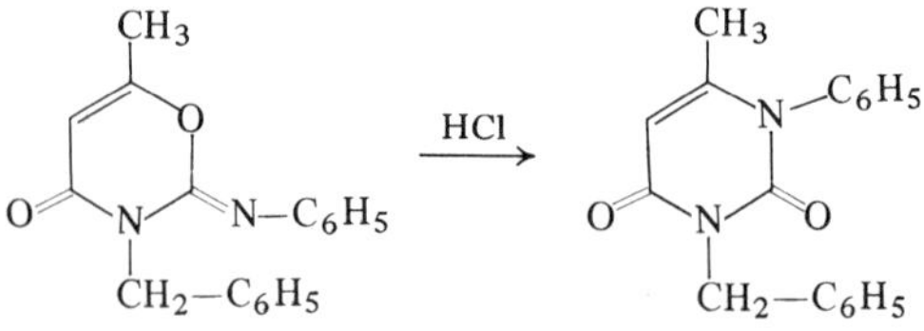

(4.958)

(4.959)

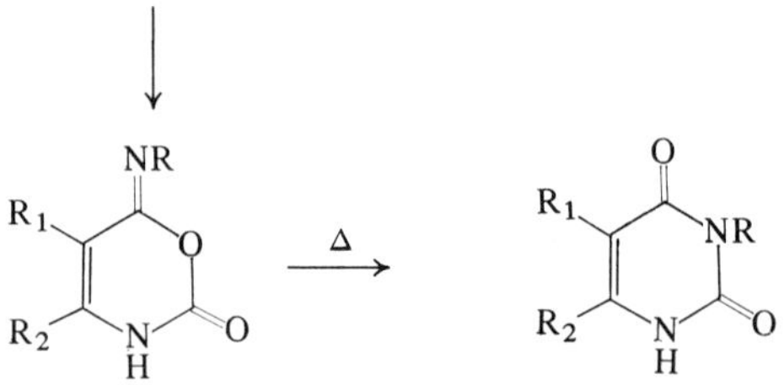

(4.960)

(4.961)

R	s-C$_4$H$_9$	s-C$_4$H$_9$	s-C$_4$H$_9$
R$_1$	H	Br	Cl
R$_2$	CH$_3$	CH$_3$	CH$_3$

of interest because bond breakage takes place at C_4–O as well as at C_2–O. This last bond fission has also been observed with the five-membered azalactones[615,616].

A novel ring contraction with a very wide scope has been observed when the 4-aryl-4H-3,1-benzoxazine (4.963a) is treated with potassium amide in liquid

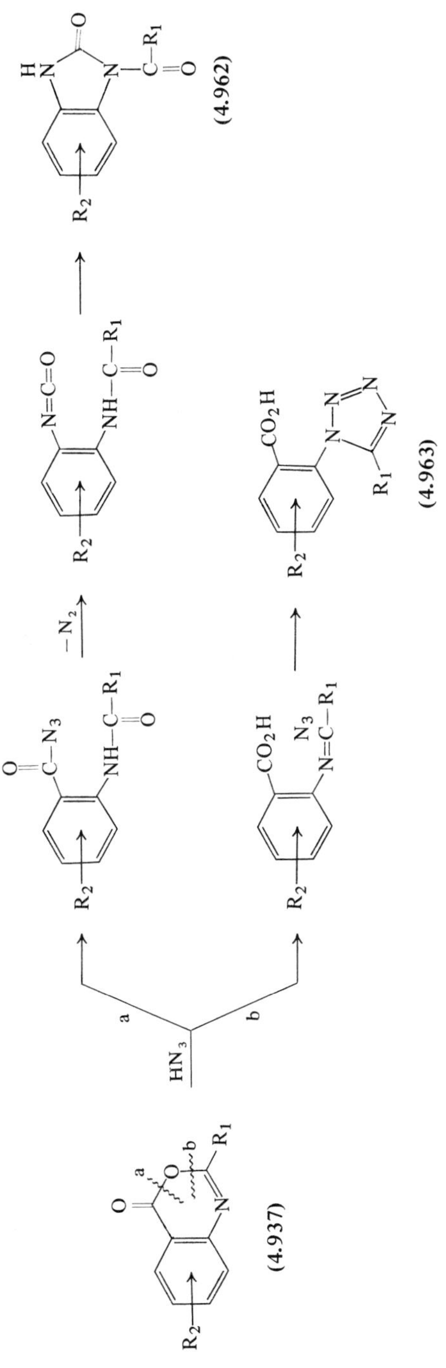

$R_2 = H; R_1 = H; CH_3; C_6H_5; p\text{-}OCH_3C_6H_4; p\text{-}NO_2C_6H_4$
$R_1 = CH_3; R_2 = 6\text{-}NO_2; 7\text{-}NO_2; 8\text{-}NO_2$

(4.962)

(4.963)

(4.937)

ammonia[616a]. 2,3-Disubstituted 3H-indol-3-ols (4.963b) are formed by a reaction which essentially involves the intramolecular addition of a primarily formed carbanion to the electrophilic iminoether. Subsequent ring opening of the oxirane ring leads to the alkoxide of the observed product.

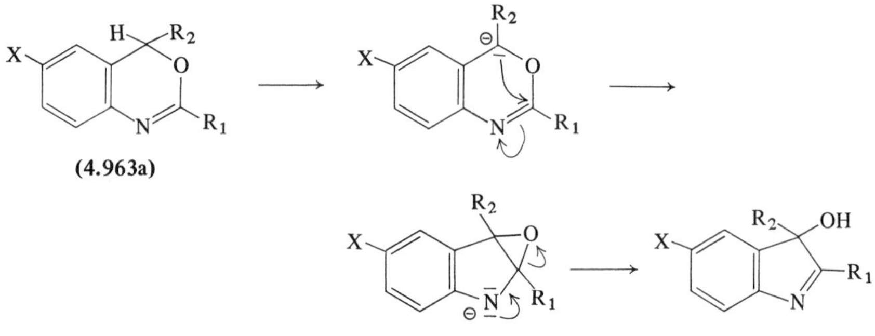

(4.963a)

(4.963b)

X   = Cl; H
$R_1 = CH_3; C_6H_5; p\text{-}CH_3C_6H_4; C(CH_3)_3$
$R_2 = C_6H_5; p\text{-}OCH_3C_6H_4; p\text{-}CH_3C_6H_4; m\text{-}CF_3C_6H_4$

b. *3-Azapyrylium salts into pyridines, pyrimidines, thiapyrylium salts and pyrazoles*

In contrast to the conversion of pyrylium salts into pyridines (see section I.A.2.a) which has been thoroughly investigated, the ring transformations of 3-aza- or 3,5-diazapyrylium salts have scarcely been studied, mainly because these compounds are less easily accessible. However, recently methods for the preparation of 3-azapyrylium salts, which are easy to carry out, have been reported. These include preparation from N-acylbenzimidchlorides and arylacetylenes in the presence of SnCl$_4$[617,618], from β-chlorovinylketones and arylnitriles[617], and from 4H-1,3-oxazines and tritylperchlorate[619]. When the 3-azapyrylium salts

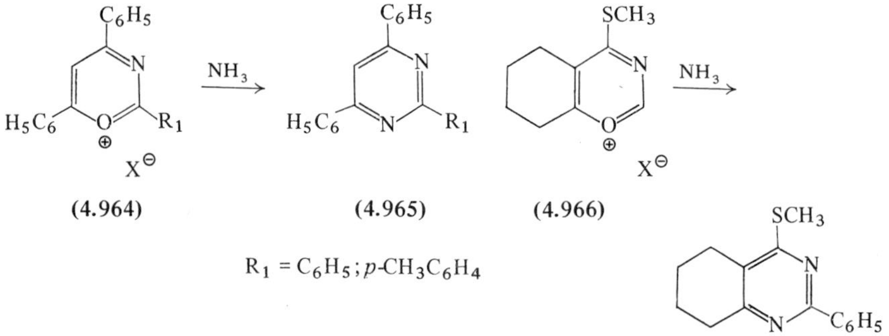

(4.964)   (4.965)   (4.966)

$R_1 = C_6H_5; p\text{-}CH_3C_6H_4$

(4.967)

(4.964) and (4.966) are treated with ammonia, just as in the case of pyrylium salts, an easy replacement of the oxygen by a nitrogen atom takes place and the pyrimidines (4.965) and (4.967) respectively are formed[617,620].

2,4,6-Trisubstituted 3-azapyrylium salts (4.968), when treated with an excess of hydrogen sulphide in methylcyanide at $0°C$ in the presence of equimolar amounts of triethylamine, give an orange-coloured solution from which, after addition of perchloric acid, the 3-azathiopyrylium salts (4.970) can be isolated[618]. The colour of the solution may be due to the formation of an intermediary compound, i.e. the β-acylaminovinylthiocarbonyl (4.969).

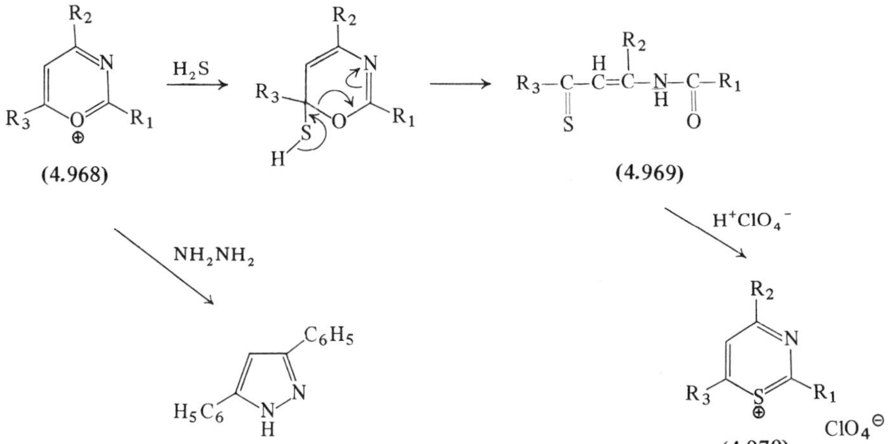

R_1	C_6H_5	$p\text{-}CH_3C_6H_4$	$p\text{-}ClC_6H_4$	C_6H_5	C_6H_5	C_6H_5	$p\text{-}ClC_6H_4$	$p\text{-}CH_3C_6H_4$
R_2	C_6H_5	C_6H_5	C_6H_5	$p\text{-}ClC_6H_4$	C_6H_5	CH_3	CH_3	CH_3
R_3	C_6H_5	C_6H_5	C_6H_5	C_6H_5	OC_2H_5	OC_2H_5	OC_2H_5	OC_2H_5

Both geometrical butadiene derivatives (4.973) were obtained by subjecting the 3-azapyrylium salts to nucleophilic attack of the carbanion of malonitrile. These can cyclize according to the route given into the corresponding 3-cyano-pyridine-2-carbonamides (4.974)[621]. It was possible to obtain evidence for the occurrence of an intermediate of type (4.972) since in one case, i.e. for (4.968) ($R_1 = R_2 = R_3 = C_6H_5$) this intermediate (4.972) ($R_1 = R_2 = R_3 = C_6H_5$) could be isolated.

Ring contraction into 3,5-diphenylpyrazole (4.971) has been reported in the reaction of (4.968) ($R = C_6H_5$) with hydrazine[617].

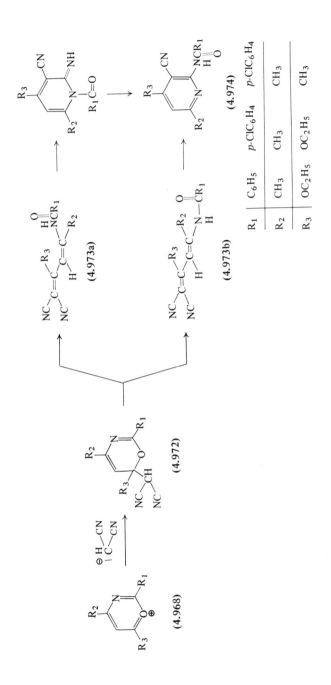

R_1	C_6H_5	p-ClC$_6$H$_4$	p-ClC$_6$H$_4$
R_2	CH_3	CH_3	CH_3
R_3	OC_2H_5	OC_2H_5	CH_3

3. RING TRANSFORMATIONS OF 1,4-OXAZINES

a. *1,4-Benzoxazines into benzoxazoles; phenoxazines into dibenzazepines*
In the 1,4-oxazine series—a class of compounds on which only a few ring transformations have been reported—it has been found that both 2,4-dihydroxy-1,4-benzoxazin-3-one (4.975) with loss of formic acid[622], and 4-hydroxy-1,4-benzoxazine-2,3-dione (4.977) with liberation of carbon dioxide[623], are converted into 2(3)-benzoxazolinone (4.976) on boiling in aqueous solution.

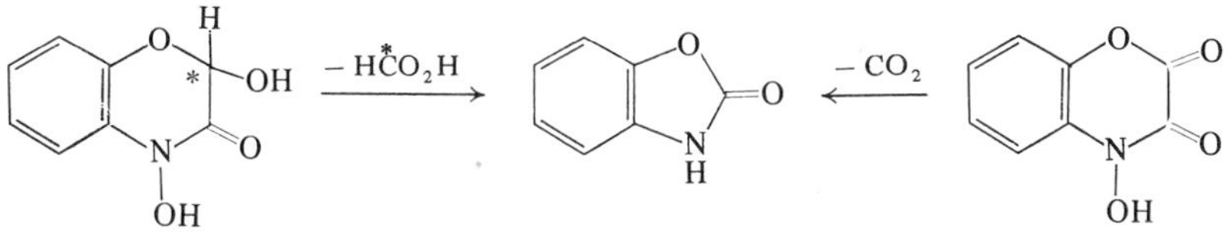

(4.975)   (4.976)   (4.977)

[14]C-Tracer experiments have shown that it is the carbon atom at position 2 in (4.975) which is eliminated during the ring transformation[624]. The authors do not present a mechanism but state that the presence of a *N*-hydroxy group and a bond which can easily be ruptured under the conditions applied, are essential for this reaction. This hypothesis is based on the fact that the compound (4.978), which contains an unreactive ether linkage, and the compound (4.979), in which no *N*-hydroxy group is present, do not react when boiled in an aqueous solution.

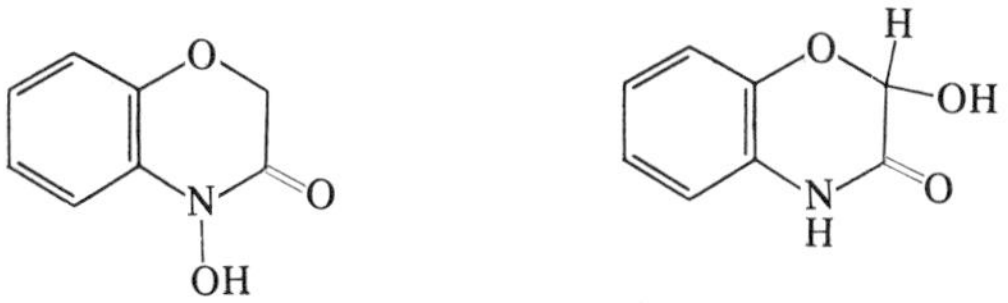

(4.978)   (4.979)

In accordance with the reported ring contraction of halocoumarins into benzofurans achieved by treatment with a base (see section I.A.1.b), it has been found[624a] that reactions of 3-chlorobenzoxazin-2-ones (4.979a) with amines give ring contraction into derivatives of benzoxazole-2-carboxylic acid (4.979b). This ring contraction also occurs with alcohols or phenols in the presence of acetic acid, yielding esters (4.979b) (R_2 = OR); with mercaptans the thiolesters (4.979b) (R_2 = SR) are obtained.

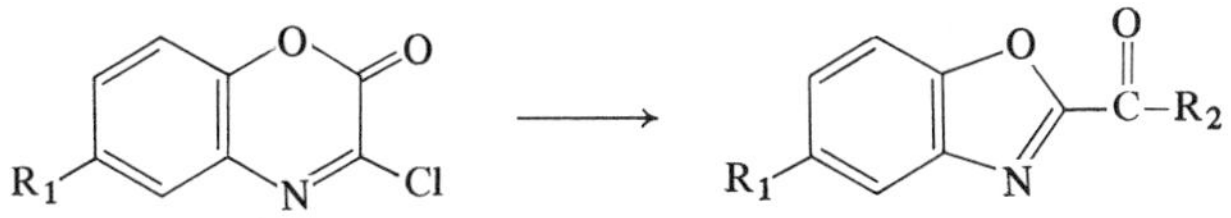

(4.979a)   (4.979b)

R_1 = H; CH$_3$; C$_6$H$_5$
R_2 = NH$_2$; NHR; NRR

The conversion of the three 5-acetylphenoxazone-2 derivatives (4.980), (4.981) and (4.982) into the 3,6,10-trihydroxydibenz[b,f]azepine (4.985) by treatment with base in methanol is another interesting case of ring expansion of 1,4-oxazines[625]. The acetyl trihydroxyphenylchinonimine (4.983) is intermediate in all these reactions in which, by condensation of the acetyl group with a hydroxyl group of the chinon, ring closure to the azepine derivative (4.984) occurs; this compound then tautomerizes further into (4.985).

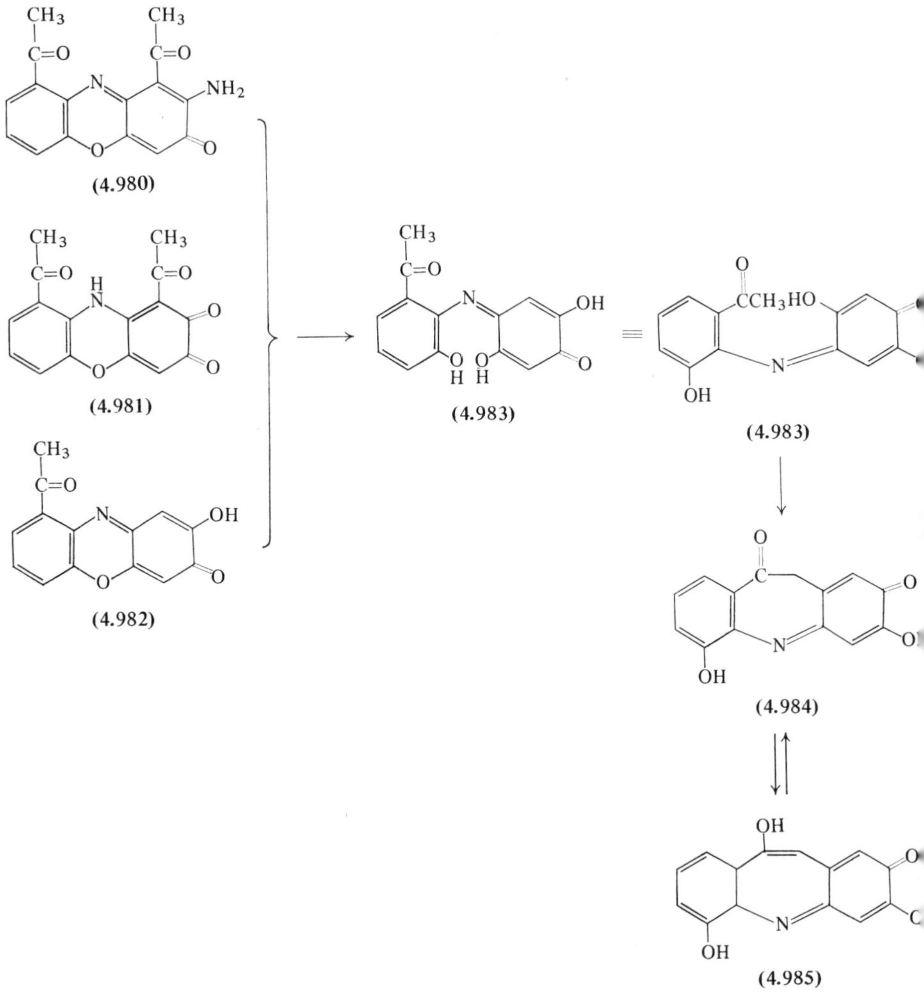

(4.980)

(4.981)

(4.982)

(4.983)

(4.983)

(4.984)

(4.985)

C. Ring transformations of six-membered heterocycles containing one sulphur and one nitrogen atom

1. RING CONTRACTIONS OF 1,2-THIAZINES

a. 1,2-Thiazines into pyrroles

In correlation with the reported ring contraction of 3,6-dihydro-1,2-oxazines into pyrroles (see section III.B.1.a), it was found that the corresponding N-aryl-3,6-dihydro-1,2-thiazine-1-oxides (4.986)—the cycloadducts of 1,3-butadienes with substituted N-sulphinylanilines—are also capable of undergoing ring contraction into the pyrroles (4.987) if they are simply heated with alcoholic potassium hydroxide[626]. This reaction seems to be of fairly general application since it has been used for the preparation of a number of substituted pyrroles. It is of interest to note that the reaction of the N-m-nitrophenyl

H	H	p-CO$_2$H	O—CH$_3$	p-CH$_3$	p-OCH$_3$	m-Cl	α-C$_{10}$H$_7$	β-C$_{10}$H$_7$	p-NH$_2$
H	CH$_3$	CH$_3$		CH$_3$	CH$_3$	CH$_3$	CH$_3$	CH$_3$	CH$_3$

derivatives of (4.986) (R$_1$ = m-NO$_2$C$_6$H$_4$, R$_2$ = CH$_3$) with base gives ring contraction into the pyrrole (4.987) (R$_1$ = m-NO$_2$C$_6$H$_4$, R$_2$ = CH$_3$) and also into 3,3'-bis(3,4-dimethyl-1-pyrrolyl)azoxybenzene (4.988)[626]; during the formation of the latter compound reduction of the nitro group apparently takes place.

A number of observations have contributed to the view that a sulphur atom is fairly readily extruded in a thermal process from a heterocycle if it leads to a new heterocycle with an aromatic structure (see, for example, the conversion of dibenzo[c,e]dithiin into dibenzothiophene (see section II.B.1.c) and the conversion of 1,4-dithiins into thiophenes (see section II.B.1.a)). The fact that this extrusion is not restricted to bivalent sulphur atoms has already been demonstrated by the preferential elimination of the oxidized sulphur in p-dithiin-1-oxide, giving thiophenes (see section II.B.1.a). In accordance with these results, it has been found that the 1,2-thiazine-1,1-dioxides (4.989), when heated at

200-225°C with copper give ready elimination of sulphur dioxide and yields the pyrroles (4.990)[627]. Sulphur dioxide extrusion is also achieved with the sulphostyril (4.991) when this compound is treated with amyl- or butylnitrite in

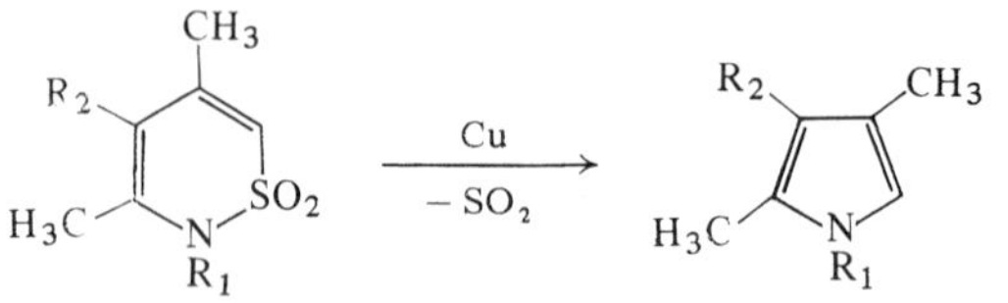

(4.989)   (4.990)

R₁	p-CH₃C₆H₄	m-N(CH₃)₂C₆H₄
R₂	CH₃	CH₃

the presence of acid. The isatine-β-oxime (4.993) is obtained; its formation occurs by a series of reactions involving the isatine-α-oxime (4.992) as intermediate which is subsequently converted by *trans* oximation into (4.993)[628]. It is not known whether the elimination reaction of sulphur dioxide occurs via a direct displacement of the α-carbon by the sulphonamide nitrogen and/or through the sulphene (4.994) acting as intermediate the formation of which is indicated by arrows.

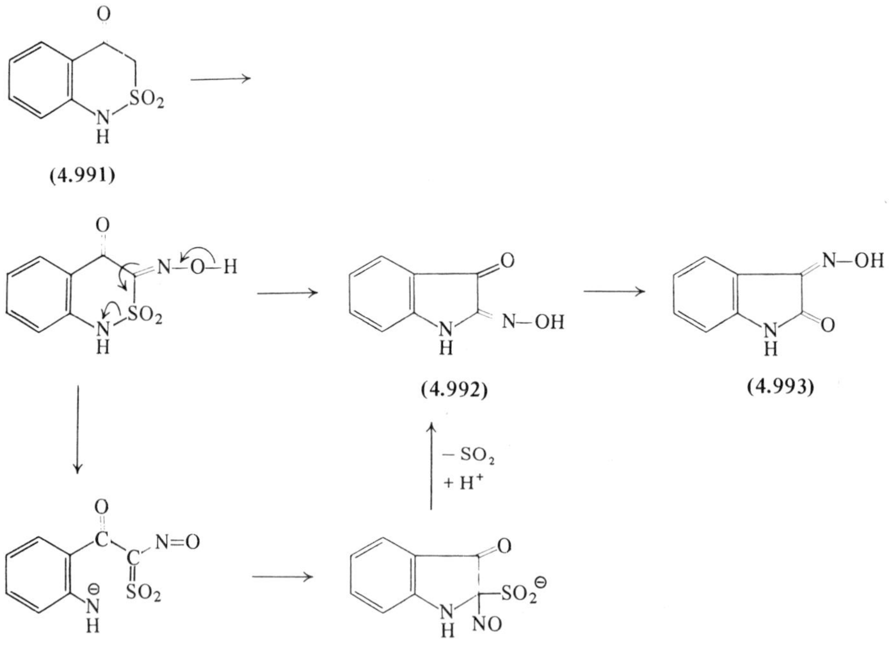

(4.991)

(4.992)   (4.993)

$- SO_2$
$+ H^+$

(4.994)

2. RING TRANSFORMATIONS OF 1,3-THIAZINES

a. *1,3-Thiazines into pyrimidines; 1,3-benzothiazines into benzisothiazoles and benzothioles; 3.1-benzothiazines into quinazolines*

Like 1,3-oxazines, 1,3-thiazines are also convertible into pyrimidines[628]. On treatment with ammonia the 2,4-dioxo-1,3-thiazine derivatives (4.995) give, with evolution of hydrogen sulphide, the uracils (4.996)[629] and the 3,1-benzothiazine-4-thione (4.997) yields with ammonia, or the appropriate alkyl- or arylamines 3*H*-quinazoline-4-thiones (4.998)[592,630]. In the case of the last reaction mentioned, a large variety of groups R_1 (aliphatic, heterocyclic or aromatic) and R_2 (alkyl, aryl, amino, hydroxyl, alkyl(aryl)amino, ureido) have been investigated. It is of interest to note that when 3-phenyl-3,4-dihydro-2*H*-1,3-benzothiazine-2-one-4-thione (4.999) is treated with methylamine, a mixture of the 1,2-dithiole (4.1000) and benzisothiazole (4.1001) is obtained[630a].

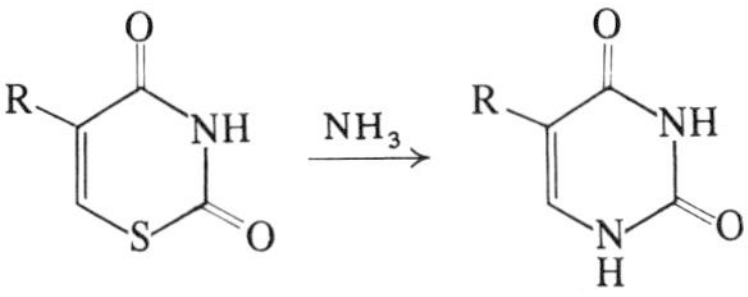

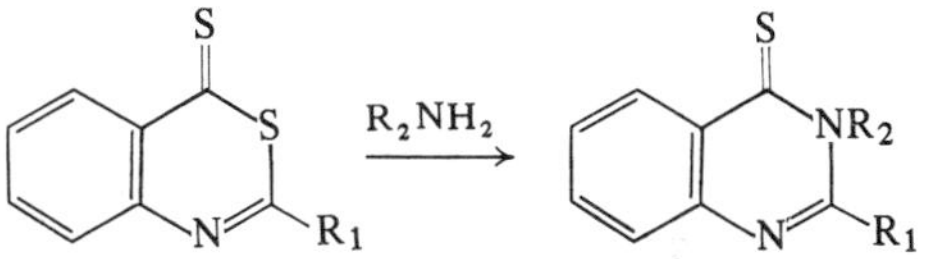

(4.995)　　　　　(4.996)　　　(4.997)　　　　　(4.998)

R = H; CH₃

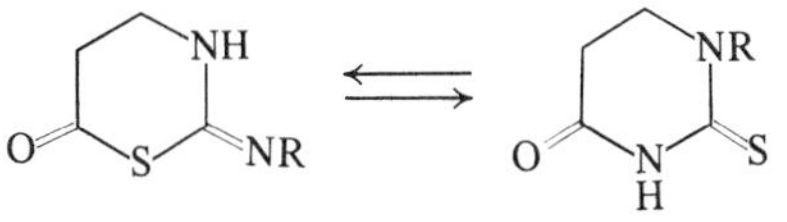

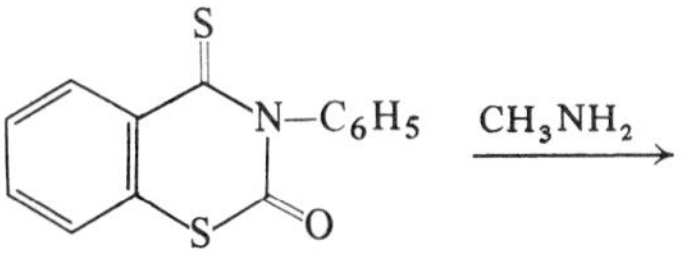

(4.1002)　　　　　(4.1003)　　　(4.999)

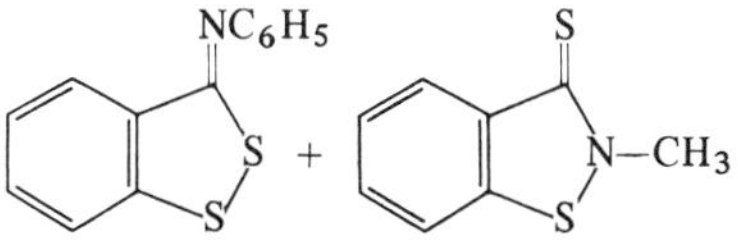

(4.1000)　　　(4.1001)

The 2-imino-6-oxo-1,3-thiazines (4.1002), when treated with dilute acetic acid or pyridine, gave the 2-thio-4-oxo-tetrahydropyrimidine (4.1003)[631]. The conversion (4.1002) → (4.1003) can also be conducted in the reverse direction. In addition, it has been found[631a] that when 4,6,6-trimethyl-3-phenyl-3,6-dihydro-2-(1*H*)-pyrimidinethione is heated with hydrochloric acid 4,4,6-trimethyl-2-phenylamino-4*H*-1,3-thiazine is formed.

Ring contraction of a 1,3-thiazine ring has been found to occur[631b] when the dihydro-1,3-benzothiazinesulphoxide (4.1003a) is refluxed with acetic anhydride (containing 1 per cent sodium acetate) leading to a mixture of the compounds (4.1003c) and (4.1003d). Probably (4.1003c) is the precursor of (4.1003d), since it has been shown[631c] that with sodium acetate–acetic anhydride a conversion of (4.1003c) into (4.1003d) can take place. The production of (4.1003c) presumably occurs via the formation of (4.1003b). In the case of the N-methyl derivative of (4.1003a), almost no ring contraction reaction takes place; the principle reaction observed being ring expansion into benzthiazepine derivatives.

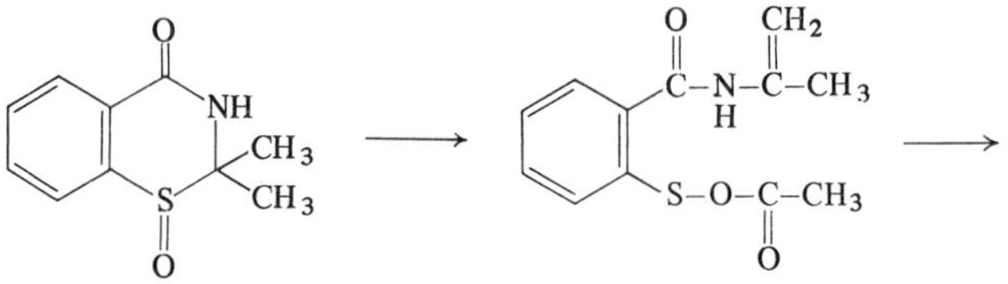

(4.1003a)   (4.1003b)

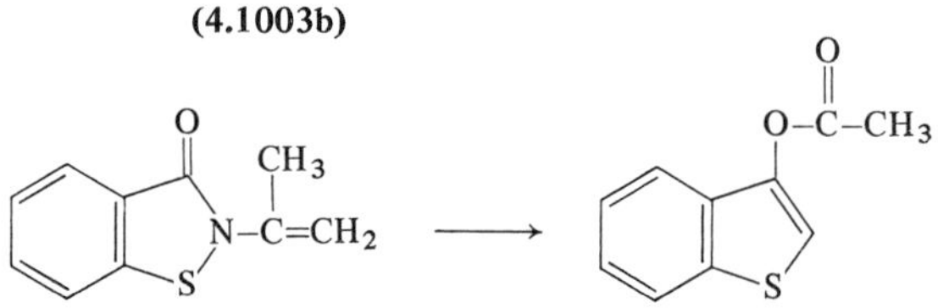

(4.1003c)   (4.1003d)

Interestingly, lithium aluminium hydride reduction of 2-phenyl-4-oxo-1,3-benzothiazine (4.1004) also gives formation, although in low yields, of benzisothiazole (4.1007) together with benzyl alcohol and 2-mercaptobenzylamine (4.1006)[632]. Apparently during the reduction of the C=O group, a reductive C–S bond fission occurs, giving the aldimine thiol intermediate (4.1005) which hydrolytically decomposes into the o-mercaptobenzylamine (4.1006) and benzaldehyde. Dehydrogenation of (4.1006) into (4.1007) can be performed by air.

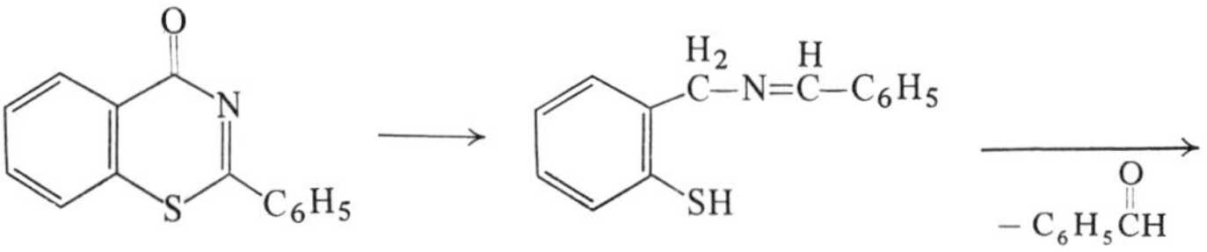

(4.1004)   (4.1005)

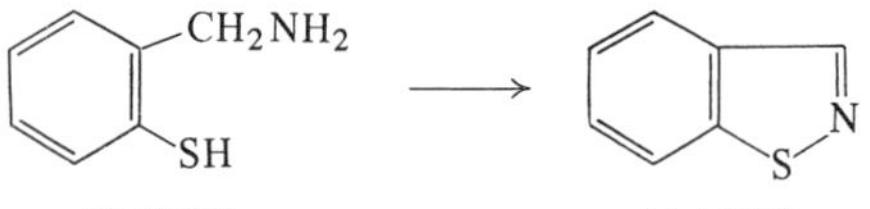

(4.1006)   (4.1007)

3. RING CONTRACTIONS OF 1,4-THIAZINES

a. *1,4-Thiazines into pyrroles and thiazolines; 1,4-benzothiazines into indoles and benzothiazoles*

Pyrolysis of 1,4-thiazines and their benzo derivatives in the presence of copper leads to removal of sulphur with formation of pyrroles and its benzo derivatives. As already discussed in section III.C.1.a, this extrusion of sulphur occurs quite readily when heteroaromatic compounds are obtained as reaction products. Thus desulphurization of phenothiazine (4.1008) leads to the carbazole (4.1009)[633], and its dibenzo derivative (4.1010) gives the corresponding 3,4,5,6-dibenzo-carbazole (4.1011)[634]. The desulphurization of a monobenzophenothiazine has also been reported[635].

(4.1008)   (4.1009)   (4.1010)

(4.1011)

N-Ethylcarbazole (4.1014) is formed from *N*-ethylphenothiazine (4.1012) by treatment with lithium in tetrahydrofuran at 50°C and subsequent carbonation[636]. At 25°C, *N*-ethyl-2-mercapto-2-carboxydiphenylamine (4.1015) is formed. The dilithio derivative (4.1013) is probably intermediate in the formation of (4.1014) as well as in that of (4.1015). Desulphurization and simultaneously dealkylation were observed when 12-methylbenzo[a]-phenothiazine was treated with lithium in tetrahydrofuran[637].

Loss of sulphur dioxide has been reported[638] during pyrolyses of the 4*H*-1,4-thiazine-1,1-dioxides (4.1016) at 300°C with reduced copper—it gives the pyrroles (4.1017)—and on treatment of the 1,4-benzothiazine-4-oxide (4.1018) with Raney nickel which leads to reductive ring contraction into the 2,3-pentamethyleneindole (4.1019)[639]. Pyrolysis of 3-phenyl-2*H*-1,4-benzothiazin-2-one at 300°C leads to decarbonylation, yielding 15-20 per cent of 2-phenylbenzothiazole[639a].

Reductive ring contraction leading to extrusion of a carbon atom outside the ring but not giving removal of sulphur is reported when the dihydro-1,4-

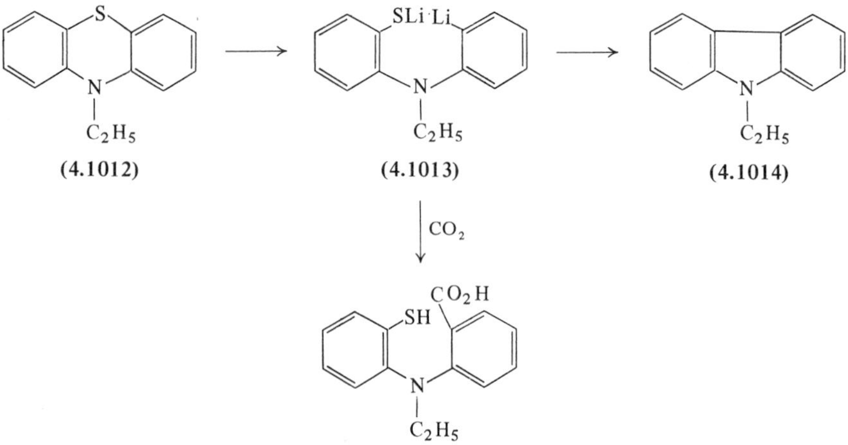

(4.1012) → (4.1013) → (4.1014)

| CO$_2$

(4.1015)

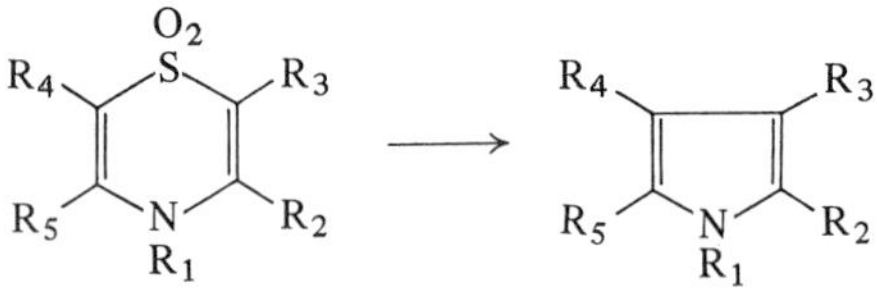

(4.1016) → (4.1017)

R_1	H	H	CH$_3$	H	H
R_2	C$_6$H$_5$	C$_6$H$_5$	C$_6$H$_5$	C$_6$H$_5$	C$_6$H$_5$
R_3	H	CH$_3$	H	C$_6$H$_5$	C$_6$H$_5$
R_4	H	CH$_3$	H	(CH)$_4$	(CH)$_4$
R_5	C$_6$H$_5$	C$_6$H$_5$	C$_6$H$_5$		

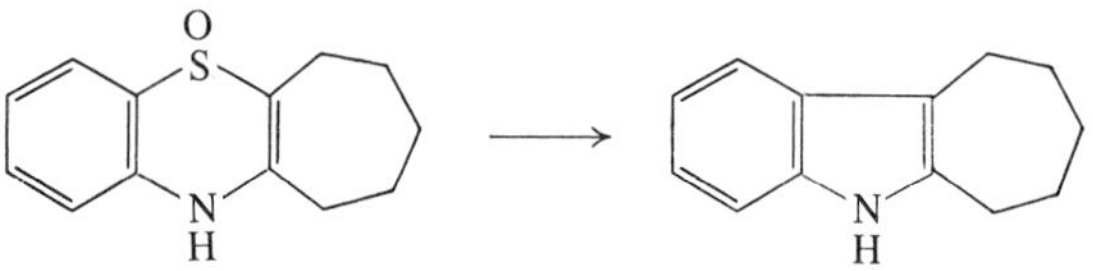

(4.1018) → (4.1019)

benzothiazine (4.1020) is boiled with zinc and acetic acid, 2-acetonylbenzo-thiazole (4.1022) being obtained[640]. In this reaction it is assumed that initially the C–S bond is hydrogenated to give the o-mercaptoacetylacetanilide (4.1021) as intermediate.

Extrusion of a carbon atom from the 1,4-thiazine ring has also been reported to occur during hydrogenation of the $2H$-1,4-benzothiazine (4.1023)[641] and of the $2H$-thiazine (4.1025)[642], yielding the 2,3-dihydrobenzthiazole (4.1024) and the Δ^3-thiazoline (4.1026) respectively.

(4.1020) (4.1021) (4.1022)

(4.1023) (4.1024)

(4.1025) (4.1026)

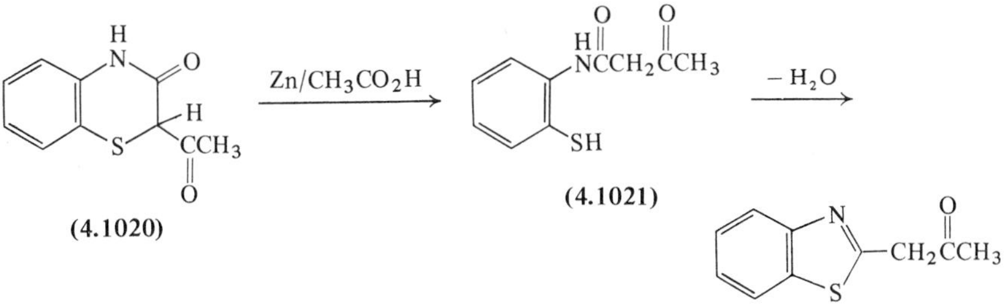

Ring contraction to a thiazole ring, involving extrusion of a ring carbon atom has been observed, when 2,3-dioxo-1,4-benzthiazine is treated with aniline or phenylhydrazine, the anilide or phenylhydrazide of benzthiazol-2-carboxylic acid being obtained[643a]. An analogous ring contraction in the benzoxazine series is discussed in section III.B.3.a.

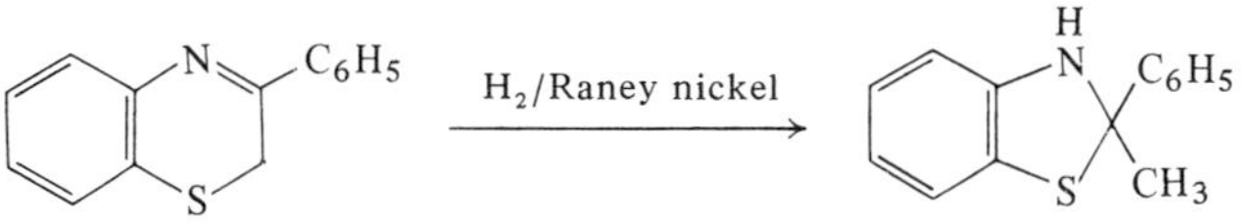

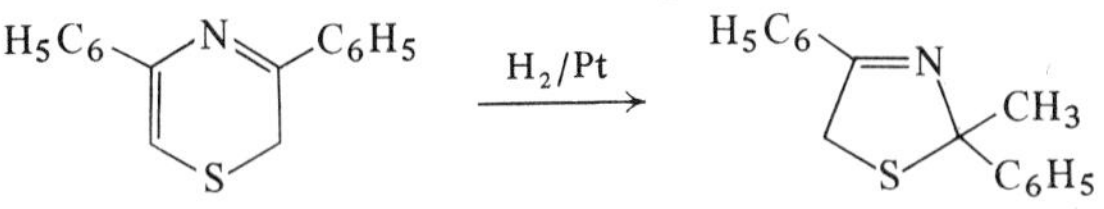

IV. RING TRANSFORMATIONS OF SIX-MEMBERED HETEROCYCLES CONTAINING THREE HETEROATOMS IN THE RING

A. Ring transformations of six-membered heterocycles containing one oxygen and two nitrogen atoms

1. 1,3,5-OXADIAZINES INTO DIHYDRO-1,3,5-TRIAZINES; 3,5-DIAZAPYRYLIUM SALTS INTO PYRIMIDINES, TRIAZINES, TRIAZOLES AND OXADIAZOLES; 1,2,4-OXADIAZINES INTO OXAZOLIDINES AND 1,3,4-OXADIAZOLES

The replacement of oxygen present in the 4,4-trifluoromethyl-1,3,5-oxadiazine (4.1027) by nitrogen takes place when (4.1028) is kept in liquid ammonia for three days in a sealed tube[643]. The presence of both trifluoromethyl groups makes the ring highly receptive to this nucleophilic attack so that the reaction occurs readily.

The ring present in the 3,5-diazapyrylium salts (4.1029), obtained by treating aryl cyanates with a complex of benzoyl chloride and $SbCl_5$, is also highly vulnerable to attack, as appears from the fact that they are easily converted into the s-triazines (4.1030)[617,644] on treatment with ammonia. However, because of this high activation of the ring the simultaneous replacement of substituents that have a strong leaving character can easily take place. To give one example,

$$F_3C \underset{H_5C_6 \diagdown O \diagdown C_6H_5}{\diagdown N \diagup N} CF_3 \xrightarrow{NH_3} F_3C \underset{H_5C_6 \diagdown N \diagup C_6H_5}{\diagdown N \diagup N} CF_3$$

(4.1027)   (4.1028)

$$R_3 \underset{O^{\oplus}}{\diagdown N \diagup N} R_1 \xrightarrow{NH_3} R_3 \underset{N}{\diagdown N \diagup N} R_1 \qquad H_2N \underset{N}{\diagdown N \diagup N} C_6H_5$$

(4.1029)   (4.1030)   (4.1031)

R_1	C_6H_5	$p\text{-}ClC_6H_4$	C_6H_5
R_2	C_6H_5	$p\text{-}ClC_6H_4$	C_6H_5
R_3	C_6H_5	$p\text{-}ClC_6H_4$	$p\text{-}CH_3C_6H_4$

$R_2 = OC_6H_5, OC_6H_4-p\text{-}CH_3$

when the 2-phenyl-3,5-diazapyrylium salts which contain aryloxy groups (4.1029) ($R_1 = C_6H_5$; $R_2 = R_3 = OC_6H_5$ or $OC_6H_4\text{-}p\text{-}CH_3$) react with ammonia, together with the formation of the corresponding s-triazines (4.1030) there is replacement of a group R_3 by an amino group and this leads to the formation of the s-triazine (4.1031)[644]. Other examples of similar ring transformations are the conversion of 2,4,6-triphenyl-3,5-diazapyrylium salts (4.1032) into the 2-hydroxy (4.1033), 2-mercapto (4.1034) and the corresponding N-methyl s-triazines (4.1036)[617].

4-Benzamido-5(R-substituted)-2,6-diphenylpyrimidines   (4.1038)   have recently been found to be formed when the pyrylium salt (4.1032) is refluxed with solutions of $CH_2(CN)_2$, $C_6H_5COCH_2CN$, or $NCCH_2CO_2CH_3$ in dioxane in the presence of triethylamine[645]. From the mechanism of all these reactions it is apparent that the introductory step is the addition of the nucleophile to position 2.

In analogy to the hydrazinolysis of 3-azapyrylium salts into pyrazoles (see section III.B.2.b) it has been found that the 3,5-diazapyrylium salt (4.1032) is converted into the 1,2,4-triazoles (4.1035) on treatment with hydrazine and methylhydrazine, and into the 1,2,4-oxadiazole (4.1037) on treatment with hydroxylamine[617].

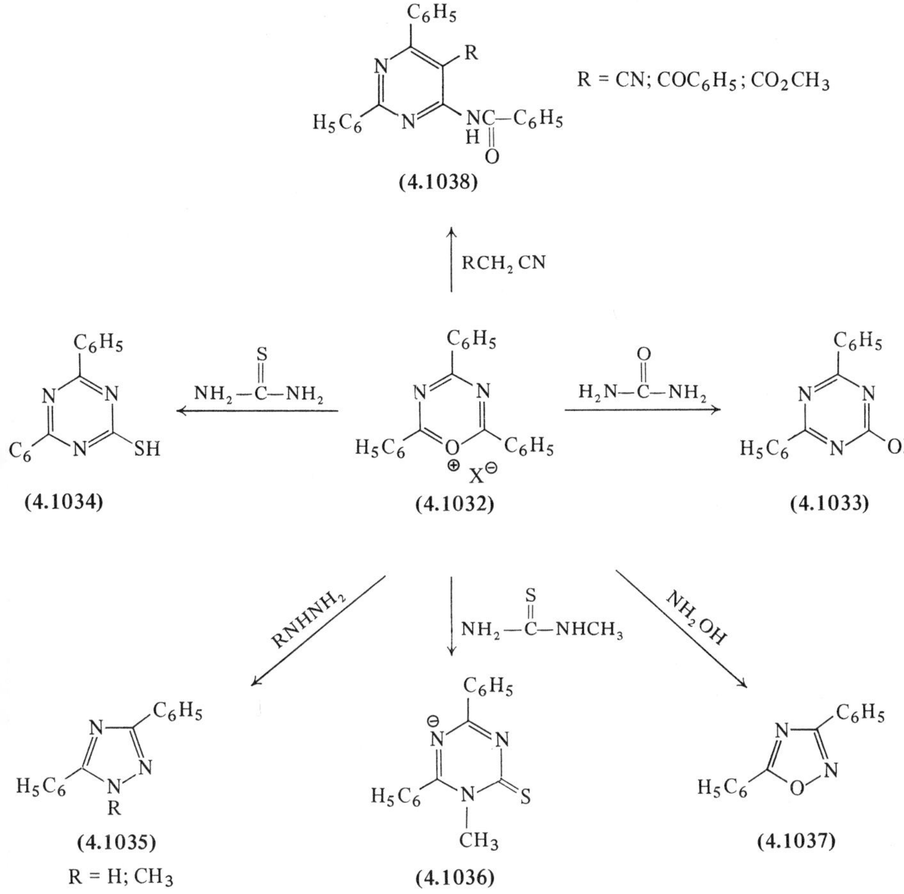

Ring contraction of an 1,2,4-oxadiazine ring into an oxazolidine ring has been found to occur when the sodium salt of 4-methyl-1,2,4-oxadiazine-3,5-dione is treated with chloroacetone[646]. No mechanism has been given.

It has been reported that ring contraction into 2,5-diphenyl-1,3,4-oxadiazole (4.1038b) occurs during treatment of 2,5-diphenyl-1,3,4-oxadiazin-6-one-4-oxide (4.103a) with alkali[696a]. The mechanism given on the next page is suggested for this reaction.

Since (4.1038a) can be prepared from the corresponding 1,4-dihydroxy-pyrazole by oxidation with peracids (see chapter 3, section 2.C.2.a) it appears possible to combine both procedures and thus produce 1,3,4-oxadiazoles by the treatment of appropriately substituted 1,4-dihydroxypyrazoles with alkaline hydrogen peroxide[646a].

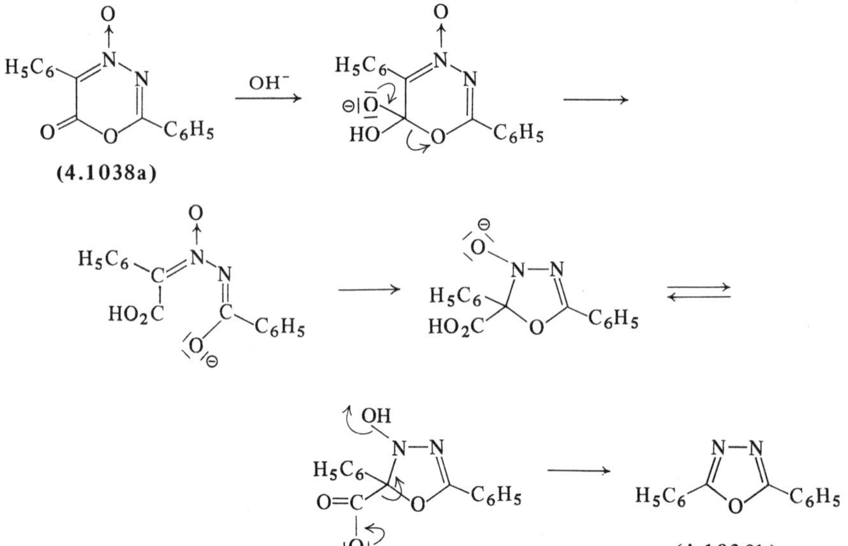

(4.1038a)

(4.1038b)

B. Ring transformations of six-membered heterocycles containing one sulphur and two nitrogen atoms

1. 1,3,4-THIADIAZINES INTO PYRAZOLES AND Δ^4-THIAZOLINES; 1,2,3-BENZOTHIADIAZINES INTO 2,1-BENZOXATHIOLES; TETRA-HYDRO-1,2,5-THIADIAZINES INTO IMIDAZOLIDINES; TETRA-HYDRO-1,3,5-THIADIAZINES INTO HEXAHYDRO-1,3,5-TRIAZINES

In the 1,3,4-thiadiazine series it was found that 2-amino-5-methyl-6-ethoxycarbonyl-1,3,4-thiadiazine (4.1040) is converted into the 3-amino-pyrazole (4.1041) in a weak acid (or neutral) medium[647]. Similarly, the 1,3,4-thiadiazines (4.1042) are transformed into the pyrazoles (4.1043) (also with removal of sulphur) when they are boiled in acetic acid or treated with ethanolic hydrogen chloride[648]. The acid-catalysed reaction is found to be fast. Possibly the acid creates a positive charge on one or both nitrogen atoms, thus considerably polarizing the C–S bond and thereby facilitating ring opening. The polarizing effect can also be brought about by the influence of an electron-attracting group, that is the ethoxycarbonyl or phenyl group in position 6. Remarkably, heating of the 1,3,4-thiadiazine (4.1040) with concentrated hydrogen chloride does not lead to liberation of sulphur but gives an intramolecular rearrangement into the five-membered thiazolon-2-imide (or

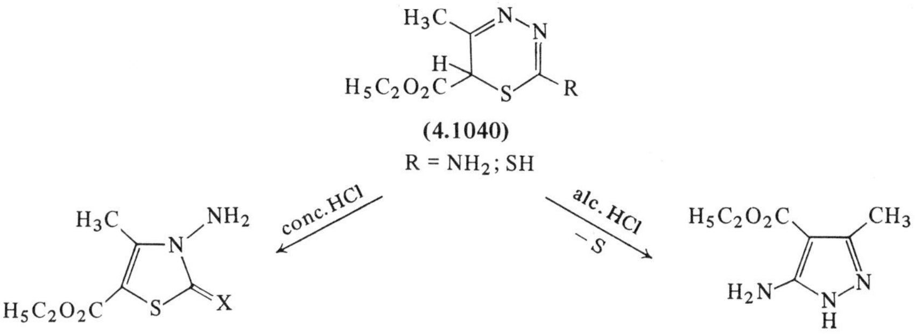

H_3C ... $H_5C_2O_2C$ ... R

(4.1040)

R = NH_2; SH

H_3C ... NH_2 ... $H_5C_2O_2C$ ... X

conc. HCl

alc. HCl — S

$H_5C_2O_2C$ ... CH_3 ... H_2N

(4.1039)

X = NH; S

(4.1041)

H_5C_6 ... N—N ... H_5C_6 ... S ... R

⟶

H_5C_6 ... R ... H_5C_6 ... N

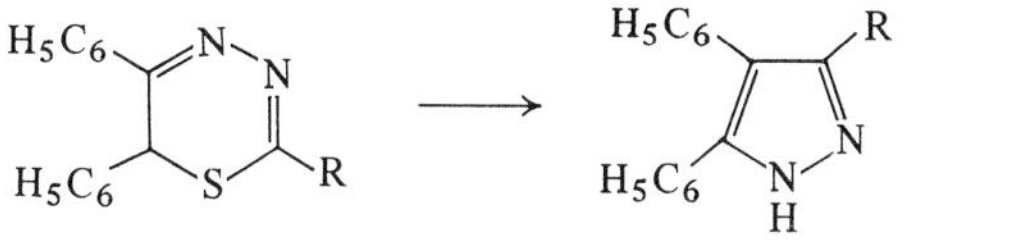

(4.1042)   **(4.1043)**

R = C_6H_5; SCH_3; $SCH_2C_6H_5$;
NH_2; $NHCH_3$; $N(CH_3)_2$

thion) (4.1039)[647,649,650]. This ring contraction has also been observed during acid treatment of 2-phenylamino-5-phenyl-1,3,4-thiadiazine (4.1044); it leads to the formation of N-amino-thiazolone-2-phenylimide (4.1045)[651] and not to the hydrazone of 3,4-diphenylthiazolone-2 (4.1046) as originally proposed[652].

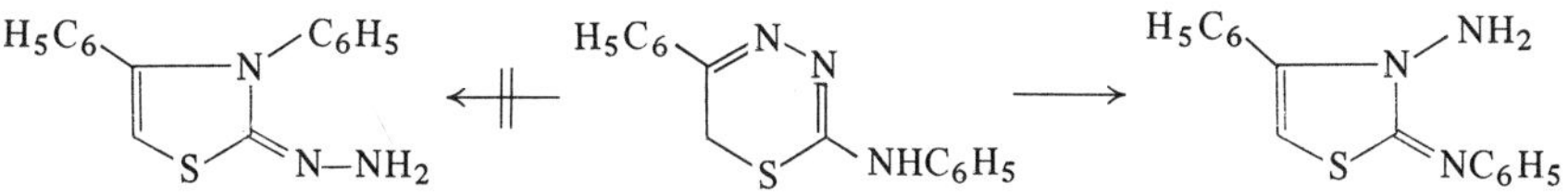

H_5C_6 ... N—C_6H_5 ... S ... N—NH_2

H_5C_6 ... N—N ... S ... NHC_6H_5

H_5C_6 ... N—NH_2 ... S ... NC_6H_5

(4.1046)   **(4.1044)**   **(4.1045)**

Because 1,3,4-thiadiazines tend readily to undergo ring contraction in acidic media, it is important to bear in mind that in all acid-catalysed condensation reactions employed for the preparation of 1,3,4-thiadiazines, ring contracted products, pyrazoles or thiazolones, are frequently obtained[647,648,650,653].

Chlorination of 2H-1,2,3-benzothiadiazine-1,1-dioxide (4.1047) with equimolar amounts of chlorine in methylene chloride gives rapid evolution of nitrogen and the formation of the somewhat unstable 3-chloro-2,1-benzoxathiole (4.1051)[654]. It is possible that an initial chlorination takes place at C_4 followed by fragmentation into the sulphene radical (4.1048) which isomerizes, via (4.1049), into the diradical (4.1050). Addition yields the benzoxathiole (4.1051). Probably because of a considerable angle strain in the transition state, ring closure of (4.1048) into a four-membered sulphone is not indicated. For more examples of the rearrangement of sulphones into cyclic sulphinate esters, see chapter 2, section I.B.1.b.

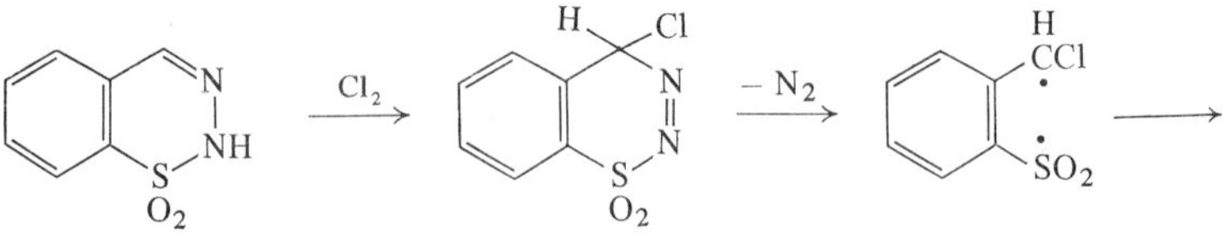

(4.1047)   (4.1048)

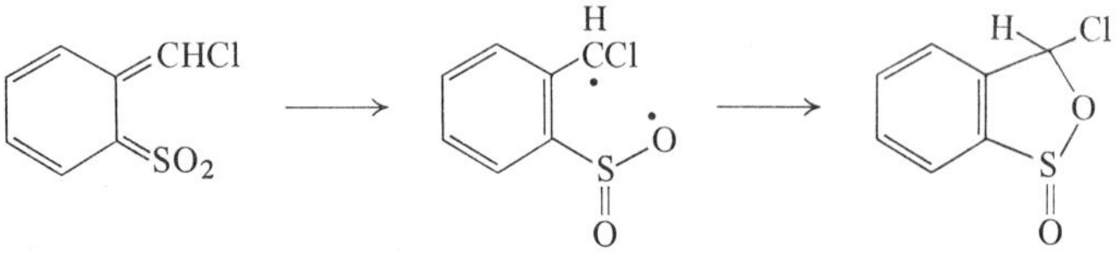

(4.1049)   (4.1050)   (4.1051)

Thermal heating of the tetrahydro-1,2,5-thiadiazine-6-thiones (4.1052) leads to extrusion of sulphur with formation of the ring-contracted product imidazolidine-2-thione (4.1053)[655,656]. A study of the stability of the 1,2,5-thiadiazine (4.1052) in relation to the size of the alkyl substituent R revealed that, if R = cyclohexyl, the compound can be stored for many weeks, while for R = i-C_3H_7, sulphur is expelled within a few days. The imidazolidines (4.1053) have also been obtained by loss of thioformaldehyde from the seven-membered 1,3,6-thiadiazepine-2-thione (4.1054).

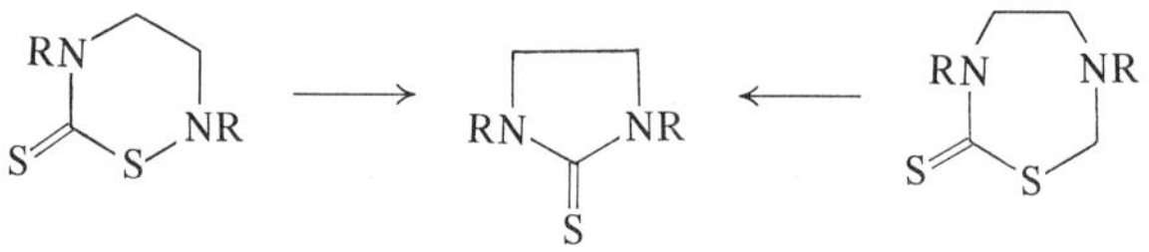

(4.1052)   (4.1053)   (4.1054)

R = C_2H_5; i-C_3H_7; n-C_4H_9- cyclohexyl; 2-C_2H_5-cyclohexyl

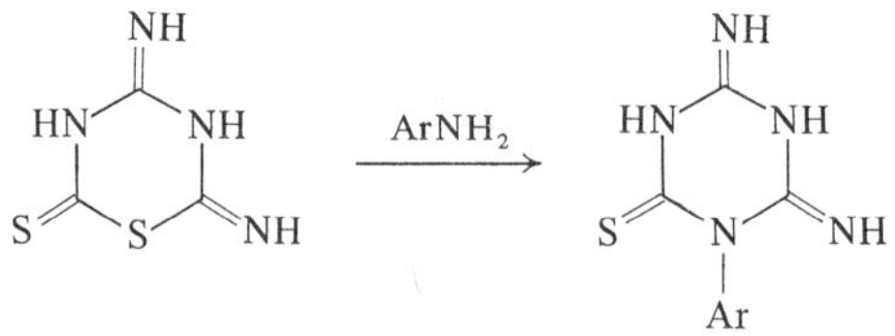

(4.1055)   (4.1056)

Sulphur-nitrogen exchange has also been reported in the reaction between 4,6-diimino-2-thio-tetrahydro-1,3,5-thiadiazine (4.1055) and aromatic amines through which 1-aryl-4,6-diimino-2-thio-hexahydro-1,3,5-triazines (4.1056) are formed[657].

C. Ring transformations of six-membered heterocycles containing three nitrogen atoms

1. RING CONTRACTIONS OF 1,2,3-, 1,2,4- AND 1,3,5-TRIAZINES

a. *1,2,4-Triazines into imidazoles, 1,2,3-triazoles and 1,2,4-triazoles; 1,3,5-triazines into imidazoles; 1,2,3-benzotriazines into indazoles; 1,2,4-benzotriazines into benzimidazoles; 1,3,5-triazine as methinylating agent for the synthesis of five-membered heterocycles*

3-Amino-5-phenyl-1,2,4-triazine-2-oxide (4.1057), which is stable in acid, gives, on heating in a 30 per cent potassium hydroxide solution, ring contraction into 4-phenyl-1,2,3-triazole (4.1058)[658]. Treatment of the deoxygenated compound (4.1059) with acetic anhydride at about 100°C gives, besides the acetyl derivative of (4.1059), the 5-acetamido-3-methyl-1,2,4-triazole (4.1060)[658]. The reaction route is obscure and no explanation for this reaction is offered.

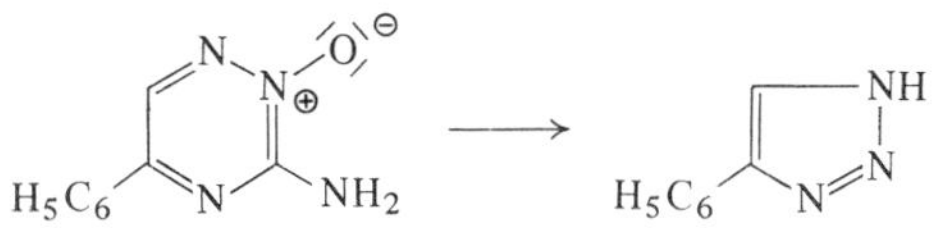

(4.1057)　　　　　(4.1058)

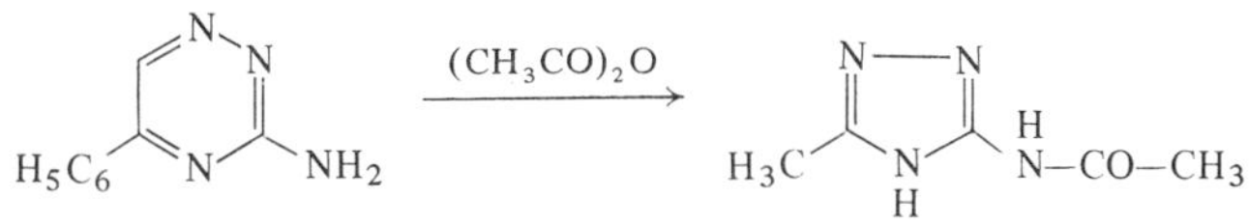

(4.1059)　　　　　(4.1060)

The 1,2,3-, 1,2,4- and 1,3,5-triazines are inclined to undergo ring contraction as a result of action by reducing reagents. Zinc in (alcoholic) alkali, zinc in acetic acid or red phosphorus in hydroiodic acid or $Na_2S_2O_4$ in acetic acid have been found to be effective reagents. Reduction of 4-oxo-1,2,3-benzotriazine (4.1061) with zinc dust and ammonium hydroxide yields indazolone (4.1062)[659]. Heating the dihydro compound (4.1063) with red phosphorus and hydroiodic acid at 180°C gives rupture of the bond between the 1 and 2 position, and subsequent ring closure into 4,5-diphenylimidazolin-2-one (4.1064)[660].

(4.1061) (4.1062)

(4.1063) (4.1064)

(4.1065) (4.1066)

R = H; $COCH_3$

(4.1067)

(4.1068)

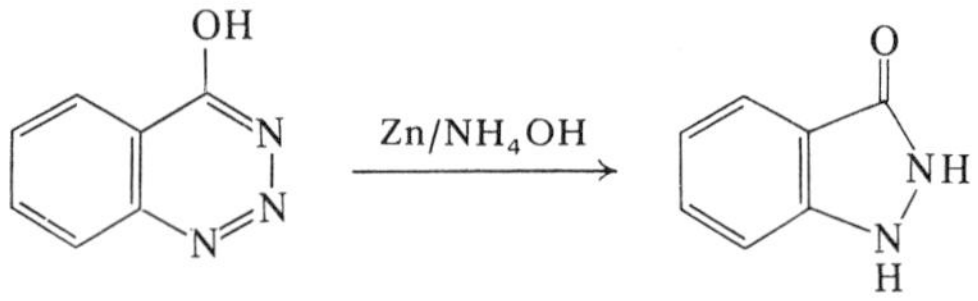

The same reductive ring contraction has been reported to occur during electrolysis of (4.1061) in both alkaline and acidic media[660a]. A similar result has been obtained during reduction of 4-methyl-1,2,3-benzotriazine-3-oxide (4.1064a) into 3-methylindazole (4.1064b) and hydroxylamine. The reaction is thought to involve primarily a two-electron reduction process, followed by fission of the nitrogen-nitrogen bond between position 2 and 3, and a subsequent two-electron reduction process leading to the formation of an arylhydrazineoxime.

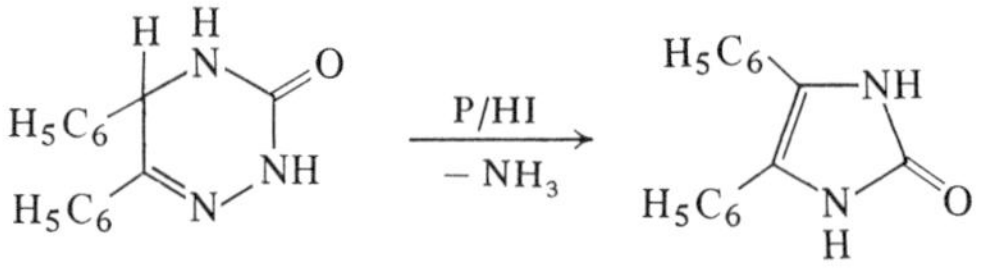

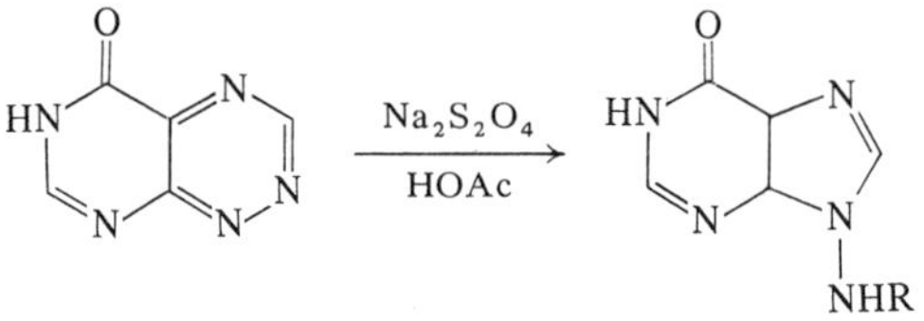

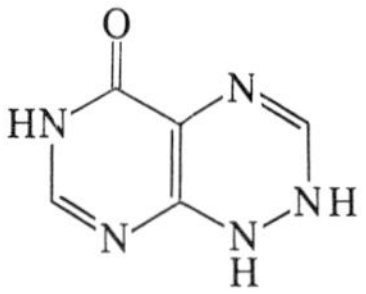

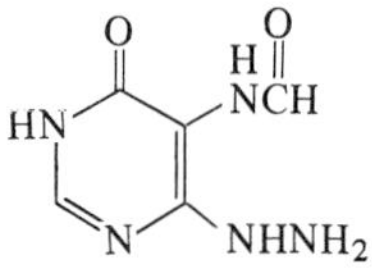

Analogous to the electrochemical reductive conversion of cinnolines into indoles (see section II.D.1.c) is the finding that the reduction of 3-phenyl-1,2,4-benzotriazine (4.1064c) in acid leads to the 2-phenylbenzimidazole cation (4.1064e)[660b]. Also in this conversion an intermediary 1,4-dihydro compound, propably (4.1064d), is proposed; however, there is as yet insufficient evidence to assign the correct structure to this compound.

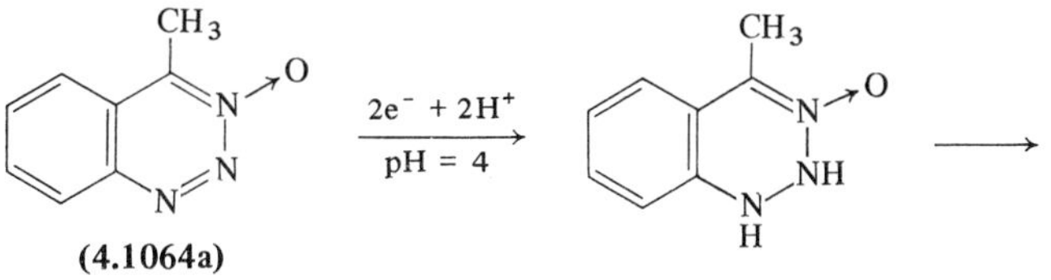

$$2e^- + 2H^+ \xrightarrow{pH = 4}$$

(4.1064a)

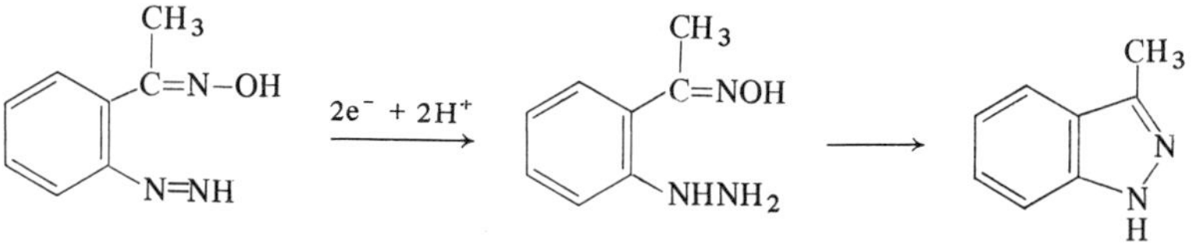

$$2e^- + 2H^+ \longrightarrow$$

(4.1064b)

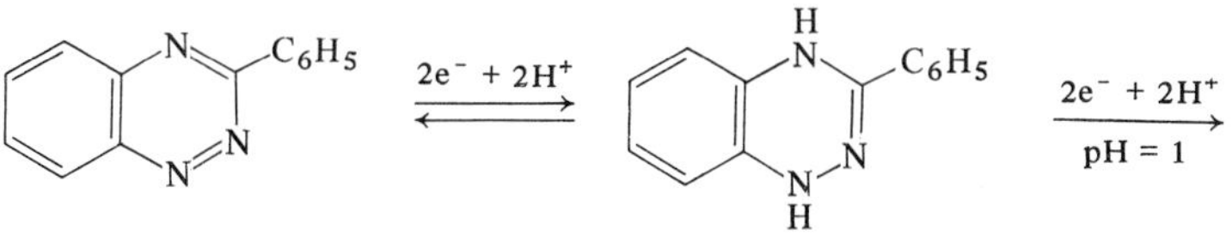

$$2e^- + 2H^+ \rightleftharpoons$$

$$2e^- + 2H^+ \xrightarrow{pH = 1}$$

(4.1064c)　　　　(4.1064d)

$$\xrightarrow{-NH_3}$$

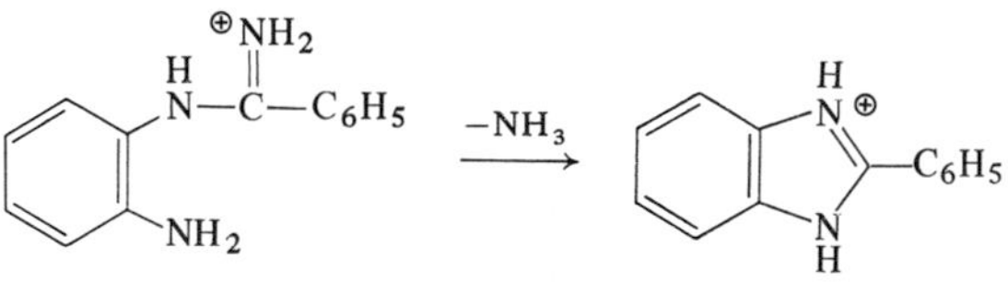

(4.1064e)

A quite similar ring contraction has been reported to occur when the 1,2,4-triazine derivative (4.1065) is reduced with $Na_2S_2O_4$ in acetic acid leading to the formation of the hypoxanthine derivative (4.1066). The reaction probably proceeds via the 1,2-dihydro-1,2,4-triazine (4.1067) and the 4-pyrimidone (4.1068)[661].

Reductive ring contractions into imidazoles have also been reported to occur in the 1,3,5-triazine series. From the 2,4,6-triaryl-1,3,5-triazines (4.1069) the 2,4,5-triarylimidazoles (4.1070) are yielded[662-665]. Reduction of the 1,3,5-triazine (4.1069) (R = CCl_2CH_3) gives, besides ring contraction, dehalogenation into a compound to which the structure (4.1070) (R = C_2H_5) has been assigned[666].

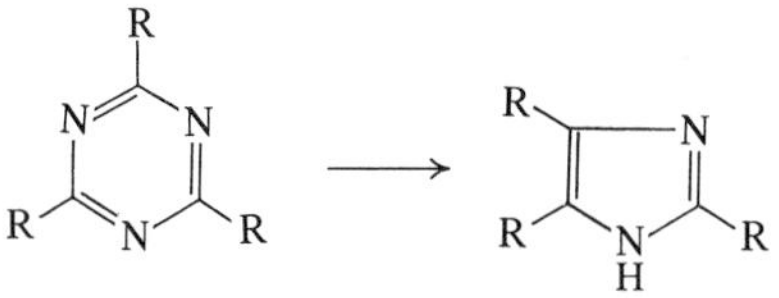

(4.1069)　　　　(4.1070)

$$R = C_6H_5; o\text{-}OHC_6H_4;$$
$$p\text{-}ClC_6H_4; p\text{-}CH_3C_6H_4$$

As a result of a successful study of the behaviour of the parent compound 1,3,5-triazine in reaction with nucleophiles, a useful method was developed for the introduction of one or more methine groups to the nitrogen of primary amines. By this reaction a primary amine (RNH_2) with 1,3,5-triazine can, for example, readily be converted into the imino compound (4.1072) which is further transformed by the excess of amine present, into N,N-disubstituted formamidine (4.1073)[666]. This reaction probably occurs via the adduct (4.1071) since in the s-triazine ring, which is characterized by a strong π-deficiency, the C=N bond is strongly polarized making addition of nucleophiles very favourable. As these methinylation reactions are easy to carry out, the yields are high and the

(4.1071) (4.1072) (4.1073)

reactions usually give very pure products as the only by-product formed is the volatile ammonia. 1,3,5-Triazine has been successfully used for the synthesis of a number of heterocycles[668]. The following compounds are obtained by this method: Δ^2-imidazoline[666], benzimidazole[666], purine[667] and purine derivatives[667], xanthine and theophylline[667], benzoxazole[667], benzothiazole[667], 2-hydroxy-, 2-amino-, and 2-thio-1,2,4-triazole[666] (or their tautomeric compounds). All these reactions are summarized in the scheme given on page 217.

Oxidation of appropriate imines with peracetic acid is a useful method for the preparation of oxaziranes. Since it is known that 1,3,5-trialkyl perhydro-1,3,5-triazines (4.1074) depolymerize in acid media into the imines (4.1075), oxidation of these compounds with peracetic acid is found to be a very attractive method for preparing N-substituted oxaziridines (4.1076)[669].

2. RING TRANSFORMATIONS OF TRIAZINES INTO OTHER SIX-MEMBERED HETEROCYCLES

a. 1,2,4-Triazines into pyridines and pyrimidines

Like the 1,2,4,5-tetrazine ring (see section V.C.1.b) the 1,2,4-triazine ring is able to react with electron-rich olefins to form dihydropyridines or pyridines[670,671]. Thus, when 5-phenyl-1,2,4-triazine (4.1077) reacts with 1-ethoxy-1-dimethylaminoethene, the dihydropyridines (4.1079) and (4.1082) and/or the pyridines (4.1080) and (4.1083) are formed. This ring conversion can be considered as an "inverse" kind of Diels–Alder-type reaction where the

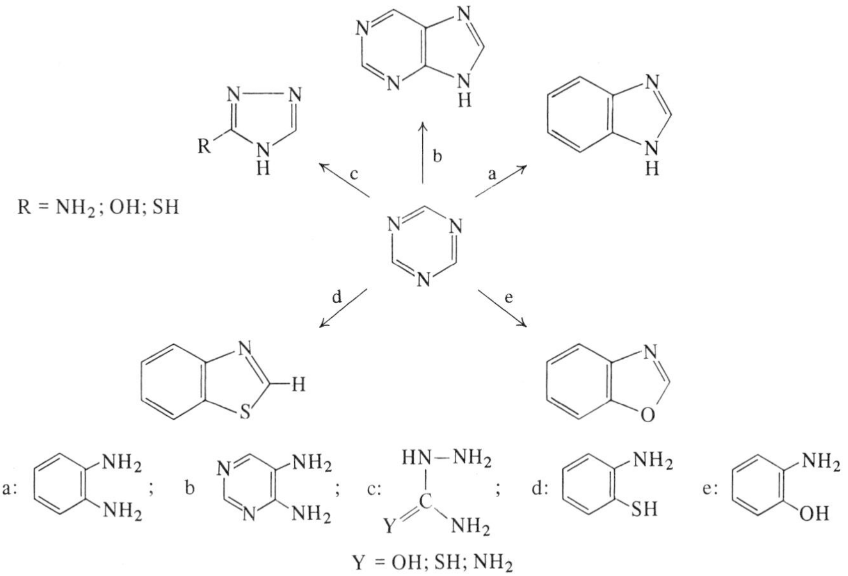

R = NH$_2$; OH; SH

a: ; b ; c: ; d: e:

Y = OH; SH; NH$_2$

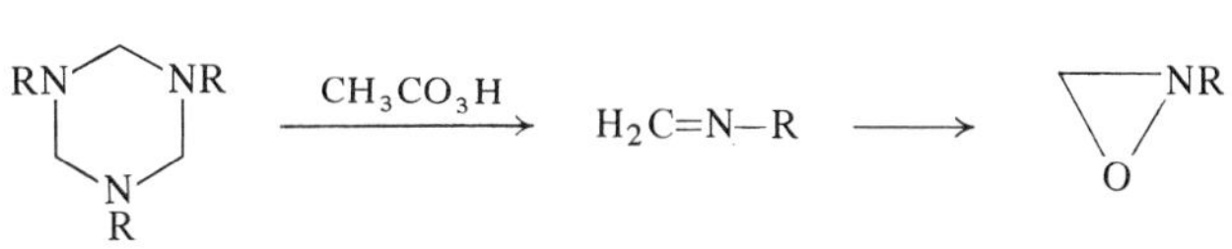

(4.1074) (4.1075) (4.1076)

dienophile is preferably electron-rich and the diene electron-deficient. In this case the 1-aza-bicyclo[2.2.2]octa-2,5-diene (4.1078) and (4.1081) are formed as intermediates. A similar ring transformation has been reported for the 1,2,4-triazines (4.1084) which react exothermally with morpholinocyclopentene to form the trimethylenepyridine (4.1085)[672]. In some cases the presence of the intermediate (4.1086) was demonstrated by nmr spectroscopy. The pyridine tricarboxylate (4.1087) was formed from (4.1084) (R = CO$_2$CH$_3$) with vinylacetate or vinylethylether.

With electron-rich acetylenes a quite different reaction takes place[671]. From the 1,2,4-triazines (4.1088) and 1-diethylaminopropyn the pyrimidines (4.1092) and not the expected pyridines are obtained[671a,b]. It was originally reported[671] that in this reaction pyridazine derivatives (4.1090) were formed but it was however proved[671a,b] that the structure assignments were incorrect and that, in fact, we deal with the formation of pyrimidine derivatives. The formation of these pyrimidines is suggested to occur via a (4 + 2) cycloaddition of the ynamine to the 1,2,4-triazine. This addition apparently does not take place on the 3,6-position—leading to the formation of (4.1089)—but on position

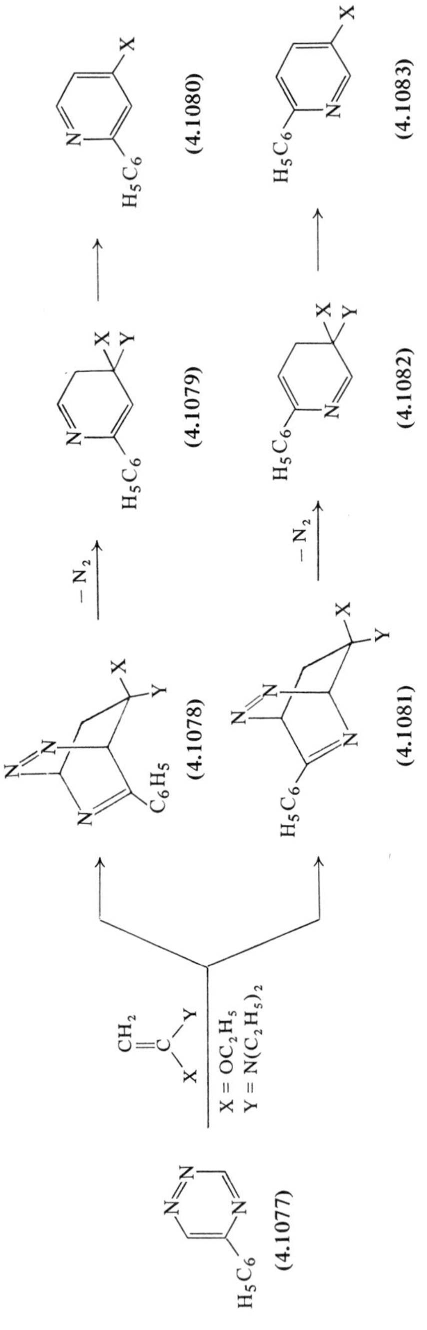
(4.1077)
$CH_2=C(X)Y$
$X = OC_2H_5$
$Y = N(C_2H_5)_2$
(4.1078)
(4.1081)
$-N_2$
(4.1079)
(4.1082)
(4.1080)
(4.1083)

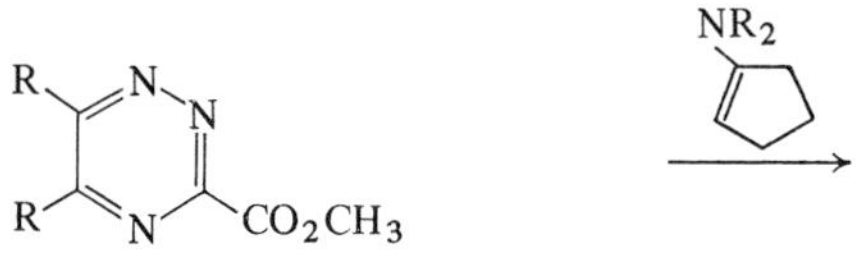

R$-$N$=$N, R$-$N, CO$_2$CH$_3$

NR$_2$

(4.1084)

R$-$, R$-$N, CO$_2$CH$_3$

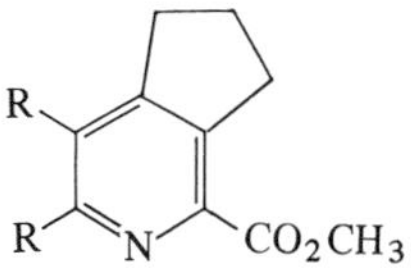

(4.1085)

R = CH$_3$; CO$_2$CH$_3$; C$_6$H$_5$; p-NO$_2$C$_6$H$_4$

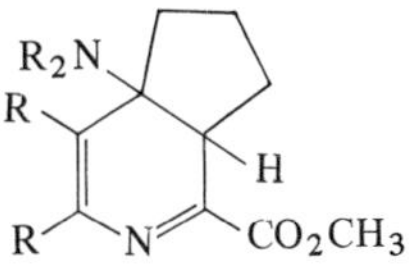

R$_2$N, R$-$, R$-$N, H, CO$_2$CH$_3$

(4.1086)

H$_3$CO$_2$C, H$_3$CO$_2$C, N, CO$_2$CH$_3$

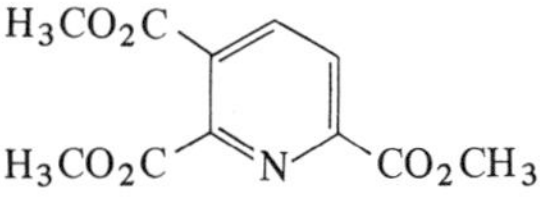

(4.1087)

2 and 5, yielding (4.1091) from which by loss of hydrogen cyanide the pyrimidines (4.1092) are formed. In accordance with the proposed mechanism, from the 6-phenyl derivative of (4.1088) (R = CO$_2$CH$_3$) the pyrimidine derivative (4.1092) (R = CO$_2$CH$_3$) is obtained with loss of benzonitrile. It was found that the position of the phenyl substituent strongly governs the mode of addition of the ynamine: addition to the 5-phenyl derivative (4.1092a) gives namely the intermediate (4.1092b) and not an intermediate of type (4.1091). This is evidenced by the fact that in this reaction the pyridine derivative (4.1092c) is formed and not a compound with a pyrimidine nucleus.

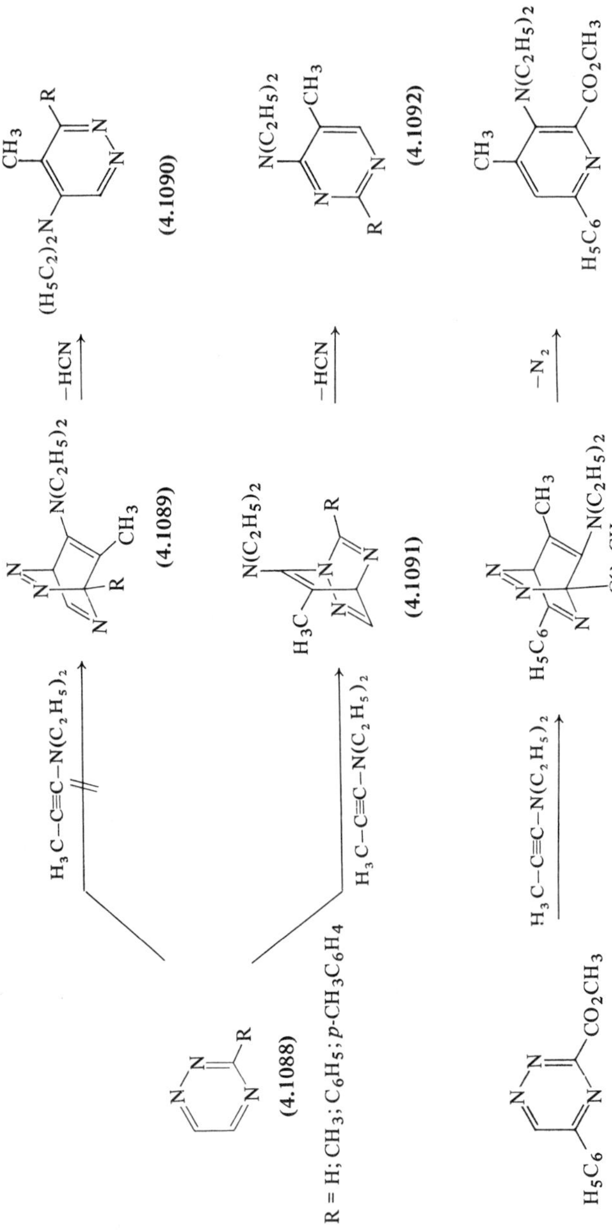

(4.1088)

R = H; CH$_3$; C$_6$H$_5$; p-CH$_3$C$_6$H$_4$

H$_3$C—C≡C—N(C$_2$H$_5$)$_2$

(4.1089)

$\xrightarrow{-HCN}$

(4.1090)

(4.1091)

H$_3$C—C≡C—N(C$_2$H$_5$)$_2$

$\xrightarrow{-HCN}$

(4.1092)

(4.1092a)

H$_3$C—C≡C—N(C$_2$H$_5$)$_2$

(4.1092b)

$\xrightarrow{-N_2}$

(4.1092c)

b. *1,3,5-Triazines into pyrimidines; 1,2,3-triazines into pyridines*
The first examples of the ring transformations of *s*-triazines into pyrimidines
were found many years ago; these were the conversions of 2,4,6-trimethyl-1,3,5-
triazine (4.1093) into the 4-aminopyrimidine (4.1094) and of the 2,4,6-triethyl-
1,3,5-triazine (4.1095) into the 4-amino-2,6-diethyl-5-methylpyrimidine
(4.1096)[673]. These reactions are not important as methods of preparation since
the required conditions are very drastic (a pressure of 8500 atm!) and the yields

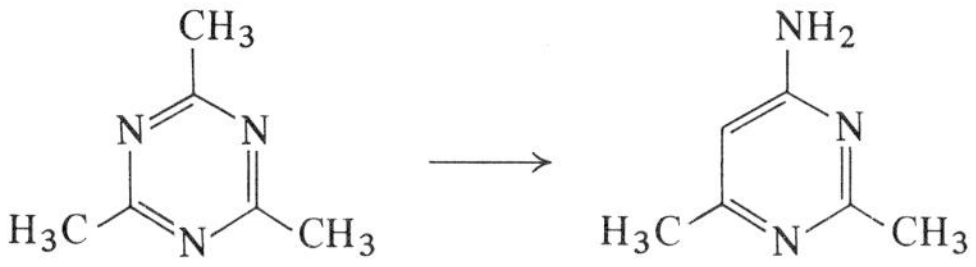

(4.1093)   (4.1094)

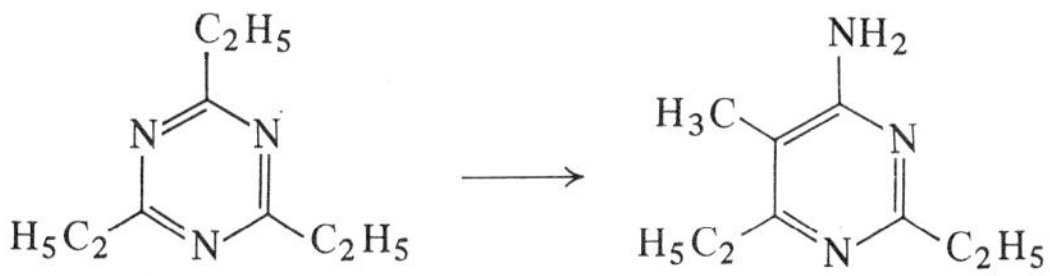

(4.1095)   (4.1096)

are inversely proportional to the required reaction conditions, that is, low. It was
consequently very surprising to find that these ring transformations occur
readily when one of the methyl groups in (4.1093) is acylated. Simple heating of
(4.1097) with water is sufficient to bring about conversion into the 4-substituted
6-acetylamino-2-methylpyrimidines (4.1098)[674,675]. This reaction is of a very
general kind as is clear from the fact that a number of 2,4-dimethyl-1,3,5-
triazines containing the acyl group in position 6 undergo these ring transformations. It is proposed that this reaction takes place in the following way (see p.

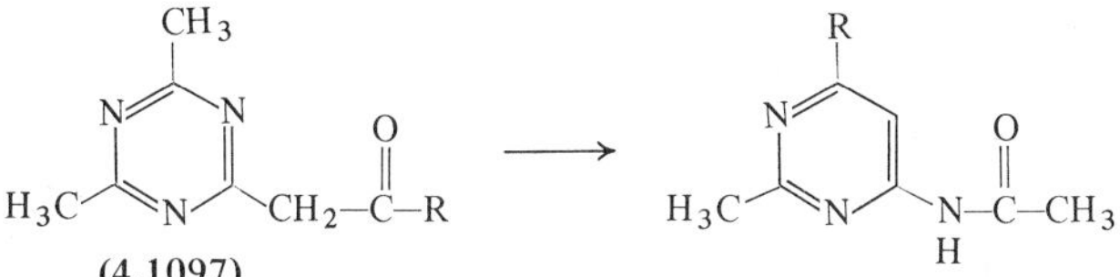

(4.1097)   (4.1098)

R = CH_3; C_6H_5; p-ClC_6H_4; p-$OCH_3C_6H_4$; 3-C_5H_4N; 2-C_5H_3S; 2-C_5H_3O

222): in the keto–enol equilibrium (4.1097)⇌(4.1099), the tautomeric enol form
is favourable to the keto form since the presence of the ring nitrogen makes
hydrogen bonding between oxygen and nitrogen possible. Because of the presence
of a partial positive charge on the ring nitrogen, the addition of water to the
azomethine linkage is promoted leading to (4.1100), in which, by an electron

shift as indicated, the open-chain intermediate (4.1101) is obtained which subsequently gives ring closure to (4.1098). The proposed mechanism suggests that a trace of acid would strongly promote the reaction; this has been shown to be the case.

(4.1097)   (4.1099)   (4.1100)

(4.1098)   (4.1101)

   The possibility of using 1,3,5-triazine as a source for generating methine groups, has already been discussed in section IV.C.1.a; this method has been satisfactorily applied for the synthesis of pyrimidines which do not contain a substituent on position 2. When 1,3,5-triazine reacts with compounds containing an active methylene group, i.e. malonitrile[677,668], malonic ester[676], benzoylacetic ester[676], arylmethyl cyanide[677-679,679a], or benzylphenylketone[679a], the 4-amino-5-cyanopyrimidine (4.1102), 4-amino-5-arylpyrimidines (4.1103), the 5-ethoxycarbonylpyrimidone-4 (4.1104), the 5-ethoxycarbonyl-4-phenylpyrimidine (4.1105) and 4,5-diphenylpyrimidine (4.1105a) respectively, are obtained. It is very interesting, however, that equimolar amounts of 1,3,5-triazine with $C_6H_5COCH_2COCH_2C_6H_5$ in the presence of triethylamine yields 3-phenyl-5-benzoylpyridone-4 (4.1105b)[679a]. The formation of the aminocyanopyrimidine (4.1102) can be postulated to take place in a series of reactions involving an initial methinylation of the malonitrile to the iminodicyano compound (4.1106) (in analogy to the formation of (4.1072)) because it is tautomeric with the aminomethylenemalonitrile, (4.1107). A renewed methinylation leads to the N-dicyanovinylformamidine (4.1108) which subsequently cyclizes into 4-amino-5-cyanopyrimidine (4.1102). The formation of the pyrimidines (4.1103), (4.1104) and (4.1105) can be explained in a similar way. As already pointed out, the importance of these pyrimidine syntheses lies in the fact that they yield only pyrimidines which are unsubstituted at position 2. This type of compound is not easily accessible by a direct synthesis and can almost only be obtained from pyrimidines which contain substituents on position 2, which are easily removable. For instance, by reduction of 2-halogenopyrimidines, desulphurization of the 2-mercapto-pyrimidines or decarboxylation of 2-carboxypyrimidines[680].

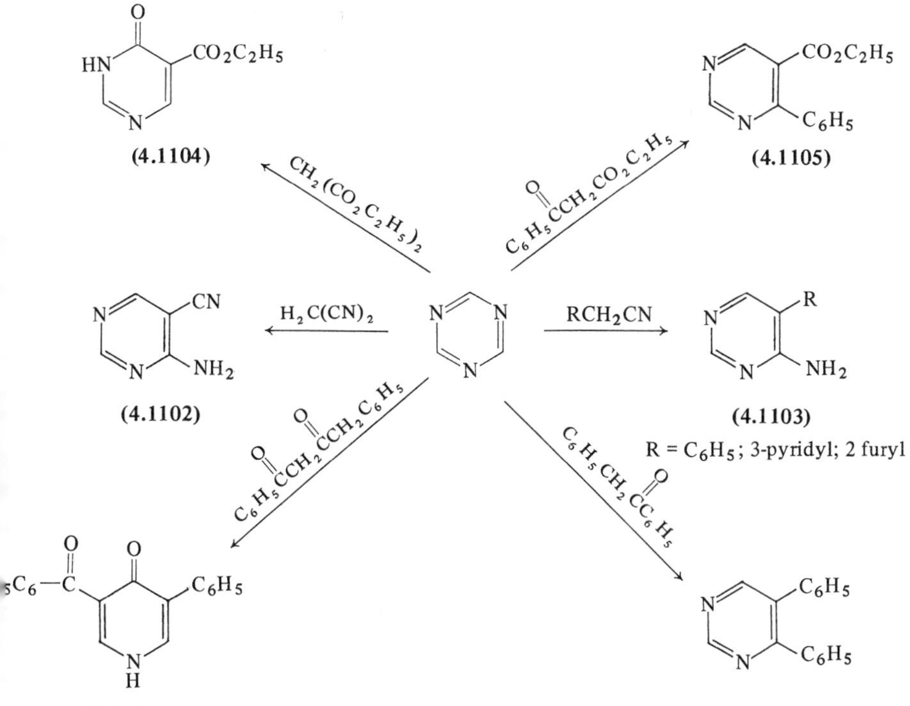

(4.1104)

(4.1105)

$CH_2(CO_2C_2H_5)_2$

$C_6H_5CCH_2CO_2C_2H_5$

(4.1102)

$H_2C(CN)_2$

RCH_2CN

(4.1103)

R = C_6H_5; 3-pyridyl; 2 furyl

$C_6H_5CCH_2CC_6H_5$

$C_6H_5CH_2CC_6H_5$

(4.1105b)

(4.1105a)

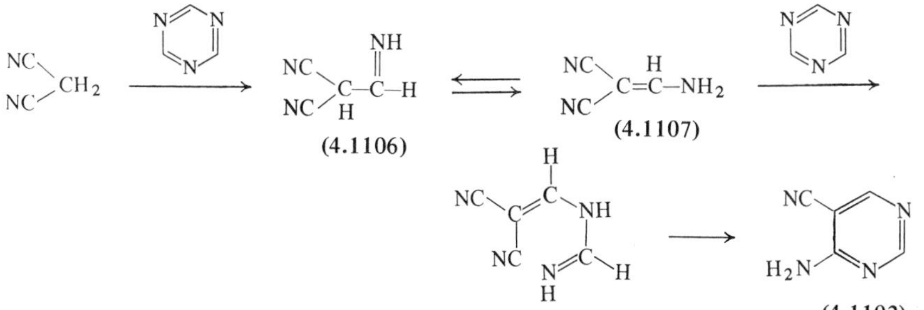

(4.1106)

(4.1107)

(4.1108)

(4.1102)

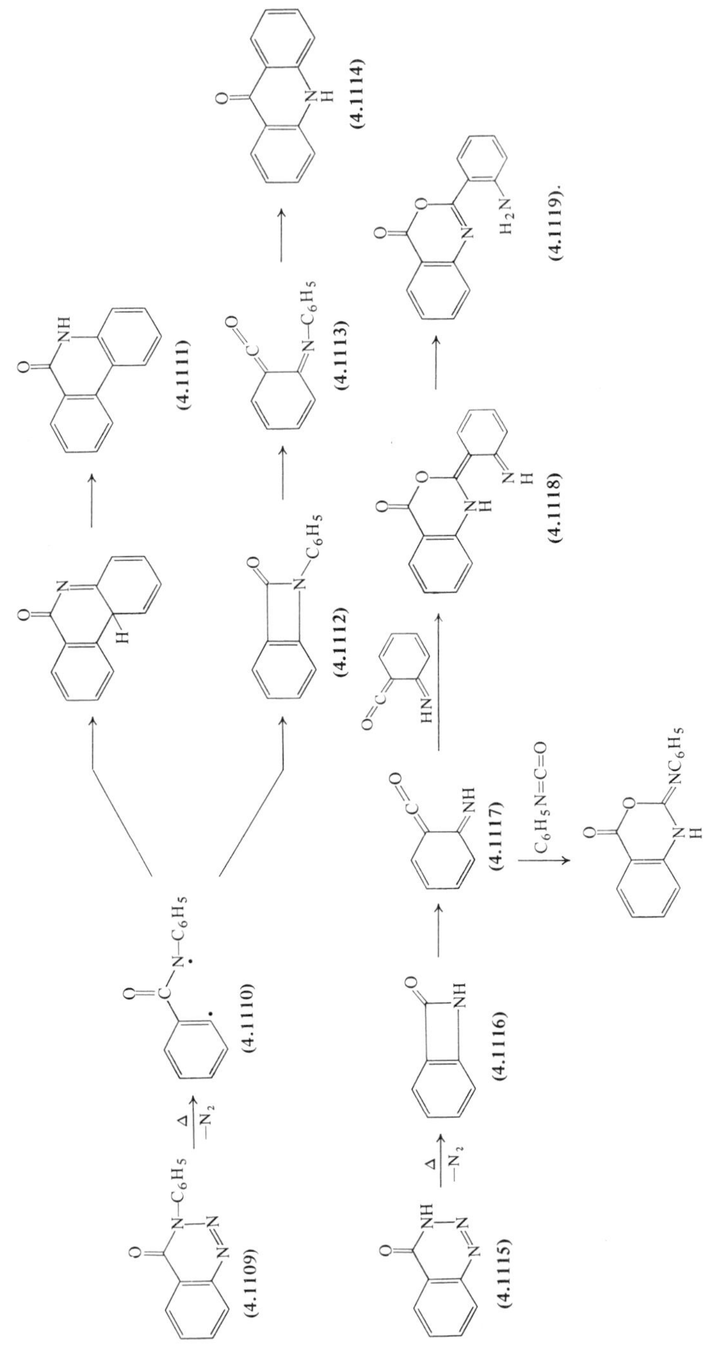

(4.1109)
(4.1110)
(4.1111)
(4.1112)
(4.1113)
(4.1114)
(4.1115)
(4.1116)
(4.1117)
(4.1118)
(4.1119).
(4.1120)
Δ
−N₂
C₆H₅N=C=O

In the reaction of acetylacetic ester with 1,3,5-triazine, surprisingly, 3,5-di(ethoxycarbonyl)-2,6-dimethylpyridine[676] is formed and not the corresponding 5-ethoxycarbonyl-4-methylpyrimidine. The mechanism of this reaction has been extensively discussed[676]. Aminoethynylation of phenylacetic esters by 1,3,5-triazine also gave pyridines[681].

Decomposition of 3-phenyl-1,2,3-benzotriazin-4-one (4.1109) at about 280°C gives an instantaneous reaction in which 9-acridone (4.1114) (31 per cent) and phenanthridin-6-one (4.1111) (14 per cent) are formed[682,683]. When the thermolysis was carried out in paraffin, benzanilide was obtained, indicating hydrogen abstraction from the solvent by the intermediate radical (4.1110) formed by ring opening and loss of nitrogen. The mechanism shown on p. 224 is suggested.

The formation of the phenanthridone (4.1111) is straightforward, but the formation of the 9-acridone (4.1114) is supposed to take place via a collapse of the 1,4-diradical (4.1110) into N-phenylbenzazetone (4.1112), ring-opening into the ketenimine (4.1113) and recyclization. The feasibility of such a heterodiene (4.1113) as an intermediate in the reaction is supported by the fact that in the thermolysis of benzotriazinone (4.1115) the most important product obtained is benzoxazinone (4.1119)[683]. The formation of this compound can reasonably be explained via the benzazetone (4.1116) and the ketene (4.1117) which in a Diels–Alder-type reaction dimerizes into (4.1118). The thermolysis of (4.1115) when carried out in the presence of phenylisocyanate[684], gives the iminobenzoxazine (4.1120) and this presents further evidence for the occurrence of a ketenimine intermediate such as (4.1117).

The photolysis of 3-phenyl-4H-benzo-1,2,3-triazine (4.1121) with ultraviolet Pyrex-filtered light gave ring contraction into (4.1123) and formation of 9,10H-phenanthridine (4.1124)[685]. Also in this case the suggestion has been made that the common 1,4-diradical (4.1122) is intermediate. When 1-methylnaphtho[1,8-de]triazine (4.1125) is irradiated in the presence of benzene, 1-aminomethyl-8-phenylnaphthalene (4.1128) is formed; in the presence of vinyl bromide, however, an N-methyl derivative (4.1130) is obtained[686]. In both reactions the radical (4.1126) is thought to be involved. This gives, with benzene, the intermediate (4.1127) which is converted into (4.1128) by hydrogen migration. However, in the presence of vinyl bromide this intermediate is converted into (4.1129) in which the ring closure effectively competes with hydrogen migration.

Irradiation of 3-phenyl-3,4-dihydronaphtho[2,3-d]1,2,3-triazin-4-one (4.1131) dissolved in THF gave, under evolution of 1 equivalent of nitrogen, formation of N-phenylnaphtho[2,3-b]azetone (4.1133) (95 per cent)[687] and benz[d]acridone (4.1135) (5 per cent). The diazonium compound is suggested as intermediate (4.1132); on losing nitrogen it yields (4.1133). The acridone (4.1135) has been shown to be formed by photolytic valence isomerization of

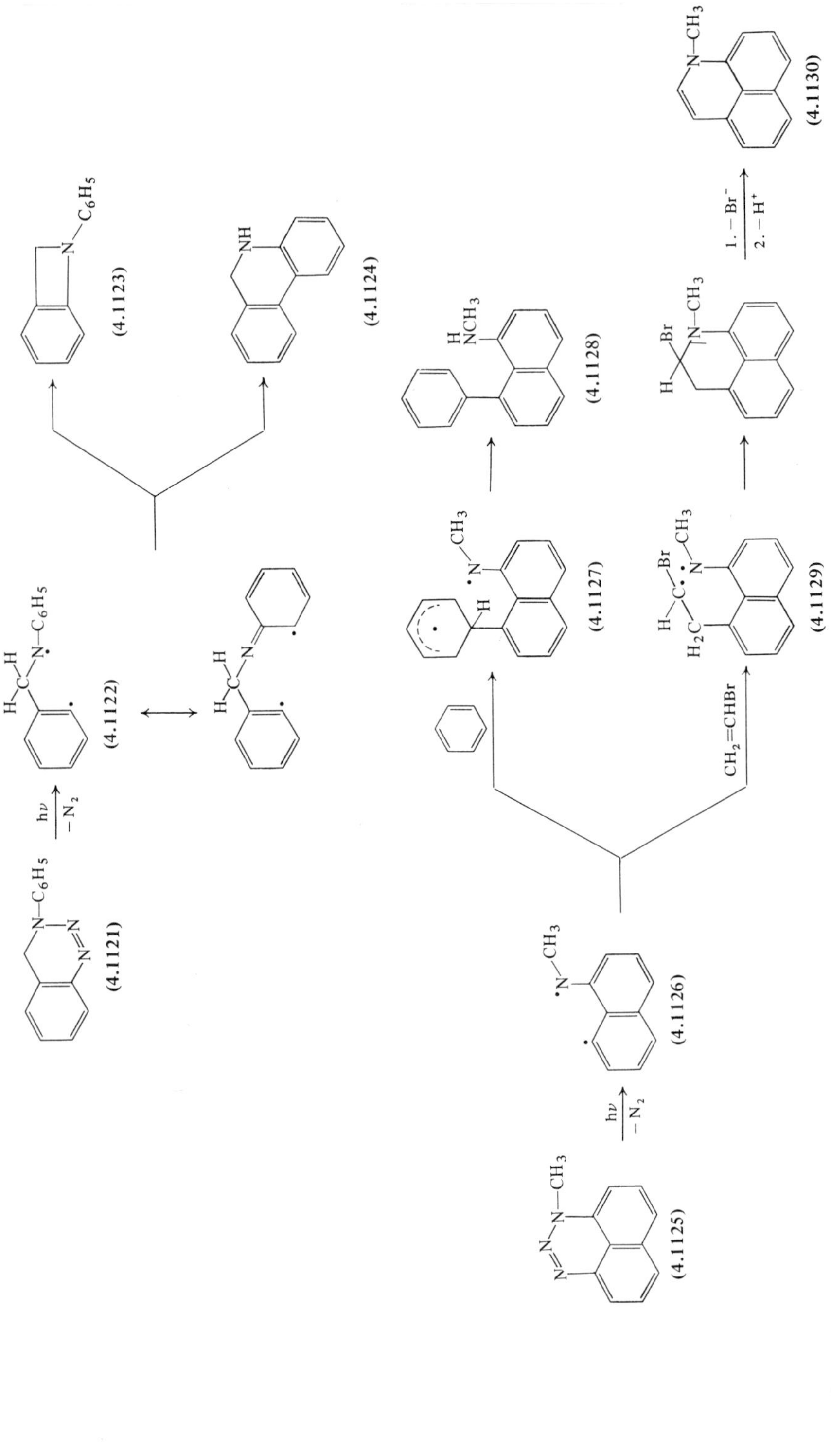
(4.1121)
hν
$-N_2$
(4.1122)
(4.1123)
(4.1124)
(4.1125)
(4.1126)
(4.1127)
(4.1128)
CH_2=CHBr
(4.1129)
1. $-Br^-$
2. $-H^+$
(4.1130)

the non-isolable ketene (4.1134). By N^{15}-labelling of the nitrogen atoms 1 and 2 in (4.1131) it could be proved that both these nitrogen atoms are cleaved off[687]. It can however, not be excluded (and it is probably more likely) that a diradical is intermediate as has been suggested in the photolysis of 3-phenyl-4H-benzo-1,2,3-triazine (4.1121).

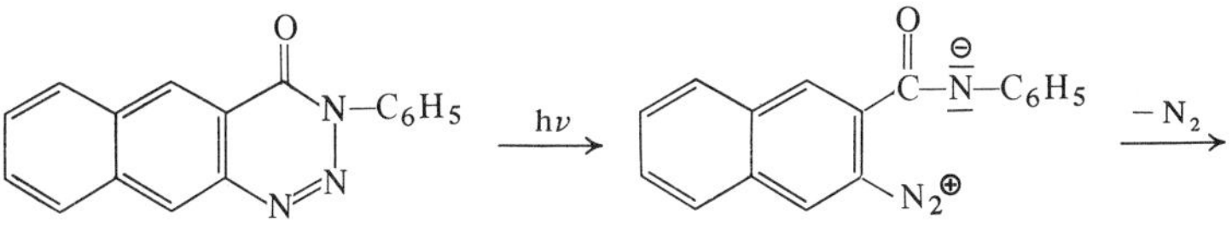

(4.1131)　　　　　　　(4.1132)

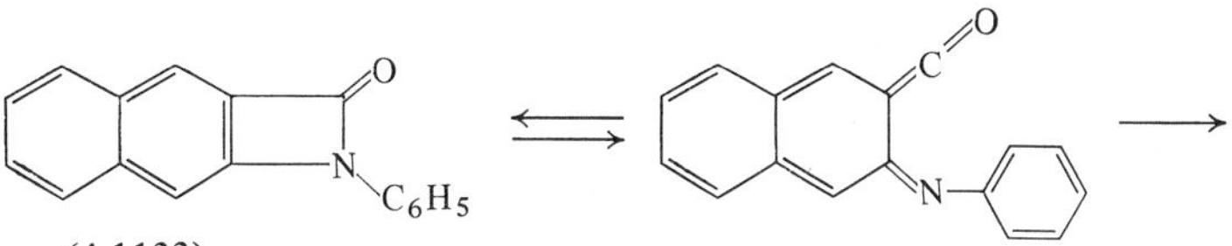

(4.1133)　　　　　　　(4.1134)

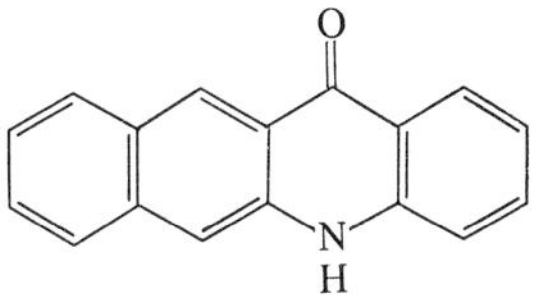

(4.1135)

V. RING TRANSFORMATIONS OF SIX-MEMBERED HETEROCYCLES CONTAINING FOUR HETEROATOMS

A. Ring transformations of six-membered heterocycles containing four sulphur atoms

1. 1,2,4,5-TETRATHIINS INTO 1,3,4-TRITHIOLES

The 1,2,4,5-tetrathiin (4.1137) formed by ring expansion of the cyclic unsaturated phosphorane (4.1136) on reaction with carbon disulphide, loses sulphur when refluxed in toluene or when treated with triethylphosphite[688]. This last reagent is known[689,690] to be particularly effective in removing sulphur from disulphides. Since the compound formed shows ir and nmr spectra closely similar to those of the starting substance, it is assumed that it is the five-membered 1,5-bis (2-oxobut-3-ylidene)-1,2,4-trithiole (4.1139) which is formed in this partial desulphurization. The formation of this 1,2,4-trithiole is believed to occur by an internal displacement of the triethylphosphorothionate groups in (4.1138) formed by addition of triethylphosphite to the disulphide (4.1137). The 1,2,4-trithiole (4.1139) can be further desulphurized into the four-membered 1,3-dithiete[688] (see chapter 3, section III.B.1).

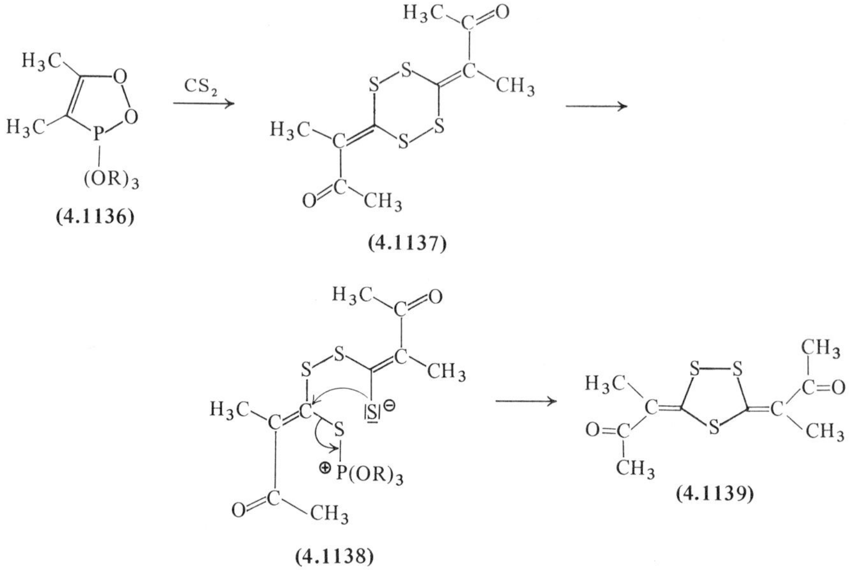

(4.1136)   (4.1137)

(4.1138)   (4.1139)

B. Ring transformations of six-membered heterocycles containing oxygen and/or sulphur and nitrogen atoms

1. 1,2,3,5-OXATHIADIAZINES INTO 1,2,4-TRIAZOLES, 1,2,4-OXADIA-ZOLES, PYRIMIDINES AND 1,3,5-TRIAZINES; 1,2,3,4-THIATRIAZINES INTO 1,2-THIAZINES

The 4,6-diaryl 1,2,3,5-oxathiadiazine-2,2-dioxides (4.1141), which are prepared by the reaction of 1 mol of sulphur trioxide with 2 mol of arylisocyanate, have been found to undergo ring transformation readily with nucleophilic reagents. Thus hydrazinolysis of (4.1141) with hydrazine or arylhydrazines give ring contraction[691] into either 3,5-diaryl 1,2,4-triazoles (4.1142) or the 3,4,5-triaryl 1,2,4-triazoles (4.1140), depending on the hydrazino compound used. When substituted hydrazines (4.1144) are used, as intermediates N-acyltriazoles, (compounds which are known to be strong acylating agents) can be advanced[694]; however, these have not been isolated. Similarly, the 3,5-diaryl 1,2,4-oxadiazoles (4.1143) are obtained[691] from (4.1141) with hydroxylamine hydrochloride.

Besides ring contraction, ring transformation of the 1,2,3,5-oxathiadiazines into pyrimidines and 1,3,5-triazines have been reported. When (4.1141) reacts

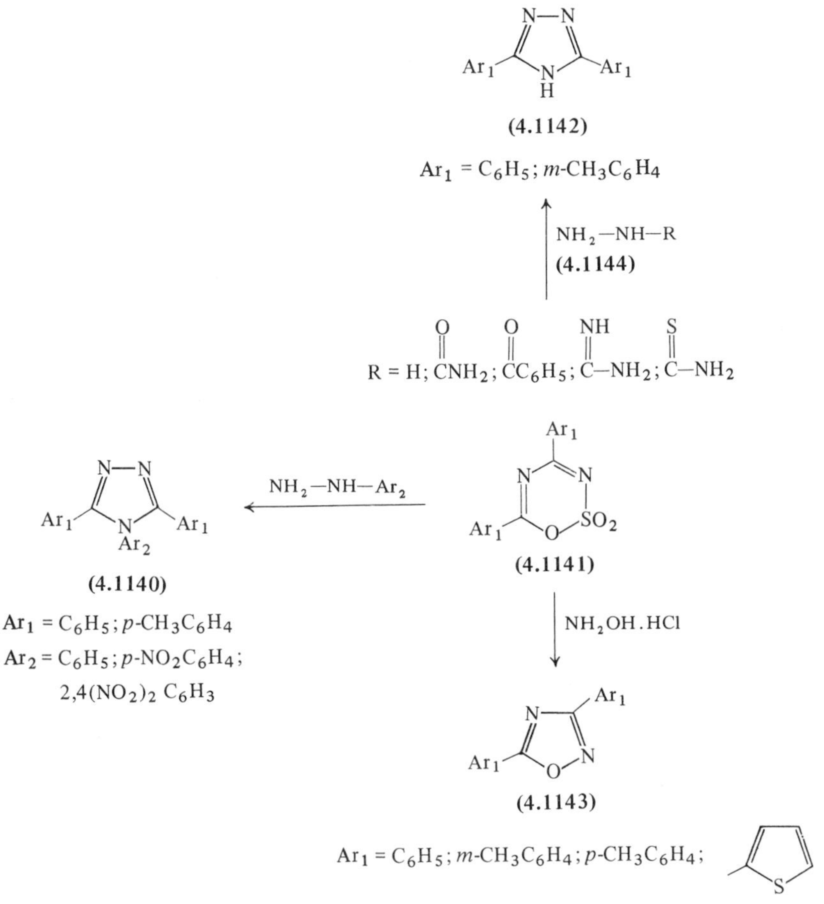

$$Ar_1 \overset{N-N}{\underset{N(H)}{\diagup\diagdown}} Ar_1$$

(4.1142)

$$Ar_1 = C_6H_5; m\text{-}CH_3C_6H_4$$

$\uparrow$

NH_2-NH-R

(4.1144)

$$R = H; \overset{O}{\overset{\|}{C}}NH_2; \overset{O}{\overset{\|}{C}}C_6H_5; \overset{NH}{\overset{\|}{C}}-NH_2; \overset{S}{\overset{\|}{C}}-NH_2$$

$$Ar_1 \overset{N-N}{\underset{N(Ar_2)}{\diagup\diagdown}} Ar_1 \quad \xleftarrow{\ NH_2-NH-Ar_2\ } \quad Ar_1 \overset{Ar_1}{\underset{O}{\diagup\diagdown}} SO_2$$

(4.1140)   **(4.1141)**

$$Ar_1 = C_6H_5; p\text{-}CH_3C_6H_4$$
$$Ar_2 = C_6H_5; p\text{-}NO_2C_6H_4;$$
$$2,4(NO_2)_2 C_6H_3$$

$\downarrow NH_2OH.HCl$

$$Ar_1 \overset{N-}{\underset{O-N}{\diagup\diagdown}} Ar_1$$

(4.1143)

$$Ar_1 = C_6H_5; m\text{-}CH_3C_6H_4; p\text{-}CH_3C_6H_4;$$

with the sodium salt of 1,3-dicarbonyl compounds[692], the sodium salt of a number of substituted acetic esters[692], or with carbonyl-stabilized sulphur ylides[693], conversion into the corresponding pyrimidine derivatives (4.1144a), (4.1146) and (4.1145) respectively occurs. When (4.1141) reacts with aliphatic or aromatic imido compounds, a smooth conversion into 2-alkyl-4,6-diaryl and 2,4,6-triaryl 1,3,5-triazines (4.1147) is achieved[691]. These reactions have been used for the formation of a great variety of 1,3,5-triazines with various substituents. With urea, thiourea or guanidine as reagent, the corresponding 1,3,5-triazines (4.1147) (R = OH; SH or NH_2) are obtained. A consideration of the mechanism leads to the suggestion that the reactions occur by a nucleophilic attack on a 1,4-dipole formed by loss of sulphur trioxide from its precursor (4.1141). On addition of, for example, the sulphur ylides, the intermediate (4.1148) is formed; ring closure then occurs giving the betaine (4.1149).

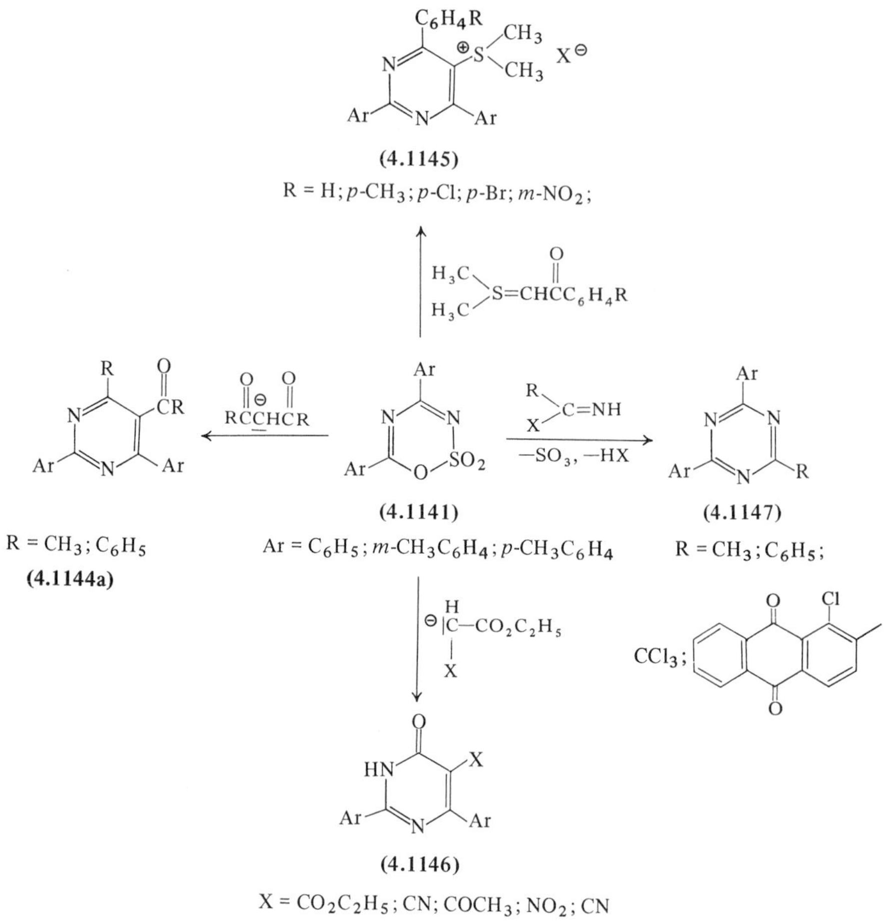

(4.1145)

R = H; p-CH$_3$; p-Cl; p-Br; m-NO$_2$;

(4.1141)

Ar = C$_6$H$_5$; m-CH$_3$C$_6$H$_4$; p-CH$_3$C$_6$H$_4$

R = CH$_3$; C$_6$H$_5$
(4.1144a)

(4.1147)

R = CH$_3$; C$_6$H$_5$;

CCl$_3$;

(4.1146)

X = CO$_2$C$_2$H$_5$; CN; COCH$_3$; NO$_2$; CN

Elimination of the hydroxide anion brings the reaction to completion forming the resonance-stabilized pyrimidine derivative. Otherwise a mechanism involving (4.1150) as intermediate has been proposed in the reaction of the sodium salts of CH-acidic compounds with (4.1141).

Thermal decomposition of the 1,2,3,4-thiatriazine-1,1-dioxide derivative (4.1151) does not lead to extrusion of sulphur dioxide but gives evolution of nitrogen with formation of the 1,2-thiazine-1,1-dioxide (4.1153)[694]. Analogous to the mechanism proposed in the somewhat similar conversion of the 1,2,3-benzotriazin-4-one (4.1109) into phenanthridone (4.1111) (see section IV.C.2.b); it can be advanced that the diradical (4.1152) is intermediate in this reaction.

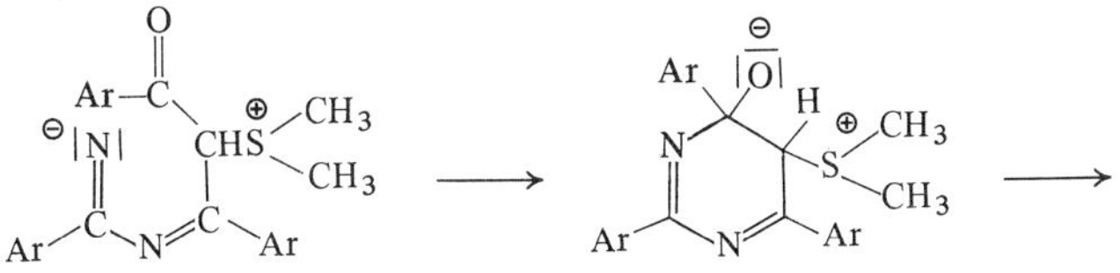

(4.1148)

(4.1149)

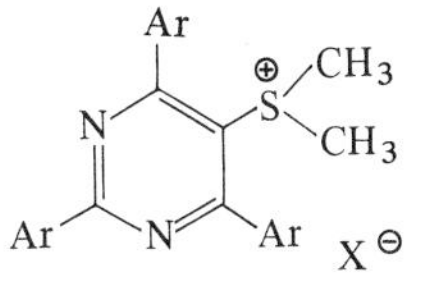

(4.1145)

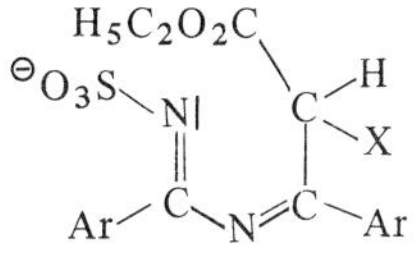

(4.1150)

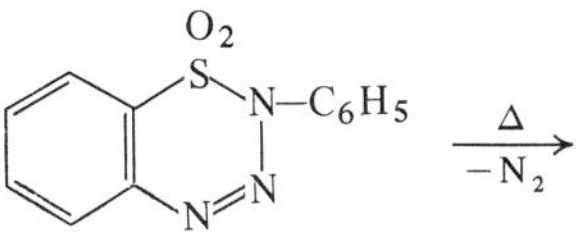

(4.1151)

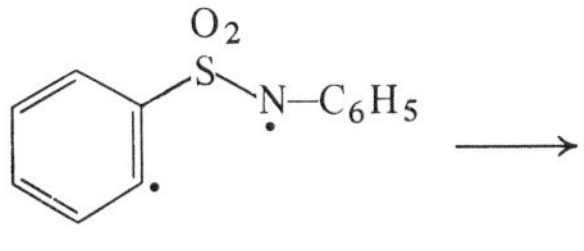

(4.1152)

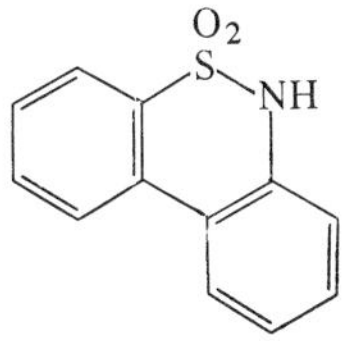

(4.1153)

Extrusion of sulphur has been reported[694a] with the 2-phenyl-4,5-diaryl 2H-1,2,3,6-thiatriazines on treatment with hydrochloric acid. The corresponding 2-phenyl-4,5-diaryl-1,2,3-triazoles are obtained.

C. Ring transformations of six-membered heterocycles containing four nitrogen atoms

1. RING TRANSFORMATIONS OF TETRAZINES

a. *1,2,4,5-Tetrazines into 1,2,4-triazoles and 1,3,4-oxadiazoles; 1,2,3,4-tetrazines into 1,2,3-triazoles*

The 1,2,4,5-tetrazine-3,6-dithione derivative (4.1154) is found to rearrange in an acidic medium into the 3,5-dimercapto-4-amino-1,2,4-triazole (4.1155)[695]. On treatment with hydrazine, this dithione (4.1154) is converted into the 4-amino-1,2,4-triazole (4.1157)[695,696].

It has been proved that the last-mentioned ring contraction does not take place via a primarily occurring ring contraction of (4.1154) into the 3,5-dimercapto-1,2,4-triazole (4.1158), which should yield (4.1157) on subsequent hydrazinolysis, but occurs via the intermediary thiohydrazinotetrazine derivative (4.1156). Quite similarly, boiling the tricyclic *as*-triazino-1,2,4,5-tetrazine derivative (4.1159) with hydrazine hydrate yields the 4-amino-3,5-dihydrazino-1,2,4-triazole

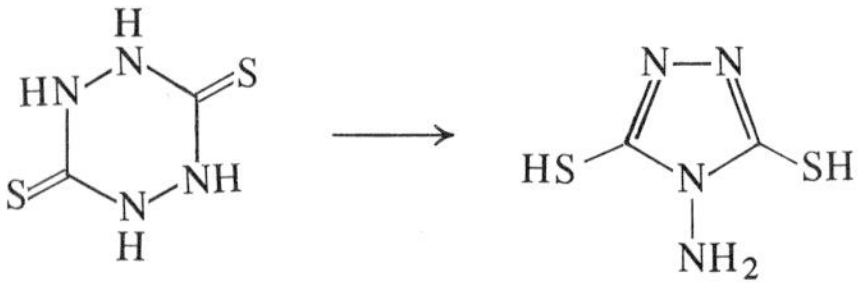

(4.1154)   (4.1155)

(4.1154) ⟶

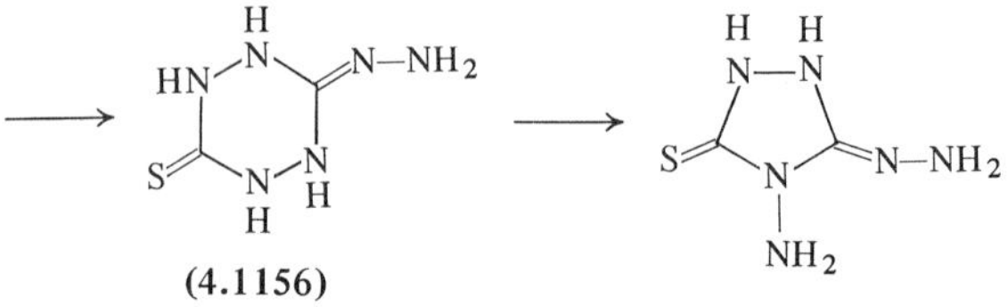

(4.1156)

(4.1157)

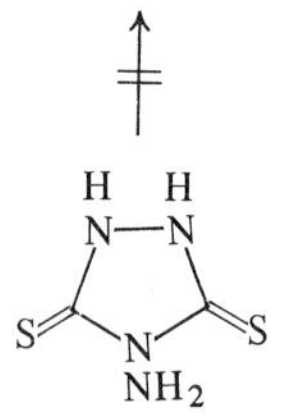

(4.1158)

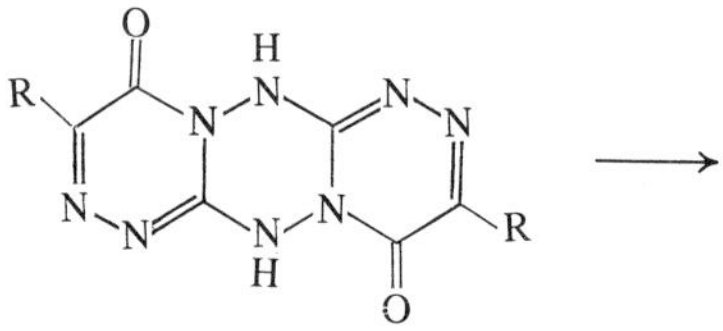

(4.1159)

R = CH_3 ; C_2H_5

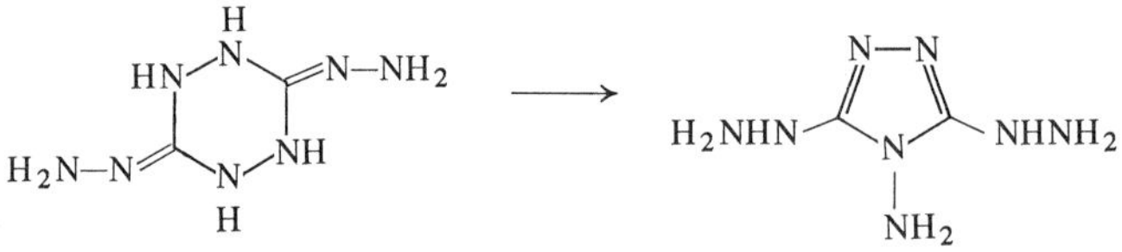

(4.1160)   (4.1161)

(4.1161)[696]. Apparently, the perhydro 1,2,4,5-tetrazine (4.1160) is formed primarily and this further isomerizes into (4.1161).

With the 1,2-dihydro-1,2,4,5-tetrazines (4.1162) analogous ring contractions were reported[697-703]. When 2,6-diphenyl- and 2,6-dipyridyl-1,2-dihydro-1,2,4,5-tetrazine (4.1162) is heated with hydrochloric acid[701-703], carbon-nitrogen bond breaking occurs and subsequent ring closure into 4-amino-3,5-

diaryl 1,2,4-triazole (4.1163) is achieved. If 1,4-diaryl 1,4-dihydro-1,2,4,5-tetrazines are heated in base, the same ring contraction has been found to occur[704]. Similarly, 3-hydroxy-1,2-dihydro-1,2,4,5-tetrazine is stated to be converted by nitrous acid into 3-hydroxy-4-amino-1,2,4-triazole[705]. Reducing agents such as zinc and acetic acid also lead to the formation of a 1,2,4-triazole, but under these conditions the 4-amino group is subsequently removed. The

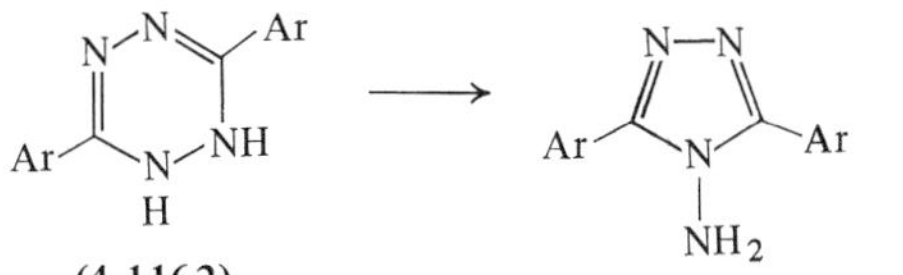

(4.1162) → (4.1163)

Ar = C_6H_5 ; C_5H_4N

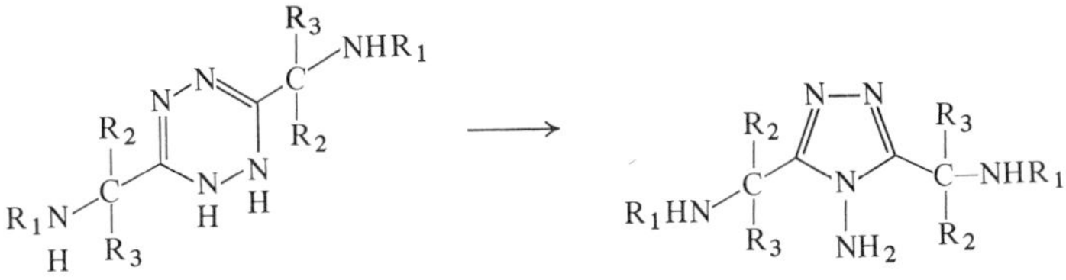

(4.1164) → (4.1165)

R_1	p-Tos	p-Tos	p-Tos	p-Tos
R_2	H	C_6H_5	C_2H_5	C_2H_5
R_3	H	H	H	C_2H_5

same ring contraction yielding (4.1165), has been reported to take place with the 3,6-bis(aminomethyl)1,2-dihydro-1,2,4,5-tetrazine (4.1164)[699]. These isomerizations which occur quite readily in acid have in the past been the cause of misleading errors in structure assignments: products previously considered to be 1,2-dihydro-1,2,4,5-tetrazines are in fact 4-amino-1,2,4-triazoles. For instance, it is apparent that condensation of N-cyanohydrazide with hydrazine does not give the expected 3-amino-1,2-dihydrotetrazine (4.1166) but the 3,4-diamino-1,2,4-triazole (4.1167)[697,698].

Oxidation of 3,6-diphenyl-1,2,4,5-tetrazine (4.1169) with a powerful oxidizing agent (40 per cent peracetic acid) does not give the corresponding N-oxide (4.1170) but instead gives the 2,5-diphenyl-1,3,4-oxadiazole (4.1168)[706]. It is not evident whether the N-oxide (4.1170) is intermediate during the oxidation reaction. Somewhat analogous ring contractions have been reported in the 2,3-dihydro-1,2,3,4-tetrazine series. Reaction of 2,3-diphenyl-

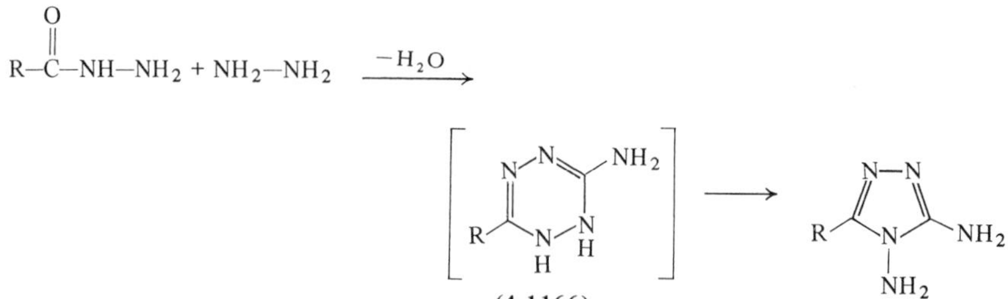

(4.1166)

(4.1167)

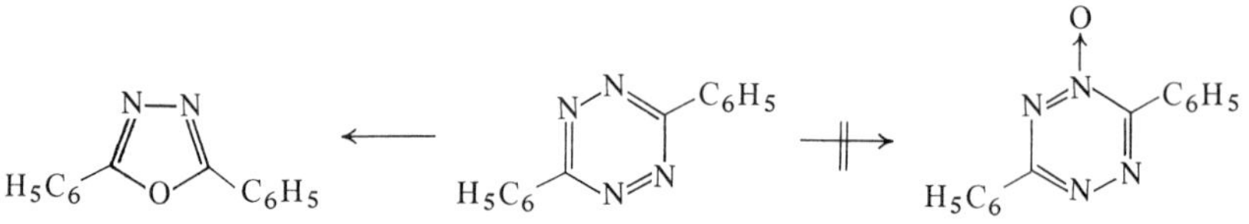

(4.1168) (4.1169) (4.1170)

2,3-dihydrotetrazine (4.1171) (R = H) with ferric chloride in hydrochloric acid[707] yields 2-phenyl-1,2,3-triazole (4.1172) (R = H); the 5,6-dimethyl derivative (4.1171) (R = CH$_3$) gives the 1,2,3-triazole (4.1172) (R = CH$_3$) on treatment with hydrochloric acid[708]. In this last reaction it is difficult to conceive how aniline, reported to be a product of the reaction, can be formed since reducing conditions are absent.

Thermal ring contraction of tetrazines into 1,2,3-triazoles has also been reported: during heating of the 2,3-dibenzoyltetrazines (4.1174) at 140-190°C, exclusive rearrangement into the 1-N,N-dibenzoylamino-1,2,3-triazole derivative (4.1175) takes place[709,710]. When treated with one equivalent of hydrochloric acid, the tetrazine (4.1174) (R = CH$_3$) gives ring contraction into the 1-benzamido derivative (4.1173) (R = CH$_3$)[711,712].

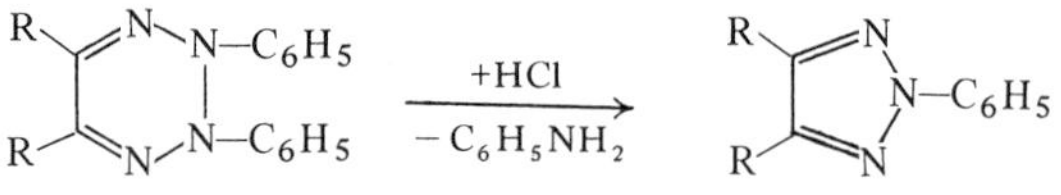

(4.1171)

(4.1172)

R = H; CH$_3$

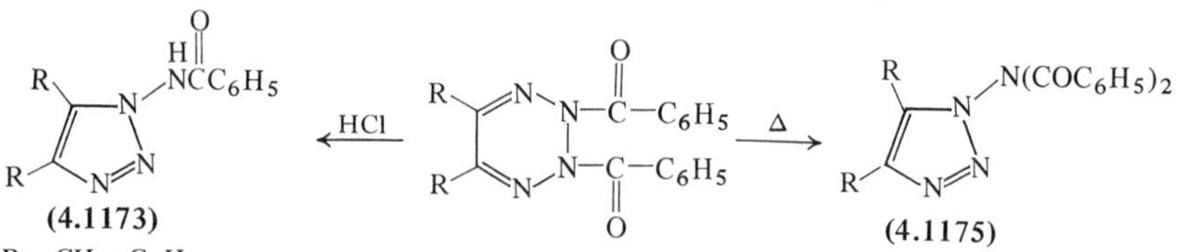

(4.1173)

R = CH$_3$; C$_6$H$_5$

(4.1174)

(4.1175)

R = CH$_3$; C$_6$H$_5$

b. *1,2,4,5-Tetrazines into pyridazines*

Olefinic hydrocarbons (ethene, propene, hexene, butadiene, cyclopentene, cyclohexene, styrene, α-methylstyrene, p-methoxystyrene and p-nitrostyrene) are readily added to the 3,6-disubstituted tetrazines (4.1176) under evolution of nitrogen; 3,4,6-trisubstituted dihydropyridazines (4.1177) are yielded. The nmr data obtained, prove unequivocally that the 1,4-dihydro structure can be assigned to these compounds[715,716].

$$R = CO_2CH_3; CHFCF_3; C_6H_5; p\text{-}NO_2C_6H_4; \alpha\text{-}C_5H_4N; CH_3$$

The influence of various substituents R in (4.1176) on the rate of the ring transformation with styrene (50°C in dioxane) shows the following reactivity order: $CO_2CH_3 \gg p\text{-}NO_2C_6H_4 > CH_3 > C_6H_5$[717]. This reactivity order indicates that stronger electron-withdrawing groups facilitate the reaction. This is also demonstrated by the finding that (4.1176) (R = CO_2CH_3, $CHFCF_3$) reacts even at room temperature with alkenes, whereas the 3,6-dimethyl or 3,6-diphenyl derivative requires higher reaction temperatures[713]. The reactivity order of various 3,6-disubstituted 1,2,4,5-tetrazines towards a number of olefins has also been established[717]. Towards (4.1176) (R = C_6H_5) (120°C in nitrobenzene), the following order has been established: p-methoxystyrene > cyclopentene > styrene > p-nitrostyrene > α-methylstyrene > cyclohexene. This reactivity order is very similar to that found in Diels–Alder reactions with hexachlorocyclopentadiene, and clearly indicates that olefins with electron-releasing substituents facilitate the reaction. In agreement with this result, (4.1176) (R = $CHFCF_3$) starts reacting at room temperature with styrene, butadiene or cyclopentene, but requires heating in order for a reaction with acrylonitrile[714] to occur.

The reaction rate between (4.1176) (R = CO_2CH_3) and styrene is found to be only slightly solvent dependent, the activation parameters (E_a = 8-14 kcal mol^{-1}; log A = 4.2-5.8 (litre $mol^{-1}s^{-1}$)) are very similar to those found in other Diels–Alder additions[716]. From studies of the effect of substituents on both 1,2,4,5-tetrazines and olefins, it would appear that the rate-determining step is the primary nucleophilic attack of the olefin on position 3 and 6 of (4.1176). This 1,4-addition can thus be considered as an "inverse" Diels–Alder-type reaction, where the dienophile is electron rich and the diene electron-deficient. It leads to the highly strained bicyclic intermediate (4.1178) which loses nitrogen in a very fast reaction with consequent relief of strain, yielding the 1,4-dihydro compound (4.1177).

(4.1176) $H_2C=CHR_1$ (4.1178) $-N_2$ (4.1177)

Extending this method to a reaction of 3,6-bis methoxycarbonyl-1,2,4,5-tetrazine (4.1176) (R = CO_2CH_3) with enol ethers[715,718], enol esters[715], enamines or ketene acetals[715], 1,4-dioxene[718] or 2-methyl-4,5-dihydrofuran[718], it appears that the 3,4,6-trisubstituted pyridazines (4.1180) can be formed[715] in excellent yield. In this elegant new synthesis of pyridazines, the 1,4-dihydro compounds (4.1179) have become transient intermediates, since they readily split off an alkanol, carboxylic acid or amino group, depending on the group R present in position 4. The reaction of (4.1176) with ethylacetimidate in dioxane leads to the addition of the C=N moiety and a further rearrangement into the 1,2,4-triazine derivative (4.1181)[718]. In a similar way the pyridazines (4.1182) and (4.1183) can now be obtained from cyclic alkenes and the tetrazine (4.1176).

The addition reaction of acetylenes[715,719] to 1,2,4,5-tetrazines (4.1176) immediately leads to pyridazines (4.1184), but the reaction rate is found to be lower than that of the olefins[715,719].

From the results discussed above it is clear that the addition reaction of appropriately substituted olefins or acetylenes to 1,2,4,5-tetrazines serves as an efficient and useful method for synthesizing substituted pyridazines. The method has a broad general scope, mainly due to the great variety of olefins and acetylenes available and the rather easy accessibility of various 1,2,4,5-tetrazines.

(4.1176) $R_2C \equiv CR_1$ (4.1184)

R = C_6H_5; CO_2CH_3

R_1	C_6H_5	OC_2H_5	$N(C_2H_5)_2$	C_6H_5	$N(C_2H_5)_2$
R_2	H	H	CH_3	$Sn(C_4H_9)_3$	$N(C_2H_5)_2$.

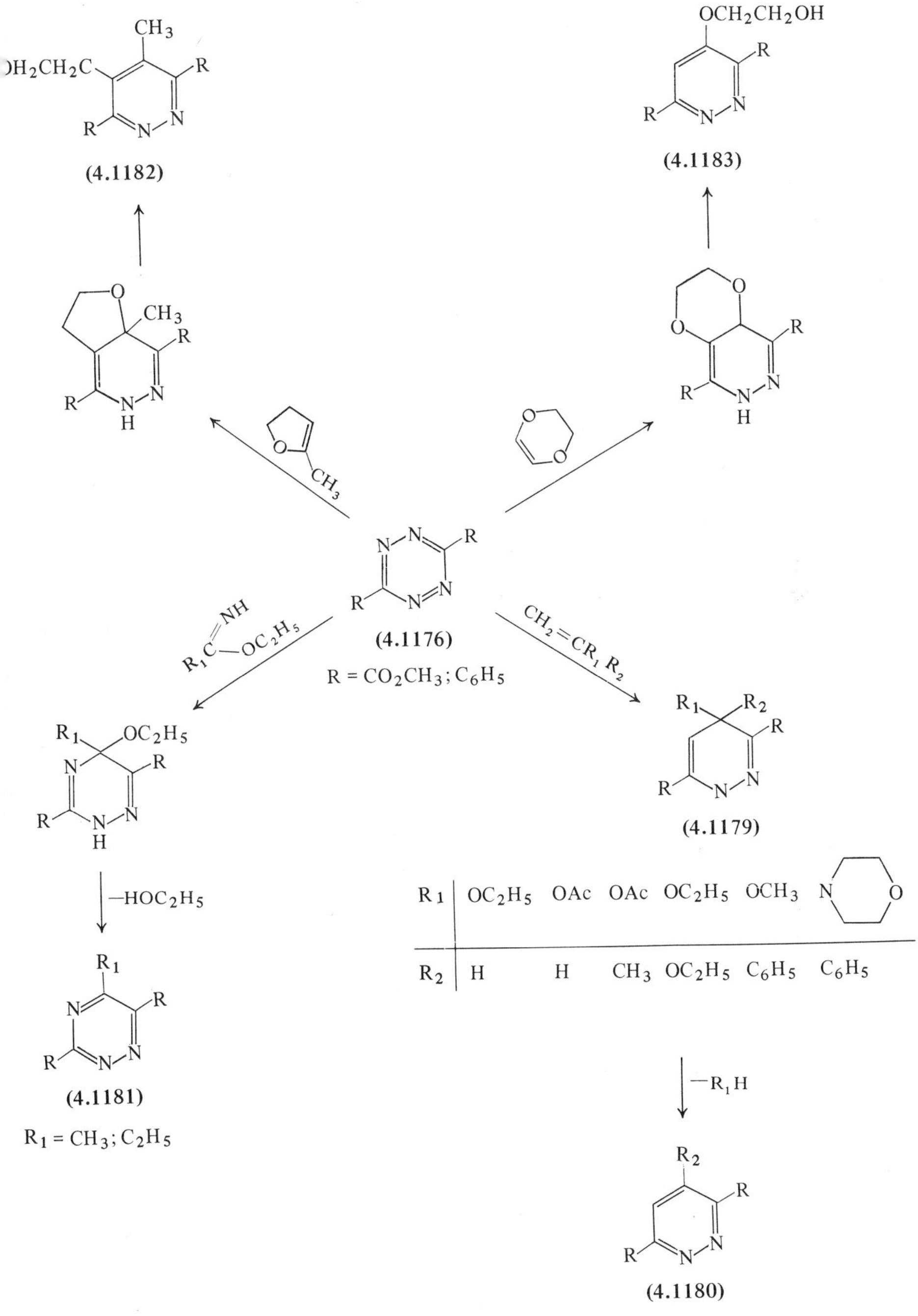

(4.1182)

(4.1183)

(4.1176)

$R = CO_2CH_3; C_6H_5$

(4.1179)

(4.1181)

$R_1 = CH_3; C_2H_5$

R_1	OC_2H_5	OAc	OAc	OC_2H_5	OCH_3	N⌒O
R_2	H	H	CH_3	OC_2H_5	C_6H_5	C_6H_5

$-R_1H$

(4.1180)

c. *1,2,4,5-Tetrazines into 1,2-diazepines*

If 3,6-dialkyl(aryl)-1,2,4,5-tetrazine (4.1176) (R = C_6H_5, CO_2CH_3, p-BrC_6H_4, CH_3, C_2F_5) is refluxed with an equimolar amount of triphenylcyclopropene, under evolution of nitrogen and with fading of the dark purple coloured solution, a compound is obtained which appears to be the corresponding 3,7-disubstituted 5H-1,2-diazepine (4.1186) (R = C_6H_5)[720,721]. The structure (4.1186) was recently confirmed by X-ray analysis[721]. The valence tautomer of (4.1176), the 3,4-diazanorcaradiene derivative (4.1185) is proposed as intermediate. Also this structure was confirmed by X-ray analysis[721]. Support for this type of intermediate has in fact been obtained, since it is possible to prepare the norcaradiene derivative (4.1187) by treating (4.1176) (R = C_6H_5)

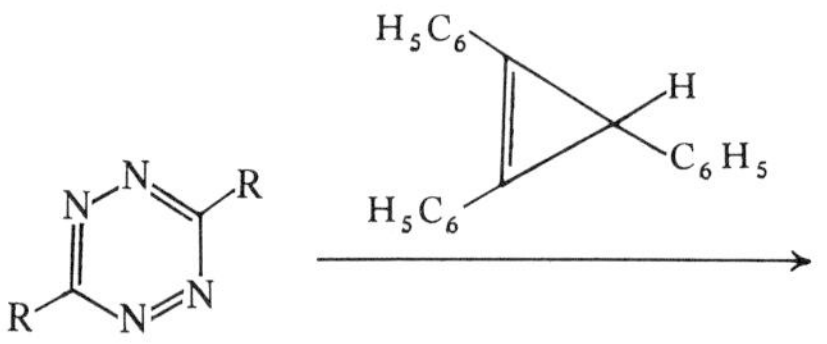

(4.1176)

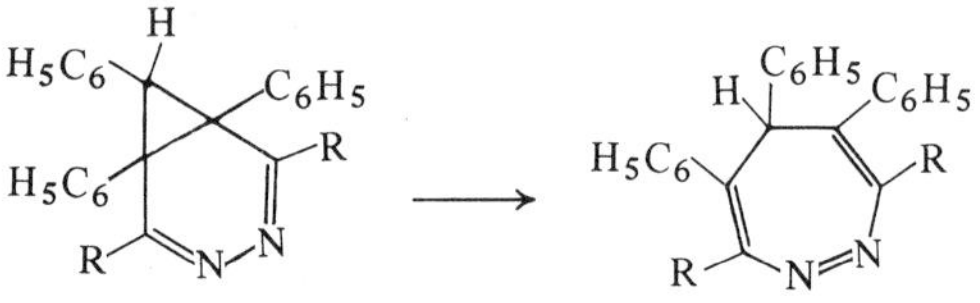

(4.1185)   (4.1186)

R = C_6H_5, CO_2CH_3, p-BrC_6H_4, CH_3, C_2F_5

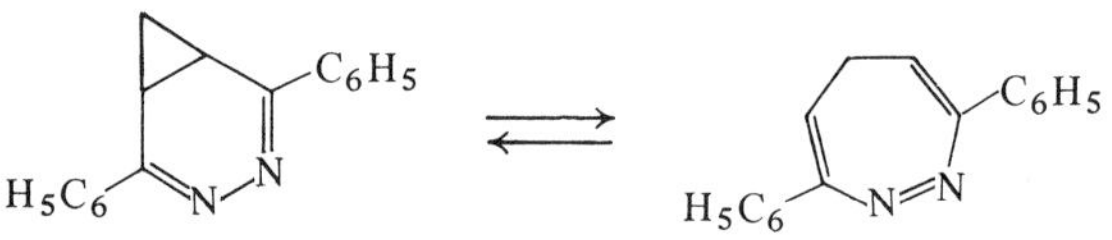

(4.1187)   (4.1188)

with cyclopropene. Nuclear magnetic resonance spectroscopy does not indicate some contribution of the 1,2-diazepine (4.1188). However, if the temperature is enhanced, the contribution of (4.1188) is found to be increased and at 150°C the equilibrium is almost entirely displaced towards (4.1188).

References

1. Dimroth, K. (1960). *Angew. Chem.* **72**, 331.
2. Dimroth, K. (1960). "Neuere Methoden der präparativen Organischen Chemie", Band III, p. 239. Verlag Chemie.
3. Schroth, W., and Fisher, G. (1964). *Z. Chem.* **4**, 281; (1964) *Chem. Abstr.* **61**, 11958.
4. Balaban, A. T. (1969). *Rev. Roum. Chim.* **14**, 1331; (1970). *Chem. Abstr.* **72**, 90194.
4a. Balaban, A. T., Schroth, W., and Fisher, G. (1969). *Advan. Heterocycl. Chem.* **10**, 241.
5. Balaban, A. T., Bedford, G. R., and Katritzky, A. R. (1964). *J. Chem. Soc.* 1646.
6. Kontecky, J. (1959). *Collect. Czech. Chem. Commun.* **24**, 1608.
7. Balaban, A. T., Sahini, V. E., and Keplinger, E. (1960). *Tetrahedron* **9**, 163.
8. Balaban, A. T., and Nenitzescu, C. D. (1960). *Chem. Ber.* **93**, 599.
9. Shuikin, N. I., Bel'skii, I. F., Balaban, A. T., and Nenitzescu, C. D. (1962). *Izv. Akad. Nauk, SSSR, Otd. Khim. Nauk* 491; (1962). *Chem. Abstr.* **57**, 15058.
10. Gird, E., Vasilescu, A., Balaban, A. T., and Barabas, A. (1968). Proc. Int. Conf. *Methods Prep. Stor. Label. Comp.* 649; (1969). *Chem. Abstr.* **71**, 38674.
11. Quint, F., Pütter, R., and Dilthey, W. (1938). *Chem. Ber.* **71**, 356.
11a. Wasserman, H. H., and Pavia, D. L. (1970). *J. Chem. Soc. D*, 1459.
11b. Yates, P., and Stout, G. H. (1954). *J. Amer. Chem. Soc.* **76**, 5110.
11c. Hall, S. S., and Chernoff, H. C. (1970). *Chem. Ind. (London)* 896.
12. Kumler, Ph. L., Pedersen, C. L., and Buchardt, O. (1968). *Acta Chem. Scand.* **22**, 2719.
13. Balaban, A. T. (1968). *Tetrahedron* **24**, 5059.
14. Dimroth, K., and Mach. W. (1968). *Angew. Chem. Int. Ed.* **7**, 460.
14a. Jurd, L. (1970). *Chem. Ind. (London)* 624.
15. Dorofeenko, G. N., Narkevich, A. N., and Zhdanov, Yu. A. (1967). *Khim. Geterotsikl. Soedin* 1130; (1968). *Chem. Abstr.* **69**, 67175.
15a. Lempert-Sreter, M., and Lempert, K. (1970). *Acta Chim. (Budapest)* **65**, 443.
15b. Snieckus, V., and Kan, G. (1970). *J. Chem. Soc. (D)*, 1208.
15c. Pedersen, C. L., and Buchardt, O. (1970). *Acta Chem. Scand.* **24**, 834.
16. Balaban, A. T. (1969). *Org. Prep. Proced.* **1**, 63.
17. Feist, F. (1892). *Chem. Ber.* **25**, 315.
18. Feist, F. (1901). *Chem. Ber.* **34**, 1992.
19. El-Kholy, I. E. S., Rafla, F. K., and Mishrikey, M. M. (1969). *J. Chem. Soc. (C)*, 1950.
20. Feist, F. (1893). *Chem. Ber.* **26**, 747.
21. Ullman, E. F. (1959). *J. Amer. Chem. Soc.* **81**, 5386.
22. Seidel, (1899). *J. Prakt. Chem.* **59**, 122; Fittig, R., and Ebert, G. (1882). *Annalen*, **216**, 162.
23. Zagorevskii, V. A., and Dudykina, N. V. (1962). *Zh. Obshch. Khim.* **32**, 2383; (1963). *Chem. Abstr.* **58**, 5617.
24. Tirodkar, R. B., and Usgaonkar, R. N. (1969). *Indian J. Chem.* **7**, 114.

25. Cheng, F. C., and Tan, S. F. (1968). *J. Chem. Soc. (C)*, 543.
26. Brent, D. A., Hribar, J. D., and DeJongh, D. C. (1970). *J. Org. Chem.* **35**, 135.
27. Orlow, N. A., and Tistschenko, W. W. (1930). *Chem. Ber.* **63**, 2948.
28. Yates, P., and Still, I. W. J. (1963). *J. Amer. Chem. Soc.* **85**, 1208.
29. Yates, P., and Jorgenson, M. J. (1958). *J. Amer. Chem. Soc.* **80**, 6150.
30. For a review on this subject see Korte, F., and Büchel, K. H. (1959). *Angew. Chem.* **71**, 709; Korte, F., and Büchel, K. H. (1960). "Neuere Methoden der präparativen Organischen Chemie", Vol. III, p. 136. Verlag Chemie.
31. Lawson, A. (1957). *J. Chem. Soc.* 144.
32. Koenigs, E., and Freund, J. (1947). *Chem. Ber.* **80**, 143.
33. Alberti, C. (1957). *Gazz. Chim. Ital.* **87**, 781; (1958). *Chem. Abstr.* **52**, 15536.
33a. Eiden, F., and Loewe, W. (1970). *Tetrahedron lett.* 1439.
34. Baker, W., and Butt, V. S. (1949). *J. Chem. Soc.* 2142.
35. Baker, W., Harborne, J. B., and Ollis, W. D. (1952). *J. Chem. Soc.* 1303.
36. Simonis, H., and Rosenberg, S. (1914). *Chem. Ber.* **47**, 1232.
37. Schönberg, A., and Sidky, M. M. (1953). *J. Amer. Chem. Soc.* **75**, 5128.
38. Eiden, F., and Haverland, H. (1968). *Arch. Pharm.* **301**. 819.
39. Mustafa, A., Hsihmat, O. H., Zayed, S. M. A. D., and Nawar, A. A. (1963). *Tetrahedron*, **19**, 1831.
40. Casini, G., Gualtieri, F., and Stein, M. L. (1969). *J. Heterocycl. Chem.* **6**, 279.
41. Hergert, H. L., Coad, P., and Logan, A. V. (1956). *J. Org. Chem.* **21**, 304.
42. Hergert, H. L., and Kurth, E. F. (1953). *J. Org. Chem.* **18**, 521.
43. Enebäck, C., and Gripenberg, J. (1957). *J. Org. Chem.* **22**, 220.
44. Oyamada, T. (1934). *J. Chem. Soc. Jap.* **55**, 755; (1935). *J. Chem. Soc. Jap.* **56**, 980; (1935). *Chem. Abstr.* **29**, 762; (1936). *Chem. Abstr.* **30**, 459; (1939). *Annalen*, **538**, 44.
45. Chadenson, M., Molho-La Croix, L., Molho, D., and Mentzer, C. (1955). *Compt. Rend.* **240**, 1362.
46. Molho, D., and Chadenson, M. (1959). *Bull. Soc. Chim. Fr.* 453.
47. Kotake, M., and Kubota, T. (1949). *Annalen*, **544**, 253.
48. Pew, J. C. (1948). *J. Amer. Chem. Soc.* **70**, 3031.
49. Geissman, T. A., and Lischner, H. (1952). *J. Amer. Chem. Soc.* **74**, 3001.
50. Bottomley, W. (1956). *Chem. Ind. (London)*, 170.
51. King, F. E., and Bottomley, W. (1954). *J. Chem. Soc.* 1399.
52. Weinges, K., and Nagel, D. (1969). *Chem. Ber.* **102**, 1592.
53. Caplin, G. A., Ollis, W. D., and Sutherland, I. O. (1967). *Chem. Commun.* 575.
53a. Crombie, L., Freeman, P. W., and Whiting, D. A. (1970). *J. Chem. Soc. (D)*, 563.
54. Bennett, P., and Donelly, J. A. (1969). *Chem. Ind. (London)*, 783.
55. Merchant, J. R., and Rege, D. V. (1969). *Tetrahedron Lett.* 3589.
56. Thyagarajan, B. S., Balasubramanian, K. K., and Rao, R. B. (1966). *Chem. Ind. (London)*, 2128.
57. Korobitsyna, I. K., and Pivnitskii, K. K. (1960). *Zh. Obshch. Khim.* **30**, 4008; (1961). *Chem. Abstr.* **55**, 22277.
58. Philips Petroleum Co. (1951). U.S. Pat. 2,731,475; Hillyer, J. C., and Edmonds J. T. Jr. (1956). *Chem. Abstr.* **50**, 12106.

59. Criegee, R., Bruyn, P. de, and Lohaus, G. (1953). *Annalen*, **583**, 19.
60. Siemiatycki, M., and Fugnitto, R. (1961). *Bull. Soc. Chim. Fr.* 538.
61. Zhungietu, G. I., Lazur'ewskii, G. V. (1968). *Zh. Vses. Khim. Obshchest.* 13, 597; (1969). *Chem. Abstr.* **70**, 47246.
62. Balaban, A. T., Farcasiu, D., and Nenitzescu, C. D. (1962). *Tetrahedron*, 18, 1075.
63. Simalty, M., Carretto, J., and Fugnitto, R. (1966). *Bull. Soc. Chim. Fr.* 2959.
64. Köbrich, G. (1961). *Annalen*, **648**. 114.
65. Bolle, J., and Tomaszewski, G. French Pat. 1,340,970; (1964). *Chem. Abstr.* **60**, 5463.
66. Tolmachev, A. I. (1963). *Zh. Obshch. Khim.* **33**, 1864; (1964). *Chem. Abstr.* **60**, 689.
67. Dorofeenko, G. N., and Safaryan, G. P. (1970). *Khim. Geterotsikl. Soedin.* 278; (1970). *Chem. Abstr.* **72**, 111216.
68. Baeyer, A., and Piccard, J. (1911). *Annalen*, **384**, 208.
69. Balaban, A. T., and Nenitzescu, C. D. (1959). *Annalen*, **625**, 74.
70. Praill, P. F. G., and Whitear, A. L. (1959). *Proc. Chem. Soc.* 312; (1961). *J. Chem. Soc.* 3573.
71. Dorofeenko, G. N., and Krivun, S. V. (1967). *Metody Poluch. Khim. Reaktivov Prep.* 17, 149; (1969). *Chem. Abstr.* **71**, 61161.
72. Balaban, A. T., and Nenitzescu, C. D. (1961). *J. Chem. Soc.* 3553.
73. Balaban, A. T., and Nenitzescu, C. D. (1960). *Tetrahedron Lett.* 2, 7.
74. Balaban, A. T., Gavăt, M., and Nenitzescu, C. D. (1962). *Tetrahedron*, 18, 1079.
75. Dilthey, W. (1922). *J. Prakt. Chem.* **104**, 28.
76. Dilthey, W., Fröde, G., and Koenen, H. (1926). *J. Prakt. Chem.* **114**, 153.
77. Dilthey, W., and Radmacher, W. (1925). *J. Prakt. Chem.* **111**, 153.
78. Dorofeenko, G. N., and Krivun, S. V. (1964). *Zh. Obshch. Khim.* **34**, 105; (1964). *Chem. Abstr.* **60**, 10641.
79. Dorofeenko, G. N., Korol'chenko, G. A., and Krivun, S. V. (1965). *Khim. Geterotsikl. Soedin.* 817; (1966). *Chem. Abstr.* **64**, 19548.
80. Dorofeenko, G. N., Krivun, S. V., and Mezheritskii, V. V. (1965). *Zh. Obshch. Khim.* **35**, 632; (1965). *Chem. Abstr.* **63**, 2947.
81. Dilthey, W. (1916). *J. Prakt. Chem.* **94**, 53.
82. Dorofeenko, G. N., and Krivun, S. V. (1962). *Zh. Obshch. Khim.* **32**, 2386; (1963). *Chem. Abstr.* **58**, 7904.
83. Bos, H. J. T., and Arens, J. F. (1963). *Rec. Trav. Chim. Pays-Bas*, **82**, 845.
84. Lombard, R., and Stephan, J. P. (1958). *Bull. Soc. Chim. Fr.* 1458.
85. Durden, J. A., and Crosby, D. G. (1965). *J. Org. Chem.* **30**, 1684.
86. Simalty-Siemiatycki, M. (1965). *Bull. Soc. Chim. Fr.* 1944.
87. Treibs, A., and Bader, H. (1957). *Chem. Ber.* **90**, 789.
88. Wizinger, R., Losinger, S., and Ulrich, P. (1956). *Helv. Chim. Acta*, **39**, 5; Wizinger, R., Grüne, A. and Jacobi, E. (1956). *Helv. Chim. Acta*, **39**, 1; Krivun, S. V., Dorofeenko, G. N., and Kovalevskii, A. S. (1970). *Chim. Geterotsikl. Soedin.* 6, 733; (1970). *Chem. Abstr.* **73**, 98769.
89. Dilthey, W. (1921). *J. Prakt. Chem.* **102**, 209.
90. Dorofeenko, G. N., and Krivun, S. V. (1967). *Metody Poluch. Khim. Reaktivov Prep.* 17, 152; (1969). *Chem. Abstr.* **71**, 12945.

91. Dorofeenko, G. N., Demidenko, E. I., and Krivun, S. V. (1967). *Izv. Vyssh. Ucheb. Zaved., Khim. Tekhnol.* **10**, 304; (1968). *Chem. Abstr.* **68**, 29529.
92. Gastaldi, C. (1922). *Gazz. Chim. Ital.* **52**, 169; (1922). *Chem. Abstr.* **16**, 2515; Gastaldi, C., and Peyretti, G. L. (1923). *Gazz. Chim. Ital.* **53**, 11; (1923). *Chem. Abstr.* **17**, 2284.
93. Diels, O., and Alder, K. (1927). *Chem. Ber.* **60**, 716.
94. Schneider, W. (1924). *Annalen,* **438**, 115.
95. Balaban, A. T. (1968). *Tetrahedron Lett.* 4643.
96. Balaban, A. T., and Nenitzescu, C. D. (1961). *J. Chem. Soc.* 3561.
97. Dorofeenko, G. N., and Zhungietu, G. I. (1965). *Zh. Obshch. Khim.* **35**, 963; (1965). *Chem. Abstr.* **63**, 9909.
98. Balaban, A. T., Genya, A., and Nenitzescu, C. D. (1961). *Izv. Akad. Nauk SSSR, Otdel. Khim. Nauk.* 1102; (1961). *Chem. Abstr.* **55**, 27288.
99. Balaban, A. T., Nenitzescu, C. D., Gavat, M., and Mateescu, G. (1961). *Izv. Akad. Nauk SSSR, Otdel. Khim. Nauk,* 3565.
100. Krivun, S. V., Dulenko, V. I., Dulenko, L. V., and Dorofeenko, G. N. (1966). *Dokl. Akad. Nauk SSSR,* **166**, 359; (1966). *Chem. Abstr.* **64**, 11153.
101. Shriner, R. L., Johnston, H. W., and Kaslow, C. E. (1948) *J. Org. Chem.* **14**, 204.
102. Balaban, A. T., and Toma, C. (1966). *Tetrahedron Suppl.* **7**, 1.
103. Dupré M., Filleux-Blanchard, M. L., Simalty, M., and Strzelecka, H. (1969). *Compt. Rend.* **268**, 1611.
104. Balaban, A. T., Gard, E., Vasilescu, A., and Barabas, A. (1965). *J. Label. Compounds,* **1**, 266; (1966). *Chem. Abstr.* **64**, 15830.
105. Shriner, R. L., and Knox, W. R. (1951). *J. Org. Chem.* **16**, 1064.
106. Mutz, G. See note 6, reference 1.
107. Mueller, A., El-Sawy, M. M., Meszaros, M., and Ruff, F. (1967). *Acta Chim. Acad. Sci. Hung.* **52**, 261; (1966). *Acta Chim. Acad. Sci. Hung.* **50**, 387; (1967). *Chem. Abstr.* **67**, 64223.
108. Dulenko, L. V., and Dorofeenko, G. N. (1967). *Metody Poluch. Khim. Reaktivov Prep.* **17**, 91; (1969). *Chem. Abstr.* **71**, 12952.
109. Balaban, A. T., and Barbulescu, N. S. (1966). *Rev. Roum. Chim.* **11**, 109; (1966). *Chem. Abstr.* **65**, 678.
110. Dorofeenko, G. N., Shdanow, Ju. A., Shungijetu, G. I., and Krivun, S. V. (1966). *Tetrahedron,* **22**, 1821.
111. Dorofeenko, G. N., Dulenko, L. V. (1969). *Khim. Geterotsikl. Soedin.* 417; (1970). *Chem. Abstr.* **72**, 3412.
112. Zhungietu, G. I. (1969). *Zh. Obshch. Khim.* **39**, 716; (1969). *Chem. Abstr.* **71**, 38831.
113. Krivun, S. V., and Dorofeenko, G. N. (1966). *Khim. Geterotsikl. Soedin.* 656; (1967). *Chem. Abstr.* **67**, 64170.
114. Krivun, S. V., and Dorofeenko, G. N. (1966). *Chem. Heterocycl. Compounds,* **2**, 499.
115. Krivun, S. V., and Dorofeenko, G. N. (1964). *Zh. Obshch. Khim.* **34**, 2091; (1964). *Chem. Abstr.* **61**, 6982.
116. Toma, C., and Balaban, A. T. (1966). *Tetrahedron Suppl.* **7**, 9.
117. Dilthey, W., and Dierichs, H. (1935). *J. Prakt. Chem.* **144**, 1.
118. Schneider, W., and Seebach, F. (1921). *Chem. Ber.* **54**, 2285.

119. Schneider, W., and Müller, W. (1924). *Annalen.* **438**, 147.
120. Schneider, W., and Riedel, W. (1941). *Chem. Ber.* **74**, 1252.
121. Dilthey, W. (1922). *Chem. Ber.* **55**, 57.
122. Balaban, A. T., and Nenitzescu, C. D. (1959). *Annalen*, **625**, 74.
123. Zhdanov, Yu. A., Dorofeenko, G. N., and Narkevich, A. N. (1963). *Zh. Obshch. Khim.* **33**, 2418; (1963). *Chem. Abstr.* **59**, 14105.
124. Narkevich, A. N., Dorofeenko, G. N., and Zhdanov Yu. A., (1967). *Dokl. Akad. Nauk SSSR.* **176**, 103; (1968). *Chem. Abstr.* **68**, 78605.
125. Khromov-Borisov, N. V., and Gavrilova, L. A. (1961). *Zh. Obshch. Khim.* **31**, 2192; (1962). *Chem. Abstr.* **56**, 2415.
126. Toma, C., and Balaban, A. T. (1966). *Tetrahedron Suppl.* **7**, 27.
127. Lombard, R., and Kress, A. (1960). *Bull. Soc. Chim. Fr.* 1528.
128. Susan, A. B., and Balaban, A. T. (1969). *Rev. Roum. Chim.* **14**, 111; (1969). *Chem. Abstr.* **71**, 61150.
129. Dimroth, K., Reichardt, C., Siepmann, T., and Bohlmann, F. (1963). *Annalen*, **661**, 1.
129a. Dorofeenko, G. N., Narkevich, A. N., Zhdanov, Yu. A., and Soroka, T. G. (1970). *Khim. Geterotsikl. Soedin.* **3**, 315; (1970). *Chem. Abstr.* **73**, 66502.
130. Anker, R. M., and Cook, A. H. (1946). *J. Chem. Soc.* 117.
131. King, L. C., and Ozog, F. J. (1955). *J. Org. Chem.* **20**, 448.
131a. Skatteboel, L. (1970). *J. Org. Chem.* **35**, 3200.
132. Kanai, K., Umehara, M., Kitano, H., and Fukui, K. (1963). *Nippon Kagaku Zasshi*, **84**, 432; (1963). *Chem. Abstr.* **59**, 13934.
133. Wizinger, R., and Ulrich, P. (1956). *Helv. Chim. Acta*, **39**, 207; (1956). *Helv. Chim. Acta*, **39**, 217.
134. Kato, H., Ogawa, T., and Ohta, M. (1960). *Bull. Chem. Soc. Jap.* **33**, 1467.
135. Märkl, G., Lieb, F., and Merz, A. (1967). *Angew. Chem.* **79**, 947; (1967). *Angew. Chem. Int. Ed.* **6**, 944.
136. Märkl, G. (1966). *Angew Chem.* **78**, 907; (1966). *Angew. Chem. Int. Ed.* **5**, 846.
137. Warren, S. G. (1968). *Angew. Chem. Int. Ed.* **7**, 606.
138. Märkl, G., Lieb, F., and Merz, A. (1967). *Angew. Chem.* **79**, 475.
139. Price, C. C., Parasaran, T., and Lakshminarayan, T. V. (1966). *J. Amer. Chem. Soc.* **88**, 1034.
140. Wiley, R. H., Smith, N. R., and Knabeschuh, L. H. (1953). *J. Amer. Chem. Soc.* **75**, 4482; Wiley, R. H. (1954). *J. Amer. Chem. Soc.* **76**, 311; Wiley, R. H., Beasly, P., and Knabeschuh, L. H. (1954). *J. Amer. Chem. Soc.* **76**, 311.
141. Stetter, H., and Schellhammer, C. W. (1957). *Chem. Ber.* **90**, 755; El Kholy, I. E. S., Rafla, F. K., and Soliman, G. (1946). *J. Chem. Soc.* 117.
141a. Chun-shan Wang (1970). *J. Heterocycl. Chem.* **7**, 389; see also the references cited therein.
142. Elkaschef, M. A. F., and Nosseir, M. H. (1960). *J. Amer. Chem. Soc.* **82**, 4344; Conrad, M., and Guthzeit, M. (1887). *Chem. Ber.* **20**, 154.
143. Borsche, W., and Bonacker, I. (1921). *Chem. Ber.* **54**, 2678; Elkaschef, M. A. F., Nossier, M. H., and Abdel-Kader, A. (1963). *J. Chem. Soc.* 4647.
144. El Kholy, I. E. S., Rafla, F. K., and Mishrikey, M. M. (1969). *J. Chem. Soc. (C)*, 1950.

145. Meislich, H. (1962). *In* "Pyridine and its Derivatives" (Ed. E. Klingsberg), Chapter 12, p. 509, Interscience Publishers, New York and London.
146. Von Pechmann, H., and Welsh, W. (1884). *Chem. Ber.* **17**, 2384, 2396; Von Pechmann, H. (1891). *Annalen*, **264**, 261.
147. Caldwell, W. T., Tyson, F. T., and Lauer, L. (1944). *J. Amer. Chem. Soc.* **66**, 1479.
148. Gault, H., Gilbert, J., and Briancourt, D. (1968). *Compt. Rend.* **266**, 131.
149. Wiley, R. H., Knabeschuh, L. H., Duckwall, A. L., and Smith, N. R. (1954). *J. Amer. Chem. Soc.* **76**, 625.
150. Korte, F., Dürbeck, H., and Weisgerber, G. (1967). *Chem. Ber.* **100**, 1305.
151. Bartholomew, D. H., Campbell, A., Oliver, D. W. H., and Dutten, B. G. Brit. Pat. 1,087,279; (1968). *Chem. Abstr.* **68**, 87183.
152. Skvortsova, G. G., Kozyrev, V. G., and Zapunnaya, K. V., USSR Pat. 215,214; (1968). *Chem. Abstr.* **69**, 59102.
153. See: Gelas, J. (1967). *Bull. Soc. Chim. Fr.* 3093, and further literature references cited therein.
154. Ichimoto, J., Fujii, K., and Tatsumi, Ch. (1967). *Agri. Biol. Chem.* **31**, 979.
155. Thomas, A. F., and Marxer, A. (1958). *Helv. Chim. Acta,* **41**, 1898; (1960). *Helv. Chim. Acta,* **43**, 469.
156. Jones, R. G., and Mann, M. J. (1953). *J. Amer. Chem. Soc.* **75**, 4048; Ainsworth, C., and Jones, R. G. (1954). *J. Amer. Chem. Soc.* **76**, 3172.
157. Marxer, A., and Thomas, A. F. (1960). *Angew. Chem.* **72**, 270.
158. Ost, J. (1879). *J. Prakt. Chem.* **19**, 203; (1881). *J. Prakt. Chem.* **23**, 441; (1883). *J. Prakt. Chem.* **27**, 257.
158a. Jaworski, T., and Kwiatkowski, S. (1970). *Rocz. Chem.* **44**, 555.
159. Kato, H., Ogawa, T., and Ohta, M. (1960). *Chem. Ind. (London),* 1300.
160. Strzelecka, H. (1966). *Ann. Chim. (Paris),* 201; (1967). *Chem. Abstr.* **66**, 2435.
161. Balaban, A. T., Frangopol, P. T., Katritzky, A. R., and Nenitzescu, C. D. (1962). *J. Chem. Soc.* 3889.
162. Kato, H., Ogawa, T., and Ohta, H. (1960). *Bull. Chem. Soc. Jap.* **33**, 1467; (1960). *Bull. Chem. Soc. Jap.* **33**, 1468; (1961). *Chem. Abstr.* **55**, 27300.
162a. Reynolds, G. A., VanAllan, J. A., and Petropoulous, C. C. (1970). *J. Heterocycl. Chem.* **7**, 1061.
163. Vartanyan, S. A., Noravyan, A. S., and Zhamagortsyan, V. N. (1966). *Arm. Khim. Zh.* **19**, 447; (1967). *Chem. Abstr.* **66**, 10634.
164. Vartanyan, S. A., Noravyan, A. S., and Zhamagortsyan, V. N. (1966). *Chem. Heterocycl. Compounds,* **2**, 510; Vartanyan, S. A., Noravyan, A. S., and Zhamagortsyan, V. N. (1967). *Arm. Khim. Zh.* **20**, 921; (1968). *Chem. Abstr.* **69**, 77081.
165. Vartanyan, S. A., Noravyan, A. S., and Zhamagortsyan, V. N. (1966). *Khim. Geterotsikl. Soedin.* 670; (1967). *Chem. Abst.* **66**, 55339.
166. Shusherina, N. P., Gladysheva, T. Kh., and Levina, R. Ya. (1968). *Vestn. Mosk. Univ. Ser.* II, **23**, 101; (1968). *Chem. Abstr.* **69**, 27191.
167. Kodrat'eva, G. V., Gunar, V. I., Ovechkina, L. F., Zav'yalov, S. T., and Krotov, A. I. (1967). *Izv. Akad. Nauk SSSR, Ser. Khim.* (3), 633-9; (1967). *Chem. Abstr.* **67**, 99689.
168. Chatterjea, J. N., Banerjee, B. K., and Jha, H. C. (1965). *Tetrahedron Lett.* 2281.

169. Ungnade, H. E., Nightingale, D. V., and French, H. E. (1945). *J. Org. Chem.* **10**, 533.
170. Desai, H. K., and Usgaonkar, R. N., *J. Indian Chem. Soc.* **41**, 821 (1964); *Chem. Abstr.* **62**, 16179 (1965).
171. Chatterjea, J. N., Jha, H. C., and Bannerjee, B. K. (1966). *J. Indian Chem. Soc.* **43**, 633; (1967). *Chem. Abstr.* **67**, 90637.
172. Ismail, A. G., and Wibberley, D. G. (1968). *J. Chem. Soc. (C)*, 2706.
173. Yagi, N., Omori, H., and Okazaki, M. (1969). *Yuki Gosei Kagaku Kyokai Shi*, **27**, 51; (1969). *Chem. Abstr.* **70**, 87526.
173a. Legrand, L., and Lozac'h, N. (1970). *Bull. Soc. Chim. Fr.* 2237.
174. Legrand, L., and Lozac'h, N. (1966). *Bull. Soc. Chim. Fr.* 3828.
175. Arndt, F., Nachtwey, P., and Pusch, J. (1925). *Chem. Ber.* **58**, 1633.
176. Arndt, F., Schwarz, R., Martius, C., and Arow, E. (1948). Revue de la Faculté de l'Université d'Istanbul A 13, 53; (1948). *Chem. Abstr.* **42**, 4176.
177. Traverso, G. (1956). Ann. Chim. (Rome), **46**. 821; (1957). *Chem. Abstr.* **51**, 6622; Traverso, G. (1957). Ann. Chim. (Rome), **47**, 3; (1957). *Chem. Abstr.* **51**, 10543.
178. Liebig, J. (1912). *J. Prakt. Chem.* **85**, 241.
179. Meyer, R., and Szanecki, J. (1900). *Chem. Ber.* **33**, 2577.
180. Schweizer, E. E., and Parham, W. E. (1960). *J. Amer. Chem. Soc.* **82**, 4085; see also Masamune, S., and Castellucci, N. T. (1965). *Chem. Ind. (London)*, 184.
181. d'Angelo, J., and Thuy, V. M. (1968). *Bull. Chem. Soc. Fr.* 2823.
182. Nerdel, F., Buddrus, J., Brodowski, W., and Weyerstahl, P. (1966). *Tetrahedron Lett.* 5385.
183. Misono, A., Osa, T., and Sanami, Y. (1968). *Bull. Chem. Soc. Jap.* **41**. 2447; (1969). *Chem. Abstr.* **70**, 19864.
184. Marcaillon, C., Fontaine, G., and Maitre, P. (1968). *Compt. Rend.* **267**, 846.
184a. Majumdar, K. C., and Thyagarajan, B. S. (1970). *J. Chem. Soc. (D)*, 1480.
185. Olsen, S., and Bredoch, R. (1958). *Chem. Ber.* **91**, 1589.
186. Whitlock, H. W. (1961). *Tetrahedron Lett.* 593; (1962). *Chem. Abstr.* **56**, 10067.
187. Andes Hess Jr., B., Bailey, A. S., and Boekelheide, V. (1967). *J. Amer. Chem. Soc.* **89**, 2746.
188. Anet, F. A. L., and Bavin, P. M. G. (1957). *Can. J. Chem.* **35**, 1084.
188a. Clinging, R., Dean, F. M., and Houghton, L. E. (1970). *J. Chem. Soc. (C)*, 897.
189. Hofmann, H., and Westernacher, H. (1969). *Chem. Ber.* **102**, 205.
190. Doherty, J. R., Keane, D. D., O'Sullivan, W. I., Philbin, E. M., and Teague, P. C. (1969). *Chem. Ind. (London)*, 586.
191. Colonge, J., and Corbet, P. (1958). *Compt. Rend.* **247**, 2144; (1958). *Chem. Abstr.* **53**, 20043.
192. Breslow, R. (1963). "Rearrangements in Small Ring Compounds", Chapter 4 of "Molecular Rearrangements" (Ed. Paul de Mayo), Interscience Publishers, New York.
193. Buchardt, O., Pedersen, C. L., Svanholm, U., Duffield, A. M., and Balaban, A. T. (1969). *Acta Chem. Scand.* **23**, 3125.
194. Balaban, A. T. (1970). *Tetrahedron*, **26**, 739.

195. Barkenbus, C., Diehl, J. F., and Vogel, G. R. (1955). *J. Org. Chem.* **20,** 871.
196. Inone, N., Yamaguchi, Sh., Shoei, I., and Suzuki, I. (1968). *Bull. Chem. Soc. Jap.* **41,** 2078.
196a. Ito, S. (1970). *Bull. Chem. Soc. Jap.* **43,** 1824.
197. Lockhart, I. M. (1968). *Chem. Ind. (London),* 1844.
198. Krapcho, J., and Turk, C. F. (1966). *J. Med. Chem.* **9,** 191; 1966. *Chem. Abstr.* **64,** 11174.
199. Gottlieb, J. (1899). *Chem. Ber.* **32,** 966.
200. Buu-Hoi, N. P. (1939). *Compt. Rend.* **209,** 321.
201. Lieck, A. (1905). *Chem. Ber.* **38,** 3853.
202. Wölbling, H. (1905). *Chem. Ber.* **38,** 3845.
203. Rosen, G., and Popp, F. D. (1969). *J. Heterocycl. Chem.* **6,** 9.
204. Whitmore, W. F., and Cooney, R. C. (1944). *J. Amer. Chem. Soc.* **66,** 1237.
205. Traverso, G. (1956). *Ann. Chim. (Rome),* **46,** 821; (1957). *Chem. Abstr.* **51,** 66221.
206. Traverso, G. (1954). *Ann. Chim. (Rome),* **44,** 1018; (1956). *Chem. Abstr.* **50,** 366.
207. Traverso, G. (1957). *Ann. Chim. (Rome),* **47,** 3; (1957). *Chem. Abstr.* **51,** 10543.
208. Lumma, Jr., W. C., and Berchtold, G. A. (1969). *J. Org. Chem.* **34,** 1566.
209. Lumma, Jr., W. C., and Berchtold, G. A. (1967). *J. Amer. Chem. Soc.* **89,** 2761.
210. Johnson, P. Y., and Berchtold, G. A. (1970). *J. Org. Chem.* **35,** 584.
211. Maheshwari, K. K., and Berchtold, G. A. (1969). *J. Chem. Soc. (D),* 13.
212. Leonard, N. J., and Figueras Jr., J (1952). *J. Amer. Chem. Soc.* **74,** 917.
213. Braun, J. V., and Weissbach, K. (1929). *Chem. Ber.* **62,** 2416.
214. Degani, I., Fochi, R., and Vincenzi, C. (1967). *Gazz. Chim. Ital.* **97,** 397; (1967). *Chem. Abstr.* **67,** 53994.
215. Hofmann, H., and Salbeck, G. (1969). *Angew, Chem. Int. Ed.* **8,** 456.
215a. Speckamp, W. N., Dijkink, J., Maassen, J. A., and Huisman, H. O. (1970). *Tetrahedron Lett.* **32,** 2743.
215b. Still, I. W. J., and Thomas, M. T. (1970). *Tetrahedron Lett.* 4225.
215c. Deckers, F. H. M., Speckamp, W. N., and Huisman, H. O. (1970). *J. Chem. Soc. (D),* 1521.
216. Suld, G., and Price, C. C. (1962). *J. Amer. Chem. Soc.* **84,** 2094.
217. Mayer, R., Broy, W., and Zahradnik, R. (1967). *In* "Advances in Heterocyclic Chemistry" (Ed. A. R. Katritzky). vol. 8, p. 219. Academic Press, New York and London.
218. Splinter, F. K., and Arold, H. (1969). *J. Prakt. Chem.* **311,** 869.
218a. Legrand, L., and Lozac'h, N. (1970). *Bull. Soc. Chim. Fr.* 2244.
219. Overberger, C. G., and Katchman, A. (1956). *J. Amer. Chem. Soc.* **78,** 1965.
220. Parham, W. E., and Huestis, L. D. (1962). *J. Amer. Chem. Soc.* **84,** 813; Parham, W. E., and Koncos, R. (1961). *J. Amer. Chem. Soc.* **83,** 4034.
221. Seidl, H., and Biemann, K. (1967). *J. Heterocycl. Chem.* **4,** 209.
222. Bergmann, E. D., and Rabinovitz, M. (1960). *J. Org. Chem.* **25,** 828.
223. Rasanu, N. (1969). *Rev. Chim. (Bucharest),* **20,** 175; (1969). *Chem. Abstr.* **71,** 38937.

223a. Wuensch, K. H., Stahnke, K. H., and Ehlers, A. (1970). *Chem. Ber.* 103, 2302.
223b. Lumma, W. C., Dutra, G. A., and Voeker, C. A. (1970). *J. Org. Chem.* 35, 3442.
224. Klingsberg, E. (1967). *Abstr. Amer. Chem. Soc. Meeting,* Sept. 1965, p. 66S; see also Popp, F. D. and Noble, A. C. (1967). *Advan. Heterocycl. Chem.* 8, 24.
225. Reitsema, R. H. (1949). *J. Amer. Chem. Soc.* 71, 2041; Biel, J. H., Hoya, W. K., and Leiser, H. A. (1959). *J. Amer. Chem. Soc.* 81, 2527.
226. Albertson, N. F., and Fillman, J. L. (1949). *J. Amer. Chem. Soc.* 71, 2818.
227. Paul, R., and Tchelitcheff, S. (1958). *Bull. Soc. Chim. Fr.* 736.
228. Crist, D. R., and Leonard, N. J. (1969). *Angew. Chem. Int. Ed.* 8, 962; Hammer, C. F., and Heller, S. R. (1966). *Chem. Commun.* 919.
229. Biel, J. H., Sprengeler, E. P., Leiser, H. A., Horner, J., Drukker, A., and Friedman, H. L. (1955). *J. Amer. Chem. Soc.* 77, 2250.
230. Ross, W. C. J. (1949). *J. Chem. Soc.* 2589.
231. Fuson, R. C., and Zirkle, Ch. L. (1948). *J. Amer. Chem. Soc.* 70, 2760.
232. Biel, J. H., Abood, L. G., Hoya, W. K., Leiser, H. A., Nuhfer, P. A., and Kluchesky, E. F. (1961). *J. Org. Chem.* 26, 4096.
233. Bondesson, G., and Dahlbom, R. (1966). *Acta Chem. Scand.* 20, 2370; (1967). *Chem. Abstr.* 66, 55449.
234. Leonard, N. J., and Ruyle, W. V. (1949). *J. Amer. Chem. Soc.* 71, 3094.
235. Leonard, N. J., and Barthel Jr., E. (1950). *J. Amer. Chem. Soc.* 72, 3632.
236. Leonard, N. J., Fuller, G., and Dryden Jr., H. L. (1953). *J. Amer. Chem. Soc.* 75, 3727.
237. Brown, R. F. C., Clark, V. M., and Todd, A. (1959). *J. Chem. Soc.* 2105.
238. Clemo, G. R., Raper, R., and Vipond, H. J. (1949). *J. Chem. Soc.* 2095.
239. Leonard, N. J., Sentz, R. C., and Middleton, W. J. (1953). *J. Amer. Chem. Soc.* 75, 1674.
240. Clemo, G. R., and Ramage, G. R. (1931). *J. Chem. Soc.* 437; Clemo, G. R., Metcalfe, T. P., and Raper, R. (1936). *J. Chem. Soc.* 1429.
241. Prelog, V., and Seiwerth, R. (1939). *Chem. Ber.* 72, 1638.
242. Leonard, N. J., and Wildman, W. C. (1949). *J. Amer. Chem. Soc.* 71, 3089.
243. Leonard, N. J., and Pines, S. H. (1950). *J. Amer. Chem. Soc.* 72, 4931.
244. Reby, C., Likforman, J., and Gardent, J. (1969). *Compt. Rend.* 269, 45.
245. Grewe, R., and Winter, G. (1959). *Chem. Ber.* 92, 1092.
246. Hatt, H. H., and Stephenson, E. F. M. (1952). *J. Chem. Soc.* 199.
247. Di Maio, G., and Tardella, P. A. (1963). *Proc. Chem. Soc. London,* 224.
248. Quin, L. D., and Pinion, D. O. (1970). *J. Org. Chem.* 35, 3130; Quan, P. M. and Quin, L. D. (1966). *J. Org. Chem.* 31, 2487.
248a. Brown, R. F. C., Subrahmanyan, L., and Whittle, C. P. (1967). *Aust. J. Chem.* 20, 339.
248b. Quin, L. D., and Pinion, D. O. (1970). *J. Org. Chem.* 35, 3134.
249. Biellmann, J. F., and Callot, H. J. (1969). *J. Chem. Soc. (D),* 140; Biellmann, J. F., Callot, H. J., and Goeldner, M. P. (1969). *J. Chem. Soc (D),* 141; Biellmann, J. F., and Callot, H. J. (1970). *Tetrahedron,* 26, 4799, 4809.
250. Ashby, J., Cort, L. A., Elvidge, J. A., and Eisner, U. (1968). *J. Chem. Soc. (C),* 2311.

251. Childs, R. F., and Johnson, A. W. (1965). *Chem. Commun.* 95; Childs, R. F., and Johnson, A. W. (1966). *J. Chem. Soc. (C)*, 1950.
252. Süs, O. (1944). *Annalen*, **556**, 65; (1953). *Annalen*, **579**, 140.
253. Süs, O., and Möller, K. (1955). *Annalen*, **593**, 91.
254. Süs, O. (1955). *Z. Wissenschaftliche Photographie*, **50**, 476.
255. Süs, O., Glos, M., Möller, K., and Eberhardt, H. D. (1953). *Annalen*, **583**, 150.
256. Süs, O., and Möller, K. (1956). *Annalen*, **599**, 233.
257. Huggett, C., Arnold, R. T., and Taylor, T. I. (1942). *J. Amer. Chem. Soc.* **64**, 3043.
258. Smith, P. A. S. (1963). "Molecular Rearrangements" (Ed. P. de Mayo), Chapter 8, vol. I, Interscience Publishers, New York and London.
258a. Wentrup, C., and Crow, W. D. (1970). *Tetrahedron*, **26**, 3965.
258b. Bristol, D. W., and Dittmer, D. C. (1970). *J. Org. Chem.* **35**, 2487.
259. Streith, J., and Sigwalt, C. (1966). *Tetrahedron Lett.* 1347.
260. Streith, J., Darrah, H. K., and Weil, M. (1966). *Tetrahedron Lett.* 5555.
261. Streith, J., Danner, B., and Sigwalt, C. (1967). *Chem. Commun.* 979.
262. Streith, J., Blind, A., Cassal, J. M., and Sigwalt, Ch. (1969). *Bull. Soc. Chim. Fr.* 948.
262a. Streith, J., and Sigwalt, C. (1970). *Bull. Soc. Chim. Fr.* 1157.
262b. For reviews of the photochemical reactions of aromatic amine oxides, see Kaneko (1968). *J. Syn. Org. Chem. Jap.* **26**, 758; also Spence, G. G., Taylor, E. C., and Buchardt, O. (1970). *Chem. Rev.* **70**, 231.
263. Ishikawa, M., Kaneko, C., Yokoe, I., and Yamada, S. (1969). *Tetrahedron*, **25**, 295.
264. Kamlet, M. J., and Kaplan, L. A. (1967). *J. Org. Chem.* **22**, 576; Splitter, J. S., and Calvin, M. (1958). *J. Org. Chem.* **23**, 651; Bonnett, R., Clark, V. M., and Todd, A. (1959). *J. Chem. Soc.* 2102; Kaminsky, L. S., and Lamchen, M. (1967). *J. Chem. Soc. (C)*, 1683. 2128); Sternbach, L. H., Koechlin, B. A., and Reeder, E. (1962). *J. Org. Chem.* **27**, 4671.
265. Alkaitis, A., and Calvin, M. (1968). *Chem. Commun.* 292.
266. Emmons, W. D. (1957). *J. Amer. Chem. Soc.* **79**, 5739.
267. Hata, N. (1961). *Bull. Chem. Soc. Jap.* **34**, 1440, 1444; (1962). *Chem. Abstr.* **56**, 4286; Hata, N., and Tanaka, I. (1962). *J. Chem. Phys.* **36**, 2076; (1962). *Chem. Abstr.* **57**, 5888.
268. Streith, J., and Cassal, J. M. (1967). *Compt. Rend.* **264**, 1307.
269. Ishikawa, M., Yamada, S., and Kaneko, C. (1965). *Chem. Pharm. Bull.* **13**, 747; (1965). *Chem. Abstr.* **63**, 8313.
270. Ishikawa, M., Yamada, S., Hotta, H., and Kaneko, C. (1966). *Chem. Pharm. Bull.* **14**, 1102; (1967). *Chem. Abstr.* **66**, 37146.
271. Kaneko, C., Yokoe, I., and Ishikawa, M. (1967). *Tetrahedron Lett.* 5237.
272. Kaneko, C. (1968). *J. Synth. Org. Chem. (Japan)*, **26**, 758.
273 Buchardt, O., Becher, J., Lohse, C., and Moeller, J. (1966). *Acta Chem. Scand.* **20**, 262; (1966). *Chem. Abstr.* **64**, 17536.
274. Buchardt, O., and Lohse, C. (1966). *Acta Chem. Scand.* **20**, 2467.
275. Buchardt, O., and Lohse, C. (1966). *Tetrahedron Lett.* 4355.
276. Buchardt, O., Becher, J., and Lohse, C. (1965). *Acta Chem. Scand.* **19**, 1120.
277. Buchardt, O., Kumler, Ph. L., and Lohse, C. (1969). *Acta Chem. Scand.* **23**, 159.

278. Nakashima, T., and Suzuki, I. (1969). *Chem. Pharm. Bull.* **17**, 2293; (1970). *Chem. Abstr.* **72**, 55145.
279. Taylor, E. C., Furth, B., and Pfau, M. (1965). *J. Amer. Chem. Soc.* **87**, 1400.
280. Kellogg, R. M., Van Bergen, T. J., and Wynberg, H. (1969). *Tetrahedron Lett.* 5211.
281. Den Hertog, H. J., Martens, R. J., Van der Plas, H. C., and Bon, J. (1966). *Tetrahedron Lett.* 4325.
282. Den Hertog, H. J., and Buurman, D. J. (1967). *Tetrahedron Lett.* 3657.
282a. Van der Lans, H. N. M., Den Hertog, H. J., and Van Veldhuizen, A. (1971). *Tetrahedron Lett.* 1875.
282b. Van der Lans, H. N. M., and Den Hertog, H. J. Unpublished results.
283. Roelfsema, W. A., and Den Hertog, H. J. (1967). *Tetrahedron Lett.* 5089.
283a. Bos, L. B., and Den Hertog, H. J. Unpublished results.
283b. Roelfsema, W. A. (1971). Thesis, Agricultural University, Wageningen, The Netherlands; see also (1972). Communications, pp. 72-75. Agricultural University, Wageningen, The Netherlands.
284. Roelfsema, W. A., and Den Hertog, H. J. Unpublished results.
285. Rees, C. W., and Sabet, C. R. (1965). *J. Chem. Soc.* 870.
286. Ueda, K. (1937). *J. Pharm. Soc. Jap.* **57**, 827; (1938). *Chem. Abstr.* **32**, 563.
287. Davies, W., Ramsay, T. H., and Stove, E. R. (1949). *J. Chem. Soc.* 2633.
288. Goldschmidt, (1888). *Monatsh. Chem.* **9**, 676.
289. Sargeson, A. M., and Sasse, W. H. F. (1958). *Proc. Chem. Soc. London*, 150.
290. Sasse, W. H. F. (1959). *J. Chem. Soc.* 3046.
291. Sasse, W. H. F. (1960). *J. Chem. Soc.* 526.
292. Baeyer, A., and Homolka, B. (1883). *Chem. Ber.* **16**, 2216; (1884). *Chem. Ber.* **17**, 985.
293. Lamberton, J. A., and Price, J. R., (1953). *Austr. J. Chem.* **6**, 173.
294. Lahey, F. N., Lamberton, J. A., and Price, J. R. (1950). *Aust. J. Sci. Res. A*, **3**, 155.
295. DRP 292 394 (1914); (1916). *Farbw. Hoechst. Frdl.* **13**, 447.
296. Ziegler, E., Kappe, Th., and Salvador, R. (1963). *Monatsh. Chem.* **94**, 453.
297. Ziegler, E., and Kappe, Th. (1963). *Monatsh. Chem.* **94**, 698.
298. Ziegler, E., and Kappe, Th. (1963). *Monatsh. Chem.* **94**, 736.
299. Ziegler, E., Salvador, R., and Kappe, Th. (1963). *Monatsh. Chem.* **94**, 941, 944; (1964). *Chem. Abstr.* **60**, 2885.
300. Ziegler, E., Foraita, H. G., and Kappe, Th. (1967). *Monatsh. Chem.* **98**, 324; (1966). *Monatsh. Chem.* **97**, 409; (1966). *Chem. Abstr.* **65**, 7137.
301. Dahn, H., and Donzel, A. (1967). See Ref. 9 *in Chimica*, **21**, 517.
302. Smith, P. A. S., and Kan, R. O. (1961). *J. Amer. Chem. Soc.* **83**, 2580.
303. Sarel, S., and Greenberger, A. (1958). *J. Org. Chem.* **23**, 330.
304. Ochiai, E., and Takahashi, M. (1968). *Chem. Pharm. Bull.* **16**, 1; (1966). *Chem. Pharm. Bull.* **14**, 1144, 1272; Ochiai, E., Kataoka, H., Dodo, T., and Takahashi, M. (1962). *Chem. Pharm. Bull.* **10**, 76; Ochiai, E., Takahashi, M., Tamai, Y., and Kataoka, H. (1963). *Chem. Pharm. Bull.* **11**, 137.

305. Ochiai, E., Kataoka, H., Dodo, T., and Takahashi, M. (1962). *Ann. Rep. Itsuu Lab.* 12, 11; Ochiai, E., and Tamai, Y. (1963). *Ann. Rep. Itsuu Lab.* 13, 5; Ochiai, E., Kataoka, H., and Takahashi, M. (1963). *Ann. Rep. Itsuu Lab.* 13, 13; Takahashi, M. (1963). *Ann. Rep. Itsuu Lab.* 13, 25; (1963). *Ann. Rep. Itsuu Lab.* 14, 23; Ochiai, E., and Dodo, T. (1965). *Ann. Rep. Itsuu Lab.* 14, 33.

306. Takahashi, M., and Tanabe, R. (1968). *Ann. Rep. Itsuu Lab.* 15, 41.

307. König, W., Coenen, M., Bahr, F., May, B., and Bassl, A. (1966). *J. Prakt. Chem.* 33, 54.

308. König, W., Coenen, M., Lorenz, W., Bahr, F., and Bassl, A. (1965). *J. Prakt. Chem.* 30, 96.

309. Davies, L. S., and Jones, G. (1969). *Tetrahedron Lett.* 1549.

310. Davies, L. S., and Jones, G. (1970). *J. Chem. Soc. (C),* 688.

311. Alberti, C. (1957). *Gazz. Chim. Ital.* 87, 720; (1958). *Chem. Abstr.* 52, 15535.

312. Koenigs, E., and Freund, J. (1947). *Chem. Ber.* 80, 143.

313. Backeberg, O. G., and Friedmann, C. A. (1938). *J. Chem. Soc.* 972; Backeberg, O. G. (1938). *J. Chem. Soc.* 1083.

314. De Stevens, G., Halamandaris, A., Bernier, M., and Blatter, H. M. (1963). *J. Org. Chem.* 28, 1336.

315. Bowie, R. A. (1970). *Chem. Commun.* 565.

315a. Kirby, G. W., Tan, S. L., and Uff, B. C. (1970). *J. Chem. Soc. (D),* 1138.

316. Den Hertog, H. J., and Van der Plas, H. C. (1964). "Advances in Heterocyclic Chemistry" (Ed. A. R. Katritzky) Part IV, p. 121.

317. Den Hertog, H. J., and Van der Plas, H. C. (1969). *In* "Chemistry of Acetylenes" (Ed. H. G. Viehe) Chapter 17. Dekker, New York.

318. Den Hertog, H. J., Van der Plas, H. C., Pieterse, M. J., and Streef, J. W. (1965). *Rec. Trav. Chim. Pays-Bas,* 84, 1569.

319. Streef, J. W., and Den Hertog, H. J. (1969). *Rec. Trav. Chim. Pays-Bas,* 88, 1391.

320. Streef, J. W., and Den Hertog, H. J. (1968). *Tetrahedron Lett.* 5945.

321. Den Hertog, H. J., and Buurman, D. J. (1967). *Rec. Trav. Chim. Pays-Bas,* 86, 187.

322. Czuba, W. Ref. 12 in reference 321.

323. Van der Plas, H. C., Zuurdeeg, B., and Van Meeteren, H. W. (1969). *Rec. Trav. Chim. Pays-Bas,* 88, 1156.

323a. Snieckus, V., and Kan, G. (1970). *Tetrahedron Lett.* 26, 2267.

324. Rees, C. W., and Yelland, M. (1969). *J. Chem. Soc. (D),* 377.

325. Nelson, N. A., Fassnacht, J. H., and Piper, J. U. (1961). *J. Amer. Chem. Soc.* 83, 206.

326. Lyle, R. E., and Lyle, G. G. (1954). *J. Amer. Chem. Soc.* 76, 3536.

327. Bergmann, E. D., and Rabinovitz, M. (1960). *J. Org. Chem.* 25, 827.

328. Craig, P. N., Lester, B. M., Saggiomo, A. J., Kaiser, C., and Zirkle, Ch. L. (1961). *J. Org. Chem.* 26, 135.

329. Whitlock, H. W. (1961). *Tetrahedron Lett.* 593.

330. Morosawa, S. (1958). *Bull. Chem. Soc. Jap.* 31, 418; (1959). *Chem. Abstr.* 53, 8160.

331. Overberger, C. G., Reichenthal, J., and Anselme, J. P. (1970). *J. Org. Chem.* 35, 138.

332. Dmitriev, L. B., Grandberg, I. I., and Moskalenko, V. A. (1968). *Izv. Timiryazev Sel'shokhoz. Akad.* 195; (1968). *Chem. Abstr.* 69, 43771.

333. Leonard, N. J., Jann, K., Paukstelis, J. V., and Steinhardt, C. K. (1963). *J. Org. Chem.* **28**, 1499.
334. Tarbell, D. S., and Noble Jr., P. (1950). *J. Amer. Chem. Soc.* **72**, 2657.
335. Grob, C. A., Hoegerle, R. M., and Ohta, M. (1962). *Helv. Chim. Acta,* **45**, 1823.
336. Ingold, C. K. (1953). Structure and mechanism, *in* "Organic Chemistry", p. 473.
337. Wittig, G., and Ludwig, H. (1954). *Annalen,* **589**, 55; Wittig, G., Koenig, G., and Clauss, K. (1955). *Annalen,* **593**, 127.
338. Jenny, E. F., and Melzer, A. (1966). *Tetrahedron Lett.* **29**, 3507; Hill, R. K., and Chan, T. H. (1966). *J. Amer. Chem. Soc.* **88**, 866.
339. Bullock, E., Gregory, B., Johnson, A. W., Brignell, P. J., Eisner, U., and Williams, H. (1962). *Proc. Chem. Soc. London.* 122; (1963). Williams, H. (1963). *J. Chem. Soc.* 4819.
340. Benary, E. (1920). *Chem. Ber.* **53**, 2218.
341. Brignell, P. J., Eisner, U., and Williams, H. (1965). *J. Chem. Soc.* 4226.
342. Anderson, M., and Johnson, A. W. (1964). *Proc. Chem. Soc. London.* 263.
343. Anderson, M., and Johnson, A. W. (1965). *J. Chem. Soc.* 2411.
344. Allgrove, R. C., and Eisner, U. (1967). *Tetrahedron Lett.* 499.
345. Ashby, J., and Eisner, U. (1967). *J. Chem. Soc. (C),* 1706.
346. Gill, G. B., Harper, D. J., and Johnson, A. W. (1968). *J. Chem. Soc. (C),* 1675.
347. Acheson, R. M., and Stubbs, J. K. (1969). *J. Chem. Soc. (C),* 2316.
348. Cromarty, A., and Proctor, G. R. (1968). *Chem. Commun.* 842.
349. Wolff, H. (1946). *Org. React.* **3**, 307.
350. Blatt, A. H. (1933). *Chem. Rev.* **12**, 215; (1933). *Chem. Abstr.* **27**, 3194.
351. Dickerman, S. C., and Lindwall, H. G. (1949). *J. Org. Chem.* **14**, 530.
352 Smith, P. A. S. (1948). *J. Amer. Chem. Soc.* **70**, 320.
353. Dickerman, S. C., and Moriconi, E. J. (1955). *J. Org. Chem.* **20**, 206; Dickerman, S. C., and Besozzi, A. J. (1954). *J. Org. Chem.* **19**, 1855.
354. Ittyerah, P. I., and Mann, F. G. (1958). *J. Chem. Soc.* 467.
354a. Fryer, R. I., Earley, J. V., Evans, E., Schneider, J., and Sternbach, L. H. (1970). *J. Org. Chem.* **35**, 2455.
354b. Wuensch, K. H., Stahnke, K. H., and Gomoll, P. (1970). *Z. Chem.* **10**, 219.
355. Mikhlina, E. E., and Rubtsov, M. V. (1963). *Zh. Obshch. Khim.* **33**, 2167; (1963). *Chem. Abstr.* **59**, 13991.
356. Rubtsov, M. V., Mikhlina, E. E., Vorob'eva, V. Ya., and Yanina, A. D. (1964). *Zh. Obshch. Khim.* **34**, 2222; (1964). *Chem. Abstr.* **61**, 9481.
357. Mikhlina, E. E. Vorob'eva, V. Ya., Shedchenko, V. I., and Rubtsov, M. V. (1965). *Zh. Org. Khim.* **1**, 1336; (1965). *Chem. Abstr.* **63**, 13257.
358. Paquette, L. A., and Scott, M. K. (1968). *J. Org. Chem.* **33**, 2379.
359. Rozantzev, E. G., and Neiman, M. B. (1964). *Tetrahedron,* **20**, 131; Bauer, L., and Hewitson, R. E. (1962). *J. Org. Chem.* **27**, 3982.
360. Collicutt, A. R., and Jones, G. (1960). *J. Chem. Soc.* 4101.
361. Fozard, A., and Jones, G. (1964). *J. Chem. Soc.* 2763.
362. Harfenist, M., and Magnien, E. (1958). *J. Amer. Chem. Soc.* **80**, 6080.
363. Godefroi, E. F., and Wittle, E. L. (1956). *J. Org. Chem.* **21**, 1163.
364. Lyle, R. E., and Troscianiec, H. J. (1955). *J. Org. Chem.* **20**, 1757.
365. Buchardt, O. (1966). *Tetrahedron Lett.* 6221.
366. Kaneko, C., Yamada, S., and Ishikawa, M. (1966). *Tetrahedron Lett.* 2145.

367. Kaneko, C., Yamada, S., and Yokoe, I. (1967). *Iyo Kizai Kenkyusho Hokoku, Tokyo Ika Shika Daigaku,* 1, 1; (1968). *Chem. Abstr.* 68, 104845.
368. Kaneko, C., and Yamada, S. (1966). *Chem. Pharm. Bull.* 14, 555; (1966). *Chem. Abstr.* 65, 10574.
369. Kaneko, C., Yamada, S., Yokoe, I., and Ishikawa, M. (1967). *Tetrahedron Lett.* 1873.
370. Kaneko, C., Yamada, S., and Yokoe, I. (1966). *Tetrahedron Lett.* 4701.
371. Buchardt, O., Lohse, C., Duffield, A. M., and Djerassi, C. (1967). *Tetrahedron Lett.* 2741.
372. Kaneko, C., and Yamada, S. (1966). *Rep. Res. Inst. Dental Materials, Tokyo, Medico-Dental University,* 2, 804.
373. Simonsen, O., Lohse, C., and Buchardt, O. (1970). *Acta Chem. Scand.* 24, 268.
374. Lohse, C. (1968). *Tetrahedron Lett.* 5625.
375. Buchardt, O., Kumler, Ph. L., and Lohse, C. (1969). *Acta Chem. Scand.,* 23, 2149; see Hoffmann, R., and Woodward, R. B. (1968). *Accounts Chem. Res.,* 1, 17.
375a. Taylor, E. C., and Spence, G. G. (1968). *Chem. Commun.,* 1037.
376. Ishikawa, M., Kaneko, C., and Yamada, S. (1968). *Tetrahedron Lett.* 4519.
377. Kaneko, C., Yokoe, I., and Ishikawa, M. (1967). *Tetrahedron Lett.* 5237; Mantsch, H., and Zanker, V. (1966). *Tetrahedron Lett.* 4211.
378. Streith, J., and Cassal, J. M. (1968). *Angew. Chem.* 80, 117; (1968). *Angew. Chem. Int. Ed.* 7, 129; Streith, J., and Sigwalt, C. (1966). *Tetrahedron Lett.* 1347; Streith, J., Danner, B., and Sigwalt, C. (1967). *Chem. Commun.* 979.
379. Streith, J., and Cassal, J. M. (1968). *Tetrahedron Lett.* 4541.
380. Sasaki, T., Kanematsu, K., and Kakehi, A. (1969). *J. Chem. Soc. (D),* 432.
381. Streith, J., and Cassal, J. M. (1967). *Compt. Rend.* 264, 1307.
382. Sasaki, T., Kanematsu, K., Kakehi, A., Ichikawa, J., and Hayakawa, K. (1970). *J. Org. Chem.* 35, 426.
383. Balasubramanian, A., McIntosh, J. M., and Snieckus, V. (1970). *J. Org. Chem.* 35, 433.
384. Snieckus, V. (1969). *J. Chem. Soc. (D),* 831.
385. Carruthers, W., and Johnstone, R. A. W. (1965). *J. Chem. Soc.* 1653.
386. Meisenheimer, J. (1919). *Chem. Ber.* 52, 1667.
387. Quin, L. D., and Shelburne, F. A. (1965). *J. Org. Chem.* 30, 3135.
388. Schmitz, E., and Ohme, R. (1962). *Chem. Ber.* 95, 2012.
389. Lednicer, D., and Hauser, C. R. (1957). *J. Amer. Chem. Soc.* 79, 4449.
390. Antonov, V. K., Shkrob, A. M., and Shemyakin, M. M. (1963). *Tetrahedron Lett.* 439.
391. Shemyakin, M. M., Antonov, V. K., Shkrob, A. M., Sheinker, Yu. N., and Senyavina, L. B. (1962). *Tetrahedron Lett.* 701.
392. Danö, M., Furberg, S., and Hassel, O. (1950). *Acta Chem. Scand.* 4, 965.
393. Böeseken, J., Tellegen, F., and Cohen Henriquez, P. (1935). *Rec. Trav. Chim. Pays-Bas,* 54, 733.
394. Gudz, V. N., Samsonova, V. G., and Vampilova, V. V. (1966). *Tr. Buryat. Kompleks. Nauch.-Issled. Inst., Akad. Nauk SSSR, Sib. Otd.* 20, 17; (1968). *Chem. Abstr.* 68, 114515.

395. Horner, L., and Brüggemann, H. (1960). *Annalen*, **635**, 22.
396. Horner, L., and Jurgeleit, W. (1955). *Annalen*, **591**, 138.
397. Greenbaum, M. A., Denney, D. B., and Hoffmann, A. K. (1956). *J. Amer. Chem. Soc.* **78**, 2563.
398. Gulbins, K., Benzing, G., Maysenhölder, R., and Hamann, K. (1960). *Chem. Ber.* **93**, 1975; Gulbins, K., and Hamann, K. (1958). *Angew. Chem.* **70**, 705.
399. Parham, W. E., and Traynelis, V. J. (1954). *J. Amer. Chem. Soc.* **76**, 4960.
400. Parham, W. E., and Traynelis, V. J. (1955). *J. Amer. Chem. Soc.* **77**, 68.
401. Parham, W. E., Nicholson, I., and Traynelis, V. J. (1956). *J. Amer. Chem. Soc.* **78**, 850.
402. Du Pont de Nemours & Co. U.S. Pat. 3,400,134.
403. Parham, W. E., Mayo, G. L. O., and Gadsby, B. (1959). *J. Amer. Chem. Soc.* **81**, 5993.
404. Draber, W. (1967). *Chem. Ber.* **100**, 1559.
405. Purrello, G., and Lo Vullo, A. (1967). *Boll. Sedute Accad. Gioenia Sci. Natur. Catania*, **9**, 46; (1969). *Chem. Abstr.* **70**, 3692.
406. Boberg, F. (1964). *Annalen*, **679**, 118.
407. Fickentscher, K. (1969). *Arch. Pharm.* **302**, 285; (1969). *Chem. Abstr.* **71**, 13072.
408. Grigg, R., Hayes, R., and Jackson, J. L. (1969). *J. Chem. Soc. (D)*, 1167.
409. Parham, W. E. Wijnberg, H., Hasek, W. R., Howell, P. A., Curtis, R. M., and Lipscomb, W. N. (1954). *J. Amer. Chem. Soc.* **76**, 4957.
410. Cullinane, N. M., Morgan, N. M. E., and Plummer, C. A. J. (1937). *Rec. Trav. Chim. Pays-Bays*, **56**, 627.
411. Gilman, H., and Swayampati, D. R. (1955). *J. Amer. Chem. Soc.* **77**, 3387.
412. Fickentscher, K. (1969). *Chem. Ber.* **102**, 2378.
413. Szmant, H. H., and Alfonso, L. M. (1957). *J. Amer. Chem. Soc.* **79**, 205.
414. Szmant, H. H., and Alfonso, L. M. (1956). *J. Amer. Chem. Soc.* **78**, 1064.
414a. Fickentscher, K. (1970). *Chem. Ber.* **103**, 3000.
415. Luettringhaus, A., Mohr, M., and Engelhard, N. (1963). *Annalen*, **661**, 84; (1963). *Chem. Abstr.* **58**, 13940.
416. Atkinson, E., Beer, R. J. S., Harris, D., and Royall, D. J. (1967). *J. Chem. Soc. (C)*, 638.
417. Beer, R. J. S., Harris, D., and Royall, D. J. (1964). *Tetrahedron Lett.* 1531.
417a. Burdon, J., and Parsons, I. W. (1971). *J. Chem. Soc. (C)*, 355.
417b. Böhme, H., and Sell, K. (1948). *Chem. Ber.* **81**, 123.
418. Schroth, W., Billig, F., and Reinhold, G. (1967). *Angew. Chem.* **79**, 685; (1967). *Angew. Chem. Int. Ed.* **6**, 698.
419. Schroth, W., Langguth, R., and Billig, F. (1965). *Z. Chem.* **5**, 352; (1964). *Chem. Abstr.* **64**, 581.
420. Bohlmann, F. (1966). *Fortschr. Chem. Forsch.* **6**, 89; Mortensen, J. T., Sörensen, J. S., and Sörensen, N. A. (1964). *Acta Chem. Scand.* **18**, 2392.
421. Armarego, W. L. F., and Turner, E. E. (1956). *J. Chem. Soc.* 1665.
422. Armarego, W. L. F., and Turner, E. E. (1957). *J. Chem. Soc.* 13.
423. Van Meeteren, H. W., and Van der Plas, H. C. (1966). *Tetrahedron Lett.* 4517.
424. Van Meeteren, H. W., and Van der Plas, H. C. (1968). *Rec. Trav. Chim. Pays-Bas.* **87**, 1089.

425. Van Meeteren, H. W., Van der Plas, H. C., and De Bie, D. A. (1969). *Rec. Trav. Chim. Pays-Bas,* **88,** 728.
426. Van Meeteren, H. W., and Van der Plas, H. C. (1969). *Rec. Trav. Chim. Pays-Bas,* **88,** 204.
427. Shiner, Jr., V. J., and Wilson, J. W. (1962). *J. Amer. Chem. Soc.* **84,** 2402.
428. Crow, W. D., and Wentrup, C. (1968). *Chem. Commun.* 1082; (1968). *Chem. Commun.* 1026.
429. Fosse, R., Hieulle, H., and Bass, L. W. (1924). *Compt. Rend.* **178,** 811.
430. Lingens, F., and Schneider-Bernlöhr, H. (1965). *Annalen,* **686,** 134.
431. Hayes, D. H., and Hayes-Baron, F. (1967). *J. Chem. Soc. (C),* 1528.
432. Levene, P. A., and Bass, L. W. (1926). *J. Biol. Chem.* **71,** 167; (1927). *Chem. Abstr.* **21,** 587.
433. Caputto, R., Leloir, L. F., Cardini, C. E., and Paladini, A. C. (1950). *J. Biol. Chem.* **184,** 333; (1950). *Chem. Abstr.* **44,** 7925.
434. Baron, F., and Brown, D. M. (1955). *J. Chem. Soc.* 2855.
435. Witzel, H. (1959). *Annalen,* **620,** 126.
436. Budovskii, E. I., Haines, J. A., and Kochetkov, N. K. (1964). *Dokl. Akad. Nauk SSSR,* **158,** 379; (1964). *Chem. Abstr.* **61,** 16341.
437. Takemura, S., and Miyazaki, M. (1959). *Bull. Chem. Soc. Jap.* **32,** 926; (1960). *Chem. Abstr.* **54,** 15487.
438. Habermann, V. (1963). *Coll. Czech. Chem. Commun.* **28,** 510; (1961). *Coll. Czech. Chem. Commun.* **26,** 3147; (1963). *Chem. Abstr.* **58,** 14352.
439. Verwoerd, D. W., and Zillig, W. (1963). *Biochim. Biophys. Acta,* **68,** 484; (1963). *Chem. Abstr.* **59,** 4264.
440. Temperli, A., Türler, H., Rüst, P., Danon, A., and Chargaff, E. (1964). *Biochim. Biophys. Acta,* **91,** 462; (1965). *Chem. Abstr.* **62,** 2945.
441. Ellery, B. W., and Symons, R. H. (1966). *Nature,* **210,** 1159.
442. Sedat, J., and Sinsheimer, R. L. (1964). *J. Mol. Biol.* **9,** 489; (1965). *Chem. Abstr.* **62,** 1982.
443. Zee-Cheng, K. Y., and Cheng, C. C. (1968). *J. Org. Chem.* **33,** 892.
444. Wiley, R. H., and Yamamoto, Y. (1960). *J. Org. Chem.* **25,** 1906.
445. Van der Plas, H. C., and Jongejan, H. (1967). *Tetrahedron Lett.* 4385.
446. Van der Plas, H. C., and Jongejan, H. (1968). *Rec. Trav. Chim. Pays-Bas,* **87,** 1065.
447. Baumbach, F., Henning, H. G., and Hilgetag, G. (1964). *Z. Chem.* **4,** 67; (1964). *Chem. Abstr.* **60,** 12006.
448. Brown, D. J., Ford, P. W., and Paddon-Row, M. N. (1968). *J. Chem. Soc. (C),* 1452.
449. Van der Plas, H. C., and Jongejan, H. (1970). *Rec. Trav. Chim. Pays-Bas,* **89,** 680.
450. Van der Plas, H. C., and Jongejan, H. (1971). *Rec. Trav. Chim. Pays-Bas,* **90,** 326.
451. Biffin, M. E. C., Brown, D. J., and Porter, Q. N. (1967). *Tetrahedron Lett.* 2029; (1968). *J. Chem. Soc. (C),* 2159.
452. Van der Plas, H. C., Vollering, M. C., and Jongejan, H. Unpublished results.
453. Kato, T., Yamanaka, H., and Yasuda, N. (1967). *J. Org. Chem.* **32,** 3593.
453a. Thompson, T. W. (1968). *Chem. Commun.* 532; Longridge, J. L. and Thompson, T. W. (1970). *J. Chem. Soc.* (C), 1658.
454. King, F. E., and Clark-Lewis, J. W. (1951). *J. Chem. Soc.* 3077.
455. Clark-Lewis, J. W. (1958). *Chem. Rev.* **58,** 63.

456. Aspelund, H. (1938). *Acta Acad. Aboensis, Math. Phys.* **10**, no. 14; (1939). *Acta Acad. Aboensis, Math. Phys.* **12**, no. 5; (1947). *Chem. Abstr.* **41**, 2414.

457. Pellizzani, G. (1887). *Gazz. Chim. Ital.* **17**, 409.

458. Aspelund, H. (1968). *Acta Acad. Aboensis, Math. Phys.* **28**; (1969). *Chem. Abstr.* **70**, 106469.

459. Aspelund, H. (1942). *Acta Acad. Aboensis, Math. Phys.* **13**, no. 1; (1943). *Acta Acad. Aboensis, Math. Phys.* **13**, no. 2; (1945). *Chem. Abstr.* **39**, 2053.

460. Wöhler, F., and Liebig, J. (1838). *Annalen,* **26**, 241; Schlieper, A. (1845). *Annalen,* **263**, 55; Richardson, G. M., and Cannan, R. K. (1929). *Biochem. J.* **23**, 68; (1929). *Chem. Abstr.* **23**, 2638; Patterson, J. W., Lazarow, A., Lemm, F. J., and Levey, S. (1949). *J. Biol. Chem.* **177**, 187; (1949). *Chem. Abstr.* **43**, 4306.

461. Biltz, H., Heyn, M., and Bergius, M. (1917). *Annalen,* **413**, 68; Biltz, H., and Kobel, M. (1921). *Chem. Ber.* **54**, 1802.

462. Clark-Lewis, J. W., and Thompson, M. J. (1959). *J. Chem. Soc.* 2401.

463. Fisher, F. R., and Day, A. R. (1955). *J. Amer. Chem. Soc.* **77**, 4894.

464. Clark-Lewis, J. W., and Edgar, J. A. (1962). *J. Chem. Soc.* 3887.

464a. Deeleman, R. A. F., and Van der Plas, H. C. (1973). *Rec. Trav. Chim. Pays-Bas,* **92**, 317; Streith, J., Leibovici, C., and Martz, P. (1971). *Bull. Soc. Chim. Fr.,* 4152.

465. Kwart, H., Spayd, R. W., and Collins, C. J. (1961). *J. Amer. Chem. Soc.* **83**, 2579; Kwart, H., and Sarasohn, I. M. (1961). *J. Amer. Chem. Soc.* **83**, 909.

466. Aspelund, H. (1939). *Acta Acad. Aboensis, Math. Phys.* **12**, no. 5; (1937). *Acta Acad. Aboensis, Math. Phys.* **11**, no. 2; (1950). Ware, E., *Chem. Rev.* **46**, 403.

467. Hepner, B., and Frenkenberg, S. (1932). *J. Prakt. Chem.* **134**, 249.

468. King, F. E., and Clark-Lewis, J. W. (1951). *J. Chem. Soc.* 3080.

469. Dudley, K. H., and Hemmerich, P. (1967). *J. Org. Chem.* **32**, 3049.

470. Otter, B. A., and Fox, J. J. (1967). *J. Amer. Chem. Soc.* **89**, 3663.

471. Barbieri, W., Bernardi, L., Palamidessi, G., and Venturi, M. T. (1968). *Tetrahedron Lett.* 2931.

472. Barbieri, W., Bernardi, R., Luini, F., and Palamidessi, G. (1969). *Farmaco, Ed. Sci.* **24**, 561; (1969). *Chem. Abstr.* **7**, 61327.

472a. Barbieri, W., Bernardi, L., Luini, F., and Palamidessi, G. (1970). *Farmaco, Ed. Sci.* **25**, 702.

472b. Barbieri, W., Bernardi, L., Luini, F., and Palamidessi, G. (1970). *Farmaco, Ed. Sci.* **25**, 694.

473. Field, G. F., and Sternbach, L. H. (1968). *J. Org. Chem.* **33**, 4438.

474. Streith, J., and Martz, P. (1969). *Tetrahedron Lett.* 4899.

475. Brown, D. J., and England, B. T. (1970). *Aust. J. Chem.* **23**, 625.

476. Van der Plas, H. C., Jongejan, H., and Geurtsen, G. (In press). *Rev. Trav. Chim. Pays-Bas.*

477. Eliel, E. L., McBride, R. T., and Kaufmann, St. (1953). *J. Amer. Chem. Soc.* **75**, 4291.

478. Brody, F., and Ruby, P. R. (1960). *In* "Pyridine and its Derivatives" (Ed. E. Klingsberg), Part I, pp. 474-489, Interscience, New York.

479. Van der Plas, H. C., and Geurtsen, G. (1964). *Tetrahedron Lett.* 2093.

480. Van der Plas, H. C. (1965). *Tetrahedron Lett.* 555.
481. Van der Plas, H. C., Smit, P., and Koudijs, A. (1968). *Tetrahedron Lett.* 9.
482. Van der Plas, H. C., Haase, B., Zuurdeeg, B., and Vollering, M. C. (1966). *Rec. Trav. Chim. Pays-Bas,* **85**, 1101.
483. Van der Plas, H. C., and Zuurdeeg, B. (1969). *Rec. Trav. Chim. Pays-Bas,* **88**, 426.
484. Van Meeteren, H. W., and Van der Plas, H. C. (1967). *Rec. Trav. Chim. Pay-Bas,* **86**, 15.
485. Van Meeteren, H. W., and Van der Plas, H. C. (1966). Proceedings Second International Conference on Methods of Preparing and Storing Lab. Comp., Brussels, p. 483.
486. Van der Plas, H. C., Zuurdeeg, B., and Van Meeteren, H. W. (1969). *Rec. Trav. Chim. Pays-Bas,* **88**, 1156.
487. Van Meeteren, H. W. (1968). Thesis, Leiden.
487a. Bunnett, J. F., and Hrutfiord, E. F. (1962). *J. Org. Chem.* **27**, 4152.
488. Van Meeteren, H. W., and Van der Plas, H. C. (1966). *Tetrahedron Lett.* 4517.
489. Van Meeteren, H. W., and Van der Plas, H. C. (1971). *Rec. Trav. Chim. Pays-Bas,* **90**, 105.
490. Taylor, E. C., and Jefford, C. W. (1962). *J. Amer. Chem. Soc.* **84**, 3744.
490a. Fisher, H., Grob, C. A., and Renk, E. (1959). *Helv. Chim. Acta* **42**, 872.
491. Taylor, E. C., Jefford, C. W., and Cheng, C. C. (1961). *J. Amer. Chem. Soc.* **83**, 1261.
491a. Delia, T. J., and Venton, D. L. (1970). *J. Heterocycl. Chem.* **7**, 1183.
491b. Pfleiderer, W., and Kempter, F. E. (1970). *Chem. Ber.* **103**, 908.
492. Sternbach, L. H., and Reeder, E. (1961). *J. Org. Chem.* **26**, 1111.
493. Archer, G. A., Sternbach, L. H. French Pat. 1,507,878; (1969). *Chem. Abstr.* **70**, 37854.
494. Sternbach, L. H., Reeder, E., Keller, O., and Metlesics, W. (1961). *J. Org. Chem.* **26**, 4488.
495. Bell, S. C., Gochman, C., and Childress, S. J. (1962). *J. Med. Pharm. Chem.* **5**, 63; (1962). *Chem. Abstr.* **57**, 12494.
496. Takacsik, T., Kosztin, P., Popa, V., and Mathe, E. Rom. 49,035; (1968). *Chem. Abstr.* **69**, 106759.
497. Farber, S., Wuest, H. M., and Meltzer, R. I. (1963). *J. Med. Chem.* **7**, 235; (1964). *Chem. Abstr.* **60**, 13243.
498. Hoffmann-La Roche. Neth. Appl. 6,413,180; (1965). *Chem. Abstr.* **63**, 14890.
499. Spencer, J. L., Lilly, Eli. and Co. U.S. Pat. 3,462,419; (1969). *Chem. Abstr.* **71**, 91549.
500. Fryer, R. I., Schmidt, R. A., and Sternbach, L. H. (1964). *J. Pharm. Sci.* **53**, 264; (1964). *Chem. Abstr.* **60**, 15872.
501. Saucy, G., and Sternbach, L. H. (1962). *Helv. Chim. Acta,* **45**, 2226.
502. Gordon, M., Pachter, I. J., and Wilson, J. W. (1963). *Arzneim.-Forsch.* **13**, 802; (1964). *Chem. Abstr.* **60**, 4147.
503. Sternbach, L. H., Fryer, R. I., Keller, O., Metlesics, W., Sach, G., and Steiger, N., (1963). *J. Med. Chem.* **6**, 261; (1963). *Chem. Abstr.* **58**, 13962.
504. Sternbach, L. H., Saucy, G., Smith, F. A., Muller, M., and Lee, J. (1963). *Helv. Chim. Acta,* **46**, 1720.

505. Keller, O., Steiger, N., and Sternbach, L. H. U.S. Pat. 3,121,075; (1964). *Chem. Abstr.* **61**, 5672.
506. Bell, S. C., Sulkowski, T. S., Gochman, C., and Childress, S. J. (1962). *J. Org. Chem.* **27**, 562.
507. Sternbach, L. H., and Reeder, E. (1961). *J. Org. Chem.* **26**, 4936.
508. Walkenstein, S. S., Wiser, R., Gudmundsen, C. H., Kimmel, H. B., and Corradino, R. A. (1964). *J. Pharm. Sci.* **53**, 1181; (1965). *Chem. Abstr.* **62**, 2114.
509. Stempel, A., Douvan, I., and Sternbach, L. H. (1968). *J. Org. Chem.* **33**, 2963.
510. Derieg, M. E., Fryer, R. I., and Sternbach, L. H. (1968). *J. Chem. Soc. (C),* 1103.
511. Stempel, A., Reeder, E., and Sternbach, L. H. (1965). *J. Org. Chem.* **30**, 4267.
512. Kaneko, C., and Yamada, S. (1967). *Tetrahedron Lett.* 5233.
513. Field, G. F., and Sternbach, L. H. (1968). *J. Org. Chem.* **33**, 4438.
513a. Wuensch, K. H., and Bajdala, H. (1970). *Z. Chem.* **10**, 144.
514. Metlesics, W., Silverman, G., and Sternbach, L. H. (1967). *Monatsh. Chem.* **98**, 633.
515. Maki, Y., Kizu, H., and Obata, K. (1963). *Yakugaku Zasshi,* **83**, 725; (1964). *Chem. Abstr.* **60**, 1742.
516. Takahashi, T., Furukawa, N., and Maki, Y. (1966). *Yakugaku Zasshi,* **86**, 867; (1966). *Chem. Abstr.* **65**, 18574.
517. Maki, Y., and Obata, K. (1963). *Yakugaku Zasshi,* **83**, 819; (1964). *Chem. Abstr.* **60**, 1742.
518. Maki, Y., Obata, K. (1964). *Chem. Pharm. Bull.* **12**, 176; (1964). *Chem. Abstr.* **61**, 650.
519. Ach, F. (1889). *Annalen,* **253**, 44.
520. Druey, J. (1963). *Pharm. Acta Helv.* **38**, 183.
521. Druey, J., Meier, K., and Staehelin, A. (1963). *Pharm. Acta Helv.* **38**, 498.
522. Dury, K. (1965). *Angew. Chem.* **77**, 282; (1965). *Angew. Chem. Int. Ed.* **4**, 229.
523. Kuhelj, F., Stanovnik, B., and Tisler, M. (1966). *Croat. Chem. Acta,* **38**, 299; (1966). *Chem. Abstr.* **66**, 55458.
523a. Maki, Y., and Beardsley, G. P. (1971). *Tetrahedron Lett.* 1507.
524. Van der Plas, H. C., and Geurtsen, G. Unpublished results.
525. Ogata, M., and Kano, K. (1967). *Chem. Commun.* 1176.
526. Kumler, Ph. L., and Buchardt, O. (1968). *J. Amer. Chem. Soc.* **90**, 5640.
527. Altman, L. J., Semmelhack, M. F., Hornby, R. B., and Vederas, J. C. (1968). *Chem. Commun.* 686.
528. Lemal, D. M., and Rave, T. W. (1963). *Tetrahedron,* **19**, 1119.
529. Pollak, A., and Tisler, M. (1964). *Tetrahedron Lett.* 253.
530. Yoneda, F., Otaka, T., and Nitta, Y. (1966). *Chem. Pharm. Bull.* **14**, 698; (1966). *Chem. Abstr.* **65**, 13673.
531. Ajello, E. (1966-1967). *Atti Accad. Sci. Lett. Arti. Palermo,* Parte I, **27**, 465; (1969). *Chem. Abstr.* **71**, 12919.
531a. Lund, H., and Lunde, P. (1967). *Acta Chem. Scand.* **21**, 1067.
532. Ikekawa, N., Honma, Y., and Kenkyusho, R. (1967). *Tetrahedron Lett.* 1197.

532a. Beak, P., and Miesel, J. L. (1967). *J. Amer. Chem. Soc.* **89**, 2375.
532b. Mager, H. I. X., and Berends, W. (1965). *Rev. Trav. Chim. Pays-Bas,* **84**, 314.
533. Lont, P. J., Van der Plas, H. C., and Koudijs, A. (1971). *Rec. Trav. Chim. Pays-Bas,* **90**, 207.
533a. Lont, P. J., Van der Plas, H. C., and Bosma, E. (1972). *Rec. Trav. Chim. Pays-Bas,* **91**, 1352; Lont. P. J., and Van der Plas, H. C. (1972). *Rec. Trav. Chim. Pays-Bas,* **91**, 850.
534. Besford, L. S., and Malcolm Bruce, J. (1964). *J. Chem. Soc.* 4037.
534a. Lund, H. (1967). *Acta Chem. Scand.* **21**, 2525.
534b. Haddlesey, D. I., Mayor, P. A., and Szinai, S. S. (1964). *J. Chem. Soc.* 5269.
534c. Ames, D. E., and Novitt, B. (1970). *J. Chem. Soc. C,* 1700.
535. Neber, P. W., Knöller, G., Herbst, K., and Trissler, A. (1929). *Annalen,* **471**, 113.
536. Malcolm Bruce, J. (1959). *J. Chem. Soc.* 2366.
537. Baumgarten, H. E., Creger, P. L., and Zey, R. L. (1960). *J. Amer. Chem. Soc.* **82**, 3977; Baumgarten, H. E., Wittman, W. F., and Lehmann, G. J. (1969). *J. Heterocycl. Chem.* **6**, 333.
538. Baumgarten, H. E., and Furnas, J. L. (1961). *J. Org. Chem.* **26**, 1536.
539. Ames, D. E., Novitt, B., Waite, D., and Lund, H. (1969). *J. Chem. Soc. (C),* 796.
540. Suzuki, I., Nakashima, T., and Nagasawa, N. (1965). *Chem. Pharm. Bull.* **13**, 713; (1965). *Chem. Abstr.* **63**, 9940.
541. Linstead, R. P., and Noble, E. G. (1937). *J. Chem. Soc.* 933.
542. Gabriel, S., and Neumann, A. (1893). *Chem. Ber.* **26**, 521.
543. Gabriel, S., and Eschenbach, G. (1897). *Chem. Ber.* **30**, 3022.
544. Gabriel, S., and Neumann, A. (1893). *Chem. Ber.* **26**, 705.
545. Daube, A. (1905). *Chem. Ber.* **38**, 206.
546. Darapsky, A., and Heinrichs, P. (1936). *J. Prakt. Chem.* **146**, 307.
547. Rowe, F. M., and Dunbar, C. (1932). *J. Chem. Soc.* 11.
548. Baloiak, S., and Domagalina, E. (1959). *Rocz. Chem.* **33**, 725; (1960). *Chem. Abstr.* **54**, 2346.
549. Drew, H. D. K., and Hatt, H. H. (1937). *J. Chem. Soc.* 16; Drew, H. D. K., and Pearman, F. H. (1937). *J. Chem. Soc.* 26; Drew, H. D. K., Hatt, H. H., and Hobart, F. A. (1937). *J. Chem. Soc.* 33.
549a. Lund, H., and Jensen, E. T. (1970). *Acta Chem. Scand.* **24**, 1867.
549b. Lund, H. (1970). *Advan. Heterocycl. Chem.* **12**, 255.
549c. Lund, H. (1965). *Collect. Czech. Chem. Commun.* **30**, 4237.
549d. Lund, H. (1970). *Advan. Heterocycl. Chem.* **12**, 224.
550. Singh, B. (1969). *J. Amer. Chem. Soc.* **91**, 3670.
551. Buchardt, O. (1968). *Tetrahedron Lett.* 1911.
552. Brand, K., and Wild, E. (1923). *Chem. Ber.* **56**, 105; Simpson, J. C. E. (1953). "Condensed Pyridazine and Pyrazine Rings." p. 188, Interscience, New York.
553. Lane, E. S. (1953). *J. Chem. Soc.* 2238; (1955). *J. Chem. Soc.* 1079.
554. Taylor, E. C., and McKillop, A. (1965). *J. Org. Chem.* **30**, 2858.
555. Lont, P. J., Van der Plas, H. C., and Verbeek, A. J. (1972). *Rec. Trav. Chim. Pays-Bas,* **91**, 949
556. Cheeseman, G. W. H., and Rafig, M. (1971). *J. Chem. Soc.* (C) 452.

556a. Albert, A., and McCormack, J. J. (1965). *J. Chem. Soc.* 6930.
557. Ahmad, Y., Habib, M. S., Mohammady, A., Bakhtiari, B., and Shamsi, S. A. (1968). *J. Org. Chem.* **33**, 201.
558. Hutzenlaub, W., Barlin, G. B., and Pfleiderer, W. (1969). *Angew. Chem. Int. Ed.* **8**, 608.
559. Kaneko, C., Yokoe, I., Yamada, S., and Ishikawa, M. (1966). *Chem. Pharm. Bull.* **14**, 1316; (1967). *Chem. Abstr.* **66**, 54822.
560. Lahmani, F., and Ivanoff, N. (1967). *Tetrahedron Lett.* 3913.
561. Lahmani, F., Ivanoff, N., and Magat, M. (1966). *Compt. Rend. Acad. Sci. Ser. (C)*, **263**, 1005.
562. Crow, W. D., and Wentrup, C. (1968). *Tetrahedron Lett.* 3115.
562a. Johnson, D. W., Austel, V., Feld, R. S., and Lemal, D. M. (1970). *J. Amer. Chem. Soc.* **92**, 7505.
563. Allison, C. G., Chambers, R. D., Cheburkov, Yu. A., MacBride, J. A. H., and Musgrave, W. K. P. (1969). *J. Chem. Soc. (D)*, 1200.
563a. Chambers, R. D., MacBride, J. A. H., and Musgrave, W. K. R. (1970). *J. Chem. Soc. D*, 739.
564. Ajello, T., Giambrone, S., and Giammanco, L. (1965). *Atti Acad. Sci. Lit. Arti, Palermo*, Part I, **26**, 321.
565. Mackay, D., Campbell, J. A., and Jennison, C. P. R. (1970). *Can. J. Cnem.* **48**, 81.
566. Franck-Neumann, M., and Leclerc, G. (1969). *Tetrahedron Lett.* 1063.
567. Morel, Th., and Verkade, P. E. (1948). *Rec. Trav. Chim. Pays-Bas*, **67**, 539.
568. Morel, Th., and Verkade, P. E. (1951). *Rec. Trav. Chim. Pays-Bas*, **70**, 35.
569. Morel, Th., and Verkade, P. E. (1949). *Rec. Trav. Chim. Pays-Bas*, **68**, 619.
570. Ferrario, (1911). *Bull. Soc. Chim. Fr.* **9**, 536.
571. Suter, C. M., McKenzie, J. P., and Maxwell, Ch. E. (1936). *J. Amer. Chem. Soc.* **58**, 717.
572. Gilman, H., Van Ess, M. W., Willis, H. B., and Stuckwisch, C. G. (1940). *J. Amer. Chem. Soc.* **62**, 2606.
573. Evans, M. L., and Mason, C. T. (1968). *J. Org. Chem.* **33**, 1643.
573a. Pedersen, A. O., Lawesson, S. O., Klemmensen, P. D., and Kolc, J. (1970). *Tetrahedron*, **26**, 1157.
573b. Chapman, O. L., and McIntosh, C. L. (1970) *J. Amer. Chem. Soc.* **92**, 7001.
574. Kresze, G., and Firl, J. (1969). *Fortschr. Chem. Forsch.* **11**, 245.
575. Kojima, S. (1954). *J. Chem. Soc. Jap.* **57**, 371; (1955). *Chem. Abstr.* **49**, 15906; (1954). *Chem. Abstr.* **57**, 819; (1955). *Chem. Abstr.* **49**, 10966.
576. Scheiner, P. (1967). *J. Org. Chem.* **32**, 2628.
577. Firl, J., and Kresze, G. (1966). *Chem. Ber.* **99**, 3695; Kresze, G., and Firl, J. (1963). *Tetrahedron*, **19**, 1329.
578. Wichterle, O., and Vavruska, M. (1952). *Chem. Listy*, **46**, 237; (1953). *Chem. Abstr.* **47**, 4330.
579. Kojima, S. (1959). *J. Chem. Soc. Jap., Ind. Chem. Sect.* **62**, 1260.
580. Kresze, G., and Braun, H. (1969). *Tetrahedron Lett.* 1743.
581. Kosinski, M. (1964). *Acta Chim. Soc. Sci. Lodz*, **9**, 93.
582. Firl, J. (1968). *Chem. Ber.* **101**, 218.
583. Meisenheimer, J. (1913). *Annalen*, **397**, 273; Franzen, V., and Otto, S. (1961). *Chem. Ber.* **94**, 1360.
584. Kloetzel, M. C. (1947). *J. Amer. Chem. Soc.* **69**, 2271.

584a. Lund, H. (1964). *Acta Chem. Scand.* **18**, 563.
585. Scheiner, P., Chapman, O. L., and Lassila, J. D. (1969). *J. Org. Chem.* **34**, 813.
586. Mecke, R., Noak, K. (1958). *Spectrochim. Acta,* **12**, 391.
587. Singh, B. (1968). *J. Amer. Chem. Soc.* **90**, 3893.
588. Banks, R. E., Haszeldine, R. N., and Matthews, V. (1967). *J. Chem. Soc. (C),* 2263.
589. Banks, R. E., Barbour, A. K., Tipping, A. E., Gething, B., Patnick, C. E., and Tatlow, J. C. (1959). *Nature,* **183**, 586.
590. Giammanco, L. (1966-67). *Atti. Accad. Sci. Lett. Arti Palermo,* Parte I, **27**, 469; (1969). *Chem. Abstr.* **71**, 49884.
591. Lacey, R. N. (1954). *J. Chem. Soc.* 845.
592. See Armarego, W. L. F. (1967). In "Quinazolines", part I, Chapter 4 of "Fused Pyrimidines" (Ed. D. J. Brown) Interscience Publishers, New York.
593. Parmar, S. S., and Arora, R. C. (1967). *J. Med. Chem.* **10**, 1182; (1966). *Can J. Chem.* **44**, 2100.
594. Zentmyer, D. T., and Wagner, E. C. (1949). *J. Org. Chem.* **14**, 967.
595. Baker, B. R., Querry, M. V., Kadish, A. F., and Williams, J. H. (1952). *J. Org. Chem.* **17**, 35.
596. Bogert, M. T., Gortner, R. A., and Amend, C. G. (1911). *J. Amer. Chem. Soc.* **33**, 949.
597. Bogert, M. T., and Cook (1906). *J. Amer. Chem. Soc.* **28**, 1449; Bogert, M. T., and May, C. E. (1909). *J. Amer. Chem. Soc.* **31**, 512.
598. Reid, W., and Valentin, J. (1968). *Chem. Ber.* **101**, 2106.
599. Staiger, R. P., and Miller, E. B. (1959). *J. Org. Chem.* **24**, 1214.
600. Kappe, Th., Steiger, W., and Ziegler, E. (1967). *Monatsh. Chem.* **98**, 214.
601. Staiger, R. P., Moyer, C. L., and Pitcher, G. R. (1963). *J. Chem. Eng. Data,* **8**, 454.
602. Steiger, W., Kappe, Th., Ziegler, E. (1969). *Monatsh. Chem.* **100**, 146.
603. Ziegler, E., Steiger, W., Kappe, Th. (1969). *Monatsh. Chem.* **100**, 150.
604. Ziegler, E., Steiger, W., Kappe, Th. (1969). *Monatsh. Chem.* **100**, 948.
605. Späth, E., and Kuffner, F. (1938). *Chem. Ber.* **71**, 1657.
606. Chan, A. K. W., and Crow, W. D. (1969). *Aust. J. Chem.* **22**, 2497.
607. Wagner, E. C. (1940). *J. Org. Chem.* **5**, 133.
608. Sheibley, F. E. (1947). *J. Org. Chem.* **12**, 743.
609. Kappe, Th., Steiger, W., and Ziegler, E. (1967). *Monatsh. Chem.* **98**, 214.
610. Steiger, W., Kappe. Th., and Ziegler, E. (1969). *Monatsh. Chem.* **100**, 528.
611. Kurihara, M., and Yoda, N. (1966). *Bull. Chem. Soc. Jap.* **39**, 1942.
612. Warrener, R. N., and Cavin, E. N. (1966). *Tetrahedron Lett.* 3231.
613. Klopping, H. L., and Loux, H. M., E. I. du Pont de Nemours & Co. U.S. Pat. 3,352,662; (1968). *Chem. Abstr.* **68**, 114610.
614. Behringer, H., and Fischer, H. J. (1961). *Chem. Ber.* **94**, 1572.
615. Behringer, H., and Grimme, W. (1959). *Chem. Ber.* **92**, 2967.
616. Awad, W. I., Fahmy, A. F. M., and Sammour, A. M. A. (1965). *J. Org. Chem.* **30**, 2222.
616a. Lednicer, D., and Emmert, D. E. (1970). *J. Heterocycl. Chem.* **7**, 575.
617. Schmidt, R. R. (1965). *Chem. Ber.* **98**, 334.
618. Schmidt, R. R., and Schwille, D. (1969). *Chem. Ber.* **102**, 269.
619. Schmidt, R. R. (1968). *Tetrahedron Lett.* 3443.

620. Hünig, S., and Hübner, K. (1962). *Chem. Ber.* **95**, 937.
621. Schmidt, R. R., Schwille, D., and Sommer, U. (1969). *Annalen*, **723**, 111.
622. Virtanen, A. I., and Hietala, P. K. (1960). *Acta Chem. Scand.* **14**, 499; Hietala, P. K., and Virtanen, A. I. (1960). *Acta Chem. Scand.* **14**, 502.
623. Honkanen, E., and Virtanen, A. I. (1960). *Acta Chem. Scand.* **14**, 1214.
624. Honkanen, E., and Virtanen, A. I. (1961). *Acta Chem. Scand.* **15**, 221.
624a. Dickoré, K., Sasse, K., and Bode, K. D. (1970). *Annalen*, **733**, 70; Gauss, W., and Heitzer, H. (1970). *Annalen*, **733**, 59.
625. Butenandt, A., Biekert, E., and Neubert, G. (1957). *Annalen*, **603**, 200.
626. Rocek, J. (1953). *Chem. Listy*, **47**, 1781; Wichterle, O., and Rocek, J. (1953). *Chem. Listy*, **47**, 1768; (1954). *Coll. Czech. Chem. Commun.* **19**, 282.
627. Rocek, J. (1954). *Coll. Czech. Chem. Commun.* **19**, 275.
628. Loev, B., and Snader, K. M. (1967). *J. Heterocycl. Chem.* **4**, 403.
629. Warrener, R. N., and Cain, E. N. (1966). *Tetrahedron Lett.* 3225.
630. Gaudin, O. P. (1966). French Pat. 1,451,163; (1967). *Chem. Abstr.* **66**, 115731.
630a. Wagner, G., and Richter, P. (1969). *Pharmazie*, **24**, 193.
631. Prosen, A., Stanovnik, B., and Tisler, M. (1964). *J. Org. Chem.* **29**, 1623.
631a. Ignatova, L. A., Ovechkin, P. L., and Unkovskii, B. V. (1970). *Zh. Vses. Khim. Obshchest.* **15**, 238.
631b. Morin, R. B., and Spry, D. O. (1970). *J. Chem. Soc. (D)*, 335.
631c. Bartlett, R. G., and McClelland, E. W. (1934). *J. Chem. Soc.* 818.
632. Bourgoin-Lagay, D., and Boudet, R. (1967). *Compt. Rend.* **264**, 1304.
633. Goske, A. (1887). *Chem. Ber.* **20**, 232; Charpentier, P. (1947). *Compt. Rend.* **225**, 306.
634. Ris, Ch. (1886). *Chem. Ber.* **19**, 2240.
635. Kym, O. (1890). *Chem. Ber.* **23**, 2458.
636. Gilman, H., Dietrich, J. J. (1958). *J. Amer. Chem. Soc.* **80**, 380.
637. Jackson, T. G., Morris, S. R., and Martin, B. W. (1969). *J. Chem. Soc. (C)*, 1728.
638. Pagani, G., and Maiorana, S. (1967). *Chim. Ind. (Milan)*, **49**, 1194; (1968). *Chem. Abstr.* **68**, 105130.
639. Carrelli, V., Marchini, P., Moracci, F. M., and Liso, G. (1967). *Tetrahedron Lett.* 3421.
639a. Rabilloud, G., Sillion, B., and De Gaudemaris, G. (1970). *C. R. Acad. Sci., Ser. C.* **270**, 2019.
640. Rogers, M. A. Th., and Sexton, W. A. (1947). *J. Chem. Soc.* 1619.
641. Wilhelm, M., and Schmidt, P. (1969). *J. Heterocycl. Chem.* **6**, 635.
642. Sica, D. (1968). *Corsi, Sem. Chim.* **11**, 104; (1970). *Chem. Abstr.* **72**, 31725.
643. Gambaryan, N. P., and Zeifman, Yu. V. (1969). *Izv. Akad. Nauk SSSR, Ser. Khim.* 2059.
643a. Zahn, K. (1923). *Chem. Ber.* **56**, 578.
644. Martin, D., and Weise, A. (1967). *Chem. Ber.* **100**, 3736.
645. Shibuya, J., and Karabayashi, M. (1969). *Bull. Chem. Soc. Jap.* **42**, 2382; (1969). *Chem. Abstr.* **71**, 91422.
646. Kornowski, H., and Delage, B. (1966). *Bull. Soc. Chim. Fr.* 3282.
647. Beyer, H., and Wolter, G. (1956). *Chem. Ber.* **89**, 1652; Beyer, H., Wolter, G., and Lemke, H. (1956). *Chem. Ber.* **89**, 2550.

648. Sandström, J. (1960). *Ark. Kemi*, **15**, 195.
649. Sandström, J. (1955). *Ark. Kemi*, **9**, 127.
650. Beyer, H. (1969). *Z. Chem.* **9**, 361.
651. Beyer, H., and Bulka, E. (1962). *Z. Chem.* **2**, 321; Bulka, E. (1959). *Angew. Chem.* **71**, 576.
652. Bose, P. K. (1925). *J. Indian. Chem. Soc.* **2**, 95; (1926). *Chem. Abstr.* **20**, 415.
653. Beyer, H., Bulka, E., and Beckhaus, F. W. (1959). *Chem. Ber.* **92**, 2593.
654. King, J. F., Hawson, A., Deaken, D. M., and Komery, J. (1969). *Chem. Commun.* 33.
655. Donia, R. A., Shotton, J. A., Bentz, L. O., and Smith Jr., G. E. P. (1948). *J. Org. Chem.* **14**, 946.
656. Donia, R. A., Shotton, J. A., Bentz, L. O., and Smith Jr., G. E. P. (1948). *J. Org. Chem.* **14**, 952.
657. Rao, Y. R., and Konher, M. V. (1969). *Indian J. Chem.* **7**, 20.
658. Sasaki, T., and Minamoto, K. (1966). *J. Org. Chem.* **31**, 3917.
659. Heller, G. (1925). *J. Prakt. Chem.* **111**, 1.
660. Biltz, H. (1905). *Annalen*, **339**, 243.
660a. Lund, H. (1970). *Advan. Heterocycl. Chem.* **12**, 256.
660b. Kwee, S., and Lund, H. (1969). *Acta Chem. Scand.* **23**, 2711.
661. Temple Jr., C., Kussner, C. L., and Montgomery, J. A. (1969). *J. Org. Chem.* **34**, 2102.
662. Radziszewski, B. (1882). *Chem. Ber.* **15**, 1493.
663. Cousin, and Volmar, (1914). *Bull. Soc. Chim. Fr.* **15**, 416.
664. Cook, A. H., and Jones, D. G. (1941). *J. Chem. Soc.* 278.
665. Piepes-Poratynski, J. (1960). *Bull. Acad. Sci. Cracovie*, 1900; (1900). *Chem. Zentrallbl.* II, 477.
666. Grundmann, C., and Kreutzberger, A. (1955). *J. Amer. Chem. Soc.* **77**, 6559.
667. Grundmann, C., and Kreutzberger, A. (1957). *J. Amer. Chem. Soc.* **79**, 2839.
668. Kreutzberger, A. (1963). *Fortschr. Chem. Forsch. Bd.* **4**, 273.
669. Emmons, W. D. (1957). *J. Amer. Chem. Soc.* **79**, 5739.
670. See note 1, reference 671.
671. Neunhoeffer, H., and Frühauf, H. W. (1969). *Tetrahedron Lett.* 3151.
671a. Steigel, A., and Sauer, J. (1970). *Tetrahedron Lett.* 3357.
671b. Neunhoeffer, H., and Frühauf, H. W. (1970). *Tetrahedron Lett.* 3355.
672. Dittmar, W., Sauer, J., and Steigel, A. (1969). *Tetrahedron Lett.* 5171.
673. Cairns, T. L., Larchar, A. W., and McKusick, B. C. (1952). *J. Amer. Chem. Soc.* **74**, 5633.
674. Osborne, D. R., and Levine, R. (1963). *J. Org. Chem.* **28**, 2933.
675. Osborne, D. R., Wieder, W. T., and Levine, R. (1964). *J. Heterocycl. Chem.* **1**, 145.
676. Kreutzberger, A., and Grundmann, C. (1961). *J. Org. Chem.* **26**, 1121.
677. Kreutzberger, A., and Abel, D. (1968). *Arch. Pharm.* **301**, 881.
678. Kreutzberger, A., and Stevens, M. F. G. (1969). *J. Chem. Soc. (C)*, 1282.
679. Kreutzberger, A., and Abel, D. (1969). *Tetrahedron Lett.* 4711.
679a. Kreutzberger, A. A. D. (1970). *Arch. Pharm. (Weinheim)*, **303**, 715.
680. Brown, D. J. (1962). "The Pyrimidines". Interscience Publishers, New York and London.

681. Kreutzberger, A., and Abel, D. (1969). *Arch. Pharm.* **302**, 362.
682. Hey, D. H., Rees, C. W., and Todd, A. R. (1968). *J. Chem. Soc. (C)*, 1028; Hey, D. H., Rees, C. W., and Todd, A. R. (1962). *Chem. Ind. (London)*, **29**,1332.
683. Smalley, R. K., Suschitzky, H., and Tanner, E. M. (1966). *Tetrahedron Lett.* 3465.
684. Herlinger, H. (1964). *Angew. Chem.* **76**, 437; (1964). *Angew. Chem. Int. Ed.* **3**, 378.
685. Flowerday, P., and Perkins, M. J. (1969). *J. Amer. Chem. Soc.* **91**, 1035.
686. Burgess, E. M., and McCullagh, L. (1966). *J. Amer. Chem. Soc.* **88**, 1580.
687. Ege, G., and Beisiegel, E. (1968). *Angew. Chem. Int. Ed.* **7**, 303.
688. Kirby, A. J. (1966). *Tetrahedron*, **22**, 3001.
689. Walling, C., and Rabinowitz, R. (1959). *J. Amer. Chem. Soc.* **81**, 1243.
690. Harvey, R. G., Jacobson, H. I., and Jensen, E. V. (1963). *J. Amer. Chem. Soc.* **85**, 1618.
691. Weidinger, H., and Kranz, J. (1963). *Chem. Ber.* **96**, 2070.
692. Weidinger, H., and Sturm, H. J. (1968). *Annalen*, **716**, 143.
693. Ueda, S., Hayashi, Y., and Oda, R. (1969). *Tetrahedron Lett.* 4967.
694. Ullmann, F., and Gross, C. (1910). *Chem. Ber.* **43**, 2694.
694a. Spasov, A., and Chemishev, B. (1970). *Dokl. Bolg. Akad. Nauk*, **23**, 791.
695. Sandström, J. (1961). *Acta Chem. Scand.* **15**, 1575.
696. Dornow, A., and Pietsch, H. (1967). *Chem. Ber.* **100**, 2585.
696a Freeman, J. P., Surbey, D. L., and Kassner, J. E. (1970). *Tetrahedron Lett.* 3797.
697. Gehlen, H., and Elchlepp, H. (1955). *Annalen*, **594**, 14.
698. Bülow, C. (1906). *Chem. Ber.* **39**, 2618, 4106.
699. Westermann, P. (1964). *Chem. Ber.* **97**, 523.
700. Curtius, Th., Darapsky, A., and Müller, E. (1915). *Chem. Ber.* **48**, 1614.
701. Müller, E., and Herrdegen, L. (1921). *J. Prakt. Chem.* **102**, 113.
702. Dallacker, F. (1960). *Monatsh. Chem.* **91**, 294; (1960). *Chem. Abstr.* **54**, 24790.
703. Pinner, A. (1893). *Chem. Ber.* **26**, 2126; (1894). *Chem. Ber.* **27**, 984, 3273.
704. Stollé, R. (1907). *J. Prakt. Chem.* **75**, 416.
705. Baker, W., Ollis, W. D., and Poole, V. D. (1950). *J. Chem. Soc.* 3389.
706. Allegretti, J., Hancock, J., and Knutson, R. S. (1962). *J. Org. Chem.* **27**, 1463.
707. Jonas, A., and Pechmann, H. von (1891). *Annalen*, **262**, 277.
708. Pechmann, H. von (1888). *Chem. Ber.* **21**, 2751.
709. Stollé, R. (1926). *Chem. Ber.* **59**, 1742.
710. Stollé, R. (1908). *J. Prakt. Chem.* **78**, 544.
711. Stollé, R. (1903). *J. Prakt. Chem.* **68**, 469.
712. Stollé, R., Münch, E., and Kind, W. (1904). *J. Prakt. Chem.* (2), **70**, 433.
713. Carboni, R. A., and Lindsey Jr., R. V. (1959). *J. Amer. Chem. Soc.* **81**, 4342.
714. Avram, M., Dinulescu, I. G., Marica, E., and Nenitzescu, C. D. (1962). *Chem. Ber.* **95**, 2248.
715. Sauer, J., Mielert, A., Lang, D., and Peter, D. (1965). *Chem. Ber.* **98**, 1435.
716. Avram, M., Bedford, G. R., and Katrizky, A. R. (1963). *Rec. Trav. Chim. Pays-Bas*, **82**, 1053.

717. Sauer, J., and Lang, D. (1964). *Angew. Chem.* **76**, 603.
718. Roffey, P., and Verge, J. P. (1969). *J. Heterocycl. Chem.* **6**, 497.
719. Neumann, W. P., and Kleiner, F. G. (1968). *Annalen,* **716**, 29.
720. Battiste, M. A., and Barton, Th. J. (1967). *Tetrahedron Lett.* 1227.
721. Heinrichs, G., Krapf, H., Schroeder, B., Steigel, A., Troll, T., and Sauer, J. (1970). *Tetrahedron Lett.* 1617.

5
Ring Transformations of
Seven-membered Heterocycles

I. RING TRANSFORMATIONS OF SEVEN-MEMBERED COMPOUNDS CONTAINING ONE HETEROATOM

An important property of unsaturated seven-membered ring systems is their tendency to transannular bridging leading to the formation of five- and/or six-membered compounds. The driving force, which determines the course of these rearrrangements is the fact that more stable compounds are formed. This tendency is especially strong in conjugated 8 electron systems, because stable aromatic 6π-electron systems are formed. These readily occurring rearrangements often lead to difficulties in the preparation of seven-membered compounds, since in many cases five- or six-membered heterocycles are found to be obtained as main products or as important by-products. Although they are beyond the scope of this book, references can be made, for example, to the ring contractions of 1-ethoxycarbonyl-, 1-cyano- and 1-methylazepines into derivatives of benzene[1-3] and cyclopentadiene[4].

A. Ring transformations of oxepines and thiepines and their hydro derivatives

1. RING CONTRACTIONS INTO FIVE- AND SIX-MEMBERED OXA- AND THIAHETEROCYCLES

The interesting fact that 2-isobutylidene-3-oxo-4,4-dimethyltetrahydrothiophene (5.4) is formed by ring contraction when 4-hydroxy-5-oxothiepanes (5.1) are treated with triphenylphosphine bromide has recently been reported[5,6]. The formation of the product (5.4) takes place via a transannular route involving the sulphur atom yielding (5.2); a subsequent carbon-sulphur bond fission yields the carbonium ion (5.3), which rearranges into (5.4). A similar type of ring contraction, involving an abnormal pinacol rearrangement has been reported to occur with 3,3,6,6-tetramethylthiepane-4,5-diol (5.5) in an acidic medium, the

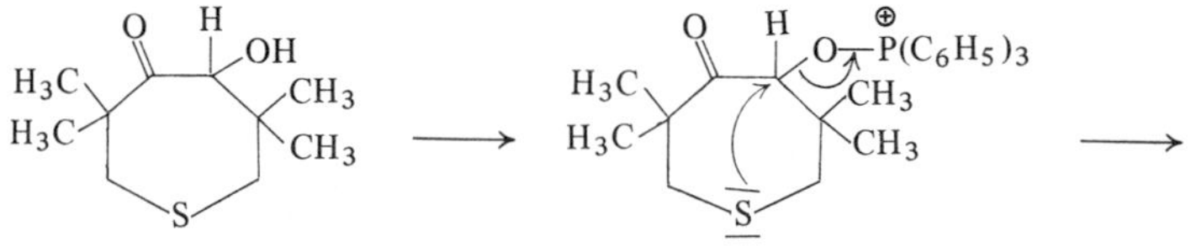

(5.1)

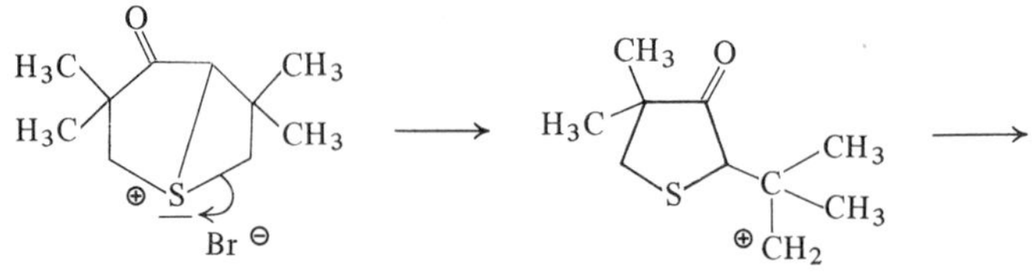

(5.2)   (5.3)

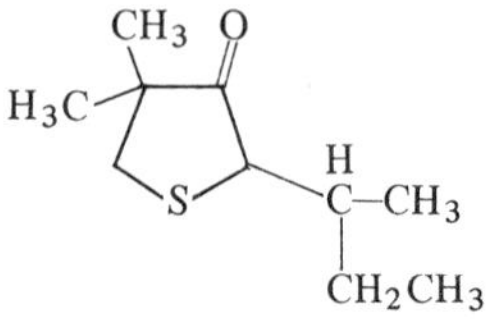

(5.4)

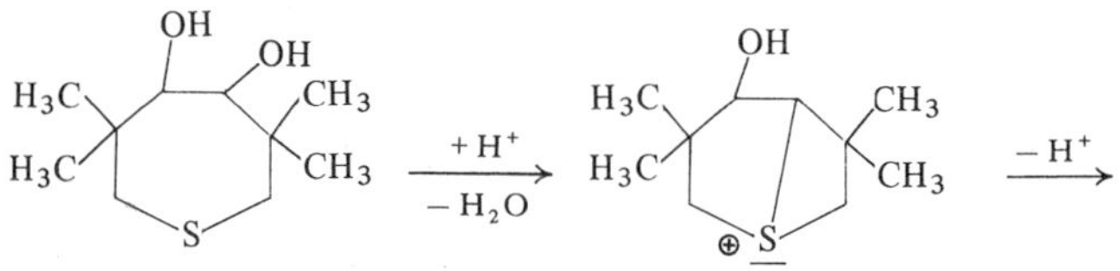

(5.5)

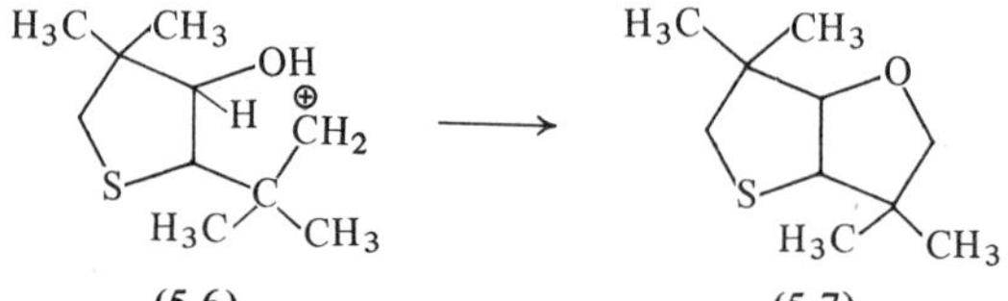

(5.6)   (5.7)

bicyclic compound (5.7) being obtained[6]. The intermediary carbonium ion (5.6) is stabilized into (5.7) through splitting off a proton.

An oxidative ring contraction of the seven-membered thiepine ring into a thiopyran ring has been reported to occur during treatment of dibenzo[b,f]-thiepine (5.8) (X = S) with an aqueous solution of potassium permanganate resulting in the formation of the 9-oxo-9H-thioxanthene (5.11) (X = S)[7]. Although a detailed mechanism has not been established, it seems reasonable to suppose that a benzilic acid rearrangement of the diketone (5.10), formed by dehydrogenation of the 1,2-diol (5.9), is involved in this ring contraction. The

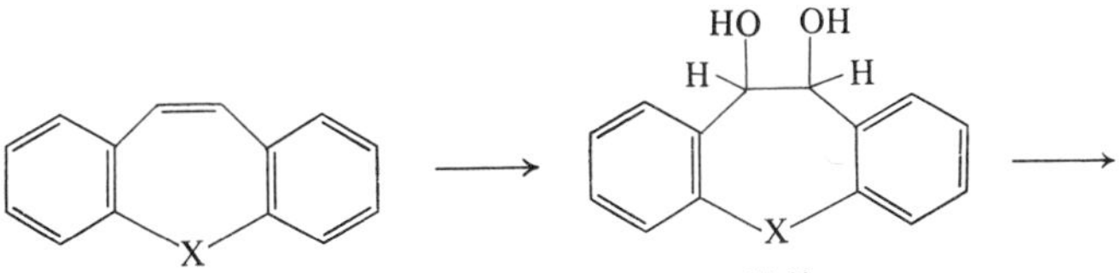

HO OH
H   H

(5.8)   (5.9)   →

O   O

HO   CO$_2$H

O

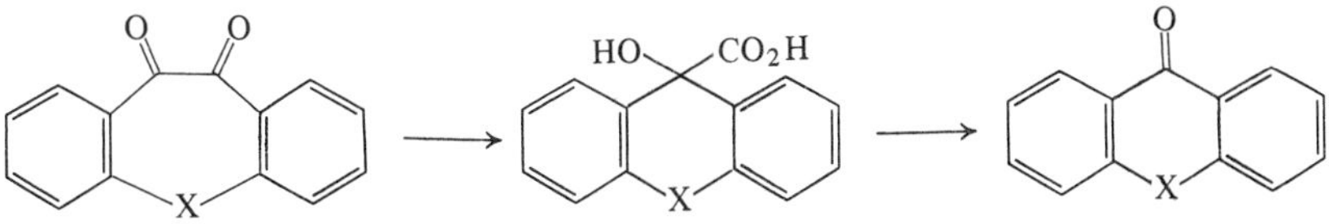

(5.10)   (5.11)

X = O, S

same ring contraction has also been reported for the benzo[b,f]oxepine (5.8) (X = O)[8]. The yield of xanthone (5.11) (X = O) is, however, low (6 per cent). In marked contrast, treatment of (5.8) (X = S) with a solution of permanganate in acetone does not yield a ring contracted product; only the thiepine ring is opened with formation of di-(o-carboxyphenyl)sulphide.

The conversion of the thiepine ring into a six-membered sulphur-containing nucleus has also been found to occur when the dibenzo[b,f]thiepine-10,11-dione monooxime (5.12) is hydrolysed with acid in aqueous ethanol[9]. Instead of the expected dione (5.10) (X = S) 9-hydroxy-9-ethoxycarbonyl-9H-thioxanthene (5.13) is obtained in a yield of more than 70 per cent. The

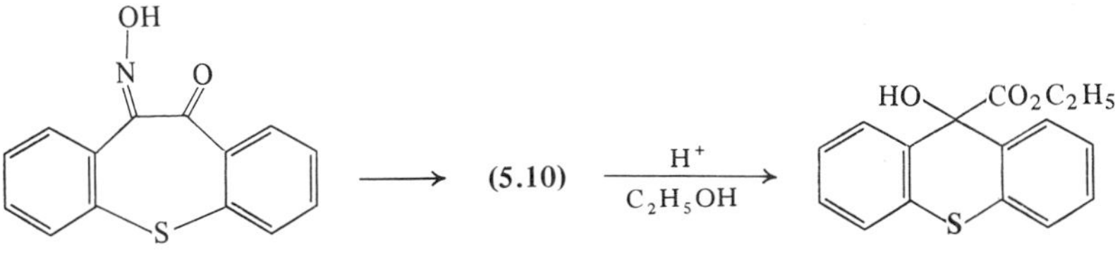

OH

N   O

HO   CO$_2$C$_2$H$_5$

(5.12)   →   (5.10) $\xrightarrow[C_2H_5OH]{H^+}$ (5.13)

diketone (5.10) (X = S), which is formed by an independent route, is also converted into (5.13) under similar conditions[9]. The fact that the S-dioxide of (5.10) (X = SO$_2$) cannot give this ring contraction, is considered as evidence for the participation of the free electron pair of the sulphur atom in (5.12) in this reaction.

B. Ring transformations of azepines and their hydro derivatives

1. RING CONTRACTIONS INTO PYRROLES AND FURANS

A dissociation-type photochemical ring contraction reaction has been found to take place during uv irradiation of the N-methylazepine (5.14) in cyclohexane[10]. In this reaction the pyrrole (5.16) is obtained, probably via the less stable intermediary bicyclic compound (5.15) formed by a transannular interaction at C_4-C_7.

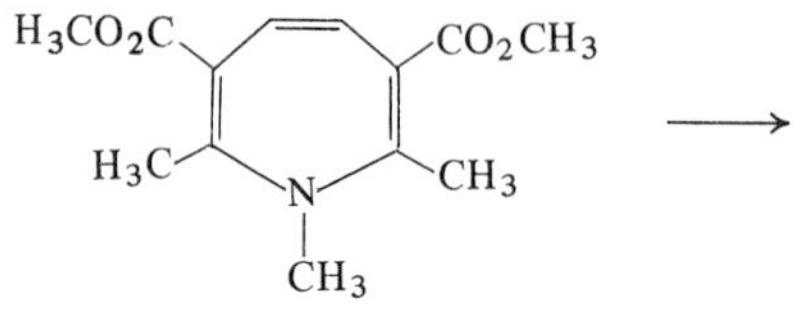

(5.14)

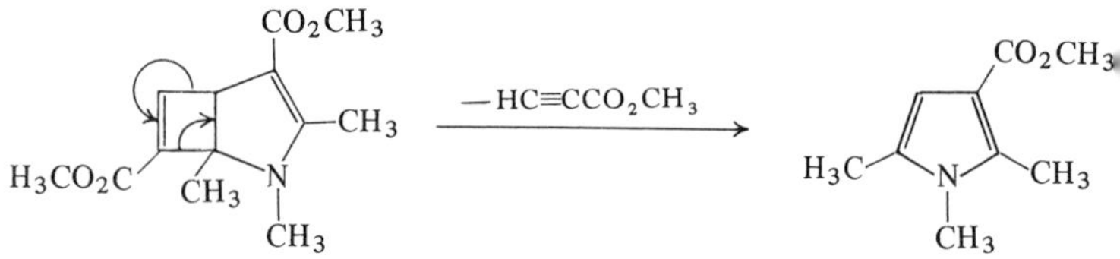

$-HC{\equiv}CCO_2CH_3$

(5.15)   (5.16)

A quite different type of ring contraction occurs in the base-catalysed conversion of the 4H-azepine (5.17) into the N-substituted pyrrole (5.19)[10]. In this reaction the transannulation takes place between nitrogen and C_5, yielding the 1-azabicyclo [3.2.0] heptene (5.18) as intermediate, and not as in the conversion (5.14) → (5.16) at C_4-C_7. In striking contrast it was found that

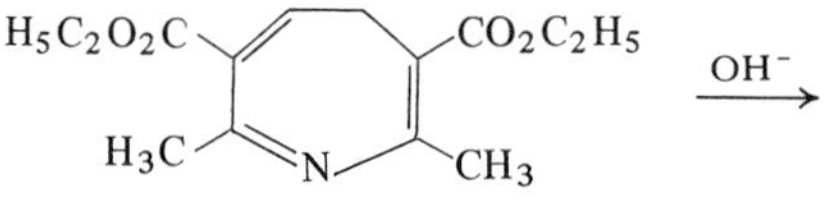

$\xrightarrow{OH^-}$

(5.17)

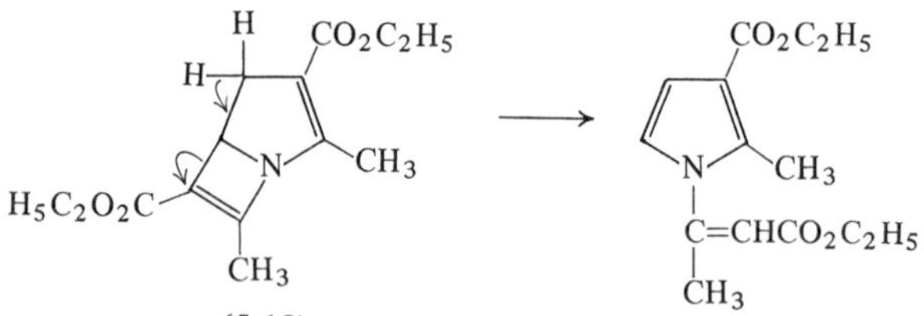

(5.18)   (5.19)

treatment of (5.17) with hydrobromic acid or hydrochloric acid leads to ring contraction into the 1,4-dihydropyridine (5.21) (X = Br or Cl). This conversion can also be performed with bromine in tetrachloromethane, but simultaneous nuclear substitution in the methyl group takes place the pentabromo compound (5.20) being formed[11].

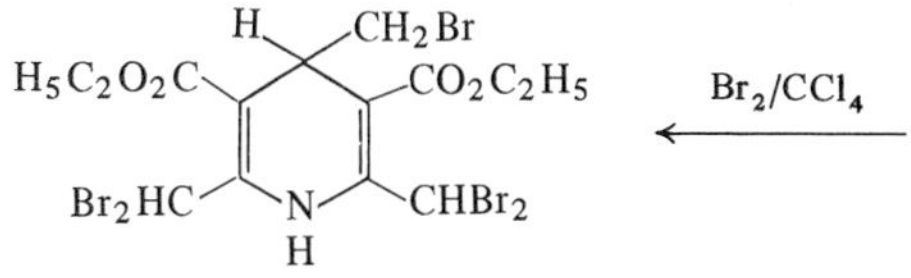

(5.20)

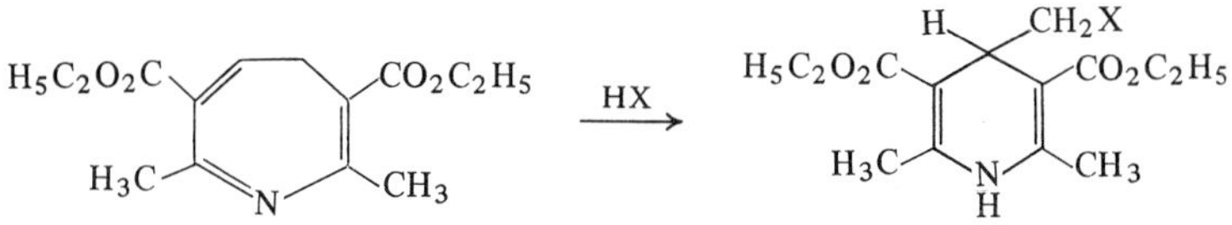

(5.17)   (5.21)

It has recently been reported that tetrahydrofuran derivatives can also be obtained from appropriately substituted azepines. When the 3H-azepines (5.22) are treated with acid, the γ-lactone (5.24) is formed[12]. The protonated azepinone (5.23) is probably the primarily formed intermediate, since it has been proved that the corresponding keto compound is also converted under identical conditions into the furan derivative (5.24). The mechanism can be depicted as follows:

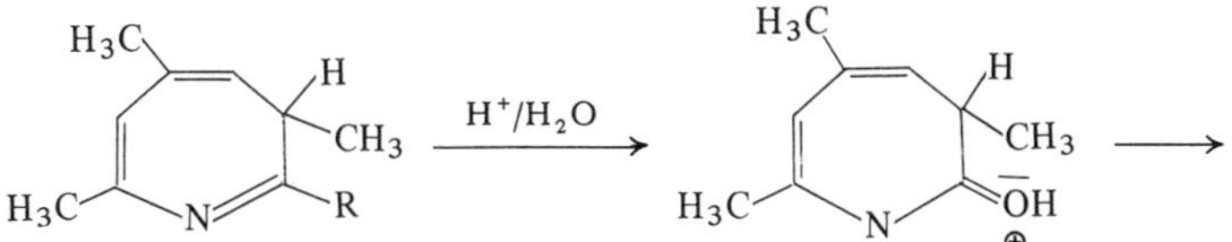

(5.22)

R = OC_2H_5 ; SC_2H_5

(5.23)

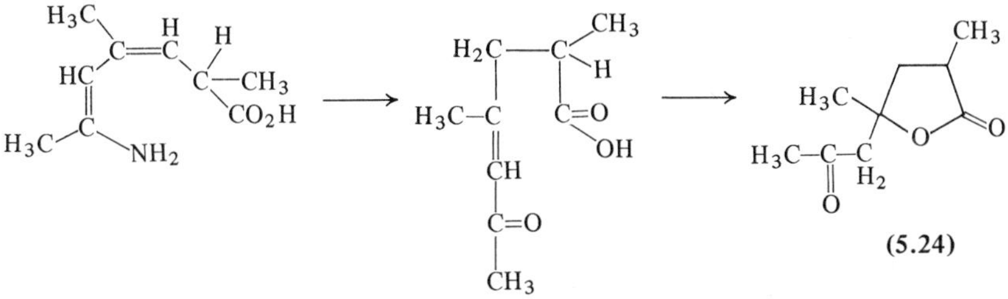

(5.24)

The behaviour of 4-cyano-4,5-dihydroazepine (5.25) towards acids and bases has been studied. This compound, which is easily prepared by means of a Wagner–Meerwein ring expansion rearrangement of 4-chloromethyl-1,4-dihydro-pyridine by the action of potassium cyanide (see chapter 4, section I.C.3.a) is found to be readily converted by base into the tetrasubstituted pyrrole (5.28) and ethylacrylate[13]. Both products are obtained by initial carbanion formation at C_4, giving the resonance stabilized (5.26), followed by transannulation at C_7, leading to the bicycle intermediate (5.27) which further decomposes by valence tautomerization into (5.28).

From nmr data and results obtained from ozonolysis and hydrogenolysis, it appears that the product obtained when (5.25) is treated with nitric acid is the furo[2,3-b]pyridine derivative (5.32)[14] and not, as originally claimed, a hydrolysis of (5.30) is apparently reduced (perhaps by complex formation), of silver nitrate, together, however, with another product, the pyrrolo[2,3-b]-pyridine (5.33). It is suggested that in the acidic medium the vinylamine part of (5.25) is easily protonated into the 4,5-dihydroazepinium ion (5.29), which, with ring opening, gives the acyclic vinylamine (5.30) that subsequently hydrolyses into (5.31). Double cyclization and oxidation results in the formation of (5.32). In the presence of silver nitrate, however, the rate of hydrolysis of (5.30) is apparently reduced (perhaps by complex formation), making possible a preferential ring-closure to the pyrrolopyridine (5.33).

Another interesting rearrangement[16] which also occurs in an acidic medium is the ring contraction of the 5-methoxycarbonyl-2,3,4,5-tetrahydro-1-benzazepinone derivative (5.34) into the oxindole (5.35); this compound can also be obtained from the oxime of 4-carboxytetralone-1 by treatment with polyphosphoric acid. It has been argued that the ring contraction of this lactam (5.34) is due to an acid-catalysed ester-amide exchange reaction rather than an

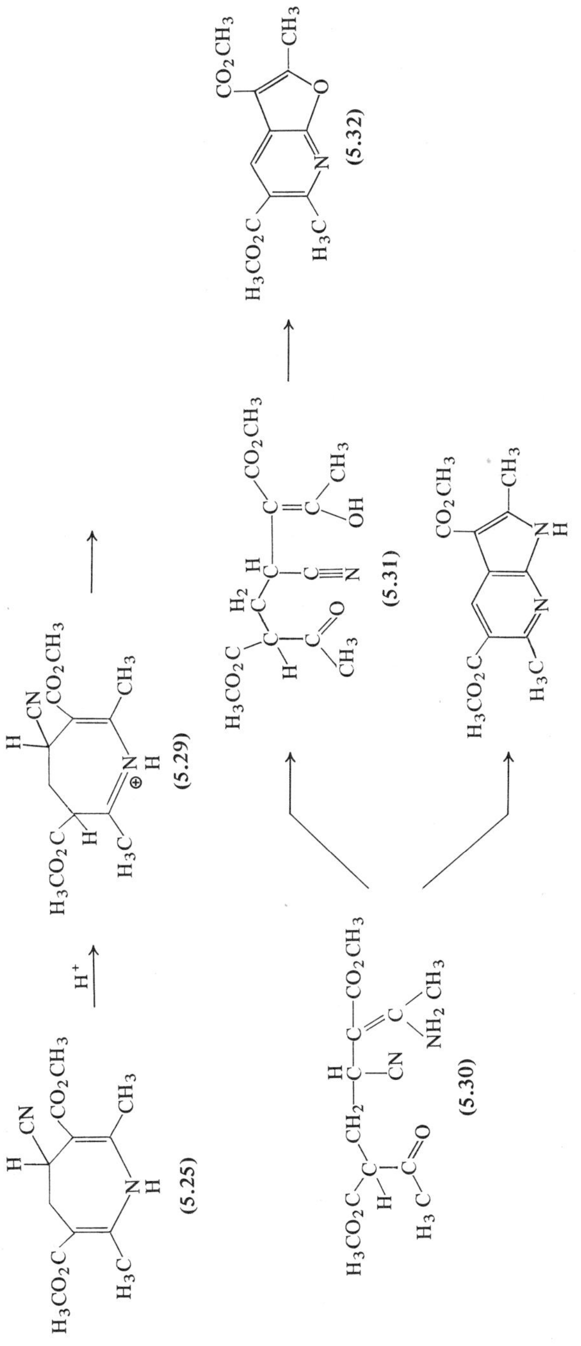

(5.25) (5.29) (5.30) (5.31) (5.32) (5.33)

initial hydrolysis to an aminodicarboxylic acid which further cyclizes into (5.35).

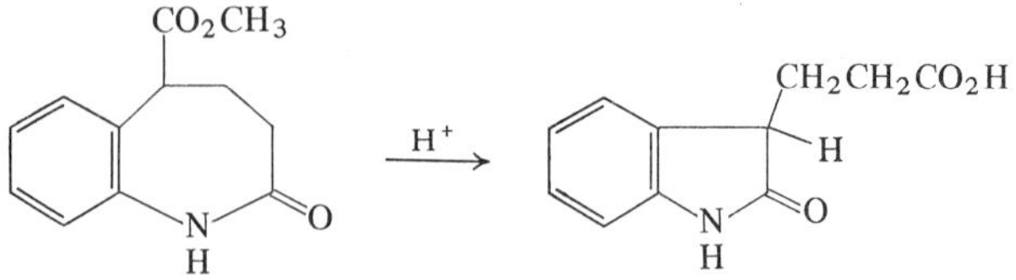

(5.34)   (5.35)

For the 1-benzazepine series, it has further been reported that the diketohydroxy compound (5.36) undergoes a rearrangement into the isatine (5.38) when acted upon by a base; the quinoline derivative (5.39) is obtained as a by-product[17]. The reaction involves ring cleavage of the seven-membered ring into the intermediary benzene derivative N-acetylisatic acid (5.37).

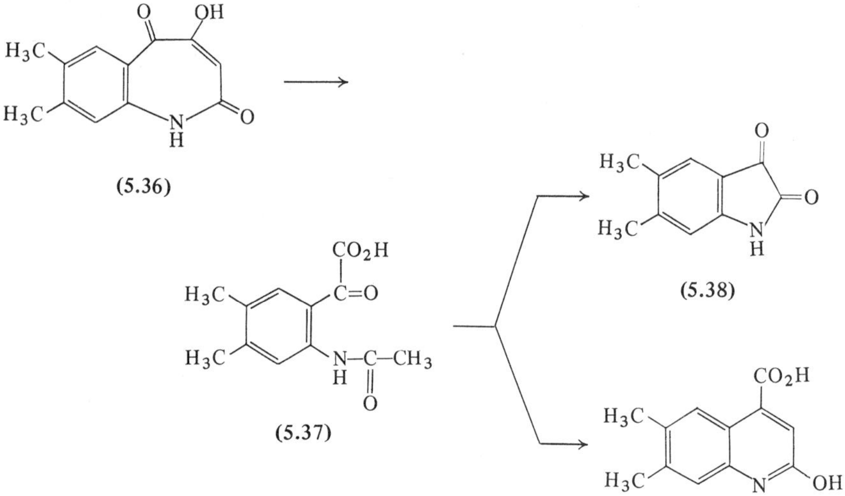

(5.36)

(5.37)

(5.38)

(5.39)

Attempts to reduce 2,7-diphenylhexahydro-4-azepinone (5.40) by the Wolff–Kishner method have failed; 2-phenyl-5-(β-phenethyl)pyrrole (5.41) was obtained[18] instead of the corresponding methylene derivative. An unusual transannular rearrangement involving the not previously reported C_α–N bond fission is necessary in order to obtain a β-phenylethyl group in position 2 of the pyrrole ring. Of the three mechanisms postulated the following seems to be the most probable:

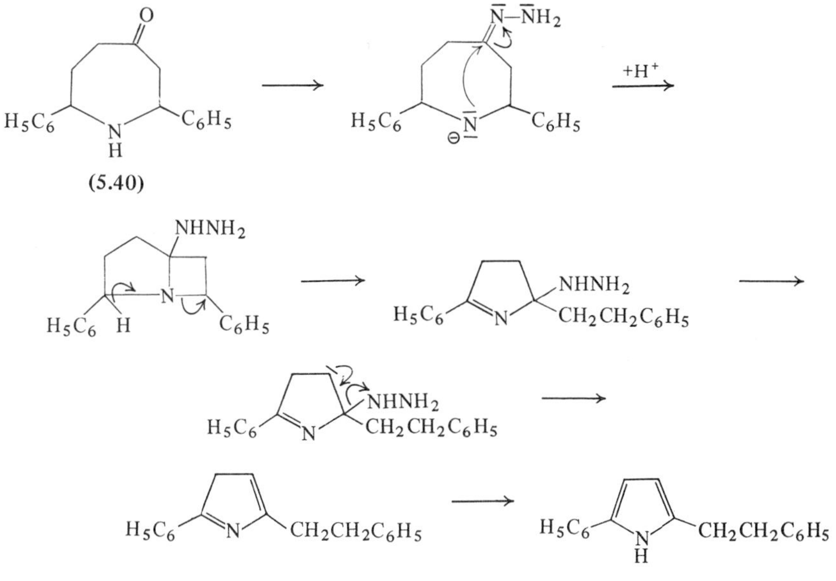

(5.40)

(5.41)

2. RING CONTRACTIONS INTO PIPERIDINES AND PYRIDINES

It has been found that the reductive ring contraction of cyclic α-aminoketones under Clemmensen conditions is not limited to the contraction of six-membered rings into five-membered rings (see chapter 4, section I.C.1.a), but also occurs with seven-membered α-aminoketones. When subjected to a reaction with amalgamated zinc and hydrochloric acid, N-methylhexahydroazepinon-3 (5.42) was found to be converted into the N-methyl-2-alkylpiperidine (5.43)[19].

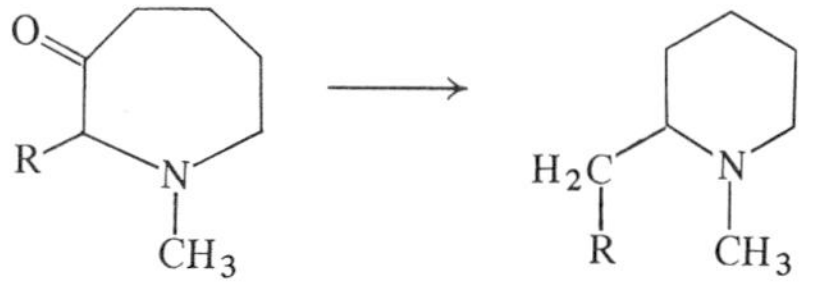

(5.42)   (5.43)

$$R = CH_3 ; C_2H_5$$

Bicyclic α-aminoketones which have a fused six- and seven-membered ring or a fused five- and seven-membered ring also show ring contraction under Clemmensen conditions[20,21]. It has been firmly established that in all these cases the ring containing the keto group contracts while the non-ketonic ring is enlarged. Hence the 3-keto-1-azabicyclo[5.4.0]undecane (5.44) gives a racemate of 4-methylquinolizidine (5.45)[20], and the ketonic 1-azabicyclo[5.3.0]decane

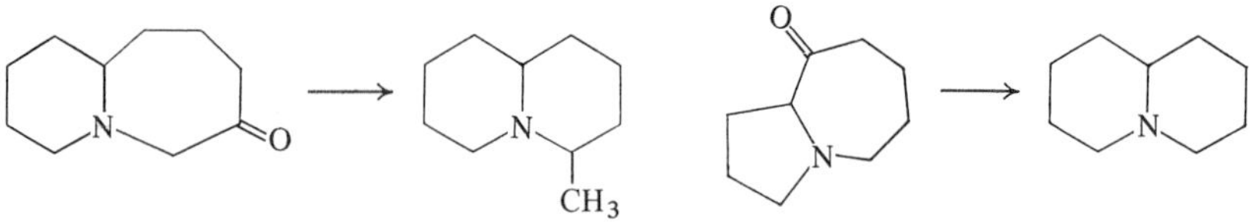

(5.44)   (5.45)   (5.46)   (5.47)

(5.46) gives the quinolizidine (5.47)[21]. The mechanism of this ring contraction is discussed in chapter 4, section I.C.1.a.

It has been found that the 2-chloro-1-benzazepine (5.48), when treated with acid, loses the arylamino group and is transformed into the 4-carboxy-2-chloro-quinoline (5.50)[23]; the course of the reaction is similar to that described for the base-catalysed conversion of benzazepinones into quinolines (see the conversion (5.36) → (5.39) discussed in section I.B.1). It has been argued[23] that this ring contraction probably occurs via the ion (5.49).

Photolytic rearrangements, involving the extrusion of a carbon atom in position 4, has been observed with the 3-benzoyl-2-methoxy-3H-azepine (5.51). Irradiation of (5.51) in methanol gives ring contraction into 7-benzoyl-2-methoxy-3-azabicyclo[2.1.0]hepta-2,4-diene (5.51a), 2-methoxy-3-phenacyl-pyridine (5.52) and 2-phenylfuro[2,3-b]pyridine (5.53)[22]. Analogous conversions have been reported with the corresponding 5-chloro- and 6-chloro derivatives of (5.51).

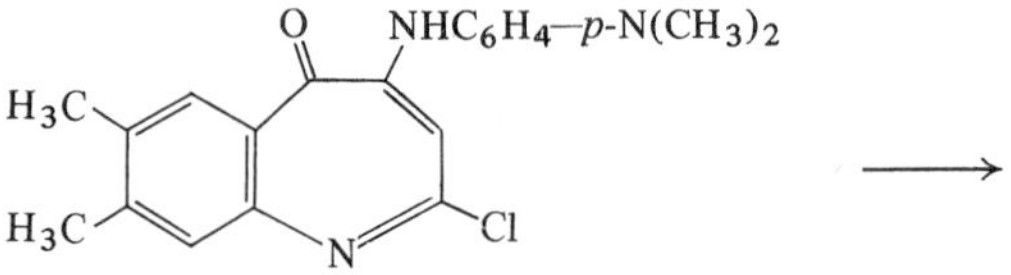

(5.48)

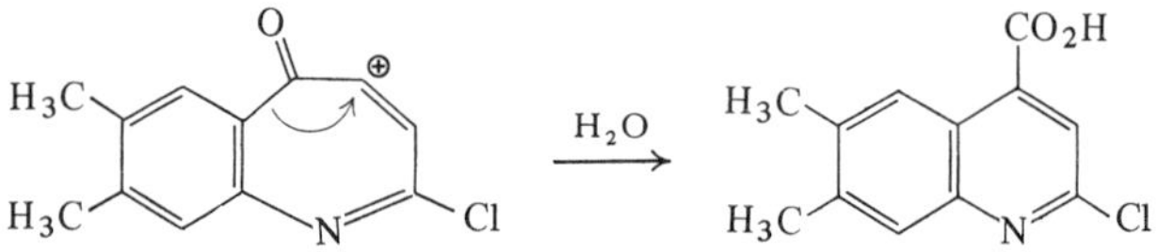

(5.49)   (5.50)

Pyrolysis gave similar pyridine derivatives. Thermal rearrangement of 2-amino-3-acyl-3H-azepines gave pyrrolo[2,3-b]pyridine derivatives[22].

Expulsion of a carbon atom at position 4 and/or 5 of the azepine ring has also been reported to occur when N-acetyl-dibenzo[b,f]azepine (5.54) is treated with hot hydrobromic acid, the 9-methylacridine (5.55) being obtained[24]. The mechanism is not clear; there is a similarity, although in the reverse sense, to the acid-catalysed ring expansion of 9-hydroxymethyl-9,10-dihydroacridine into a

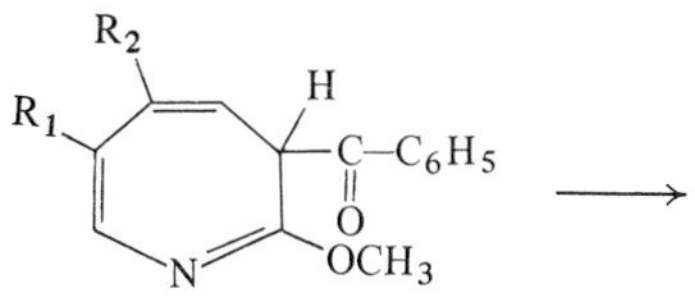

(5.51)

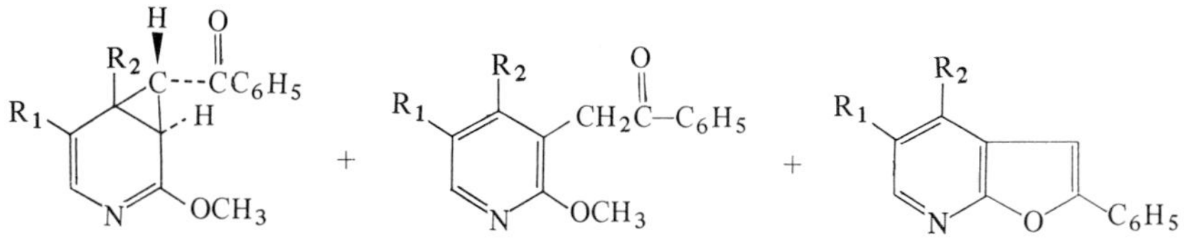

(5.51a)   (5.52)   (5.53)

R_1	H	Cl	H
R_2	H	H	Cl

dibenzoazepine discussed in chapter 4, section I.C.3.a. Treatment of the dibenzo[b,f]azepine (5.56) with potassium nitrosodisulphonate (Fremy's salt) also gives extrusion of a C_4 carbon atom since acridine-9-carboxaldehyde (5.57) is formed[25]. No mechanism is proposed.

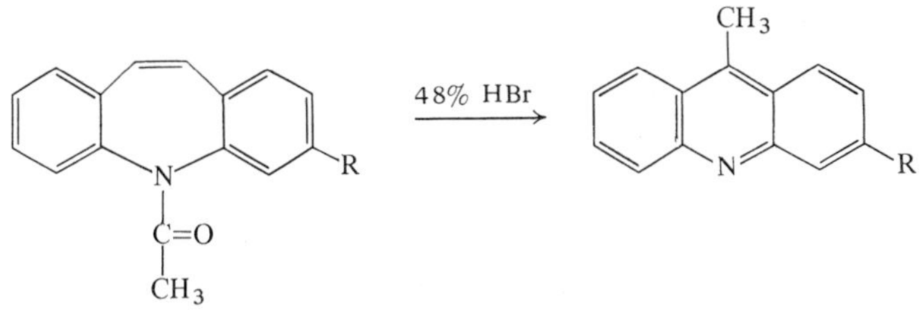

(5.54)   (5.55)   R = H; C_2H_5

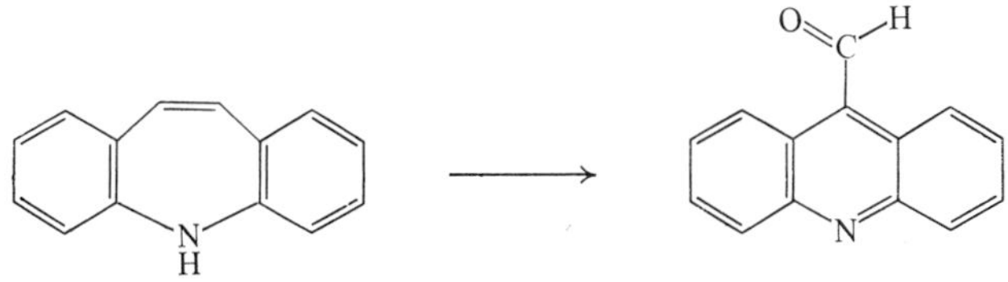

(5.56)   (5.57)

Loss of a carbon atom at position 2 takes place during thermolysis or during lithium aluminium hydride reduction of 3-oxo-dibenz[b,d]azepine (5.58), which leads to the formation of phenanthridone-2 (5.59)[26]. It has not proved possible to demonstrate that this interesting ring contraction is monomolecular and

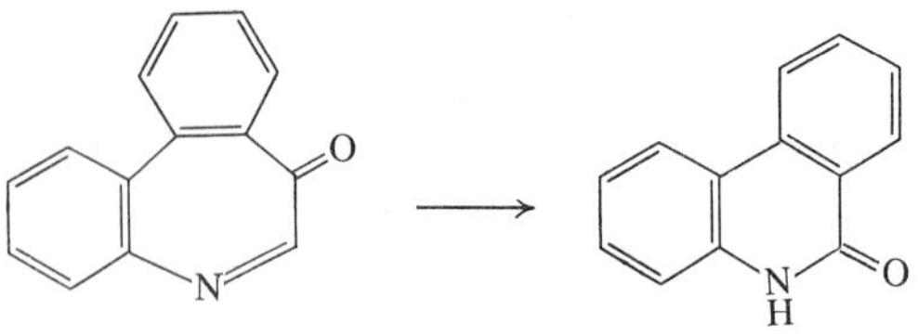

(5.58)   (5.59)

involves the extrusion of a single carbon atom. As the yields are always about 50 per cent, a disproportionation reaction cannot be excluded.

Acid-catalysed amide-ester exchange, which has been found to be responsible for the conversion of 5-methoxycarbonylbenzotetrahydroazepinone-2 (5.34) into the oxindole (5.35) also takes place during ring contraction of 4-ethoxycarbonylbenzotetrahydroazepinone-2 (5.60) into the dihydro-carbostyril derivative (5.61)[27]. Closely related ring contractions are the

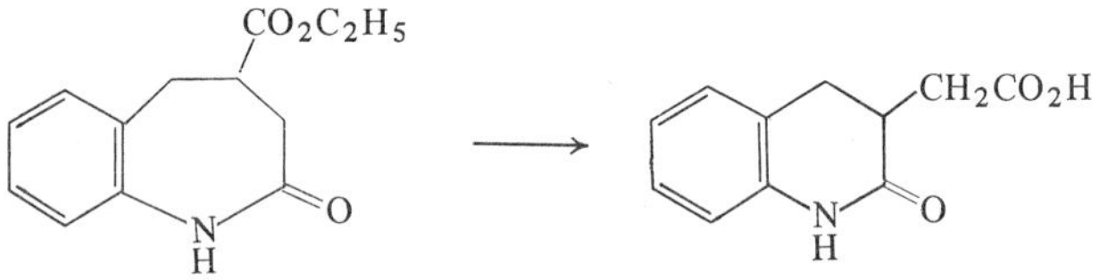

(5.60)   (5.61)

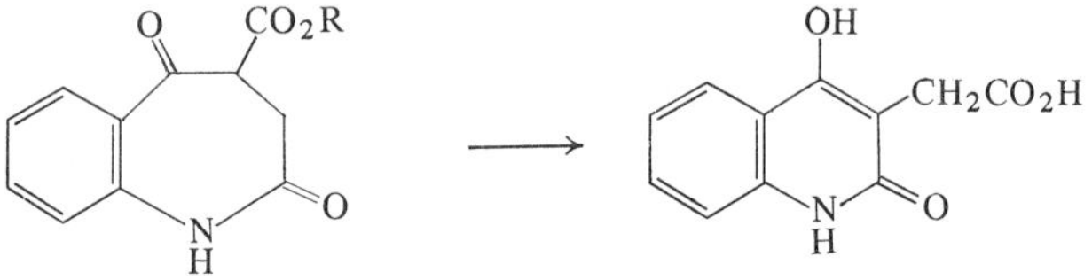

(5.62)   (5.63)

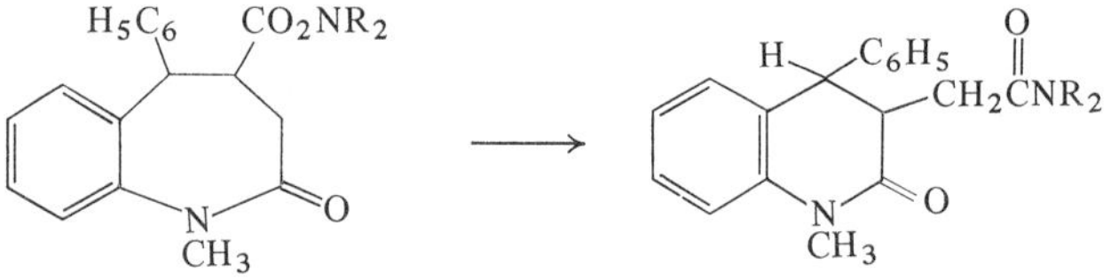

(5.64)   (5.65)

conversion of 2,5-diketo-4-ethoxycarbonyl-1-benz-2,3,4,5-tetrahydroazepine (5.62) into 2,4-dihydroxy (or its tautomeric ketoform) quinoline-3-acetic acid (5.63) by treatment with aqueous acid or base[28] and the thermal rearrangement of the tetrahydroazepine-4-carbonamides (5.64) into *cis* and *trans* derivatives of tetrahydroquinolines (5.65)[28a].

3. RING EXPANSIONS INTO OXAZOCINES

The 1-oxide of 1-methyl-2-phenylhexahydroazepine (5.66), on heating at 180°C, gives the deoxygenated compound (5.67) as major product, and in addition the unsaturated hydroxylamine (5.69) and 2-methyl-8-phenylhexahydro-oxazocine (5.68); the last product is formed by a Meisenheimer rearrangement of the nitrogen to oxygen[29]. This migration reaction is competitive with the attack of oxygen on the β-hydrogen atom; the seven-membered ring is apparently sufficiently flexible to make this attack possible. Similar ring expansions are also observed with the 2-arylpiperidine-*N*-oxides (see chapter 4, section I.C.3.b) and with pyrrolidine-*N*-oxides (see vol. 1, chapter 3, section I.C.5.c).

(5.66)

(5.67)   +   (5.68)   +   (5.69)

II. RING TRANSFORMATIONS OF SEVEN-MEMBERED HETEROCYCLES CONTAINING TWO HETEROATOMS

A. Ring transformations of seven-membered heterocycles containing two nitrogen atoms

1. RING CONTRACTIONS OF DIAZEPINES AND BENZODIAZEPINES AND THEIR HYRO DERIVATIVES INTO FIVE-MEMBERED HETEROCYCLES

a. *1,2-Diazepines and their dihydro derivatives into furans, pyrroles and pyrazoles*
Ring contraction has frequently been encountered in reactions with diazepines that have both nitrogen atoms present in positions 1 and 2. Transannular bridging is the process assumed to take place in many of these ring contractions. In a few cases these bicyclo intermediates have indeed been isolated. 3,4-Dihydro-2*H*-1,2-diazepinol-4 (5.70), when treated with cold 6*N* hydrochloric acid, gives the furfurylhydrazine (5.73)[30]; this reaction is thought to occur via the bridged bicyclic intermediate oxadiazabicyclo[3.2.1]octene (5.71), which, after protonation at N_1 (i.e. (5.72)) probably collapses in the way shown.

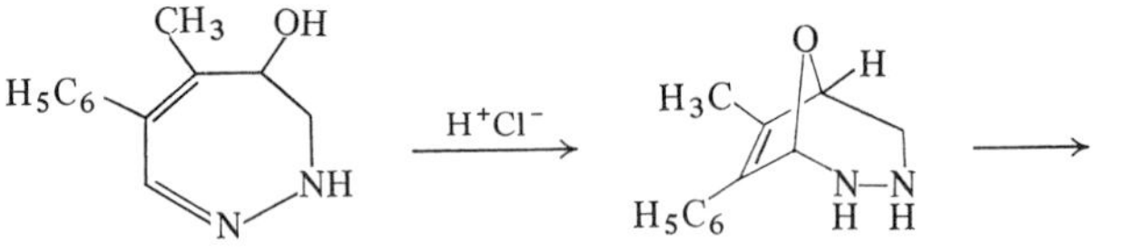

(5.70)   (5.71)

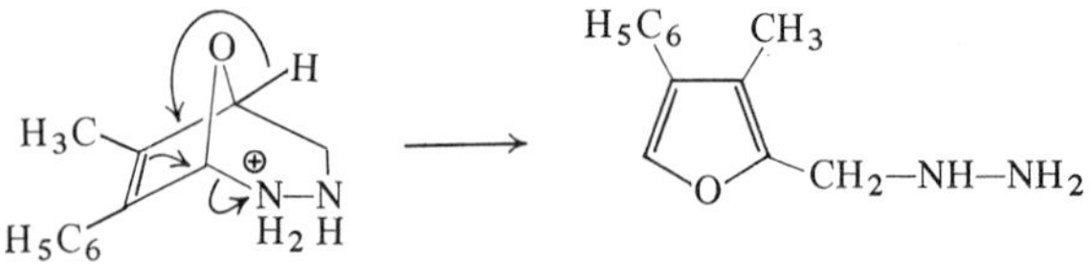

(5.72)   (5.73)

Support for the occurrence of a bicyclic compound as intermediate is provided by the fact that in the reaction of (5.70) with acetic anhydride the acetyl derivative of (5.71), i.e the bicyclo[3.2.1]octene (5.74), can be isolated and that this substance by further treatment with acid can be converted into the furfurylhydrazino compound (5.75)[30].

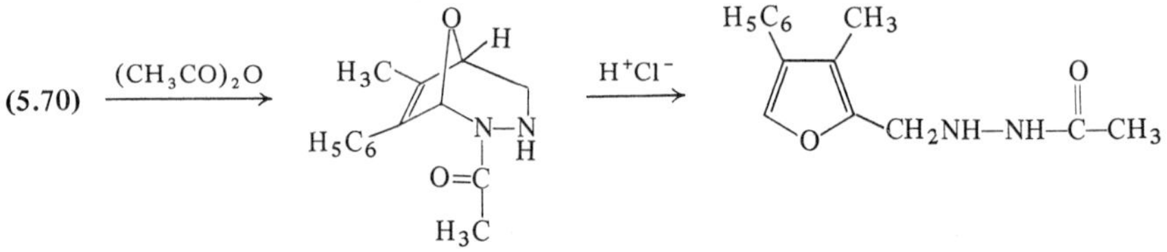

(5.74)   (5.75)

It is proposed that the bicyclic compound (5.77), which is similar in structure to (5.71), is formed as intermediate in the conversion of the 1,2-diazepinone-4 (5.76) into N-hydroxy-Δ^3-pyrrolin-2-one (5.78b) by treatment with hydroxylamine[31]. However, in contrast to (5.71), the bicyclic intermediate (5.77) does not contain a proton at C_4 which would result in a different type of fragmentation involving carbon-carbon bond fission leading to hydrazine, formaldehyde and (5.78a), which is in tautomeric equilibrium with (5.78b).

The product N-acetyl-7-methoxy-1,2-diazepinone-4 (5.81), obtained by the reaction of the 1,2-diazepinone-4 (5.76) and acetyl chloride in pyridine–methanol, when treated with base gives ring contraction into the pyrrole derivative (5.82)[32]. Although not strictly relevant to this section, it is worth mentioning in this connection that treatment with base of the 1,2-diazepinone-4 (5.76), leads to a mixture of both 2- and 6-aminopyridines (5.79) and (5.80)[32,33] (see section II.A.2.a).

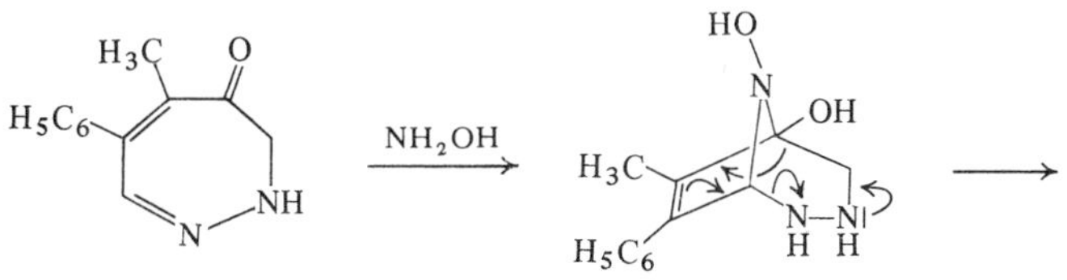

(5.76)

(5.77)

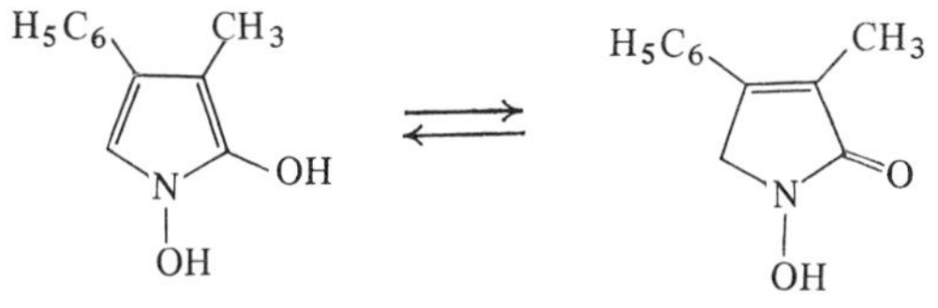

(5.78a)

(5.78b)

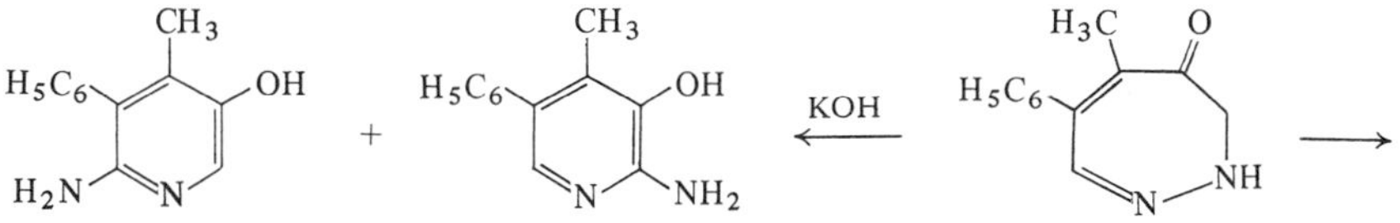

(5.79)

(5.80)

(5.76)

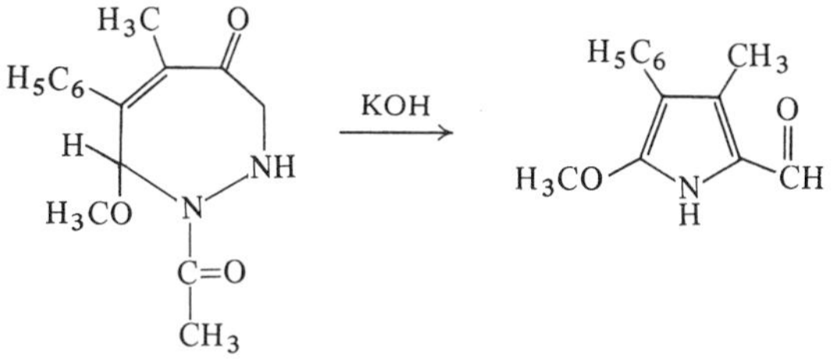

(5.81)

(5.82)

We must endeavour to find out why there is such a striking difference between the 1,2-diazepine (5.76) and its *N*-acetyl derivative (5.81) in their behaviour towards base. From the mechanism of the ring contraction it would seem that the acyclic precursor (5.83) is formed in basic media from (5.81); the presence of the acetyl group considerably decreases the nucleophilic character of the nitrogen originally present at position 1[32]. The geometry and the juxtaposition of the functional groups appear to be more suitable for the formation of the bicyclic compound (5.84), which further dehydrates into (5.85) and subsequently into the pyrrole (5.82), than for the initial formation of the six-membered compound (5.86).

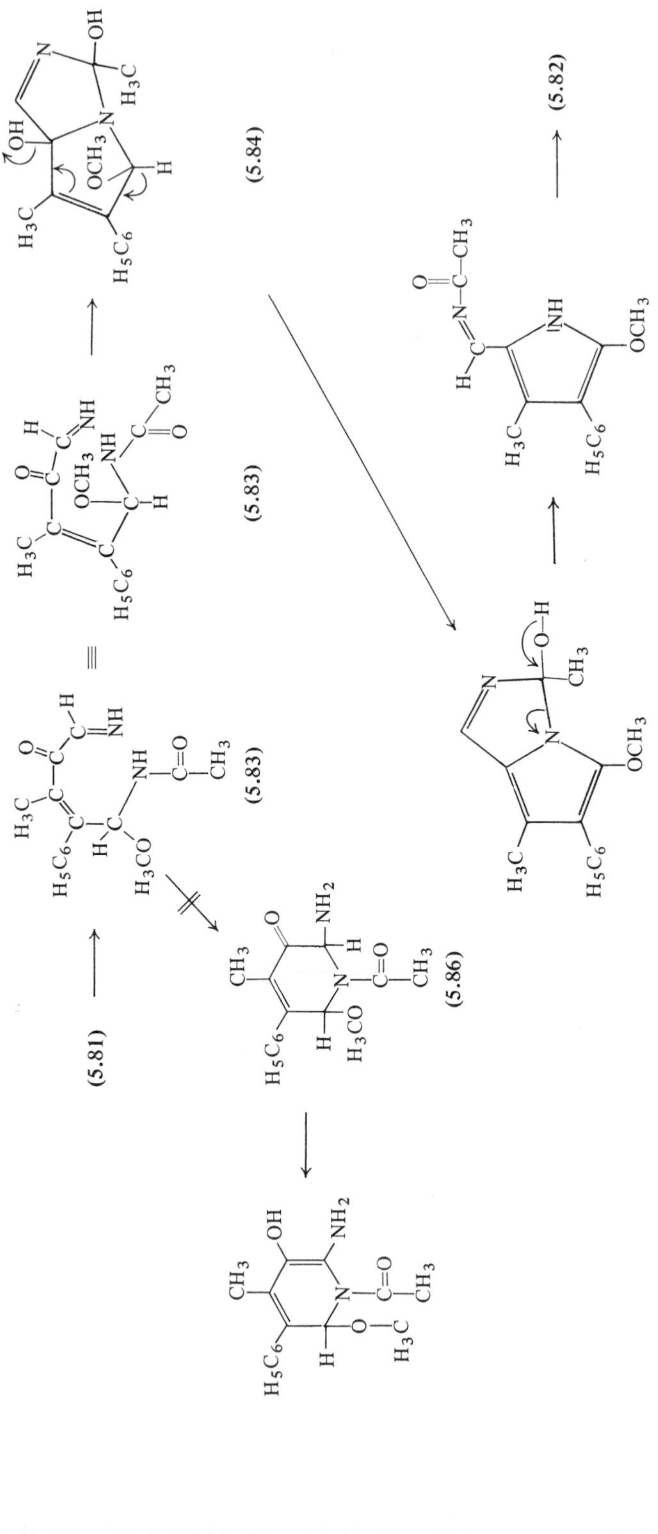

(5.81)
(5.83)
(5.83)
(5.84)
(5.85)
(5.82)
(5.86)

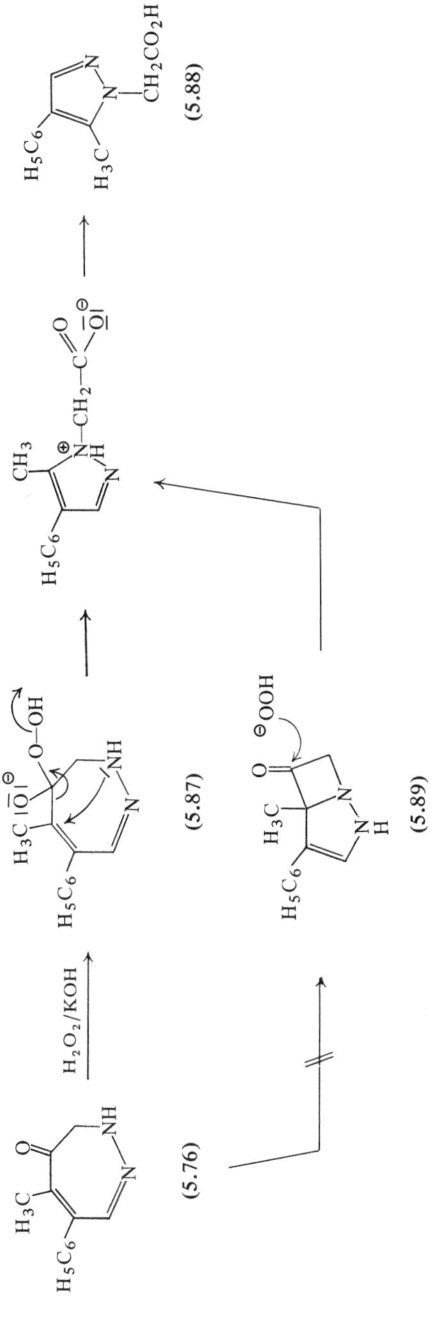

(5.76) (5.87) (5.88) (5.89)

The interesting properties of the 1,2-diazepinone-4 (5.76) are also manifested when the compound is oxidized with an alkaline solution of hydrogen peroxide; ring contraction is observed, 5-methyl-4-phenylpyrazole-1-acetic acid (5.88) being obtained[34]. It is probable that this reaction occurs by transannular bridging between N_2 and C_5 in the hydroperoxide (5.87), a participation reaction which has been observed in other reactions with this compound. An alternative pathway, the attack of the hydroperoxide anion on the bicyclic species (5.89) is also a possibility, but at present there is no direct evidence to suggest that this intermediate (5.89) plays an important role.

Ring contraction into the tetraphenylpyrrole (5.92) has been observed during thermolysis of the pentaphenyl-4H-1,2-diazepine (5.90) at 235-245°C. When a lower temperature is used the bicyclic diazabicyclo[3.2.0]heptadiene (5.91) is isolated[35,36]; this reaction is a characteristic example of a 1,4-transannular

(5.90)   (5.91)   (5.92)

(5.93)

(5.94)

R = H; CH₃; COCH₃; COC₆H₅

bridging. Some support for the existence of a bicyclic intermediate, characterized by a five-membered ring fused over the carbon-nitrogen bond with a four-membered ring, is provided by the fact that the intermediate like (5.93) is probably present when 7-benzyl-2,5-diphenyl-3,4,7-triaza-2,4-norcaradiene is heated briefly to produce ring contraction into 1-benzyl-4-phenylimidazole with loss of benzonitrile[37] and by the very recent finding[38] that on photolysis of the 1,2-diazepin-4-one (5.76) the bicyclic isomer 1,2-diazabicycloheptenone (5.94), which can revert thermally into the starting compound (5.76)[38,39], is isolated.

b. *Benzo-1,4- and -2,4-diazepines and their dihydro derivatives into indoles, isoindoles, isoindolines and indazoles*

Since the discovery of the biological activity of chlorodiazepoxide and diazepam as anti-anxiety agents (both compounds are derivatives of the 1,4-benzodiazepine ring system), a considerable amount of work has been devoted to the study of the syntheses and physical and chemical properties of compounds of this kind. In these studies a number of reactions were discovered in which 1,4-benzodiazepine derivatives are converted by ring contraction into five-membered and six-membered heterocycles.

The 1,4-benzodiazepin-2-one (5.95) is found to be converted into the isoindole carboxamide (5.99) by treatment with sodium hydride in non-aqueous media[40]. The reaction is initiated by a base-catalysed proton abstraction at C_3, and a subsequent conversion into the intermediates (5.96) and/or (5.98). Opening of the four-membered ring takes place by an anionic rearrangement of the tricyclic intermediate (5.96). The same rearrangement is exhibited by the 1,4-dimethyldiazepinium salt, yielding the corresponding N-methyl derivative of (5.99)[40], and by the thione (5.100) and tetrahydropyrimidobenzodiazepine (5.102), yielding (5.101) and (5.103) respectively[41]. The compound (5.95) (R = H), containing a hydrogen atom at N_1 instead of a methyl group, can be converted into an isoindole on treatment with acetic anhydride in pyridine[42].

A tricyclic intermediate (5.105) has been proposed in the conversion of 1-amino-7-nitro-5-phenyl-1,3-dihydro-2H-1,4-benzodiazepin-2-one (5.104) into 3-phenyl-6-nitroindazole (5.106)[43]. Further illustrations of the great variety of

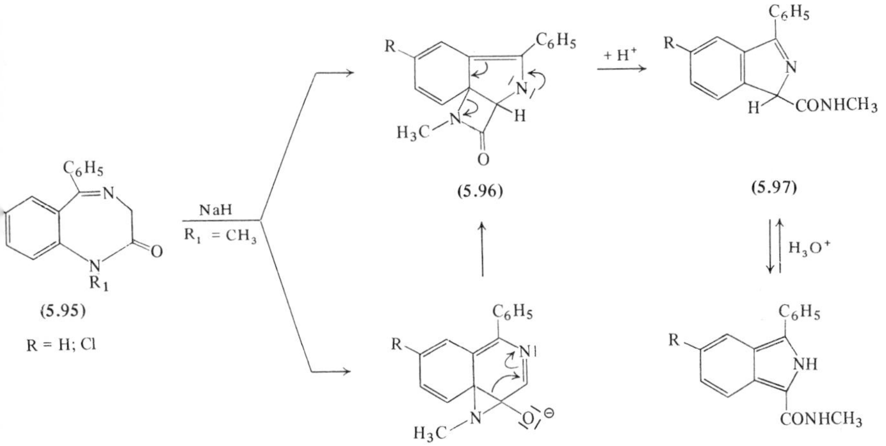

(5.96)   (5.97)

(5.95)

R = H; Cl

(5.98)   (5.99)

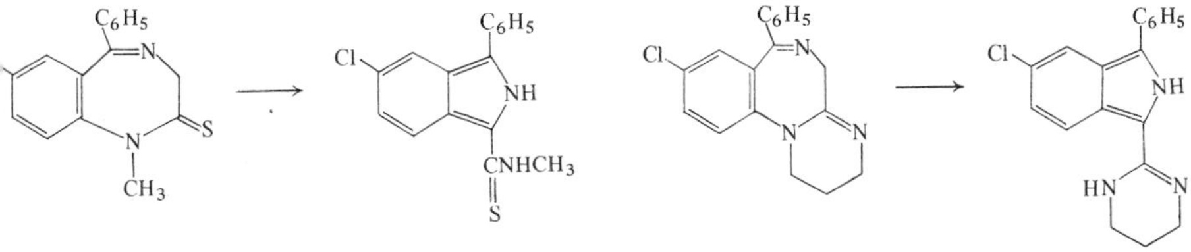

(5.100)   (5.101)   (5.102)   (5.103)

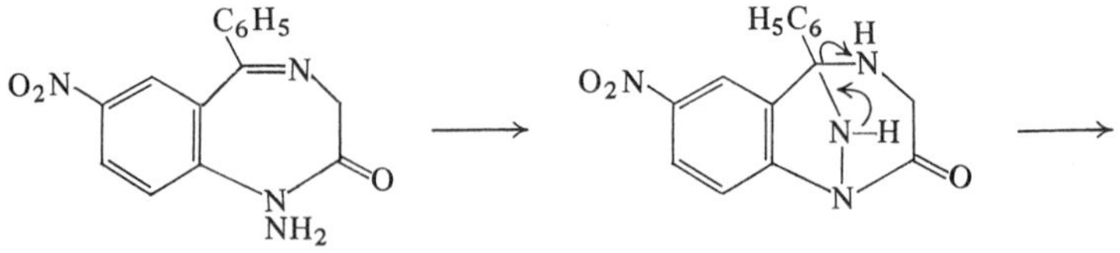

(5.104)   (5.105)

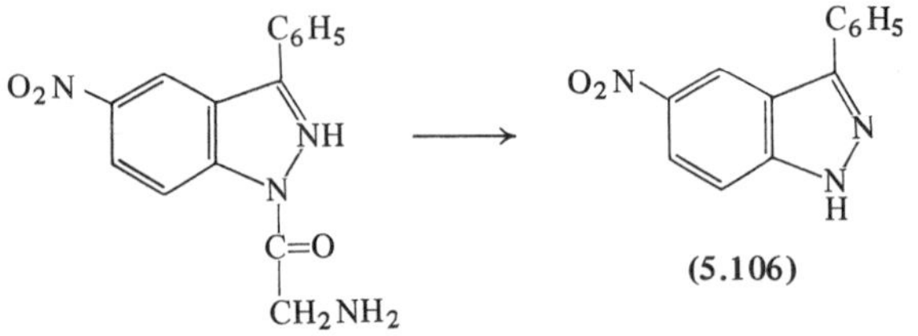

(5.106)

ring contractions of 1,4-benzodiazepine derivatives are the important ring contraction of the 3-acetoxy-1,4-benzodiazepine (5.107)—or the corresponding 3-hydroxy compound—into the indole-2-carboxaldehyde diacetale (5.108) during refluxing with a solution of sodium hydroxide in aqueous methanol[44], and also the acid-catalysed conversion of (5.109) into a mixture of the indoles (5.110) and (5.111).

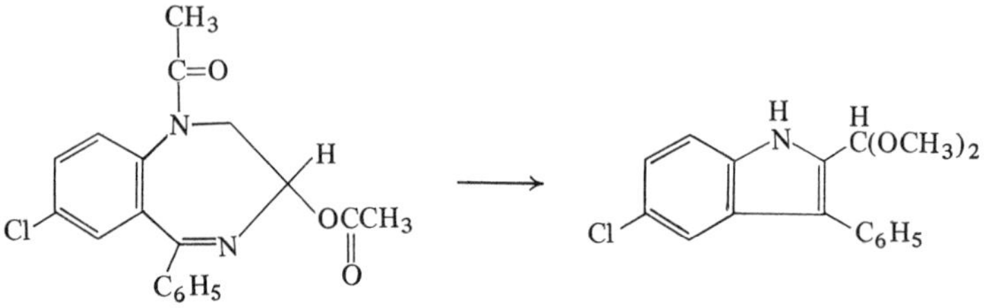

(5.107)   (5.108)

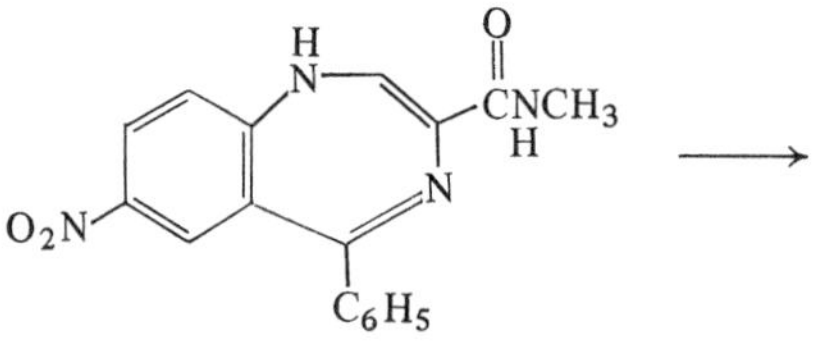

(5.109)

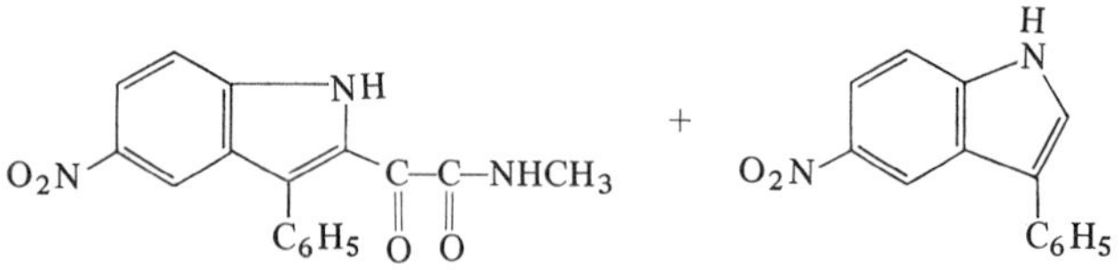

(5.110)   (5.111)

An example of ring contraction in the 2,4-benzodiazepine series, is that of the diazepin-3-one (5.112) into 1-oxo-2-isoindoline-2-carboxamide (5.114) during oxidation with chromic acid[45]. Evidence for the 2,4-benzodiazepinedione (5.113) as intermediate is provided by the fact that the 2,4-dibenzyl derivative of (5.112) is, under these conditions, converted into the 2,4-dibenzyl derivative of (5.113), which can be isolated. Although several reports have appeared in the

(5.112)   (5.113)   (5.114)

literature describing the preparation of 2,4-benzodiazepin-1,3,5-trione (5.115), more recent work[46] has shown that this compound is, in fact, the phthalimide derivative (5.116), indicating the occurrence of a ring contraction very similar to that in the conversion of (5.112) → (5.114) discussed above.

(5.115)   (5.116)

c. 1,5-Benzodiazepines into benzimidazoles and pyrazoles

2,4-Dimethyl-1,5-benzodiazepine (5.117) (R_1 = R_2 = CH_3) is converted on treatment with a mineral acid into 2-methylbenzimidazole (5.118) (R = CH_3) and acetone; the 2-methyl-4-phenyl analogue (5.117) (R_1 = CH_3, R_2 = C_6H_5) gives a mixture of (5.118) (R = CH_3) and (5.118) (R = C_6H_5) together with acetone and acetophenone[47-49]. By a similar reaction, the naphthoimidazole (5.120) was obtained from the 1,5-naphthodiazepinium salt (5.119)[50]. In this respect it is worthwhile to mention that it has been observed that pyrolysis of benzodiazepinium salts leads to the formation of benzimidazolium salts, suggesting that the melting points recorded in this series may not be particularly significant.

The ring contraction reaction (5.117) → (5.118) is thought to occur by a primary combination of o-phenylene diamine (5.121) and the diketo compound (5.122) forming the dihydrobenzimidazole (5.123) which aromatizes by loss of a carbanion into the benzimidazole; both compounds (5.121) and (5.122) must be present in a rapidly established equilibrium with the 1,5-benzodiazepine (5.117).

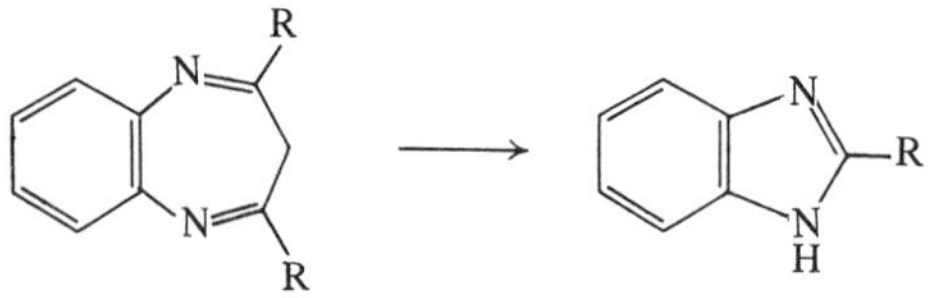

(5.117)   (5.118)

R = CH$_3$; C$_6$H$_5$

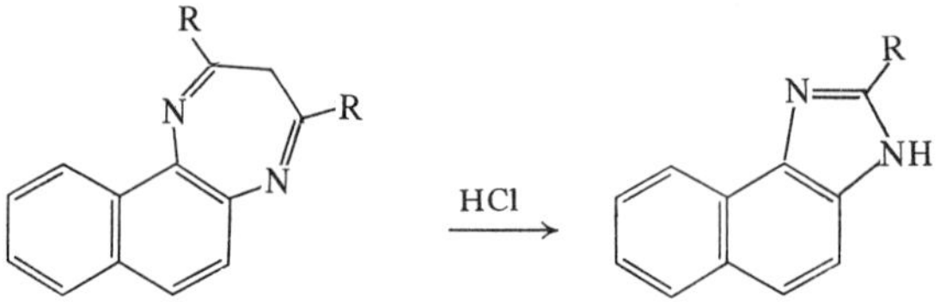

(5.119)   (5.120)

R = CH$_3$; C$_6$H$_5$

However, a mechanism involving initial hydrolysis to the open-chain compound (5.124), followed by ring closure into (5.123) and an internal displacement of the carbanion seems a more acceptable pathway[49].

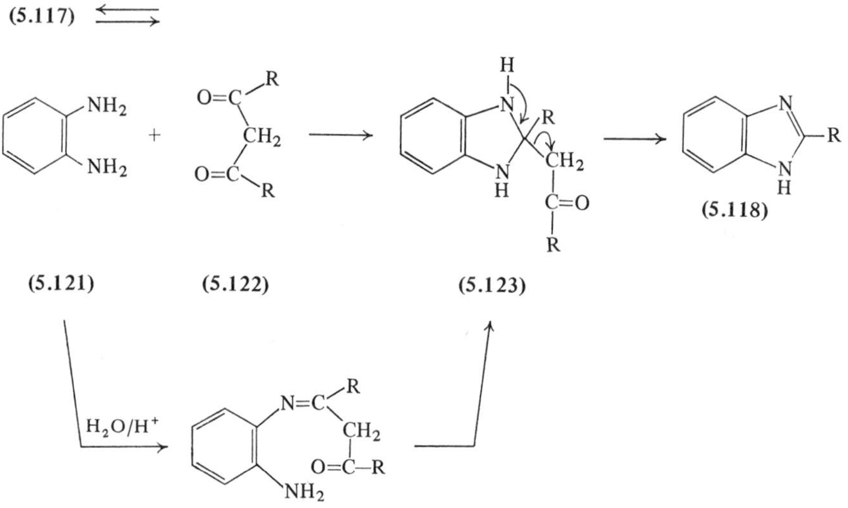

(5.117) $\rightleftarrows$

(5.121)   (5.122)   (5.123)

(5.124)

The hydrochloride (5.125) (R = CH$_3$) is converted into *N*-phenylpyrazole (5.126) (R = CH$_3$) when treated with phenylhydrazine, a reaction in which very probably the C$_2$–C$_3$–C$_4$ fragment of (5.125), i.e. (5.122) (R = CH$_3$) is involved[47,49,52]; this suggests that there is an equilibrium between (5.117),

(5.125)   (5.126)   (5.127)

(5.121) and (5.122). The open-chain compound (5.127), which is similar to (5.124), is proposed as intermediate.

Acid-catalysed ring contraction into benzimidazoles has also been found to occur with 1,5-benzodiazepin-2(3H)-ones. Boiling (5.129) (R = CH$_3$) with dilute sulphuric acid gave a low yield of 2-methylimidazole (5.118) (R = CH$_3$); the phenyl analogue (5.129) (R = C$_6$H$_5$) gives 2-phenylimidazole (5.118) (R = C$_6$H$_5$)[53,54].

(5.128)   (5.129)   (5.118)

R = CH$_3$; C$_6$H$_5$

Treatment of (5.129) (R = CH$_3$) with sodium 2-ethoxyethanolate gave 1-isopropenyl-2-benzimidazolone (5.128) (R = CH$_3$)[56]. But this ring transformation does not take place when (5.129) (R = CH$_2$CO$_2$C$_2$H$_5$) is treated with this sodium salt[51,55]. The 4-phenyl-1,5-benzodiazepinone (5.129) (R = C$_6$H$_5$) rearranges to benzimidazolone as soon as it is heated above its melting point[56]. Because these ring contractions occur so readily, alkenylbenzimidazolones[57,58] are obtained as an additional product when 1,5-benzodiazepines are prepared from β-ketoesters and o-phenylene diamine. It has recently been suggested that this thermal ring contraction reaction can in fact be effectively used to differentiate between the two theoretically possible dihydro-1,5-benzodiazepinones obtained by condensation of unsymmetrical unsubstituted o-aryl diamines with β-ketoesters[59].

The 1,5-benzodiazepine-2,4-dione (5.132) has been found to be converted when heated with o-phenylene diamine hydrochloride into the 2,2$'$-methylene bis imidazoles (5.133), while its dimethyl derivative (5.131) when boiled in sulphuric acid gives the open-chain compound (5.130)[60] and not the dihydrobenzimidazole (5.134) as originally suggested[61].

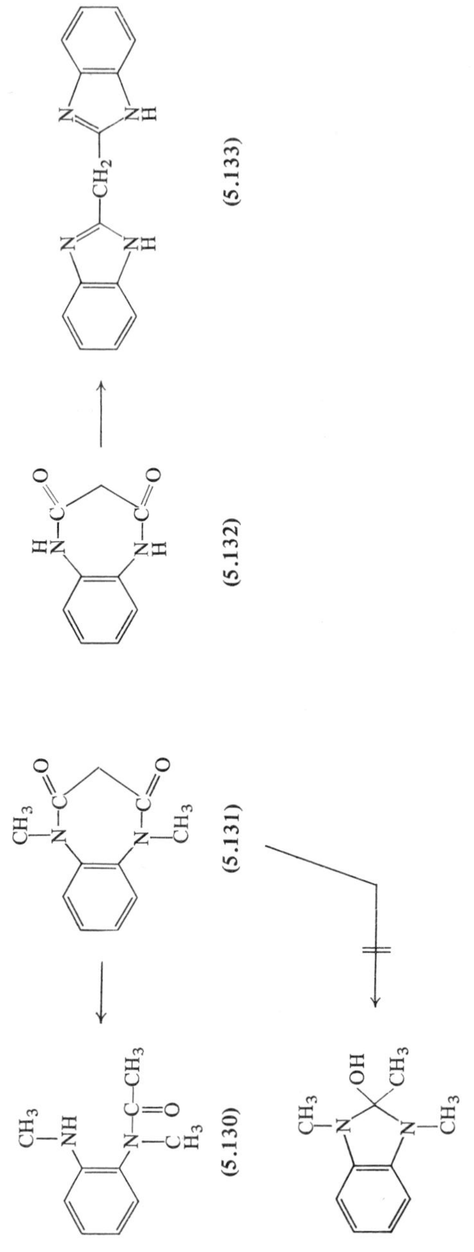

(5.133)

(5.132)

(5.131)

(5.130)

(5.134)

2. RING CONTRACTIONS OF DIAZEPINES INTO SIX-MEMBERED HETEROCYCLES

a. *1,2-Diazepines into pyridines*

As already reported in section II.A.1.a, ring contraction reactions have been reported with 5-methyl-6-phenyl-2,3-dihydro-4*H*-1,2-diazepin-4-one (5.76) in acidic as well as in basic media. Treatment of (5.76) with 6*N* hydrochloric acid at 50°C leads to the formation of the *N*-amino-3-hydroxy-5-phenylpyridinium chloride (5.135)[31,62], while in aqueous sodium hydroxide a mixture of about equal amounts of the 6-amino-3-hydroxy- (5.79) and 2-amino-3-hydroxypyridines (5.80) is obtained[33].

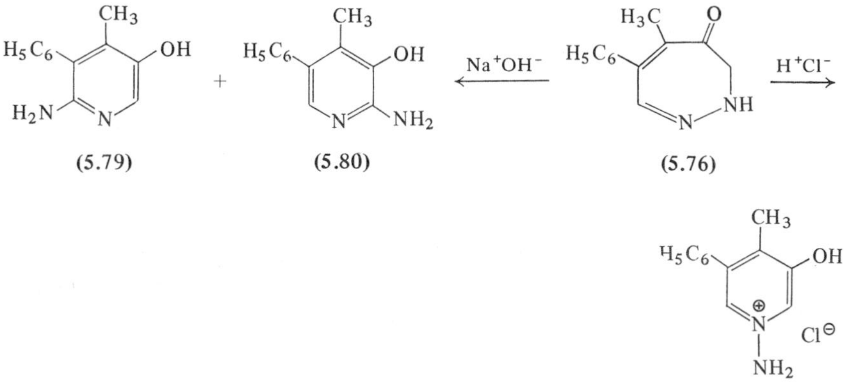

(5.79)   (5.80)   (5.76)

(5.135)

Let us first consider the behaviour of (5.76) and some of its derivatives in acidic media. It was found[31,62] that the 2-methyl derivative (5.136) does not give this ring contraction in acidic media; however, in contrast, the 1-methyl derivative (5.137), which is present as a diazepinium salt in this acid medium, reacts to form the *N*-methylamino-pyridinium salt (5.138). This result unequivocally leads to the conclusion that (*a*) it is the N_1 atom which is extruded during the ring contraction and (*b*) that the hydrogen at the N_2 atom in (5.76) plays an important role in this ring contraction reaction. In analogy to the initial

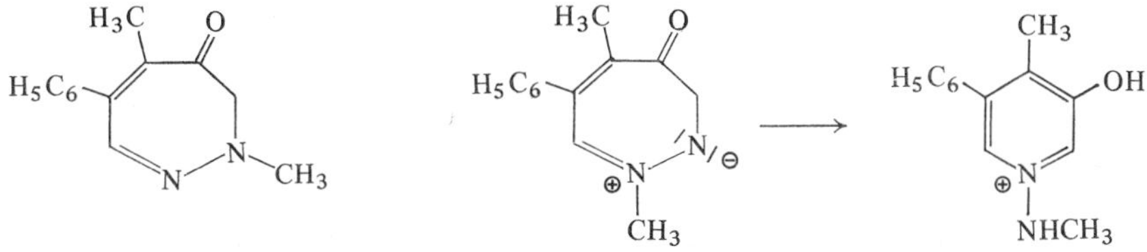

(5.136)   (5.137)   (5.138)

step proposed in the hydrolysis of hydrazones, the mechanism of the ring contraction must involve protonation of the C=N bond in (5.76), yielding (5.139). Subsequent C_7-N_2 bridging gives the bicyclic compound (5.140). The N-aminopyridinium salt (5.141) is formed by the electron shift indicated and is then converted by prototropy into (5.135). From this mechanism it is

(5.76) $\longrightarrow$ (5.139)   (5.140)

(5.141)   (5.135)

clear that blocking of the hydrogen at N_2 by a methyl group precludes this ring contraction. That a bicyclic intermediate containing a diaziridino ring can exist is proved by the fact that the N-methyl derivative of (5.135), i.e. (5.138)[31,62,63], has indeed been isolated. It is evident that one has to be aware that in acid-catalysed reactions, which should lead to compounds of type (5.76), N-aminopyridinium compounds are obtained as by-products[66]. In a study on the mechanism of the ring contraction of (5.76) in a basic medium, it appears that the rate of the reaction (5.76) → (5.79) + (5.80) is first order in hydroxyl ion concentration and that hydrogen–deuterium exchange (which occurs faster at C_3 than at C_7) takes place more quickly than the disappearance of (5.76)[64]. Moreover it has been found that the 2-methyl derivative (5.142) (R = H) and the 2-cyanoethyl derivative (5.142) (R = CH_2CN)[65] give exclusively the 2-methylaminopyridine (5.143) (R = H) and the β-[2-pyridylamino]- propionic acid (5.143) (R = CH_2CO_2H) respectively; both results indicate that it is very likely the nitrogen at position 2 which is extruded during the base-catalysed ring contraction of (5.76) into (5.79) and (5.80) respectively.

The results obtained from rate studies and an investigation of the behaviour of the N-alkylated derivatives, suggest a mechanism in which, in a rate-determining step, the carbinolamine (5.144) is obtained, first leading to the formation of the enolide carbanion (5.145). A subsequent β-elimination leads to the acyclic diimine (5.146) which can cyclize into the 2-amino compound (5.80) (route A), as well as the 6-amino compound (5.79) (route B). The supposition

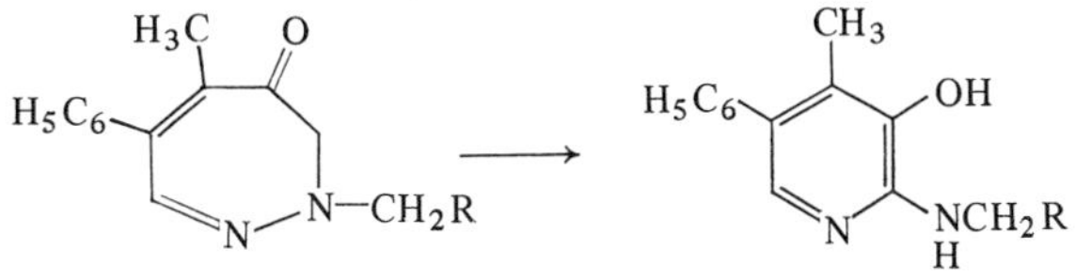

(5.142)   (5.143)

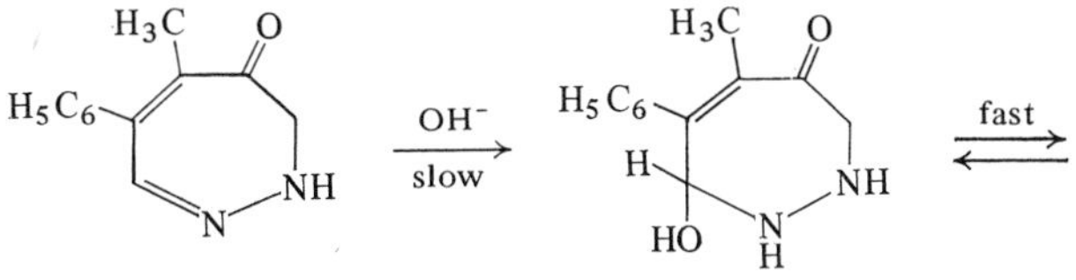

(5.76)   (5.144)

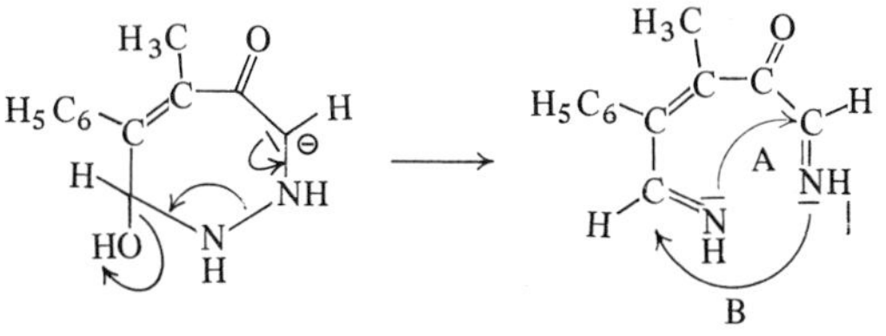

(5.145)   (5.146)

route A: (5.146) $\longrightarrow$

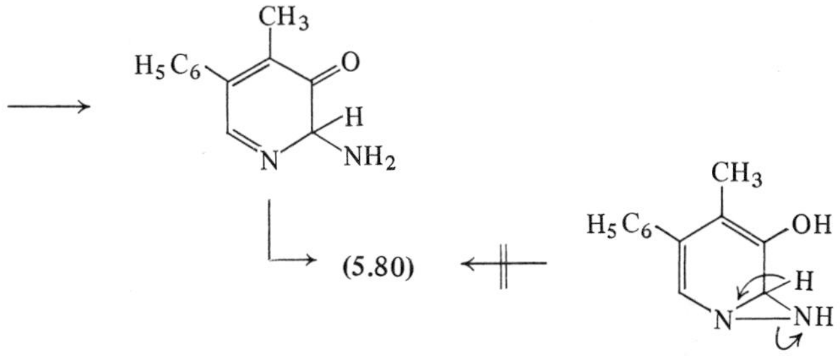

(5.80)

(5.147)

route B: (5.146) $\longrightarrow$

(5.79)

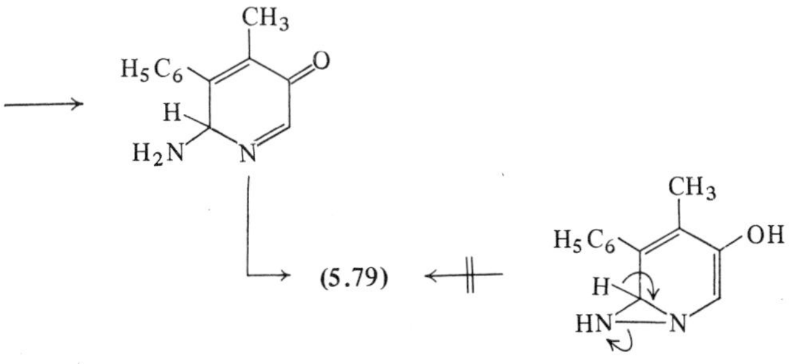

(5.148)

that the formation of (5.80) and (5.79) occurs via the diazabicycloheptanes (5.147) and (5.148) respectively[33] is incompatible with the finding that a mixture of 2- and 6-aminopyridines exclusively is obtained from (5.147) and (5.148) in base without any indication of the presence of trace amounts of N-aminopyridinium salts.

Evidence for the existence of an intermediate containing the diaziridine ring is provided by the finding that during the uv light induced ring contraction of the 1-methylbetaine (5.137) into the 2-methylaminopyridine (5.150) at −80°C, in addition to (5.150), its precursor, 4,7-dimethyl-5-phenyl-1,7-diazabicyclo-[4.1.0]-4-hepten-3-one (5.149), can be isolated[63]; the last compound (5.149) is formed by a concerted 4π electrocyclic reaction of the azomethine imine

(5.137)   (5.149)   (5.150)

system. In reverse correlation with the reaction mentioned above is the recently reported photo conversion of 1-ethoxycarbonyliminopyridinium betaine into 1-ethoxycarbonyl-1,2-diazepine[67] (for further details see chapter 4, section I.C.3.b).

A few other examples of acid-catalysed ring contractions of 1,2-diazepines into derivatives of pyridine which we will mention but not discuss in detail, include the following conversions: the 1,2-diazepinone (5.151) into the lactam (5.152)[68]; the tetramethylene-4,5-dihydro-4H-1,2-diazepine (5.153) into the 5,6,7,8-tetrahydroquinoline (5.154)[69]; (5.155) into the N-anilinoisoquinolone (5.156)[70]; the 2,3-benzodiazepine (5.157) into the N-amino-3,4-dihydroiso-quinolinium salt (5.158)[71], and the conversion of (5.159) into (5.160)[72,73]. The action of concentrated hydrochloric acid and zinc causes a similar ring contraction to occur with (5.159) (m-tolyl is replaced by a phenyl group; however, in this case, loss of the N-amino group takes place[74]). The conversion (5.153) → (5.154) is thought to occur via the 1,2-diazepinium ion (5.161) and the 1,4-dihydropyridine (5.162).

b. *1,4-Benzodiazepines into quinazolines, dihydroquinazolines, dihydro-quinoxalines and quinolines; 1,5-benzodiazepines into quinoxalines*

3-Hydroxy-1,3-dihydro-2H-1,4-benzodiazepin-2-one, which is stabile as a solid or in neutral solution, is found to be converted into dihydroquinazolines by treatment with alkali or with acid. Hence the compound (5.163), when heated vigorously with base, forms after acidification the 3,4-dihydroquinazoline-2-

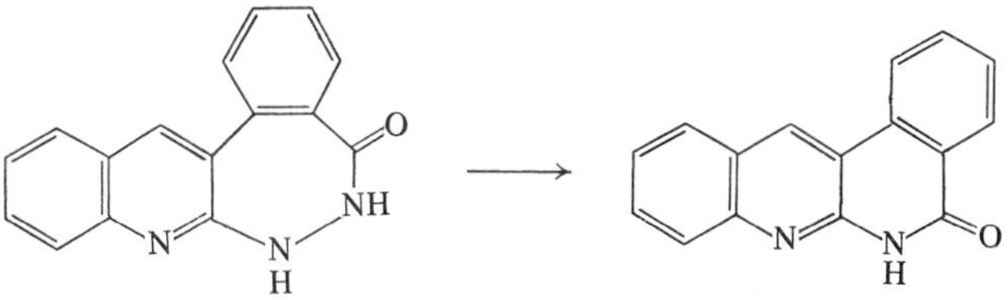

(5.151)   (5.152)

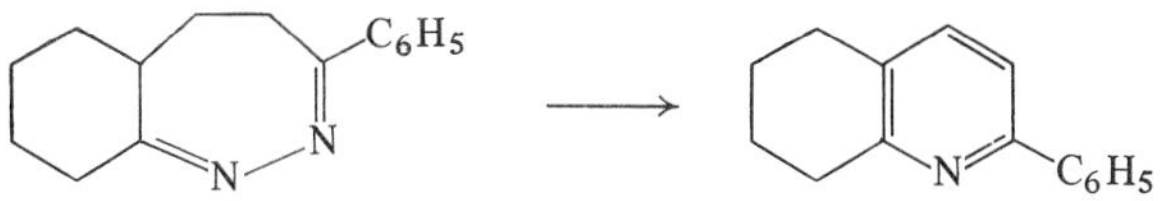

(5.153)   (5.154)

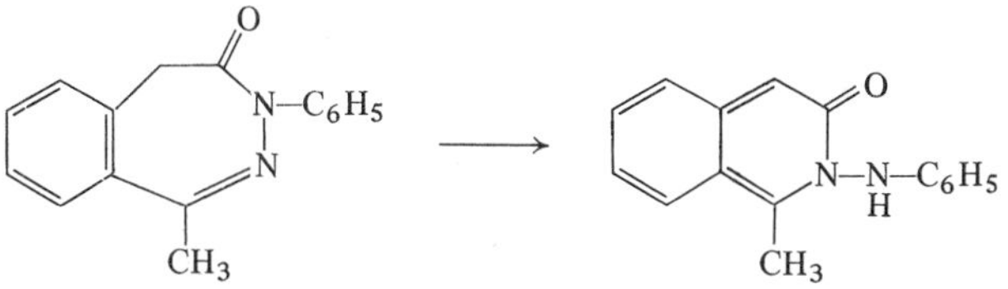

(5.155)   (5.156)

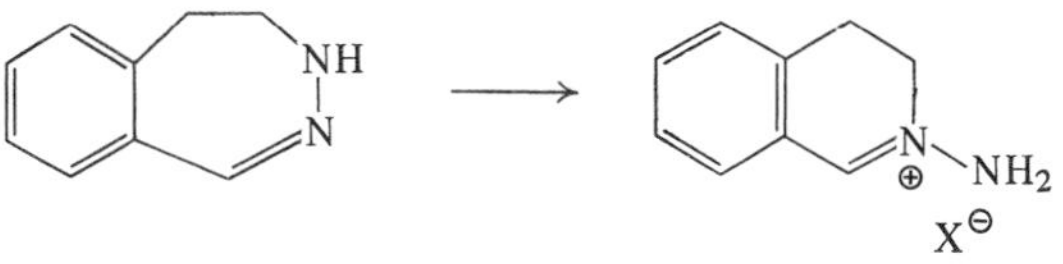

(5.157)   (5.158)

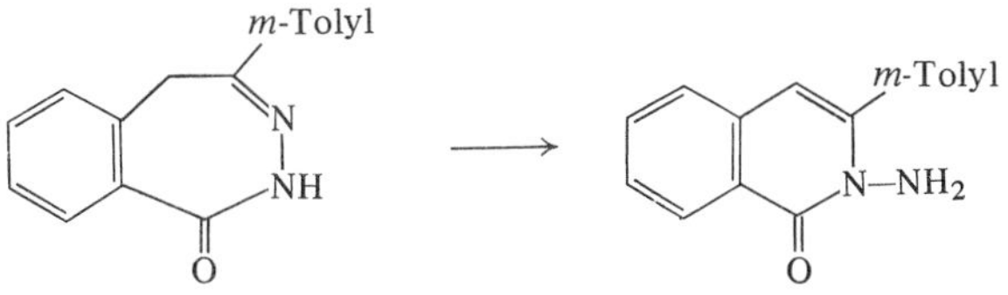

(5.159)   (5.160)

carboxylic acid (5.165)[75]. The 2,3-dione (5.164) is the primarily formed product since under somewhat milder basic conditions it is possible to isolate this and, under more strenuous conditions it is found to be converted into (5.165). Parallel rearrangements take place with the 1-methyl derivative of (5.163) which yields the 1,4-dihydroquinazoline (5.166). Similar rearrangements occur in the acid-catalysed conversions of the 3-acetoxy compound

(5.153) $\longrightarrow$

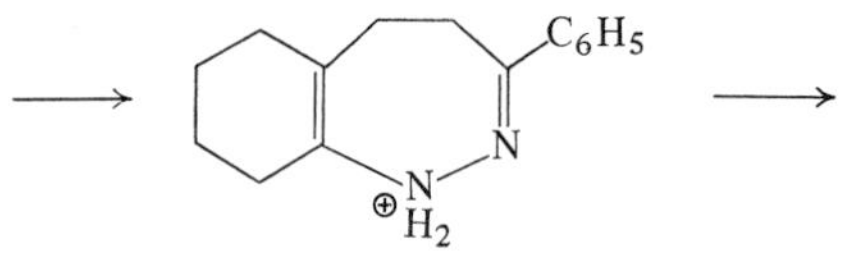

(5.161)

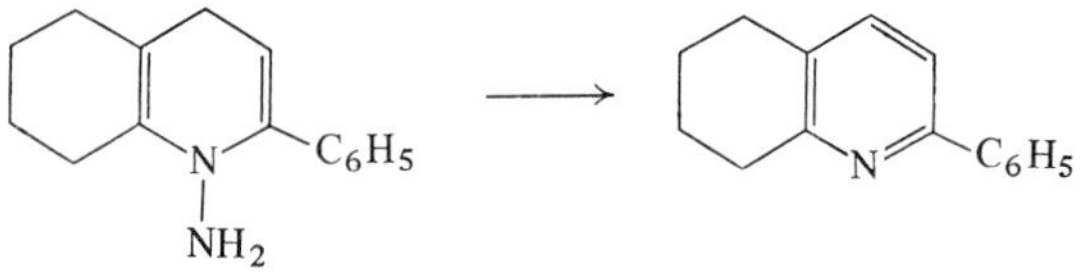

$\longrightarrow$

(5.162)

(5.154)

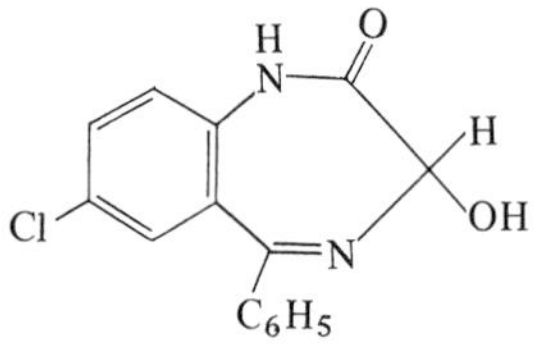

(5.163)

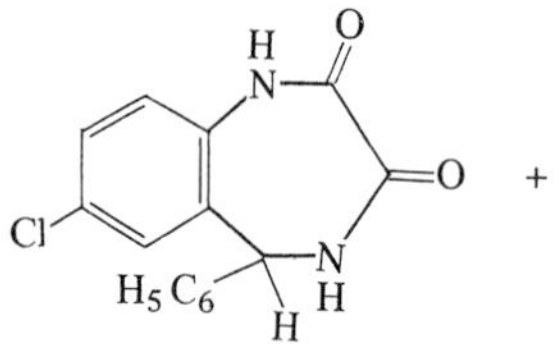

(5.164)   +

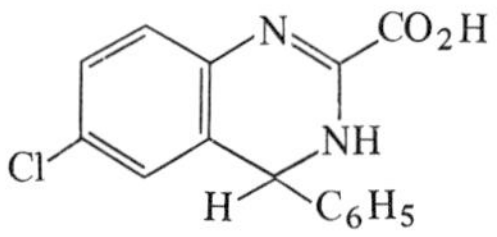

(5.165)

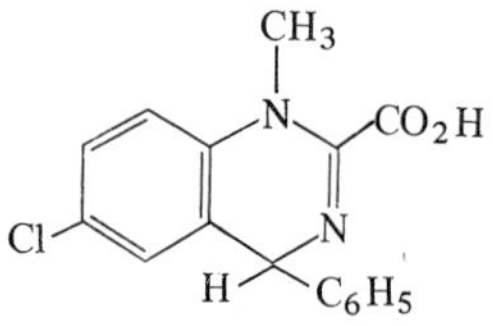

(5.166)

$(5.167)^{76,77}$ and the 2-methylamino-1,4-benzodiazepine $(5.169)^{77,78}$ yielding the quinazoline-2-carboxaldehyde (5.168). With hydrazine or primary amines the corresponding hydrazone and aldimine of (5.168) are obtained. Although it is evident that in all these ring transformation reactions the intermediates are open-chain compounds, it is not quite clear whether in the formation of these intermediates ring opening occurs between the N_1 and C_2- or the C_3 and N_4-position of the seven-membered ring[75].

Ring contraction into quinazolines has also been reported during treatment with acid of 2-acetylamino-5-aryl-3H-1,4-benzodiazepin-3-one (5.171); the 4-phenylquinazolines (5.170) and (5.172) are obtained[78].

Oxidation of the 1,4-benzodiazepine-3-carboxamide (5.173) with chromic acid also results in the formation and isolation of a quinazoline derivative, i.e. $(5.176)^{80}$. The reaction course probably involves oxidation into the open-chain benzene derivative (5.174) which cyclizes into the 1,2-dihydroquinazoline (5.175), and by a subsequent loss of formic acid, yields the product (5.176).

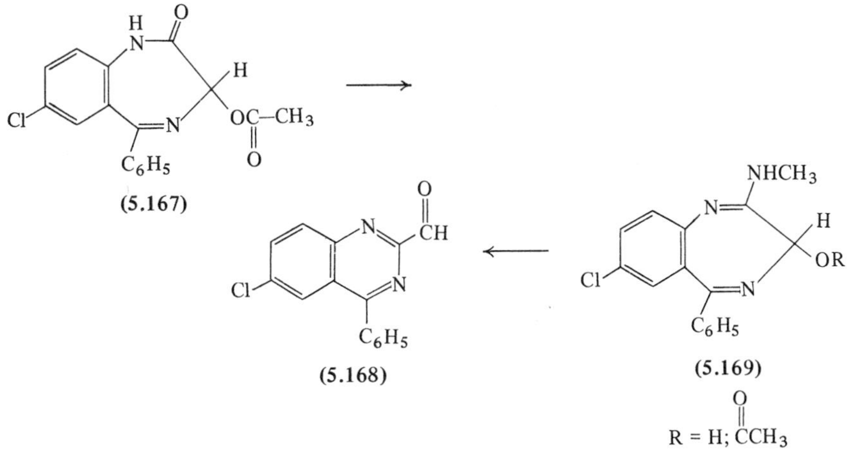

(5.167)

(5.168)

(5.169)

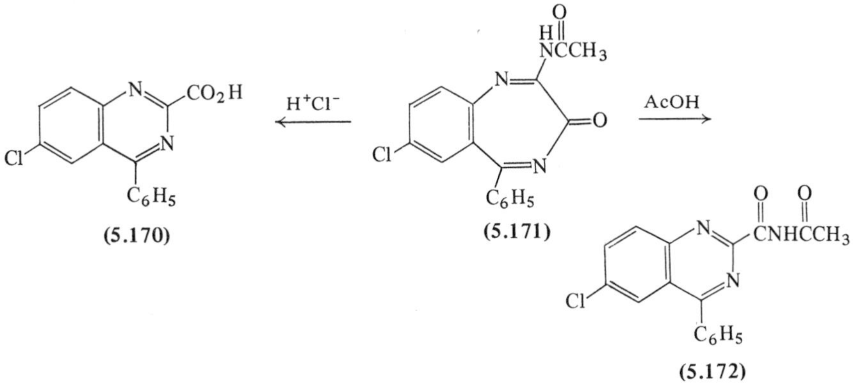

$$R = H; \overset{O}{\overset{\|}{C}}CH_3$$

(5.170) H^+Cl^- (5.171)   AcOH

(5.172)

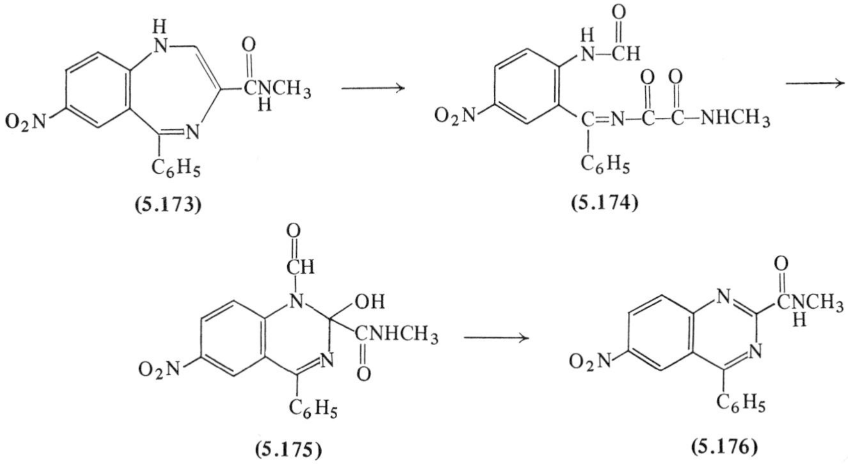

(5.173)

(5.174)

(5.175)

(5.176)

Whereas all the reactions previously mentioned in this section are concerned with ring contractions of 1,4-benzodiazepines into quinazolines, it is interesting that there are also reactions known in which 1,4-benzodiazepines are converted into quinoxalines or quinolines. It appears that the action of *p*-toluenesulphonyl chloride or phosphoryl chloride causes 1,4-benzodiazepin-2-one-4-oxides (5.178) to undergo a Beckmann-type rearrangement into 4-benzoyl-6-chloro-3,4-dihydroquinoxaline-2-(1*H*)-one (5.179)[81], and not into a carboxylic acid or a dione, as has been reported[75] for 1,4-benzodiazepines, which do not contain an

(5.177)

R = NH₂; CH₃

(5.178)

(5.179)

N-oxide function. On the contrary, treatment of (5.178) with acetic acid gave the quinazoline-2-aldehyde (5.168); with hydrazine or methylamine the corresponding hydrazone or methylimino derivative of (5.168), i.e. (5.177), was obtained.

A ring contraction into a quinoline derivative is found to occur when the diazepinone (5.180) is refluxed with acetic anhydride containing a trace of sulphuric acid[82]. The 3-acetylaminoquinolone-2 (5.183) is obtained; it is probably formed via the *N*-acetyl-1,4-benzodiazepinium acetate (5.181) and the aziridino derivative (5.182). In the absence of a *N*-1 substituent, further cyclization into an oxazoloquinoline has been observed[82].

Interestingly, it has been found[83] that electrochemical reduction of the 7-chloro-2-methylamino-5-phenyl-3*H*-4,5-dihydro-1,4-benzodiazepine (5.184) with mercury electrodes leads to loss of methylamine and ring contraction to 2-methyl-4-phenyl-6-chloro-3,4-dihydroquinazoline (5.187). It is reasonable to suppose that this reaction occurs via the tetrahydro derivative (5.185), which, because it is unstable, facilitates elimination of the methylamino group (5.186).

(5.180)   (5.181)

(5.182)   (5.183)

A nucleophilic attack of the N_4 on the electropositive carbon at C_2 and a subsequent hydride shift to C_3 leads to extrusion of C_3 resulting in the formation of the methyl substituent on position 2 in (5.187). It has been suggested[84] that a reductive ring opening into the substituted amidine (5.188) would provide an alternative route for this ring contraction; in this case ring cyclization will occur into (5.187) by loss of methylamine. Electrochemical reduction of the 1,4-benzodiazepine-4-oxide (5.190) into (5.187) has been the subject of many investigations[85-88,83]; it has been proved that these reductions take place via (5.189) and (5.184).

Ring transformation of (5.190) into the quinoxaline derivative (5.193) and the eight-membered oxadiazocine (5.192) has been found to occur during photolysis[89]. These rearrangements are believed to involve the oxaziridino compound (5.191) as the primary isomerization product. This oxaziridino compound can in fact be isolated[90] when (5.190) is irradiated in sunlight; it can serve as an experimental indication for the presence of an oxaziridine as an intermediate in the photoisomerization of the hetero-aromatic N-oxides. The formation of the N-acetylquinoxaline from (5.191) is analogous to the conversion of the pyrroline-1-oxide (5.194) into N-acetylazetidine (5.195)[91].

It is of interest that the compound 7-chloro-4,5-epoxy-5-phenyl-1,3,4,5-tetrahydro-1,4-benzodiazepin-2-one (5.195a) (R = H) undergoes with facility a ring contraction in alcohol producing good yields of 3-alkoxymethyl-6-chloro-4-hydroxy-4-phenyl-3,4-dihydroquinazolin-2(1H)-ones (5.195d)[91a]. Since this ring contraction does not occur with the 1-methyl derivative (5.195a) (R = CH_3), it can reasonably be supposed that the iminoisocyanate (5.195b) and the cyclic acylated imines (5.195c) are intermediates. Addition of the alcohol, which is known[91b] to occur readily on acylated imines, yields (5.195d).

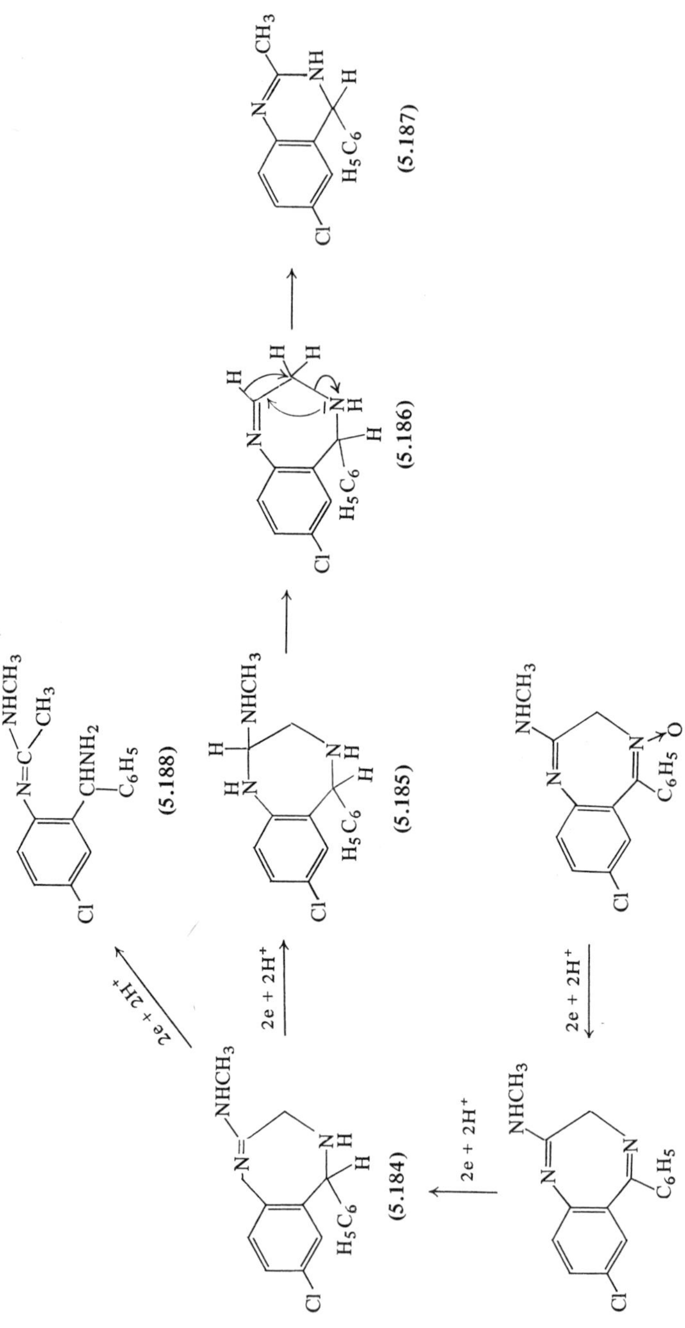

(5.184) (5.185) (5.186) (5.187) (5.188) (5.189) (5.190)

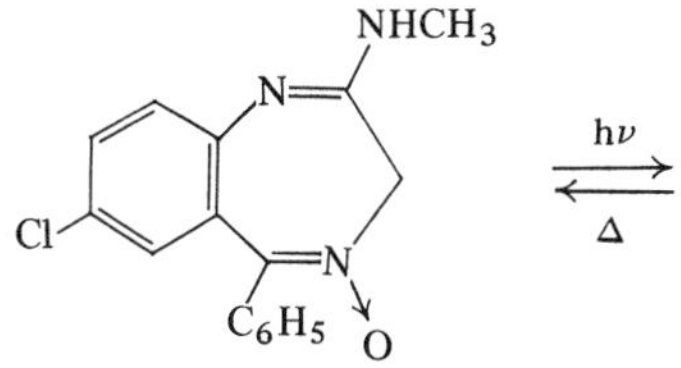

(5.190)

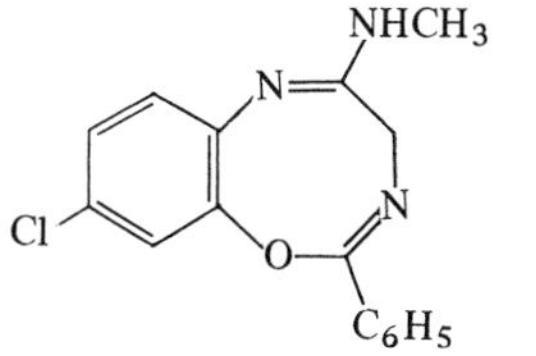

(5.191)

(5.192)

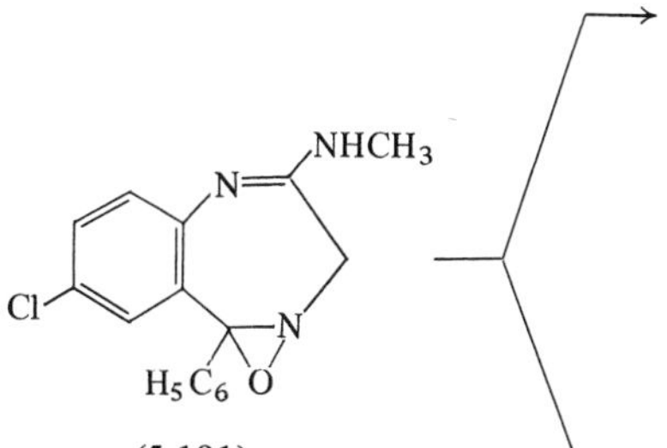

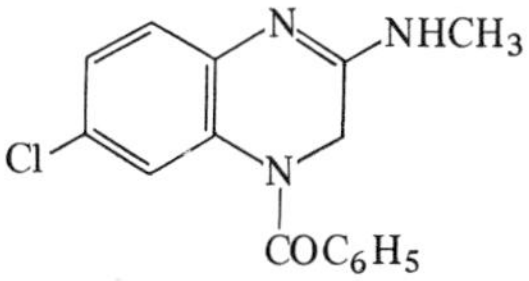

(5.193)

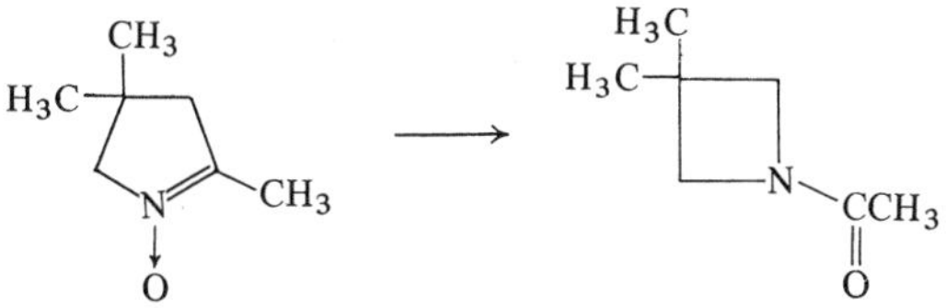

(5.194)

(5.195)

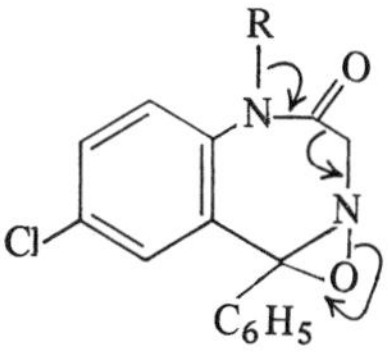

(5.195a)

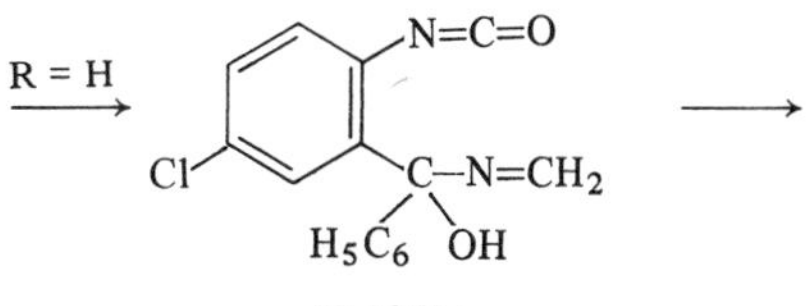

R = H

(5.195b)

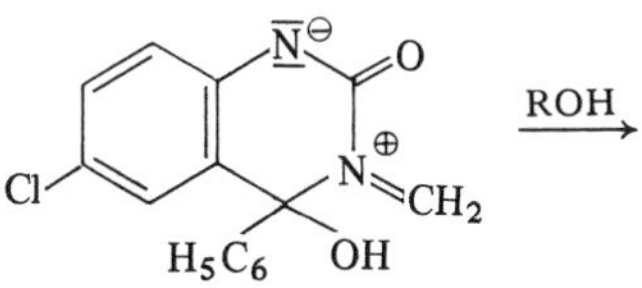

(5.195c)

ROH

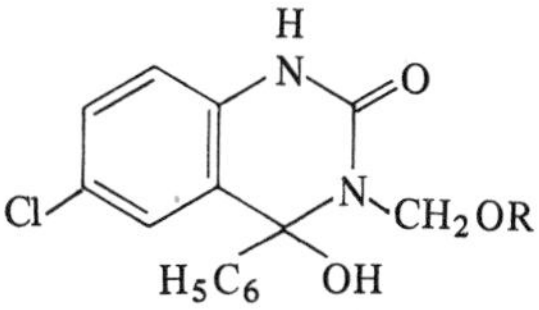

(5.195d)

A different type of ring contraction occurs when the 3-hydroxyiminobenzo-1,5-diazepine (5.196) is treated with acid; it leads to formation of 2-acetyl-3-methylquinoxaline oxime (5.197)[92].

Oxidation of the 1,5-benzodiazepine (5.198)[49] by peracetic acid, did not lead to the expected 3,6-diazabenzotropone (5.199), but resulted in a further rearrangement of (5.199) into the 2-acetyl-3-methylquinoxaline (5.200). The same product is also formed when (5.198) is photo-oxidized in a non-aqueous

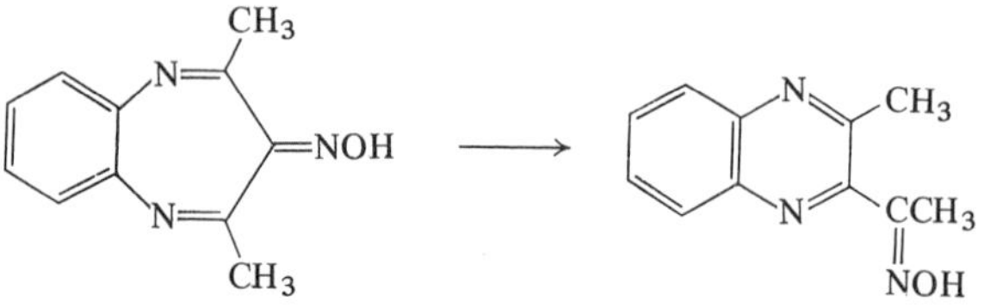

(5.196)   (5.197)

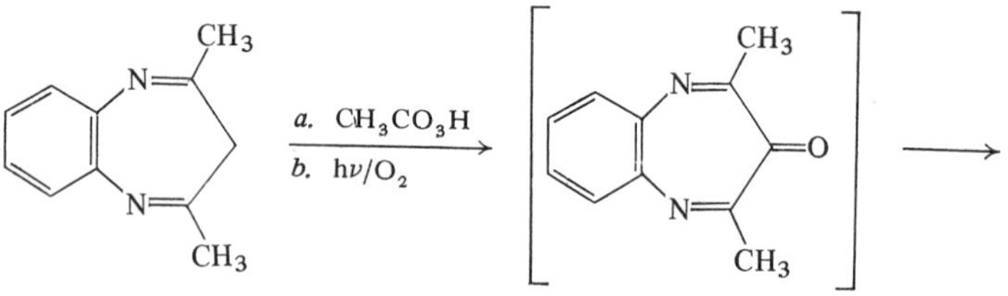

(5.198)   (5.199)

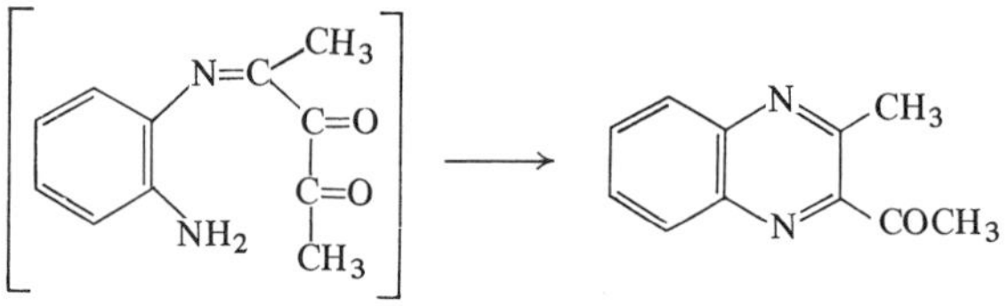

(5.200)

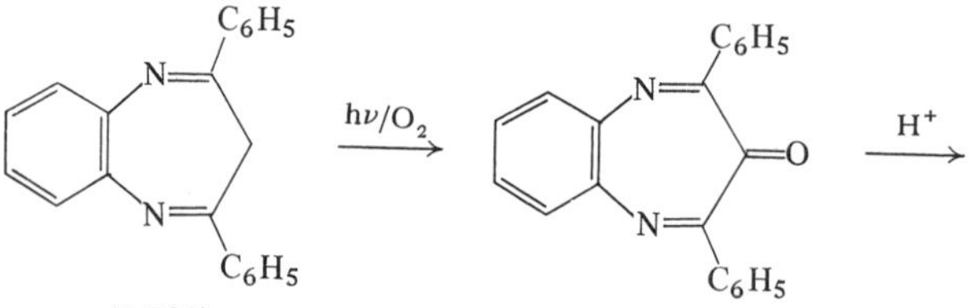

(5.201)   (5.202)

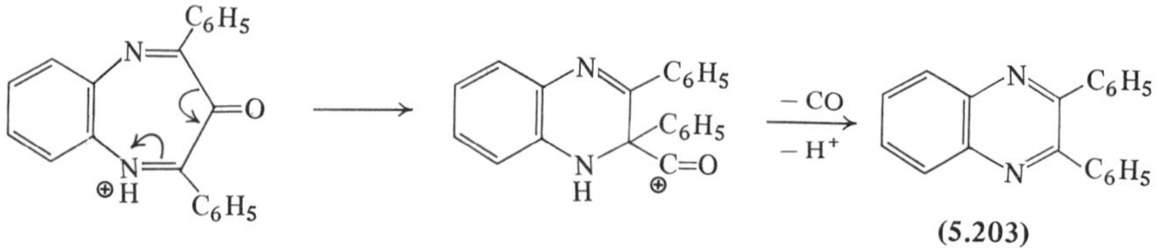

(5.203)

medium[93a]. However, it is of interest that it has been demonstrated that a 3,6-diazatropone (5.202) is obtained when 2,4-diphenyl-1,5-benzodiazepine (5.201) is photo-oxidized[93]. This compound is, however, rather unstable and appears to be easily converted into 2,3-diphenylquinoxaline (5.203) in warm concentrated sulphuric acid or acetic acid. It is proposed that this decarbonylation reaction occurs as indicated on page 300.

c. 1,2-Diazepines into pyridazines

The ring contractions exhibited by 5-methyl-6-phenyl-2,3-dihydro-4H-1,2-diazepin-4-one (5.76), and several of its derivatives are both diverse and interesting; they result in conversion into furans, pyrroles, pyrazoles (see section II.A.1.a), N-aminopyridines or 2(6)-aminopyridines (see section II.A.2.a). The conversion of 2-acetyl-3-benzylidene-3,4-dihydro-2H-1,2-diazepin-4-one (5.204) when treated with methanolic hydrochloride into benzyl-3-(4-methyl-5-phenylpyridazinyl)ketone (5.208) is another instance of its divergent behaviour[94]. It is probable that this ring contraction occurs via the intermediary diazatropone (5.205) which by a covalent hydration across the C=N bond, yields the carbinolamine (5.206). A retro-aldol reaction leads to extrusion of the C_3-atom outside the ring and formation of (5.207). Dehydration of (5.207) gives

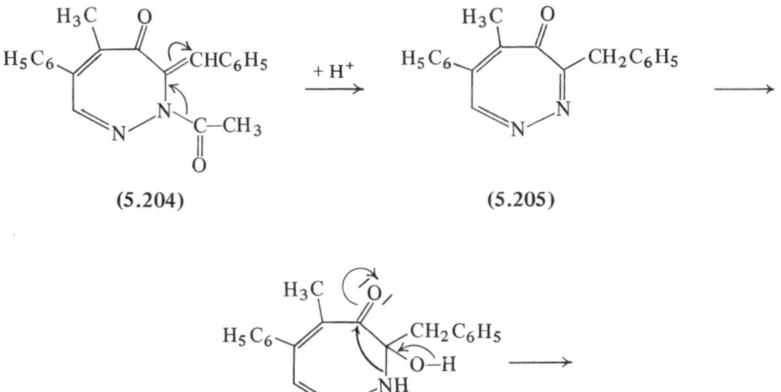

(5.204)   (5.205)

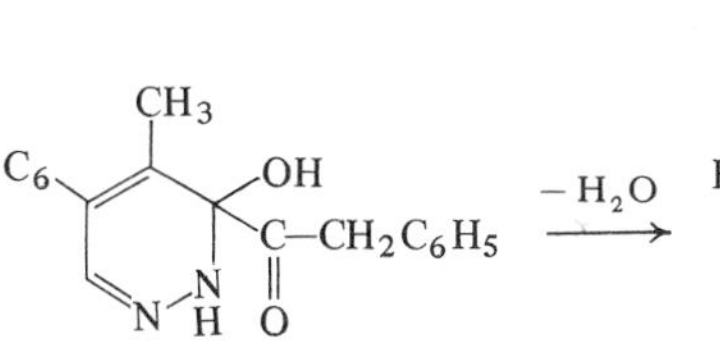

(5.206)

(5.207)   (5.208)

the pyridazine (5.208). That it is definitely the C_3-atom, which is extruded has been confirmed by ^{14}C-labelling experiments[94]. The N-substituted dihydro-1,2-diazepine (5.210) has been found to be converted in acid media into the 1,6-dihydropyridazine-3-carboxaldehyde (5.212); in basic media the well-known conversion into a pyridine compound, the 6-methylamino-3-hydroxypyridine derivative (5.150), has been achieved[95]. The course of these ring contractions can be postulated to take place as follows:

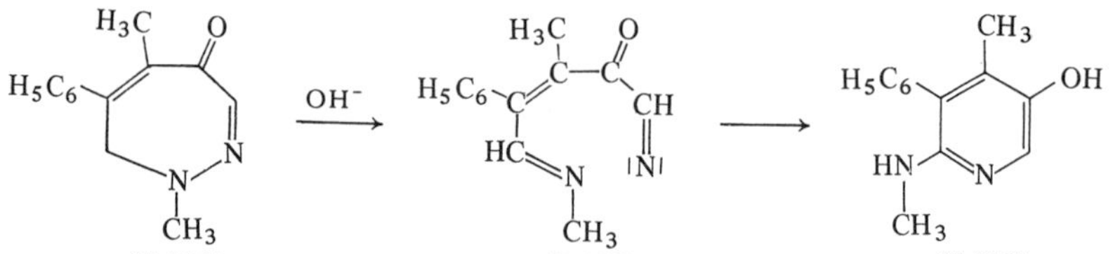

The conversion of (5.210) into (5.212) involves a covalent hydration across the C=N bond forming (5.211) in which a solvolytic displacement occurs in a manner analogous to that described[94] for the conversion of (5.206) → (5.207). In the base-catalysed reaction the acyclic anionic diimine (5.209) can be proposed as intermediate[95] because this reaction is analogous to the mechanism given for the ring contraction of 1,2-diazepines into pyridines as discussed at length in section II.A.2.a.

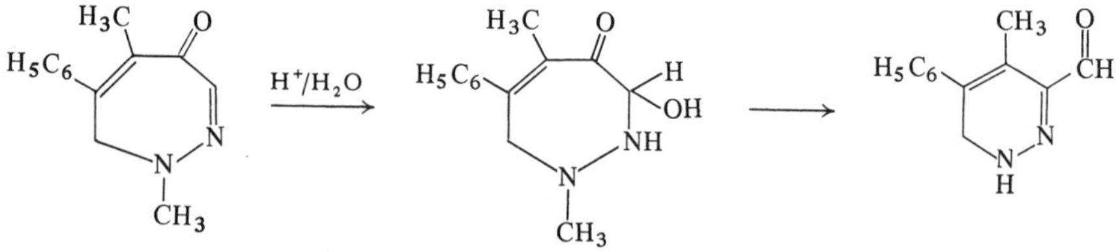

Treatment of the 6H-3,7-diphenyl-4,5-dihydro-1,2-diazepine (5.213) (R = H) with N-bromosuccinimide did not give oxidation into the corresponding 4H-1,2-diazepine as expected, but instead gave formation of the pyridazine (5.214) (R = H) together with the 3,4-diazanorcaradiene derivative (5.215) (R = H)[96,97]. With N-chlorosuccinimide a similar reaction was observed: from the compound (5.213) (R = CH$_3$, C$_6$H$_5$) has been formed the norcaradiene derivative (5.215) (R = CH$_3$, C$_6$H$_5$), in a much higher yield than with N-bromosuccinimide. In addition, the pyridazine derivatives (5.214) (R = CH$_3$, C$_6$H$_5$) and pyridazines, chlorinated on the α-position of the side-chain have been obtained. These norcaradiene derivatives are intermediates in these ring contraction reactions, as

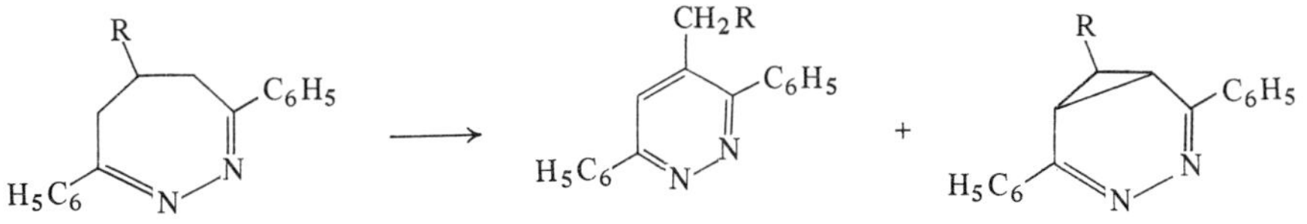

(5.213)   (5.214)   (5.215)

demonstrated by the fact that they are easily converted into the corresponding pyridazines in ethanolic hydrogen chloride. Moreover, if the course of the ring contraction is followed by uv spectrometry there are two isobestic points in the curves obtained; in the rearrangement of (5.213) (R = C_6H_5) the first point is observed at 314 mμ during the early stages of the reaction, and a second at 285 mμ, indicating that initially an intermediate is produced which then proceeds to the end-product.

By carrying out the conversion of (5.213) (R = C_6H_5) in the presence of deuterated hydrogen chloride an incorporation of about 65 per cent of deuterium into position 5, i.e. (5.219) (R = C_6H_5) has been measured (nmr and mass spectrometry). This result has led to the conclusion that by the addition of deuterium chloride to (5.216), the carbonium ion (5.218) is formed and not the carbonium ion (5.217). That there is not 100 per cent incorporation is considered to be due to a difference of C–H and C–D bond strength[98].

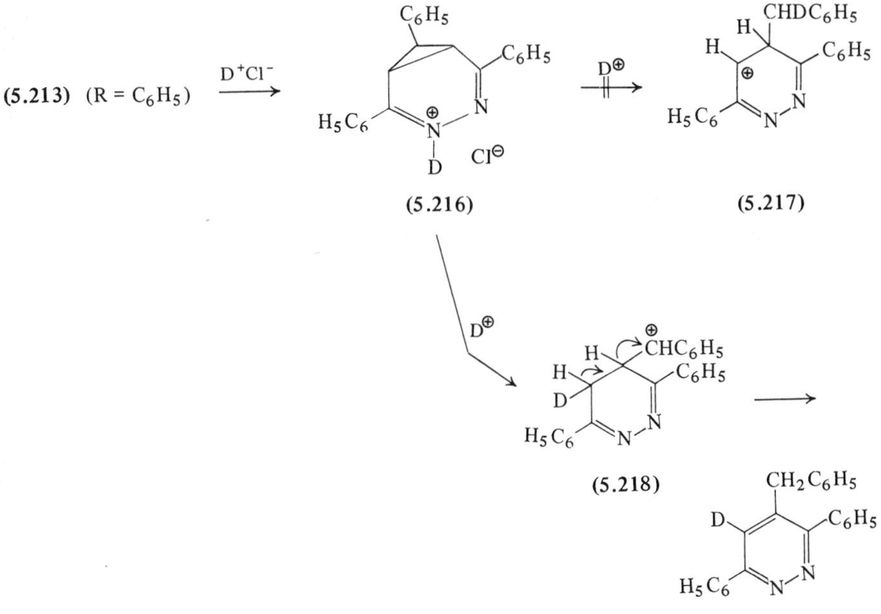

(5.216)   (5.217)

(5.218)

(5.219)

Very recently, a quite unexpected and so far almost unexplored ring contraction has been reported to occur when the 3,8-dihalogenobenzo[c,f] 1,2-diazepin-11-one (5.220) reacts with ethyl-4-(diethoxyphosphinyl)crotonate (5.221) (R:CH = CH $CO_2C_2H_5$) or with sodium hydride in acetone or in benzene, yielding the 3,8-dichlorobenzo[c] cinnoline (5.222)[99,100]. Attempts

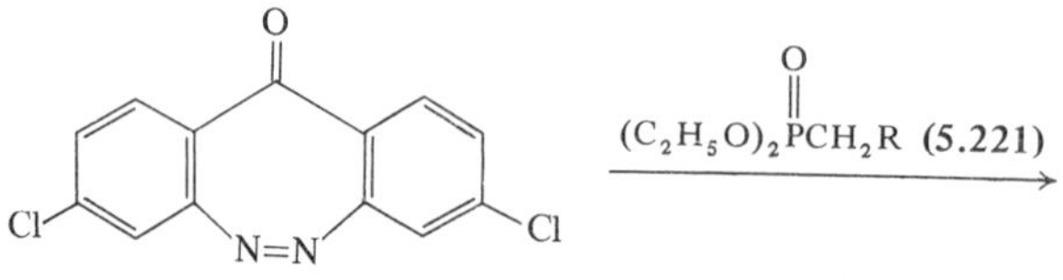

(5.220)

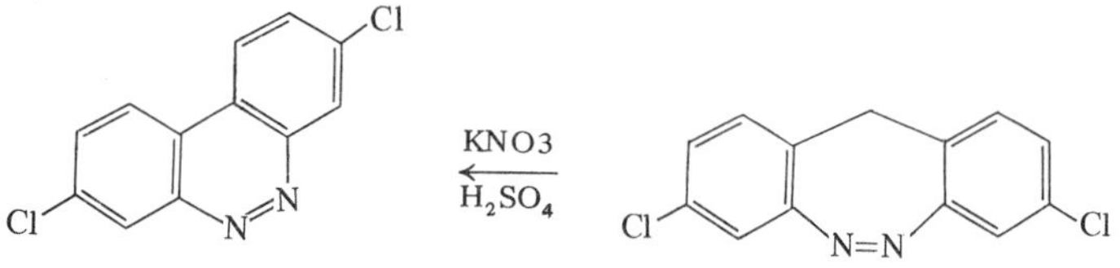

$(C_2H_5O)_2\overset{O}{\overset{\|}{P}}CH_2R$ (5.221)

(5.222)

KNO3
←
H₂SO₄

(5.223)

to bring about the reaction with (5.221) (R = $CO_2C_2H_5$, CN, $CONR_2$ or COC_6H_5) were unsuccessful. It has recently been reported that treatment of 11-H-benzo[c,f] 1,2-diazepine (5.223) with nitrate salts in sulphuric acid also leads to the formation of (5.222)[101].

B. Ring transformations of seven-membered heterocycles containing two sulphur atoms

1. RING CONTRACTIONS OF 1,4-DITHIEPANES INTO 1,4-DITHIANES

6-Hydroxy-1,4-dithiepane (5.224) is converted with thionyl chloride in chloroform into 2-chloromethyl-1,4-dithiane (5.227)[102]. It is believed that the primarily formed 6-chloro derivative (5.225) rearranges into the thiiranium intermediate (5.226) which, by a nucleophilic attack on the chloride ion, is

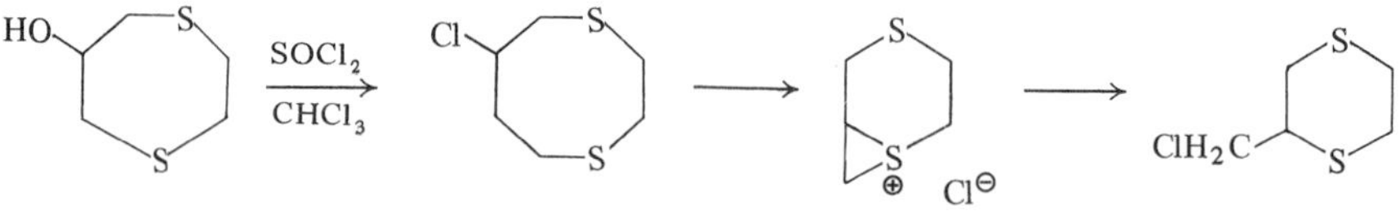

(5.224)

SOCl₂
→
CHCl₃

(5.225)

(5.226)

(5.227)

converted into (5.227). This mechanism is quite similar to that proposed for the conversion of β-hydroxyisopropylsulphide into the 2-chloro-n-propylsulphides which also involves a cyclic sulphonium intermediate[103].

C. Ring transformations of seven-membered heterocycles containing two different heteroatoms

1. RING CONTRACTIONS OF BENZ[d]1,3-OXAZEPINES INTO INDOLES; BENZ[f]1,3-OXAZEPINES INTO COUMARINS; BENZ[f]1,4-OXAZEPINES INTO ISOQUINOLINES; BENZ[f]1,5-OXAZEPINES INTO BENZOXAZOLES

6-Methylbenz[d] 1,3-oxazepine-2-carbonitrile (5.230), which is obtained during photolysis of 2-cyano-4-methylquinoline-N-oxide (5.229) ((5.230) was first erroneously assigned the structure of the oxaziridine (5.228) (see chapter 4, section I.C.3.b)), is converted[104,105] into the 3-methylindole (5.232) by prolonged heating in dilute sulphuric acid. Compound (5.230) is transformed by photolysis into the 2-cyano-3-methylindole (5.231)[107], and in the presence of methylamine into 2-methylamino-3-amino-4-methylquinoline (5.233)[106], which subsequently isomerizes. The isomerization of 2-cyanobenzoxazepines by the action of acetylchloride into 3-hydroxyquinolines has also been reported[107a].

A similar rearrangement has been found to occur if 2-phenylbenz[d] 1,3-oxazepine (5.234) is boiled in methanol, N-dimethoxyphenylmethylindole (5.236) and 2-phenylindole (5.238)[108] being formed. It thus seems probable that the formation of (5.236) and (5.238) takes place by a mechanism involving transannular bridging between nitrogen and carbon, i.e. (5.235), and between two carbon atoms, i.e. (5.237), as indicated on page 307.

When the benz[f] 1,4-oxazepine (5.239) is treated with sodamide in dioxane a mixture of products is formed[109], all of these contain the isoquinoline ring, that is (5.241), (5.242) and (5.243) respectively. The compound (5.243) is obtained by an acyloin reaction from (5.242). In the scheme proposed it has been suggested that bridging between C_2 and C_7 is the primary step forming the oxiranoisoquinoline derivative (5.240). A ring contraction into a six-membered coumarin derivative (5.245) has been reported to occur when 2-phenyl-4-methoxycarbonyltetrafluorobenz[f] 1,3-oxazepine (5.244) is refluxed with a solution of acetic acid and concentrated hydrogen chloride[110]. The compound is isolated in the form of a molecular compound with benzoic acid.

It was recently reported that 2,3-tetramethylene-2,3-dihydro-1,5-benzoxazepine (5.246) rearranges when treated with phosphorus pentachloride into the benzoxazole derivative, 2-(1-cyclohexenyl)benzoxazole (5.247)[117]. A chloroimidate and the anion of o[chloro(1-cyclohexenyl)methyleneimino] phenol are the probable intermediates.

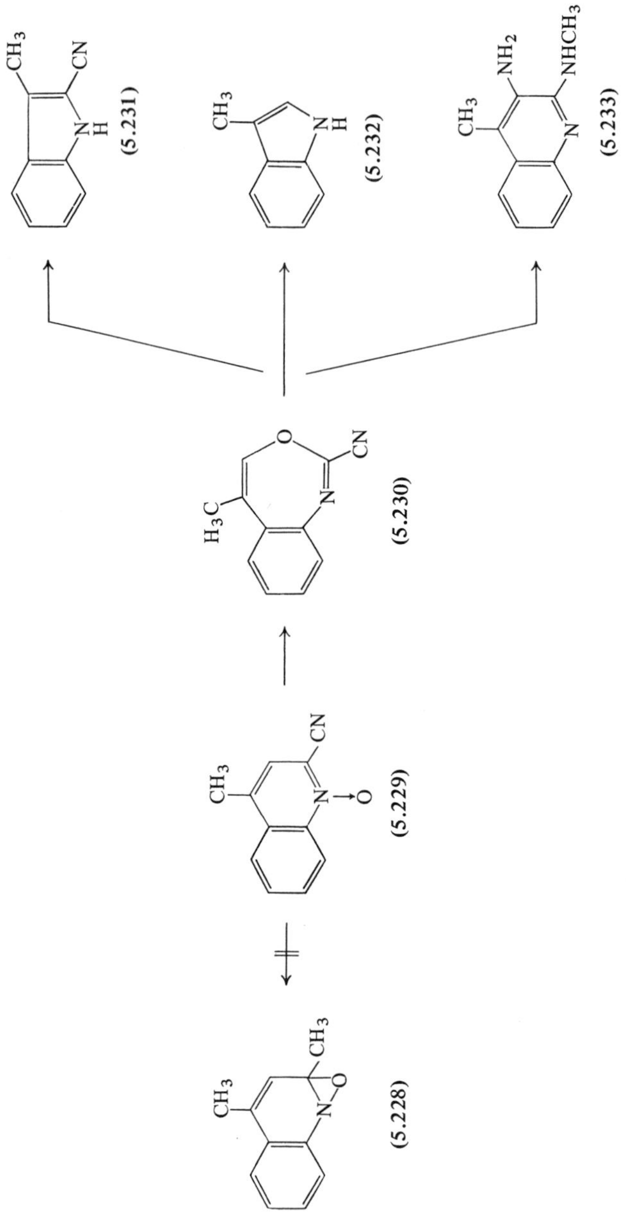

(5.228) (5.229) (5.230) (5.231) (5.232) (5.233)

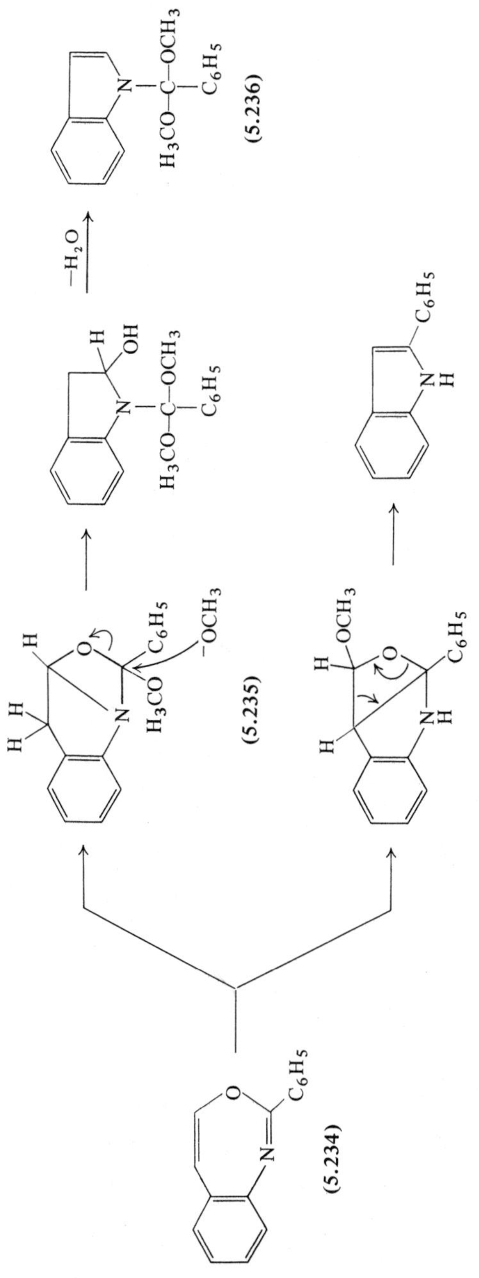

(5.236)

(5.235)

(5.238)

(5.237)

(5.234)

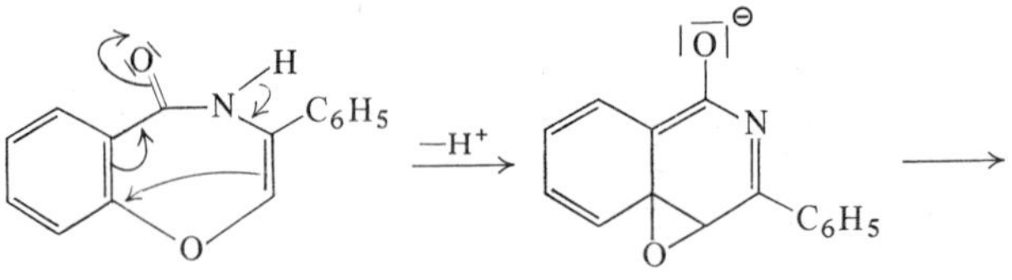

(5.239)   (5.240)

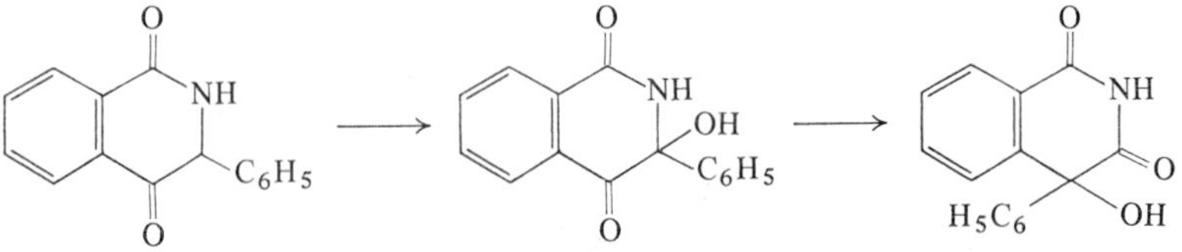

(5.241)   (5.242)   (5.243)

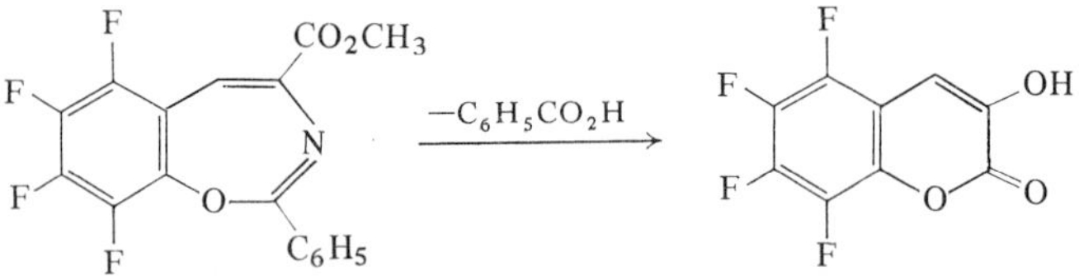

(5.244)   (5.245)

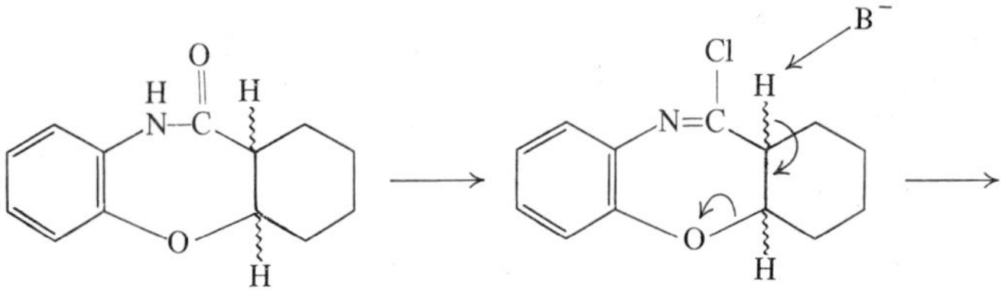

(5.246)

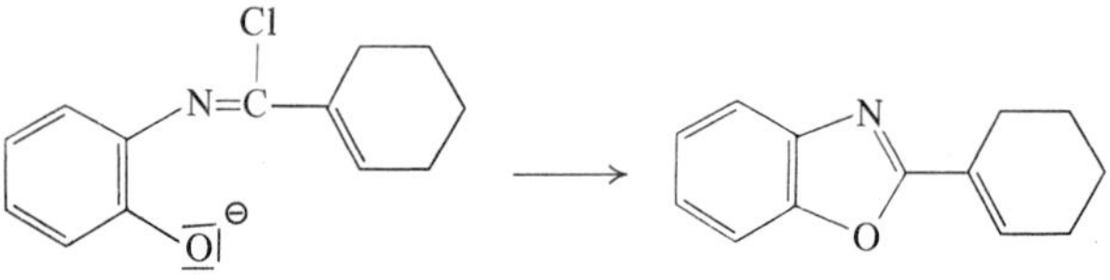

(5.247)

2. RING CONTRACTIONS OF BENZO-1,4-THIAZEPINES INTO BENZO DERIVATIVES OF THIAZOLES, ISOTHIAZOLES, INDOLES AND PYRIDINES; 1,4-THIAZEPINES INTO ISOTHIAZOLES AND THIAZOLINES

As already discussed in vol. 1, chapter 3, section II.B.1.a., the extrusion of a sulphur atom occurs readily in cases where the ring contraction leads to a new ring with an aromatic structure.

In the 1,4-thiazepine series a few examples of this type of ring contraction have been reported which support this hypothesis. Ring contraction of benzo-[b] 1,4-thiazepines into quinoline derivatives has been reported to occur during treatment with a base[111a]. Thus, the benzo[b] 1,4-thiazepine (5.247a) gives, with morpholine in boiling isopropanol, the 2-methylthio-4-phenylquinoline (5.247c); with sodium ethanolate (but not with weaker bases such as amines) the thione (5.247d) yields the thiocarbostyril (5.247e). The thiiranequinoline derivative (5.247b) is postulated as intermediate. Thermolytic extrusion of sulphur has also been observed. Hence dibenzo[b,f] 1,4-thiazepine gives the corresponding phenanthridine derivative when heated with copper in diethylphthalate[111]. From an examination of the effect of a number of substituents of different character present in several positions in phenyldibenzo[b,f] 1,4-thiazepines (5.248) upon the thermolytic formation of the corresponding phenanthridines (5.250) under different reaction conditions it appears that there

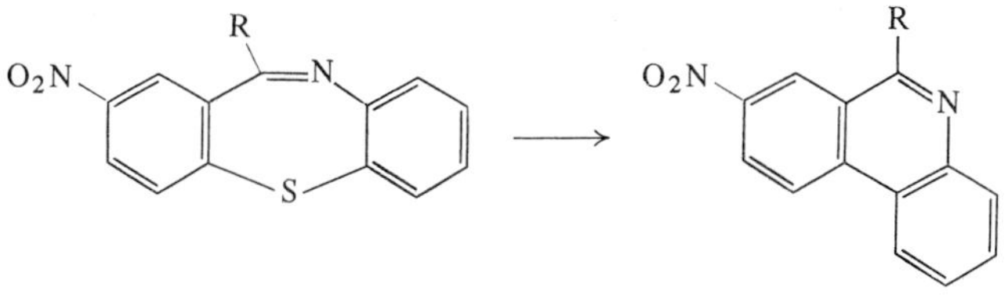

$R = H; CH_3; C_6H_5; m\text{-}NO_2C_6H_4$

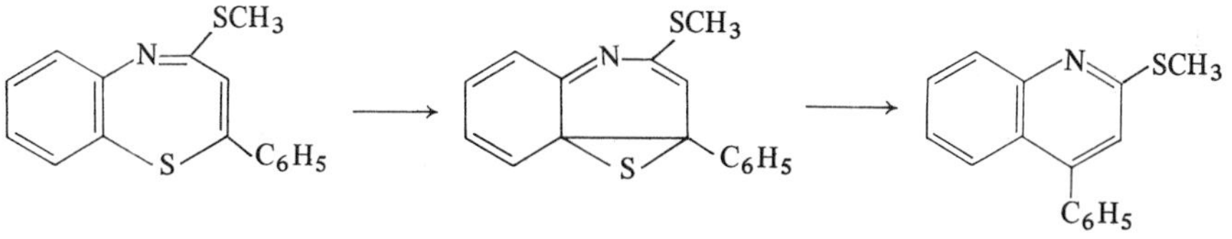

| (5.247a) | (5.247b) | (5.247c) |

| (5.247d) | (5.247e) |

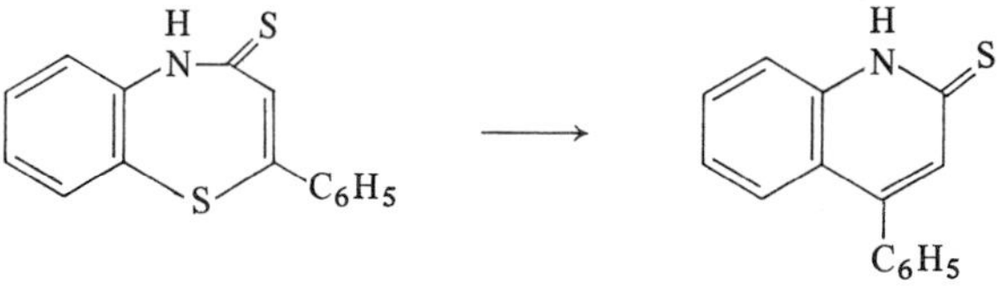

is justification for assuming that the intermediate is the tetracyclic compound (5.249)[112] containing a thiirane ring, which, by loss of sulphur aromatizes into (5.250).

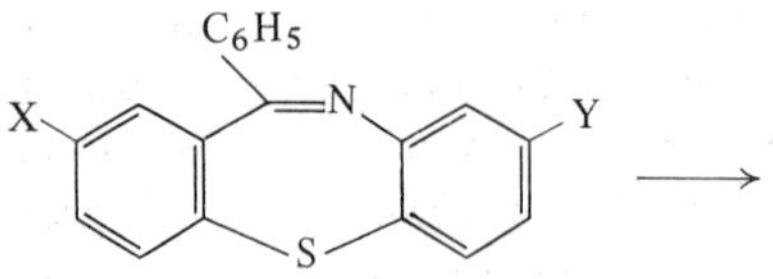

(5.248)

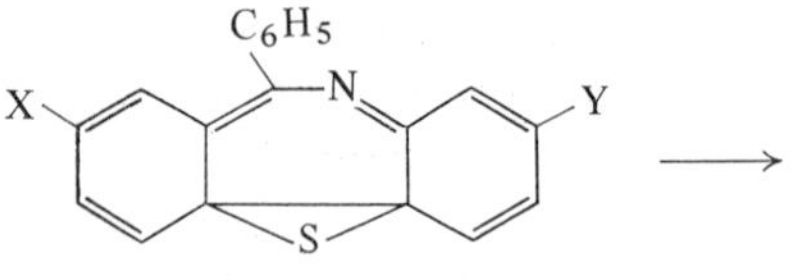

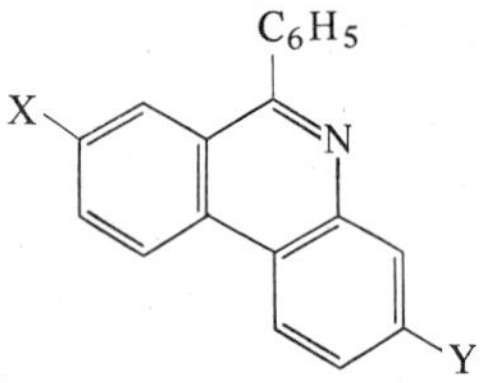

(5.249)   (5.250)

Similarly, the 1,4-thiazepine (5.251) loses its sulphur atom, yielding (5.252)[113], on treatment with hydrogen and Raney nickel or by oxidation with hydrogen peroxide in acetic acid. This method has also been successfully applied for ring contraction of the salt (5.253) into the benzo[a]quinolizinium salt (5.254)[114].

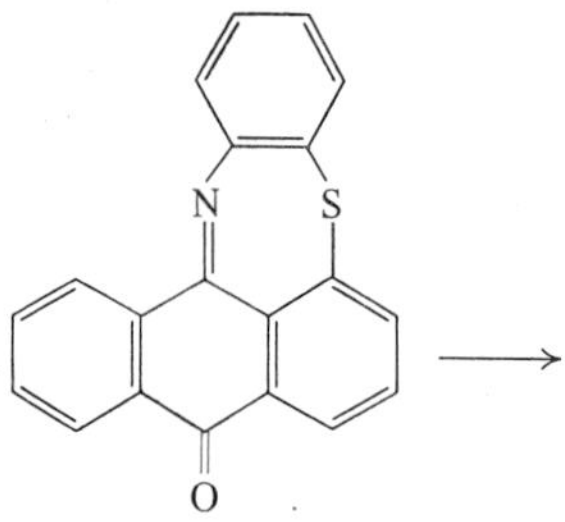

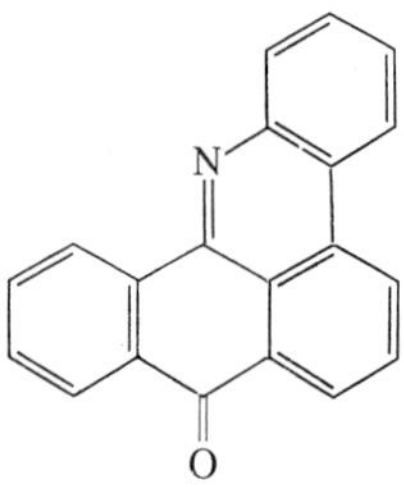

(5.251)   (5.252)

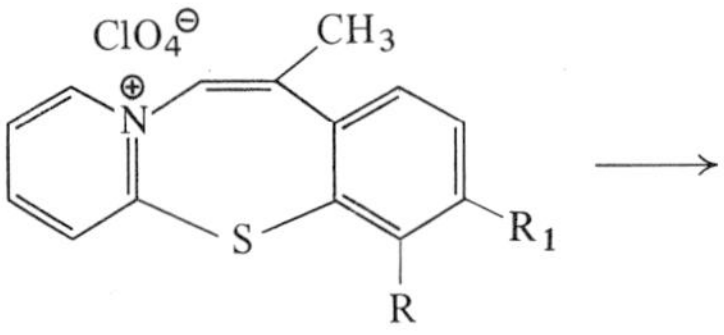

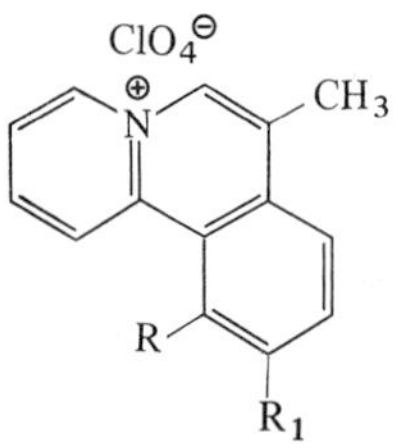

(5.253)   (5.254)

Loss of sulphur dioxide has been reported to occur when the benzo-1,4-thiazepine-S-dioxide (5.255) is treated with lithium aluminium hydride, the indole derivative (5.256) being formed[115]. This reaction is not base catalysed.
An interesting ring transformation was found[111a] when the 2,3-dihydro-

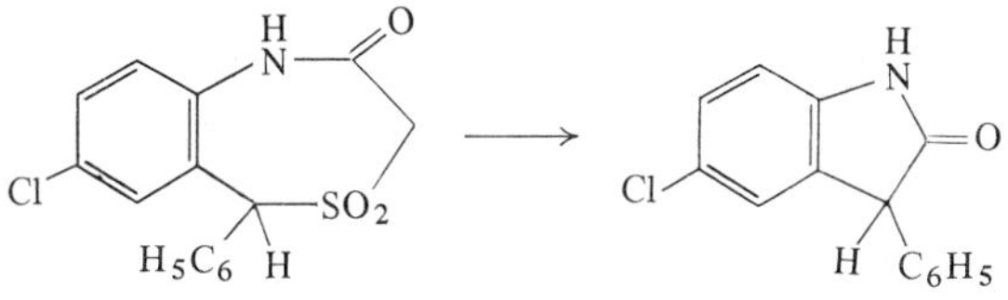

(5.255)   (5.256)

derivative of the benzo[b] 1,4-thiazepine, that is (5.256a), was treated with a base: ring contraction occurred with elimination of methylmercaptane, yielding the 2-styrylbenzothiazole (5.256c). Transannulation yielding the sulphonium ion (5.256b) is probably the initial step in this ring contraction. These results suggest that tetrahydro- and hexahydro-1,4-thiazepines can also show the type of ring contraction discussed above. This has in fact been demonstrated[111a]; 2,7-diphenylhexahydro-1,4-thiazepin-5-one, on heating in polyphosphoric acid, eliminates water and yields 2-styryl-5-phenyl-Δ^2-thiazoline.

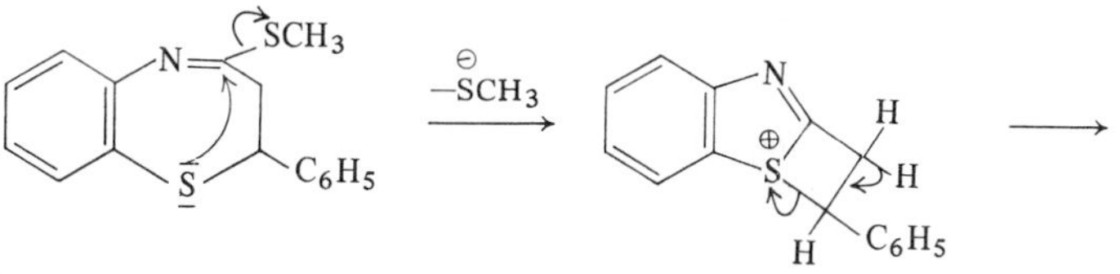

(5.256a)   (5.256b)

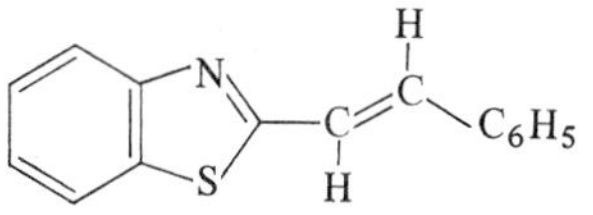

(5.256c)

Another interesting ring transformation has been found to occur when the hexahydro-1,4-thiazepine (5.257) reacts with chlorine at low temperature ($-10°C$). Besides oxidation to the tetrahydro-1,4-thiazepine (5.258), ring contraction into the isothiazol-3-ones (5.259) and (5.260) takes place[116]. The interrelation between (5.259) and (5.260) was shown by the isomerization of (5.259) into (5.260) on treatment with triethylamine at room temperature. It is suggested that the probable intermediate in this reaction sequence is the vinylsulphenyl chloride (5.261) which is converted into (5.259) via the open-chain carbonamide (5.262).

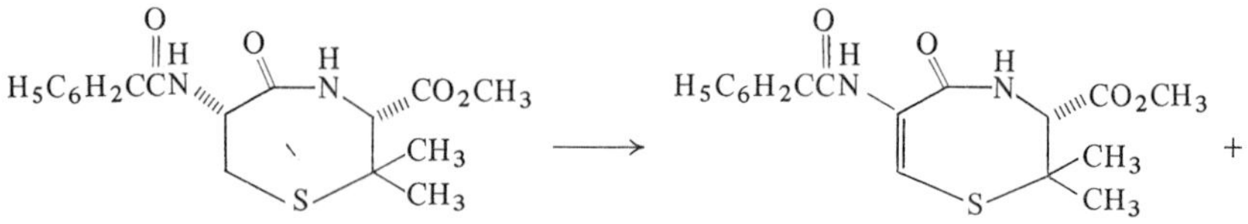

$$H_5C_6H_2CCN \quad (5.257) \longrightarrow \quad H_5C_6H_2CCN \quad (5.258) \quad +$$

(5.257)   (5.258)

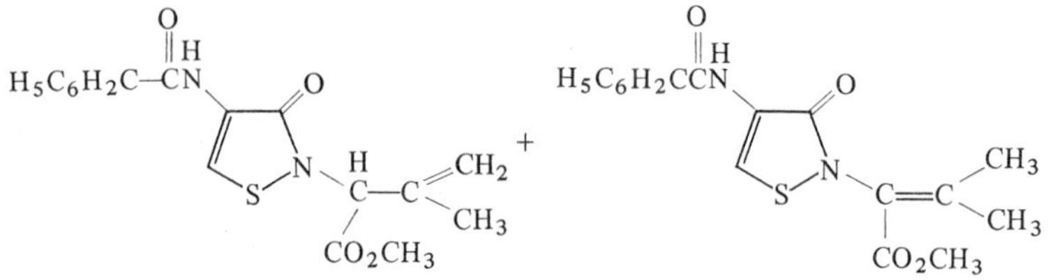

$$H_5C_6H_2C-CN \quad (5.259) \quad + \quad H_5C_6H_2CCN \quad (5.260)$$

(5.259)   (5.260)

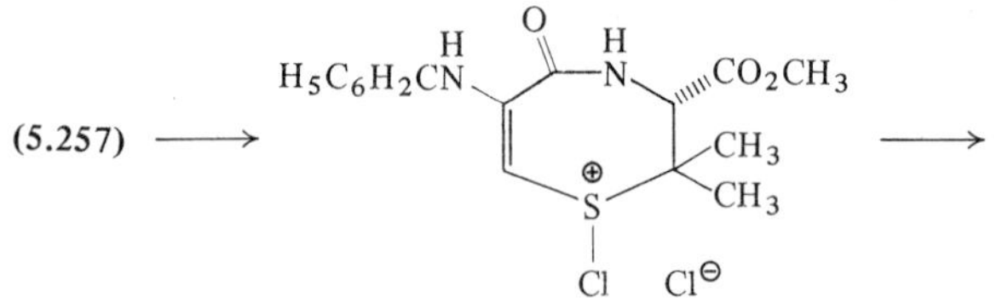

$$(5.257) \longrightarrow \quad H_5C_6H_2CN \quad (5.261) \quad \longrightarrow$$

(5.261)

$$H_5C_6H_2CCN \quad (5.262) \quad \xrightarrow{-HCl} \quad H_5C_6H_2CCN \quad (5.259)$$

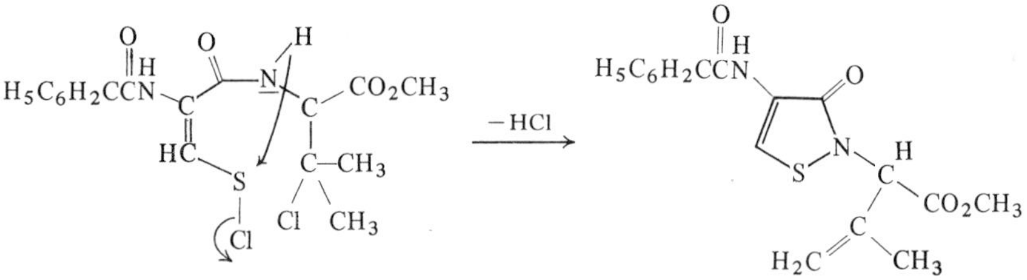

(5.262)   (5.259)

III. RING TRANSFORMATIONS OF SEVEN-MEMBERED HETEROCYCLES CONTAINING THREE HETEROATOMS

A. Thiadiazepines into pyridazines and pyrazoles; oxadiazepines into pyrazoles; benzoxadiazepines into indazoles, 3,1-benzoxazines and quinazolines; dioxa- thiepanes into tetrahydrofurans; benzotriazepines into quinoxalines

On treatment with a halogenating reagent (*N*-bromosuccinimide in tetrachloromethane, bromine in acetic acid or sulphuryl chloride with sodium iodide in acetone), 3,6-diphenyl-2,7-dihydro-1,4,5-thiadiazepine (5.263) is converted into 3,6-diphenylpyridazine (5.265)[118]. In this reaction, as in the conversion of (5.248) into (5.250), the most likely intermediate is a thiirane derivative (5.264), probably formed by an initial halogenation at the carbon adjacent to the sulphur, followed by a dehydrohalogenation process. On thermolysis of (5.263) the same ring transformation reaction was observed.

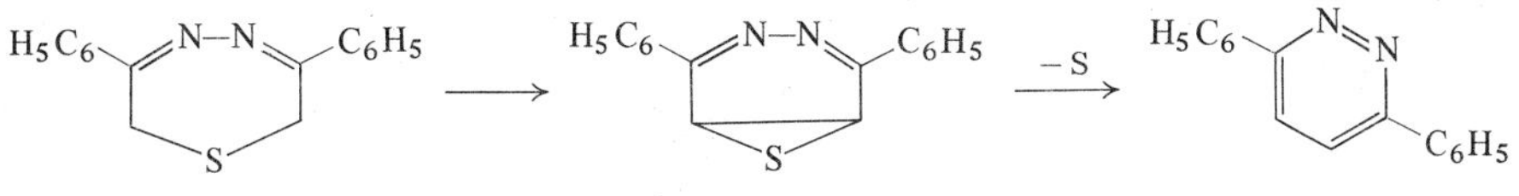

(5.263)   (5.264)   (5.265)

When heated in acetic acid the S-dioxide of (5.263) also yields the pyridazine (5.265)[118]. Quite similarly, ring contraction by sulphur monoxide extrusion, leading to the benzocinnoline (5.267c), has been observed during thermolysis of the benzo[b,f],[1,4,5] thiadiazepine-1-oxide (5.267a) in refluxing benzene[118a]. It was possible to trap the SO moiety when the thermolysis was carried out in the presence of 2,3-diphenylbutadiene, 3,4-diphenyl-2,5-dihydrothiophene-1-oxide being yielded. It is reasonable to advance the thiirane-S-oxide (5.267b) as intermediate; it is formed by the thermally initiated disrotatory electrocyclization in accordance with the Woodward Hoffmann rules[118b]. Compound (5.267c) is also found to be formed from the 1,4-dioxide (5.267d) when this compound is heated to its melting point. There is a close resemblance between the activation energies of the decomposition of (5.267a) in toluene whether or not a 1,3-diene is present, and this demonstrates that the generation of transient SO is independent of the presence of a diene. Surprisingly it is found that the S-monoxide (5.266) gives, on heating in acetic acid, ring contraction into 3,5-diphenylpyrazole (5.269)[118] and does not lead to the formation of a

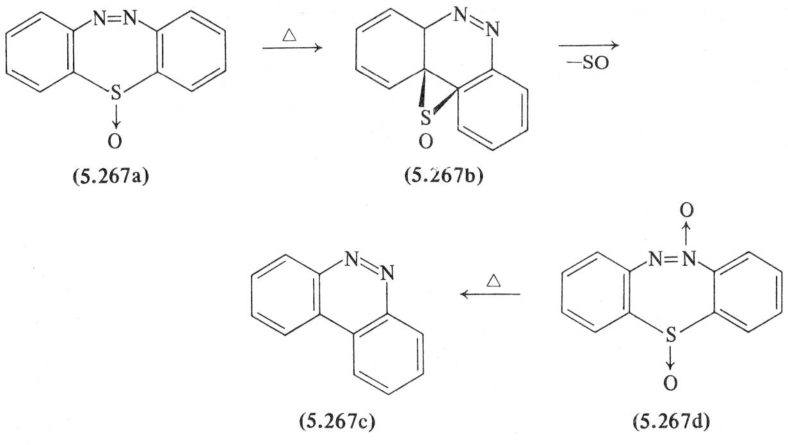

(5.267a)   (5.267b)

(5.267c)   (5.267d)

pyridazine derivative. Transannulation between C_3 and C_7 in (5.267), giving (5.268), apparently seems the step most favoured in this reaction. The further conversion that occurs is outlined below; however, there is no step analogous to (5.268) → (5.269).

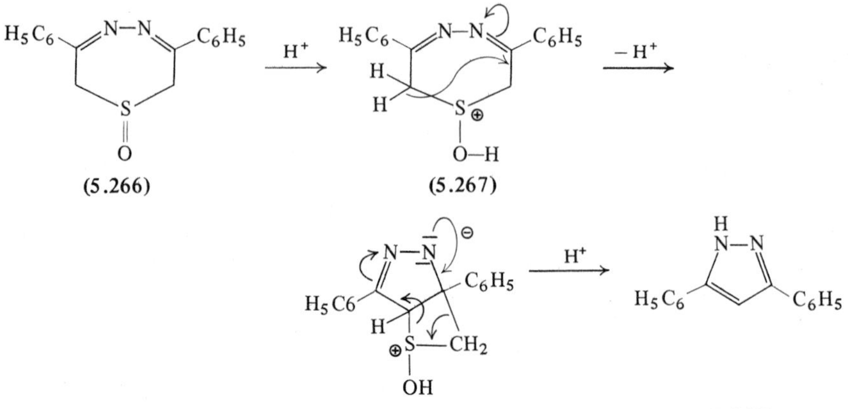

(5.266)   (5.267)

(5.268)   (5.269)

The interesting and quite often unexpected chemical behaviour of the 1,4,5-thiadiazepine system is further demonstrated by the remarkable ring contraction which has been observed when the S-dioxide (5.270) reacts with sodium ethanolate in ethanol[118]. 3-Methyl-4,5-diphenylpyrazole (5.274) is obtained. The formation of this compound has also been explained by transannulation at C_3 and C_7, (5.271) → (5.272), giving the intermediate (5.272) which is structurally analogous to (5.268). The course of the subsequent reaction apparently depends now on the oxidation level of the sulphur atom: whereas the conversion (5.268) → (5.269) does not involve a phenyl migration, (5.272) is able to undergo a phenyl migration, as shown below:

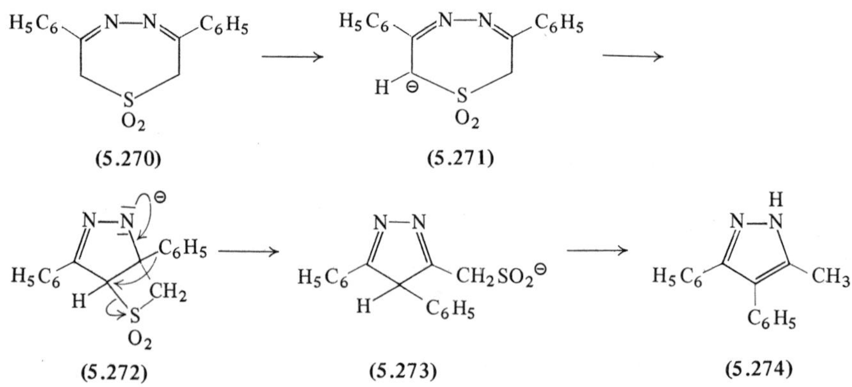

(5.270)   (5.271)

(5.272)   (5.273)   (5.274)

In the 1,4,5-oxadiazepine series, a number of useful ring contraction reactions have been reported with compounds containing an oxo group in position 2. The compounds (5.275), which are obtained by a 1:1 addition of ketene to a chinon(1,2)diazide-2, when heated above their melting point or placed in a high boiling solvent, lose carbon dioxide thus leading to the production of indazoles (5.276) in high yields[119-121].

(5.275)   (5.276)

R	H	H	H	H	H	H	H	CH$_3$
R$_1$	H	CH$_3$	CH$_3$	C(CH$_3$)$_3$	Cl	H	CO$_2$CH$_3$	Cl
R$_2$	H	H	CH$_3$	H	H	Cl	H	CH$_3$
R$_3$	H	H	H	C(CH$_3$)$_3$	H	H	H	H

A wide variety of interesting ring contractions were reported with the 7-chloro-2-methylamino-5-phenyl-3,1,4-benzoxadiazepine (5.278)[122]. Exposure of (5.278) to light, or heating (5.278) at 160–180°C, in vacuo, yields the 5-chloro-3-phenyl-1-indazolecarboxamide (5.279)[122]. A Beckmann rearrangement of (5.278) with an ethereal solution of hydrogen chloride leads to the formation of the 4-phenylimino-4H-3,1-benzoxazine derivative (5.277)[122]. Heating (5.277) gives rearrangement into the 4-quinazolone derivative (5.280)[122]. From (5.278), with boron trifluoride-etherate the 4-phenylimino-2-quinazolone derivative (5.281) is obtained; this ring contraction probably occurs via (5.277) since this compound also yields (5.281) when the same reagent is used.

A similar rearrangement has been reported with the 7-chloro-4-phenyl-3,1,4-benzoxadiazepin-2-one, yielding 6-chloro-3-phenyl-2,4-(1H,3H)-quinazolinedione[123]; although a 3,1-benoxazine derivative, has been suggested as intermediate, this compound has not been isolated.

The ring contraction of 1,2,5-benzotriazepine (5.282) has unexpectedly been shown by chemical and physical methods of analysis to lead to the N-iminoquinoxalinium ylide (5.285)[124] It has been further demonstrated that this reaction is quite general for all 3-unsubstituted benzotriazepines and can be considered to occur via the protonated form (5.283) which collapses via the diaziridine (5.284) into (5.285). A similar rearrangement leading to an

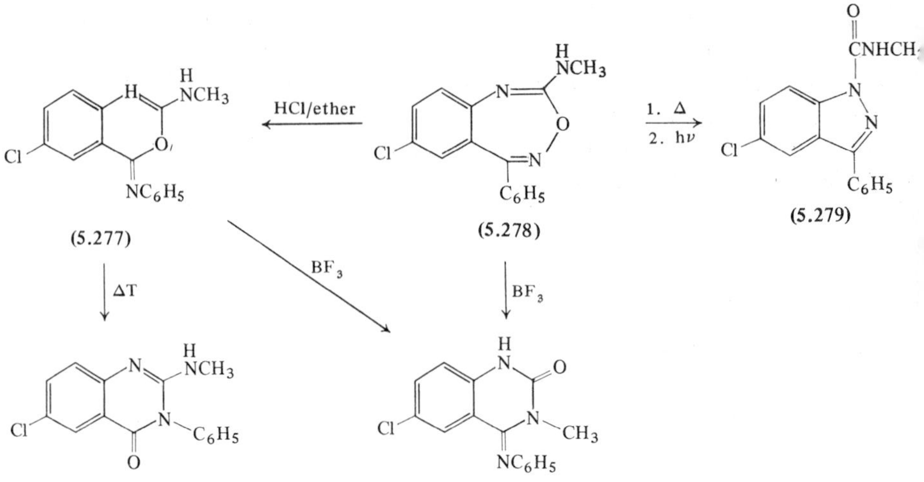

(5.277)   (5.278)   (5.279)

(5.280)   (5.281)

1-aminopyridine derivative has now been reported with a 1,2-diazepin-4-one[31,62] (see section II.A.2.a).

The 1,3,2-dioxathiepane-S-dioxide (5.287), when heated with triethylamine at 180°C for 9 h, gives, with loss of sulphur dioxide, tetrahydrofuran (5.289) in high yield (96 per cent)[125]; no indication was obtained for the cyclic product (5.286), the product usually obtained when a dialkyl sulphite[126] is decomposed. Attempts to perform this ring contraction with the six-membered 1,3,2-dioxathiane-S-monoxide were unsuccessful, the formation of acroleine and

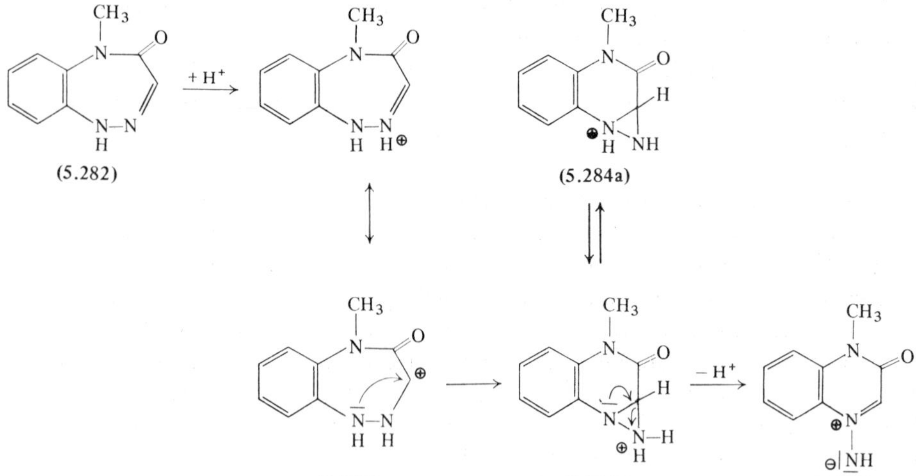

(5.282)   (5.284a)

(5.283)   (5.284b)   (5.285)

propionaldehyde only being observed. Nucleophilic attack of the amine at the S—O bond certainly causes ring opening of the seven-membered ring in the formation of (5.288). The open-chain compound (5.288) readily loses sulphur dioxide and recyclizes further with elimination of triethylamine.

(5.286)   (5.287)

$$(C_2H_5)_3\overset{\oplus}{N}-(CH_2)_4-OSO_2^{\ominus} \xrightarrow{-SO_2} (C_2H_5)_3\overset{\oplus}{N}-(CH_2)_4-\underline{O}|^{\ominus} \longrightarrow$$

(5.288)

(5.289)

References

1. Hafner, K. (1963). *Angew. Chem.* **75**, 1041; Hafner, K., Zinser, D., and Moritz, K. L. (1964). *Tetrahedron Lett.* 1733.
2. Marsh, F. D., and Simmons, H. E. (1965). *J. Amer. Chem. Soc.* **87**, 3529.
3. Childs, R. F., and Johnson, A. W. (1965). *Chem. Commun.* **5**, 95.
4. Paquette, L. A. (1964). *J. Amer. Chem. Soc.* **86**, 4096; Childs, R. F., Grigg, R., and Johnson, A. W. (1967). *J. Chem. Soc. (C)*, 201.
5. Groot, Ae, de, Boerma, J. A., and Wijnberg, H. (1969). *Rec. Trav. Chim. Pays-Bas*, **88**, 994.
6. Groot, Ae. de, Boerma, J. A., and Wijnberg, H. (1968). *Tetrahedron Lett.* 2365.
7. Bergmann, E. D., and Rabinovitz, M. (1960). *J. Org. Chem.* **25**, 828.
8. Manske, R. H. F., and Ledingham, A. E. (1950). *J. Amer. Chem. Soc.* **72**, 4797.
9. Lüttringhaus, A., and Creutzburg, G. (1968). *Angew. Chem. Int. Ed.* **7**, 128.
10. Childs, R. F., and Johnson, A. W. (1966). *J. Chem. Soc. (C)*, 1950.
11. Anderson, M., and Johnson, A. W. (1965). *J. Chem. Soc.* 2411.
12. Paquette, L. A. (1964). *J. Amer. Chem. Soc.* **86**, 4096.
13. Brignell, P. J., Bullock, E., Eisner, U., Gregory, B., Johnson, A. W., and Williams, H. (1963). *J. Chem. Soc.* 4819.
14. Bullock, E., Gregory, B., and Johnson, A. W. (1964). *J. Chem. Soc.* 1632.
15. Bullock, E., Gregory, B., and Johnson, A. W. (1962). *J. Amer. Chem. Soc.* **84**, 2260.
16. Lloyd, H. A., and Horning, E. C. (1954). *J. Amer. Chem. Soc.* **76**, 3651.
17. Rees, A. H. (1962). *J. Chem. Soc.* 3097.
18. Overberger, C. G., Reichenthal, J., and Anselme, J. P. (1970). *J. Org. Chem.* **35**, 138.
19. Leonard, N. J., and Barthel, E. (1949). *J. Amer. Chem. Soc.* **71**, 3098; (1950), *J. Amer. Chem. Soc.* **72**, 3632.

20. Leonard, N. J., and Nicolaides, E. D. (1951). *J. Amer. Chem. Soc.* **73,** 5210; Leonard, N. J., and Pines, S. H. (1950). *J. Amer. Chem. Soc.* **72,** 4931.
21. Leonard, N. J., and Wildman, W. C. (1949). *J. Amer. Chem. Soc.* **71,** 3100.
22. Ogata, M., Matsumoto, H., and Kano, H. (1969). *Tetrahedron,* **25,** 5217.
23. Rees, A. H. (private communication). "Heterocyclic Compounds" Part 9, p. 285 Elderfield; Rees, A. H., and Simon, K. (1969). *Can. J. Chem.* **47,** 1227.
24. Schindler, W., and Blattner, H. (1961). *Helv. Chim. Acta,* **44,** 753.
25. Haque, K. E., and Proctor, G. R. (1968). *Chem. Commun.* 1412.
26. Paterson, W., and Proctor, G. R. (1962). *J. Chem. Soc.* 3468.
27. Lloyd, H. A., Matternas, L. U., and Horning, E. C. (1955). *J. Amer. Chem. Soc.* **77,** 5932.
28. Geissman, T. A., and Cho, A. K. (1959). *J. Org. Chem.* **24,** 41.
28a. Vogel, A., Troxler, F., and Lindenmann, A. (1969). *Helv. Chim. Acta,* **52,** 1929.
29. Carruthers, W., and Johnstone, R. A. W. (1965). *J. Chem. Soc.* 1653.
30. Moore, J. A., Medeiros, R. W., and Williams, R. L. (1966). *J. Org. Chem.* **31,** 52; Moore, J. A., and Medeiros, R. W. (1959). *J. Amer. Chem. Soc.* **81,** 6026.
31. Moore, J. A., and Binkert, J. (1959). *J. Amer. Chem. Soc.* **81,** 6029.
32. Wineholt, R. L., Wyss, E., and Moore, J. A. (1966). *J. Org. Chem.* **31,** 48.
33. Moore, J. A., and Zoll, E. C. (1964). *J. Org. Chem.* **29,** 2124.
34. Moore, J. A., and Habraken, C. L. (1965). *J. Org. Chem.* **30,** 1889.
35. Battiste, M. A., and Barton, Th. J. (1967). *Tetrahedron Lett.* 1227.
36. Sauer, J., and Heinrichs, G. (1966). *Tetrahedron Lett.* 4979.
37. Heine, H. W., and Irving, J. (1967). *Tetrahedron Lett.* 4767.
38. Derocque, J. L., Theuer, W. J., and Moore, J. A. (1968). *J. Org. Chem.* **33,** 4381.
39. Woodward, R. B., and Hoffmann, R. (1969). *Angew. Chem. Int. Ed.* **8,** 781.
40. Fryer, R. I., Earley, J. V., and Sternbach, L. H. (1966). *J. Amer. Chem. Soc.* **88,** 3173.
41. Fryer, R. I., Earley, J. V., and Sternbach, L. H. (1969). *J. Org. Chem.* **34,** 649.
42. Fryer, R. I., Brust, B., Earley, J. V., and Sternbach, L. H. (1967). *J. Chem. Soc. (C),* 366.
43. Metlesics, W., Tavares, R. F., and Sternbach, L. H. (1965). *J. Org. Chem.* **30,** 1311.
44. Metlesics, W., Silverman, G., and Sternbach, L. H. (1964). *J. Org. Chem.* **29,** 1621.
45. Felix, A. M., and Fryer, R. I. (1968). *J. Heterocycl. Chem.* **5,** 291.
46. Grdenić, D., and Bezjak, A. (1953). *Arkiv. Kemi.* **25,** 101; (1954). *Chem. Abstr.* **48,** 11146; Hahn, V., Hammes, P., and Gerić, Z. (1954). *Experientia,* **10,** 11; (1955). *Chem. Abstr.* **49,** 3898; Kesler, M. (1955). *Arkiv. Kemi.* **27,** 67; (1955). *Chem. Abstr.* **49,** 15313.
47. Thiele, J., and Steimmig, G. (1907). *Chem. Ber.* **40,** 955.
48. Lloyd, D., McDougall, R. H., and Marshall, D. R. (1965). *J. Chem. Soc.* 3785.
49. Barltrop, J. A., Richards, C. G., Russell, D. M., and Ryback, G. (1959). *J. Chem. Soc.* 1132.

50. Ried, W., and Höhne, W. (1954). *Chem. Ber.* **87**, 1801.
51. Merz, K. W., Haller, R., and Müller, E. (1963). *Naturwissenschaften,* **50**, 663; (1964). *Chem. Abstr.* **60**, 4148.
52. Finar, I. L. (1958). *J. Chem. Soc.* 4094.
53. Davoll, J. (1960). *J. Chem. Soc.* 308.
54. Wigton, F. B., and Joullié, M. M. (1959). *J. Amer. Chem. Soc.* **81**, 5212.
55. Müller, E., Haller, R., and Merz, K. W. (1966). *Annalen,* **697**, 193.
56. Barchet, R., and Merz, K. W. (1964). *Tetrahedron Lett.* **33**, 2239.
57. Rossi, A., Hunger, A., Kebrle, J., and Hoffmann, K. (1960). *Helv. Chim. Acta,* **43**, 1298; Sexton, W. A. (1942). *J. Chem. Soc.* 303; Davoll, J. (1960). *J. Chem. Soc.* 308.
58. Wigton, F. B., and Joullié, M. M. (1959). *J. Amer. Chem. Soc.* **81**, 5212; Ried, W., and Stahlhofen, P. (1957). *Chem. Ber.* **90**, 825; Ried, W., and Storbeck, W. (1962). *Chem. Ber.* **95**, 459.
59. Israel, M., and Jones, L. C. (1969). *J. Heterocycl. Chem.* **6**, 735.
60. Smith, C. W., Rasmussen, R. S., and Ballard, S. A. (1949). *J. Amer. Chem. Soc.* **71**, 1082.
61. Shriner, R. L., and Boermans, P. G. (1944). *J. Amer. Chem. Soc.* **66**, 1810.
62. Moore, J. A. (1955). *J. Amer. Chem. Soc.* **77**, 3418.
63. Pleiss, M. G., and Moore, J. A. (1968). *J. Amer. Chem. Soc.* **90**, 4738.
64. Moore, J. A., Kwart, H., Wheeler, G., and Bruner, H. (1967). *J. Org. Chem.* **32**, 1342.
65. Theuer, W. J., and Moore, J. A. (1967). *J. Org. Chem.* **32**, 1602.
66. Moore, J. A., Marascia, F. J., Medeiros, R. W., and Wineholt, R. L. (1966). *J. Org. Chem.* **31**, 34.
67. Streith, J., and Cassal, J. M. (1968). *Angew. Chem. Int. Ed.* **7**, 129; Streith, J., Blind, A., Cassal, J. M., and Sigwalt, C. (1960). *Bull. Soc. Chim. Fr.* 948; Balasubramanian, A., McIntosh, J. M., and Snieckus, Y. (1970). *J. Org. Chem.* **35**, 433.
68. Diesbach, H., Gross, J., and Tschannen, W. (1951). *Helv. Chim. Acta,* **34**, 1050.
69. Gill, N. S., James, K. B., Lions, F., and Potts, K. T. (1952). *J. Amer. Chem. Soc.* **74**, 4923.
70. Halford, J. O., Raiford Jr., R. W., and Weissmann, B. (1961). *J. Org. Chem.* **26**, 1898.
71. Schmitz, E., and Ohme, R. (1962). *Chem. Ber.* **95**, 2012.
72. Lieck, A. (1905). *Chem. Ber.* **38**, 3853.
73. Wölbling, H. (1905). *Chem. Ber.* **38**, 3845.
74. Brody, F., and Ruby, P. R. (1960). "The Chemistry of Heterocyclic Compounds, Pyridine and its Derivatives", Part I, p. 268. Interscience Publishers, New York.
75. Bell, S. C., and Childress, S. J. (1962). *J. Org. Chem.* **27**, 1691.
76. Bell, S. C., McCaully, R. J., Gochman, C., Childress, S. J., and Gluckman, M. I. (1968). *J. Med. Chem.* **11**, 457; (1968). *Chem. Abstr.* **69**, 36088.
77. Sternbach, L. H., Reeder, E., Stempel, A., and Rachlin, A. I. (1964). *J. Org. Chem.* **29**, 332.
78. Bell, S. C., Gochman, C., and Childress, S. J. (1963). *J. Org. Chem.* **28**, 3010.
80. Fryer, R. I., Earley, J. V., and Sternbach, L. H. (1967). *J. Org. Chem.* **32**, 3798.

81. Bell, S. C., and Childress, S. J. (1964). *J. Org. Chem.* **29**, 506.
82. Fryer, R. I., and Sternbach, L. H. (1965). *J. Org. Chem.* **30**, 524.
83. Oehlschlaeger, H., and Hoffmann, H. (1967). *Arch. Pharm.* **300**, 817.
84. See Lund, H. (1970). "The Chemistry of the Carbon-Nitrogen Double Bond" (Ed. S. Patai), Chapter 11, p. 540, Interscience Publishers, New York.
85. Oehlschlaeger, H. (1963). *Arch. Pharm.* **296**, 396.
86. Sankowski, B. Z., Levin, M. S., Urbigkit, J. R., and Wollick, E. G. (1964). *Anal. Chem.* **36**, 1991.
87. Oehlschlaeger, H., Volke, J., and Hoffmann, H. (1966). *Collect. Czech. Chem. Commun.* **31**, 1264; (1966). *Chem. Abstr.* **64**, 15718.
88. Oehlschlaeger, H., Volke, J., Hoffmann, H., and Kurek, E. (1967). *Arch. Pharm.* **300**, 250.
89. Field, G. F., and Sternbach, L. H. (1968). *J. Org. Chem.* **33**, 4438.
90. Sternbach, L. H., Koechlin, B. A., and Reeder, E. (1962). *J. Org. Chem.* **27**, 4671.
91. Kaminsky, L. S., and Lamchen, M. (1966). *J. Chem. Soc. (C)*, 2295.
91a. Ning, R. Y., Douvan, I., and Sternbach, L. H. (1970). *J. Org. Chem.* **35**, 3243.
91b. Bell, S. C., McCaully, R. J., and Childress, S. J. (1968). *J. Org. Chem.* **33**, 216.
92. Barltrop, J. A., and Richards, C. G. (1957). *Chem. Ind. (London)*, 466.
93. Yonezawa, T., Matsumoto, M., and Kato, H. (1968). *Bull. Chem. Soc. Jap.* **41**, 2543; (1969). *Chem. Abstr.* **70**, 37797.
93a. Matsumoto, M., Matsumura, Y., Iio, A., and Yonezawa, T. (1970). *Bull. Chem. Soc. Jap.* **43**, 1496.
94. Bly, R. K., Zoll, E. C., and Moore, J. A. (1964). *J. Org. Chem.* **29**, 2128.
95. Moore, J. A., and Theuer, W. J. (1965). *J. Org. Chem.* **30**, 1887.
96. Amiet, R. G., Johns, R. B., and Markham, K. R. (1965). *Chem. Commun.* 128.
97. Amiet, R. G., and Johns, R. B. (1968). *Aust. J. Chem.* **21**, 1279.
98. Dewar, M. J. S., and Narayanaswami, K. (1964). *J. Amer. Chem. Soc.* **86**, 2422.
99. Dubois, R. J., and Popp, F. D. (1968). *Chem. Commun.* 675.
100. Popp, F. D., Dubois, R. J., and Casey, A. C. (1969). *J. Heterocycl. Chem.* **6**, 285.
101. Dubois, R. J., and Popp, F. D. (1969). *Chem. Ind. (London)*, 620.
102. Fuson, R. C., and Speziale, A. J. (1949). *J. Amer. Chem. Soc.* **71**, 1582.
103. Fuson, R. C., Price, C. C., and Burness, D. M. (1946). *J. Org. Chem.* **11**, 475.
104. Kaneko, C., and Yamada, S. (1966). *Chem. Pharm. Bull.* **14**, 555; Kaneko, C., Yamada, S., and Ishikawa, M. (1966). *Tetrahedron Lett.* 2145.
105. Kaneko, C., and Yamada, S. (1967). *Chem. Pharm. Bull.* **15**, 663.
106. Kaneko, C., Yokoe, I., and Ishikawa, M. (1967). *Tetrahedron Lett.* 5237.
107. Ishikawa, M., Kaneko, C., and Yamada, S. (1968). *Tetrahedron Lett.* 4519.
107a. Yamada, S., and Kaneko, C. (1969). *Iyo Kizai Kenkyusho Hokoku, Tokyo Ika Shika Daigaku*, **3**, 75; (1970). *Chem. Abstr.* **73**, 98765.
108. Kaneko, C., Yamada, S., and Yokoe, I. (1966). *Tetrahedron Lett.* 4701.
109. Schenker, K. (1966). *Chimica*, **20**, 157.

110. Petrova, T. D., Mameav, V. P., Yakobson, G. G., and Vorozhtsov Jr., N. N. (1968). *Khim. Geterotsikl. Soedin.* 777; (1969). *Chem. Abstr.* **71**, 12949.
111. Jarrett, A. D., and Loudon, J. D. (1957). *J. Chem. Soc.* 3818.
111a. Wilhelm, M., and Schmidt, P. (1970). *Helv. Chim. Acta,* **53**, 1697.
112. Galt, R. H. B., and Loudon, J. D. (1959). *J. Chem. Soc.* 885.
113. Galt, R. H. B., Loudon, J. D., and Sloan, A. D. B. (1958). *J. Chem. Soc.* 1588.
114. Bradsher, C. K., and McDonald, J. W. (1961). *Chem. Ind. (London),* 1797.
115. Kuch, H., Seidl, G., and Schmitt, K. (1967). *Arch. Pharm.* **300**, 299.
116. Leonard, N. J., and Wilson Jr., G. E. (1964). *J. Amer. Chem. Soc.* **86**, 5307.
117. McEvoy, F. J., Allen Jr., G. R. (1970). *J. Org. Chem.* **35**, 1183.
118. Loudon, J. D., and Young, L. B. (1963). *J. Chem. Soc.* 5496.
118a. Chow, Y. L., Tam, J. N. S., Blier, J. E., and Szmant, H. H. (1970). *J. Chem. Soc. D,* 1604.
118b. Hoffmann, R., and Woodward, R. B. (1968). *Accounts Chem. Res.* **1**, 17.
119. Ried, W., and Dietrich, R. (1960). *Naturwissenschaften,* **47**, 445.
120. Ried, W., and Dietrich, R. (1963). *Annalen,* **666**, 135; (1963). *Angew. Chem.* **75**, 476.
121. Ried, W., and Wagner, K. (1965). *Annalen,* **681**, 45.
122. Metlesics, W., Silverman, G., and Sternbach, L. H. (1967). *Monatsh. Chem.* **98**, 633.
123. Sulkowski, T. S., and Childress, S. J. (1962). *J. Org. Chem.* **27**, 4424.
124. Rissi, S., Pirola, O., and Selva, F. (1968). *Tetrahedron,* **24**, 6395.
125. Gilles, R. G. (1960). *J. Org. Chem.* **25**, 651.
126. Bissinger, W. E., Kung, F. E., and Hamilton, C. W. (1948). *J. Amer. Chem. Soc.* **70**, 3940.

Author Index

Numbers in brackets are reference numbers and are included to assist in locating references in which the authors' names are not mentioned in the text. Numbers followed by asterisks indicate the page on which the reference is listed.

A

Abdel-Kader, A., 34 (143), 243*
Abel, D., 222 (677, 679), 225 (681), 262*, 263*
Abood, L. G., 59 (232), 247*
Ach, F., 148 (519), 257*
Acheson, R. M., 97 (347), 251*
Ahmad, Y., 167 (557), 170 (557), 259*
Ainsworth, C., 36 (156), 244*
Ajello, E., 152 (531), 257*
Ajello, T., 173 (564), 259*
Albert, A., 167 (556a), 259*
Alberti, C., 9 (33), 10 (33), 84 (311), 240*, 250*
Albertson, N. F., 58 (226), 247*
Alder, K., 24 (93), 28 (93), 242*
Alfonso, L. M., 112 (413, 414), 253*
Alkaitis, A., 69 (265), 248*
Allegretti, J., 233 (706), 263*
Allen, G. R., Jr., 305 (117), 321*
Allgrove, R. C., 93 (344), 251*
Allison, C. G., 173 (563), 259*
Altman, L. J., 151 (527), 257*
Amend, C. G., 190 (596), 260*
Ames, D. E., 158 (534c), 160 (539), 258*
Amiet, R. G., 302 (96, 97), 320*
Anderson, M., 93 (343), 94 (342, 343), 251*, 269 (11), 317*
Andes Hess, B., Jr., 43 (187), 245*
Anet, F. A. L., 43 (188), 245*
d'Angelo, J., 41 (181), 245*
Anker, R. M., 24 (130), 34 (130), 243*

Anselme, J. P., 89 (331), 250*, 272 (18), 317*
Antonov, V. K., 108 (390, 391), 252*
Archer, G. A., 141 (493), 256*
Arens, J. F., 23 (83), 241*
Armarego, W. L. F., 115 (421, 422), 190 (592), 192 (592), 203 (592), 253*, 260*
Arndt, F., 40 (175, 176), 245*
Arnold, R. T., 65 (257), 248*
Arold, H., 53 (218), 246*
Arora, R. C., 190 (593), 260*
Arow, E., 40 (176), 245*
Ashby, J., 64 (250), 94 (250), 95 (250, 345), 247*, 251*
Aspelund, H., 125 (456, 458, 459), 128 (459), 129 (456, 466), 130 (456), 255*
Atkinson, E., 113 (416), 253*
Austel, V., 173 (562a), 259*
Avram, M., 235 (714, 716), 263*
Awad, W. I., 194 (616), 260*

B

Backeberg, O. G., 84 (313), 250*
Bader, H., 23 (87), 241*
Baeyer, A., 22 (68), 24 (68), 28 (68), 79 (292), 241*, 249*
Bahr, F., 82 (307, 308), 250*
Bailey, A. S., 43 (187), 245*
Bajdala, H., 145 (513a), 257*
Baker, B. R., 191 (595), 260*
Baker, W., 10 (34, 35), 233 (705), 240*, 263*
Bakhtiari, B., 167 (557), 170 (557), 259*

323

Subject Index